The
Evolving Continents

The Evolving Continents

2nd Edition

Brian F. Windley
Department of Geology
University of Leicester

JOHN WILEY & SONS

Chichester · New York · Brisbane · Toronto · Singapore

Reprinted June 1978
Reprinted October 1978
Reprinted August 1979
Reprinted May 1982
Reprinted September 1984

Library of Congress Cataloging in Publication Data:

Windley, B. F. (Brian F.)
  The evolving continents.

  Bibliography: p.
  Includes index.
  1. Continental drift. 2. Geodynamics. I. Title.
QE511.5.W56 1984   551.1′36   83-21597

ISBN 0 471 90376 0

ISBN 0 471 90390 6 (pbk.)

British Library Cataloguing in Publication Data:

Windley, B. F.
  The evolving continents.—2nd ed.
  1. Plate tectonics
  I. Title
  551.1′36   QE511.4

ISBN 0 471 90376 0
ISBN 0 471 90390 6 (pbk.)

Typeset by Preface Ltd, Salisbury, Wilts.
Printed by Pitman Press Ltd, Bath, Avon.

To Judith
with whom I have shared much joy on many continents

Contents

Preface to First Edition

In Universities today students tend to be taught separate courses in self-contained packages by different members of staff. In order to avoid a pigeon-holed view of the earth sciences, there is therefore an increasing need for an integrating course in which many themes are woven together. Before we, the staff, let the students graduate at the University of Leicester we give them such a course in which the integrating medium is earth history. This book has grown out of that course.

In the last decade or so two major developments have taken place. Firstly, the well-known revolution in earth sciences during the 1960s was based on new geophysical knowledge of the world's ocean floor and plate mosaic. Since the beginning of the 1970s there has been a growing attempt to work out the structure and evolution of Phanerozoic fold belts in the light of the current concepts of sea-floor spreading and continental drift. Secondly, there has been a great advance, mostly in the 1970s, in the understanding of the Precambrian Shield regions, assisted considerably by improvements in age dating techniques. The combination of these two factors means that whilst the 1960s was the decade of the oceans, the 1970s is the decade of the continents. I have been fortunate in that these advances took place during the preparation of this text.

Important aspects of this book are as follows. Firstly, most books on earth history pay little attention to the Precambrian. It seems to me that sufficient is now known about the Archaean and Proterozoic for them to play a major role in any such text. For this reason the first half of this volume is a review of the Precambrian. Secondly, many books on plate tectonics tend to concentrate on geophysics and the oceanic crust and the principles of plate motion; few deal largely with the 'geology' of plate tectonics from a continental standpoint. With this in mind the second half contains a review of the geology of continental break-up and key Phanerozoic fold belts according to plate tectonic theory. In the last chapter I have brought together the reasons for many of the long-term changes in the rock record—they tell the story of the evolving continents. And, finally, I have summarized in all chapters the types of mineralization characteristic of different tectonic belts and periods of earth history because so often metallogeny is ignored by academics. Thus throughout I have emphasized those subjects that the reader will not easily find in synthesized book form elsewhere.

I have not made an exhaustive review of all shields and fold belts throughout the world; rather, I have selected classic regions which are representative of the rock groups or tectonic belts developed at particular times in earth history. In this way it is possible to characterize with well-documented examples the successive stages in development of the continental crust.

On the whole I have tried not to present just a descriptive account of 'what is where', but rather to bring out as many ideas about genetic processes and tectonic models as possible—facts without ideas can be very dull. There is current debate about the interpretation of much data and so I have pin-pointed many such areas of controversy because they enable the student, undergraduate or graduate, to focus in on the debates of today.

It seemed to me repetitious to give a description of each fold belt or continent in turn because so often the geology is broadly similar, and therefore I have synthesized the relevant contemporaneous component parts of fold belts, cratons or time periods as much for the Precambrian (such as greenstone belts, dyke swarms and tillites) as for the Phanerozoic (such as island arcs, palaeoclimatic indicators and Pangaea). But for many chapters I found that there are no recent adequate or suitable syntheses in the literature and so I have brought together in the form of new reviews data and ideas on, for example, Archaean high-grade regions, models for Archaean and Proterozoic crustal evolution, Proterozoic belts and geosynclines, and the break-up of Pangaea. Often different authors have produced their own models for a particular fold belt, or they have synthesized just one rock component such as tillites, Cordilleran volcanics, island arc sediments, evaporites, anorthosites or alkaline complexes. So I have endeavoured to bring together the models and syntheses for the different rock components of the various fold belts and time periods in order to portray in an interdisciplinary way a comprehensive picture of continental development.

For the Phanerozoic I have taken a frank plate tectonic approach. For the Precambrian I have trod more cautiously, documenting opposing ideas and points of view which reflect the state of the science today, but at the same time trying to emphasize those areas of Precambrian geology which are amenable to plate tectonic interpretation.

The book is intended for senior undergraduates who have already received a grounding in the main subject of geology, post-graduate students who may be able to see their own specialization in a better perspective, and professional earth scientists who are interested in an interdisciplinary overview of recent advances in understanding of continental evolution. I have not hesitated to give a generous list of references as they contain in their data and ideas the building blocks with which the science has developed; they may save the student, young and old, many hours of library search.

I have benefited from discussions over the years with many friends who have assisted me in understanding the intricacies of a variety of subjects, but in particular I wish to thank the following colleagues who read and improved individual chapters: F. B. Davies, J. D. Hudson, C. H. James, M. A. Khan, M. J. Le Bas, P. C. Sylvester-Bradley and J. H. McD. Whitaker. However, I alone am responsible for any shortcomings or omissions. I wish to acknowledge with thanks the receipt in 1975–76 of a Research Fellowship from the Leverhulme Trust Fund which enabled me to complete the writing free of lecturing duties. Last, but by no means least, for the patient typing of seemingly endless drafts and for her indispensable editorial assistance, I wish to thank my wife, Judith.

Leicester   Brian F. Windley
September, 1976

Preface to Second Edition

In the seven years since I finished writing the first edition there have been many important developments in the Earth Sciences, several with far-reaching implications. The aim of this second edition is to provide an updated synthesis of all those developments that are relevant to crustal evolution. In order to do this I have slimmed down the older or less important parts of the first edition and added a very large amount of new data and ideas on innumerable subjects. It has to be admitted that much of the first edition is now out-of-date, but with this new edition the reader will have an insight into all the recent and current advances.

There is a new introductory chapter which presents relevant time-scale information and a general guide for the reading of the book. Also, there is a new chapter on the Himalayan belt about which there is much growing interest.

The five chapters on the Proterozoic have been substantially revised and a new one on Late Proterozoic mobile belts added; the chapter on 'Crustal Evolution in the Proterozoic' has been entirely rewritten.

There are new sections, or parts of sections, on the following Archaean subjects: early active continental margins, crustal evolution, greenstone belts in India and W Australia, banded iron formations in greenstone belts, and granulite-gneiss belts in India and the Limpopo.

For the Proterozoic there are new sections on the Svecokarelides of the Baltic Shield, the Wopmay Orogen, the Wollaston Lake fold belt, phosphorites, evaporites, sedimentation types and their environments, early plate tectonic settings, anorthosites, Sn mineralization in rapakivi granites, W in limestones, and U-V-Cu in clastics, the Grenville, Dalslandian and Pan-African mobile belts, and Proterozoic greenstone belts and granulite–gneiss belts.

Turning to the Phanerozoic, new sections have been added on phosphorites and oceanic upwelling, faunal extinctions, the Iapetus ocean, the tectonic evolution of the Appalachian and Hercynian fold belts, the elevation and climate of the Pangaea supercontinent, the structure of mid-oceanic ridges, the contrast between the arcs of the W and E Pacific, hot spot tracks and oceanic islands, mineralization in the oceanic environments, the suspect terrains and metamorphic core complexes of the Cordilleran belt of western N America, the structure and mineralization of the Andes, the Coast Range plutonic complex of British Columbia, Palaeo- and Neo-Tethys, Miocene evaporites and ore deposits of the Alpine belt; tectonic accretion, metamorphic belts, fore-arcs and back-arcs of island arcs; the structure and evolution of the Himalayas, and indentation tectonics in Asia.

The final chapter on 'The Evolving Continents' has been substantially revised to take into account, in an integrated manner, this new information. New sections are on the early Precambrian oxygenic atmosphere, uranium mineralization in the early Proterozoic and the Cenozoic, the isotopic composition of Phanerozoic sediments with time, models for the chemical evolution and magmatism of the continental crust together with continental growth models, the thermal regimes of plate tectonic environments, heat production and geothermal gradients in crustal evolution, early Precambrian base metal sulphide deposits, and the Archaean–Proterozoic boundary.

There are 103 new figures and tables, in place of several older ones, and about 1250 new references to papers that have appeared since late 1976 whilst an equivalent number of the older references have been deleted. I have kept to this style of detailed documentation of references, which are the building-blocks of the science, because of the widespread enthusiastic response to this aspect of the first edition.

It is amazing how quickly the Earth Sciences are evolving on a broad front. I hope I have captured in these pages the essence of these recent advances and that the reader will enjoy an integrated perspective with an evolutionary flavour.

Chapter 1

Introduction

Before launching into 3800 Ma of time travel we should have an introductory briefing. The aim of this book is to provide an up-to-date synthesis of earth history as indicated by the geological record; thus no attempt is made to consider the first 800 million missing years between the formation of the planet and the appearance of the earliest rocks (Smith, 1981). But in order to discuss the remaining 3800 Ma we should first define the geological time scale and, in particular, the differences in subdivision of the Precambrian and Phanerozoic throughout the world.

There are two framework scales for expressing time (Harland, 1975; Harland *et al*. 1982). The chronostratigraphic scale is constructed on the superposition principle of supracrustal rocks with boundaries based on unconformities and palaeontologically defined divisions. The chronometric scale defines geological time boundaries by numbered multiples of duration of a standard year (in SI units—Système International d'Unités: 10^0 yr = 1 a; 10^3 yr = 1 ka; 10^6 yr = 1 Ma; 10^9 yr = 1 Ga). The scale takes advantage of isotopic methods for defining time boundaries and is particularly useful for subdividing Precambrian time. In 1976 the IUGS Subcommission on Geochronology recommended a standard set of decay constants and isotopic abundances for U, Th, Rb, Sr, K and Ar and these are now widely used in geochronology (Steiger and Jäger, 1977). Most ages quoted in this second edition are based on the new constants.

The Precambrian Time Scale

Of course there is no definition of the time of the beginning of the Precambrian—that just depends on the advances in geochronological research on the oldest rocks. The International Union of Geological Sciences has approved the terms 'Archaean' and 'Proterozoic' (James, 1978) but their mutual boundary time varies slightly from continent to continent: 2500 Ma in Canada and the USA, and 2600 Ma in the USSR and China (Table 1.1). This boundary is conveniently selected to separate Archaean rocks, which are commonly highly deformed and metamorphosed, from the little-deformed, unconformable, shelf-type Proterozoic rocks. However, crustal evolution has been diachronous with the result that in southern Africa undeformed cover sequences of 'Proterozoic-like' rocks began to appear about 3000 Ma in the Pongola Series. The term Katarchaean has only been taken up widely in the USSR. Table 1.1 gives a breakdown of the subdivisions and regional names of the Proterozoic. The terms Aphebian, Helikian and Hadrynian are used for eras in Canada (Douglas, 1980), and W, X, Y, Z by the Geological Survey in the USA (H. L. James, 1972), whereas the more general divisions, early, middle and late Proterozoic, are used throughout the USA and Mexico (Harrison and Peterman, 1980). The Subcommission on Precambrian Stratigraphy of IUGS agreed in 1979 to divide the Proterozoic eon

Table 1.1 A selection of Precambrian time scales modified after W. B. Harland *et al.* 1982. *A Geologic Time Scale*, reproduced by permission of Cambridge University Press

Eon	Ma	USSR Western	China central	Australia	Africa Southern	Canada (GSC)	USA (USGS)	USA and Mexico	Ma
Phanerozoic (Ph)	~590				(Fish river)				570
Proterozoic	~670	Vendian ~680 Kudash R4 ~700	Sinian	Marinoan	Nama	Hadrynian	Z Zedian	Late Prot	700
	~800	Karatan	Qingbaikou	Sturtian / Torrensian	Gariep				800
		Riphean		Willouran	Nosib				900
	~1050 R3	Yurmatin	Jixian ~1000		Koras ~1080	1000 (Neo-)	Y Yovian	Middle Prot	1000
	~1350 R2	Burzyan	Nankou 1400			Helikian			
	~1650 R1		Changcheng	Carpentarian	Waterberg	(Paleo-) 1600			1500
	~1900	Karelian	~1800	~1800		1800			
	2100	Futuo		Nullaginian	~2070	Aphebian L M	X Xenian	Early Prot	2000
	2400	~2300	Wutai	~2300	Transvaal / Griqualand West	E			
(Pt) 2500	2630 2800	~2600 Byelomorian	~2600 Dantazi	~2630 Sham Vaian ~2800 Bulawayan ~3060	Ventersdorp Witwatersrand Dominion Reef Pongola	Late Ar 2500 Middle Ar 2900	W Weltian	Late Arch 2500 Middle Arch	2500 3000
Archean	3060		Qianxi	Yilgarn / Pilbara	Moodies Fig Tree Onverwacht	3400			3500
(Ar)	3800 4000	Katarchean	~3800	Sebakwian	~3750	Early Ar		Early Arch	4000
				[Zimbabwe]	[South Africa]				4500

into three eras with age boundaries at 1600 Ma and 900 Ma (Sims, 1980). It also recommended dividing the Archaean eon at 2900 Ma and 3500 Ma. In the USSR the subunits of the Proterozoic are the Karelian and the Riphean, the latter being broken down into R_1–R_4. The latest Proterozoic unit is the post-680 Ma Vendian (Keller, 1979; Semikhatov, 1981). There are also type reference sequences (Supergroups and Groups) for the Archaean and Proterozoic divisions in the European (Baltic Shield) and Asiatic (Aldan Shield) parts of the USSR (Kratz and Mitrofanov, 1980). In China the Precambrian is divided into the Sinian and Presinian (the units shown in Table 1.1 are after Wang,

1980). Note that for the early–mid–Precambrian Ma and Wu (1981) have the Qianxi, Fuping, Wutai and Luliang stages separated at 3000 Ma, 2500 Ma, 2050 Ma and 1750 Ma. The data from Australia in Table 1.1 are from Rutland (1981), and those from southern Africa are from Harland *et al.* (1982) which are comparable with those from Tankard *et al.* (1982).

The Phanerozoic Time Scale

This classification is based on subdivisions such as eons (e.g. Phanerozoic), eras (e.g. Palaeozoic), periods (e.g. Cretaceous or Neogene), epochs (e.g. Early, Middle and

Late Cretaceous) and ages (e.g. Albian). There is a current attempt on a global basis to define boundaries by reference to type sections agreed upon by the international community, e.g. the Silurian-Devonian boundary (McLaren, 1977).

The Precambrian–Cambrian boundary, ranging from 570 Ma to 600 Ma, has yet to be defined precisely, although 590 Ma is a likely choice (Harland *et al*. 1982). The position of the boundary is relevant to the Ediacaran fauna of soft-bodied organisms which developed in latest Precambrian time (Chapter 8); the term Ediacaran may, in fact, be elevated to the status of an epoch. For classification and definition of the time boundaries of all the eras, periods, epochs and ages of Phanerozoic time, the reader is referred to the excellent review by Harland *et al*. (1982). Tables 1.2 and 1.3 summarize the main time

data for the Cambrian–Permian and Permian–Holocene, respectively.

The Framework of the Continents

We have evidence that rocks have been forming on earth for the last 3750 Ma, and Fig. 1.1 shows their mutual distribution. Archaean rocks tend to occur as cratons within Proterozoic belts and it has been thought that these were Archaean nuclei around which the Proterozoic belts accreted whilst the Phanerozoic belts, in turn, accreted around the Proterozoic cratons. However, such a simple picture hides a more complicated distribution of rocks that are too small to be represented on this map. Clearly there are numerous small remnants of Archaean rocks in many Proterozoic and Phanerozoic belts and of Proterozoic rocks in Phanerozoic

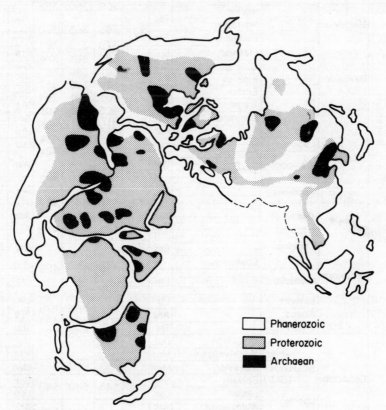

Phanerozoic
Proterozoic
Archaean

Fig. 1.1   A Permian pre-drift map of the continents showing the distribution of Archaean regions within Proterozoic cratons surrounded by Phanerozoic mobile belts

Table 1.2 Chronostratigraphic scale with chronometric calibrations for the Cambrian to the Permian. Modified after Harland and coworkers 1982. Reproduced by permission of Cambridge University Press

Eon	Era	Period Sub-period	Epoch	Age			Abbreviation	Van Eysinga, 1975	Armstrong, 1978	Pickton in Hambrey and Harland, 1981	Harland et al., 1982 Ma	Age	Duration
Phanerozoic	Palaeozoic	Permian P₁		Asselian	C		Ass	280	288	290	286	Ass	9
		Carboniferous (C) — Pennsylvanian (Pen)		Gzelian	Noginskian	Stephanian	Gze					Gze	?
				Kasimovian	B		Kas					Kas	?
					Krevyakinsk. A								
				Moscovian	Myachkov. Ctb	Westphal	Mos		307	300	296	Mos	? / 34
					D / C / B / A					308			
				Bashkirian	Yeadonian C / B	Namurian	Bsh		330	(320)	315	Bsh	?
					A					(328)	320?		
		Mississippian (Mis)		Serpukhovian			Spk		341	(335)	333	Spk	13 / ?
								325					
			Visean	Holkerian			Vis					Vis	19 / 40
									356	(350)	352		
			Tournaisian	Ivorian			Tou		368	365	360	Tou	8
				Hastarian				345					
		Devonian (D)	D₃	Famennian			Fam		379	373	367	Fam	7
				Frasnian			Frs		385	381	374	Frs	7
			D₂	Givetian			Giv		391	388	380	Giv	6 / 48
				Eifelian			Eif	370	396	395	387	Eif	7
			D₁	Emsian			Ems		401	401	394	Ems	7
				Siegenian			Sig		406	406	401	Sig	7
				Gedinnian			Ged	395	(417)	411	408	Ged	7
									(425)				
		Silurian (S)	Pridoli				Prd			415	414	Prd	6
			Ludlow				Lud		432	422	421	Lud	7 / 30
			Wenlock	Sheinwoodian			Wen	423	440	428	428	Wen	7
			Llandovery				Lly	435	446	440	438	Lly	10
		Ordovician (O)	Ashgill				Ash		445	447	448	Ash	10
			Caradoc	Costonian			Crd	450	465	469	458	Crd	10 / 67
			Llandeilo				Llo		477	477	468	Llo	10
			Llanvirn				Lln		492	485	478	Lln	10
			Arenig				Arg	500	500	495	488	Arg	10
			Tremadoc				Tre		510	505	505	Tre	17
		Cambrian (€)	Merioneth (Mer)	Dolgellian			Dol					Dol	9
				Maentwrogian			Mnt	515	524	520	523	Mnt	9
			St. Davids (St D)	Menevian			Men					Men	9 / 85
				Solvan			Sol	540	545	540	540	Sol	8
			Caerfai (Cr f)	Lenian			Len					Len	15
				Atdabanian			Atb			570	570	Atb	15
(Ph)	(Pz)			Tommotian			Tom	570	574	590	590	Tom	20

Table 1.3 Chronostratigraphic scale with chronometric calibrations for the Permian to the Holocene. Modified after W. B. Harland *et al*. (1982). *A Geologic Time Scale*. Reproduced by permission of Cambridge University Press

Eon	Era	Sub-era / Period	Epoch	Age	Abbreviation	Van Eysinga, 1975	Armstrong, 1978	Pickton in Hambrey and Harland, 1981	Harland et al., 1982 — Ma	Harland — Age	Harland — Duration
Phanerozoic (Ph)	Cenozoic (Cz)	Quaternary (Q) or Pleistogene	Holocene		Hol	0.01		0.01	0.01	Hol 0.01	2.0
			Pleistocene		Ple	1.8		2.0	2.0	Ple 1.99	
		Tertiary (TT) — Neogene (Ng) [22.6]	Pliocene (Pli) 2	Piacenzian	Pia					Pia	3.1
			Pliocene (Pli) 1	Zanclian	Zan	5		5	5.1	Zan	
			Miocene 3	Messinian	Mes					Mes	6.2
			Miocene 3	Tortonian	Tor				11.3	Tor	
			Miocene 2	Serravallian	Srv					Srv	3.1
			Miocene 2	Langhian – Late	Lan$_2$				14.4	Lan$_2$	
			Miocene 2	Langhian – Early	Lan$_1$					Lan$_1$	
			Miocene 1	Burdigalian	Bur					Bur	10.2
			Miocene 1	Aquitanian	Aqt	22.5		22.5	24.6	Aqt	
		Paleogene (Pg) [40.4]	Oligocene 2	Chattian	Cht				32.8	Cht	8.2
			Oligocene 1	Rupelian	Rup	38		37	38.0	Rup	5.2
			Eocene 3	Priabonian	Prb				42.0	Prb	4
			Eocene 2	Bartonian	Brt					Brt	8.5
			Eocene 2	Lutetian	Lut				50.5	Lut	
			Eocene 1	Ypresian	Ypr	55		53.5	54.9	Ypr	4.4
			Paleocene (Pal) 2	Thanetian	Tha				60.2	Tha	5.3
			Paleocene (Pal) 1	Danian	Dan	65	65	65	65	Dan	4.8
	Mesozoic (Mz)	Cretaceous (K) — K2 Senonian [79]		Maastrichtian	Maa		65	65	65	Maa	8
				Campanian	Cmp		72	73	73	Cmp	10
				Santonian	San		84	84	83	San	4.5
				Coniacian	Con		88	86	87.5	Con	1
				Turonian	Tur	80	90	88	88.5	Tur	2.5
				Cenomanian	Cen	100	92	90	91	Cen	6.5
		K1 Neocomian		Albian	Alb		106	95	97.5	Alb	15.5
				Aptian	Apt		116	107	113	Apt	6
				Barremian	Brm	118	123	115	119	Brm	6
				Hauterivian	Hau		127	121	125	Hau	6
				Valanginian	Vlg		130	126	131	Vlg	7
				Berriasian	Ber	141	136	131	138	Ber	6
		Jurassic (J) — J3 Malm [69]		Tithonian	Tth		143	135	144	Tth	6
				Kimmeridgian	Kim		149	141	150	Kim	6
				Oxfordian	Oxf	160	157	143	156	Oxf	7
		J2 Dogger		Callovian	Clv		162	149	163	Clv	6
				Bathonian	Bth		166	156	169	Bth	6
				Bajocian	Baj	176		165	175	Baj	6
				Aalenian	Aal		177	171	181	Aal	7
		J1 Lias		Toarcian	Toa		188	178	194	Toa	6
				Pliensbachian	Plb		198	183	200	Plb	6
				Sinemurian	Sin			189	206	Sin	6
				Hettangian	Het	195	211	192	213	Het	7
		Triassic (Tr) — Tr3 [35]		Rhaetian	Rht		220	(197)	219	Rht	6
				Norian	Nor		228	(202)	225	Nor	6
				Carnian	Crn		234	(207)	231	Crn	6
		Tr2		Ladinian	Lad		238	(214)	238	Lad	7
				Anisian	Ans		242	(221)	243	Ans	5
		Scythian (Scy) — Tr1	Olenekian	Spathian	Spa			(224)		Spa	1¼
			Olenekian	Smithian	Smi			(228)		Smi	1¼
			Induan	Dienerian	Die			(231)		Die	1¼
			Induan	Griesbachian	Gri	230	247	235	248	Gri	1¼
	Paleozoic	Permian (P) — P2 [38]		Tatarian	Tat				253	Tat	5
				Kazanian	Kaz		252	(239)		Kaz	2½
				Ufimian	Ufi	251	259	(242)	258	Ufi	2½
		P1		Kungurian	Kun			(255)	263	Kun	5
				Artinskian	Art		269	(266)	268	Art	5
				Sakmarian	Sak		278	277		Sak	9
				Asselian	Ass		288	290	285	Ass	9

belts; some of these are reworked fragments within younger high deformation zones and others are uplifted segments of widespread basement that typically underlies many orogenic belts. For example, the fact that most of the Himalayas is underlain by reworked Proterozoic sediments and gneisses can be demonstrated frequently in places where the uplift has been very high. Yet the formation of the Alpine-Himalayan belt clearly involved very little accretion of new material: in the Alpine belt there is a remarkable lack of granitic rocks (Chapter 20), in the Himalayas there is only one island arc (Kohistan-Ladakh) and one Andean-type batholithic belt (Kangdese), and the orogenic belt is characterized more by erosion of older rocks and re-deposition as sediments, intrusion of granites formed by partial melting of lower continental crust and by tectonic reworking (metamorphism and deformation) of Precambrian basement (Chapter 21). Thus it can be said that the formation of Mesozoic–Cenozoic orogenic belts by plate tectonic processes typically involved more plate destruction than accretion. Significantly, the evolution of some Proterozoic mobile belts is characterized more by plate destruction than accretion (Kröner, 1977b) and so the relationships cannot be used to suggest lack of lateral plate motions in Precambrian times (Chapter 10). Also, the formation of some Andean-type belts has involved tectonic erosion at the site of subduction with the result that Precambrian crystalline basement extends as far as the present coastline (e.g. Peru, Shackleton *et al*. 1979). So on all scales we are seeing relicts of older rocks within younger rocks and a basement of older rocks beneath younger cover sequences. This means that the distribution of Archaean, Proterozoic and Phanerozoic rocks depicted in Fig. 1.1 is not a direct reflection of the ratio in which they formed. This subject is related to that of 'Continental Growth Models' reviewed in Chapter 22.

Another important aspect to consider when looking at Fig. 1.1 is that differential erosion and uplift are considerable. The Archaean terrains described in Chapters 2 and 3 formed

at deep- and shallow-crustal levels, respectively, and are thus invariably separated by faults. Turning to the youngest orogenic belt in the world, the vertical tectonic level along the Himalayas also varies enormously. In southern Tibet the uplift is so little that Palaeogene sediments are still preserved on the mountain tops, whereas in the Karakorum-Kohistan ranges of N Pakistan recent uplift has been so rapid that most of the Phanerozoic pile of sediments has been removed by erosion. Thus, again, basement is as important as cover, a fact that should be borne in mind when reading many chapters in this book.

In Perspective—Forwards of Backwards?

The subject of how the continents have evolved through geological time is currently of much interest. The aim of this book is to reflect that interest by synthesizing data and ideas in such a way that the reader gets an evolutionary feel about the process of events, and in order to obtain such a perspective the reader should start at the beginning. However, so that the many threads found in earlier chapters may be brought together, the final chapter reviews in a more integrated way some of the key aspects bearing on the subject of crustal evolution. With such an approach the reader will find that an understanding of Archaean geology will help immeasurably when beginning to dip into early Proterozoic tectonics. Likewise, late Proterozoic conditions are useful in the appreciation of the environment in which early Palaeozoic sediments and life forms evolved. Conversely, we could ask two questions: is it possible to appreciate the tectonics of the late Palaeozoic without understanding the plate motions of the early–mid Palaeozoic, or is it helpful in understanding the evolution of the Himalayas to know something about the drift of continents following the break-up of Pangaea? Clearly the answers are no and yes, respectively.

However, such an approach has certain disadvantages, and this concerns the subject of plate tectonics. This is a conceptual model which is unifying and revolutionizing the earth sciences, but it is based only on the geology of the last 200 Ma, the age of the oldest extant ocean floor. All pre-Mesozoic geology is interpretative with respect to the plate tectonics cycle of events; therefore, in order to understand the geology of pre-Mesozoic orogenic belts in plate tectonic terms, or rather if we wish to see how far back in time we can apply such a model, we have to have a basic comprehension of the relevant aspects of plate tectonics. Following this logic when reading this book, if the reader feels he/she wishes to consider these questions rather than the evolutionary theme referred to earlier, it would be sensible first to read Chapters 17–21 (in that order) and then the chapters on the Palaeozoic, Proterozoic and Archaean. This would demonstrate how progressively difficult it is to interpret the geological record in plate tectonic terms further back in time.

Perhaps this is one of the first books that can be read equally advantageously from back to front as from beginning to end!

Chapter 2

Archaean Granulite–Gneiss Belts

The Archaean (pre-2500 Ma) regions of the world contain two types of terrain: gneiss-dominated belts metamorphosed largely to a high metamorphic grade, the subject of this chapter, and the well-preserved, low-grade, volcanic-dominated greenstone belts of Chapter 3. Some regions contain both types and their mutual relationships, as well as the general problems of their possible evolution, are reviewed in Chapter 4.

Most high-grade regions went through a major supracrustal-plutonic event in the period 3100–2800 Ma and a few, i.e. West Greenland, the Limpopo belt, several shields in the USSR (Shuldiner, 1982) and Labrador, have a history that started 3800–3600 Ma.

The commonest rocks are granulite to upper amphibolite facies gneisses; special problems are attached to their mode of origin and to the geochemistry of orthopyroxene-bearing types that might be termed charnockites or felsic granulites. The gneisses contain the remains of some of the earliest sediments and volcanics, about which little is known in detail, and also of layered igneous complexes with calcic anorthosites. The rocks are typically folded into large-scale interference patterns and traversed by major shear belts.

The most important recent advances in the understanding of the evolution of Archaean high-grade regions have been in the North Atlantic craton (Fig. 2.1) that includes NW Scotland, Greenland and the Labrador coast (Bridgwater et al., 1973, 1978). In this chapter, therefore, data from these regions will be prominent.

Distribution

The Archaean high-grade rocks occur in a wide variety of environments: extensive areas such as the North Atlantic craton and the Aldan Shield, linear belts like the Limpopo mobile belt of southern Africa, small areas between greenstone belts (e.g. Zimbabwe), and in small isolated remnants such as in the Lofoten Islands of Norway.

When plotted on a pre-Mesozoic, precontinental drift map the occurrences are grouped into several areas. The regions concerned are surrounded by younger mostly Proterozoic belts, often formed by remobilization of the Archaean rocks, or partially covered by younger deposits; it is certainly safe to regard the present Archaean areas as just minimal remains of once larger regions.

Rock Units

There is a succession of different rock types in these high-grade regions. It is usual to find that major layers are mutually conformable and thus it is possible to map out an intercalation of rock units. In some parts of the world the gneisses are regarded as recrystallized sediments or volcanics and thus the present succession is thought to represent a remnant supracrustal pile. However, the results of V. R. McGregor (1973) from West Greenland demonstrate that the intercalation there is due not to supracrustal deposition but to deformation and conformable intrusion. A comparable picture has emerged from Lab-

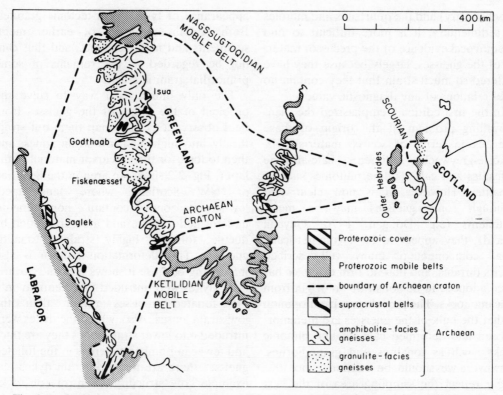

Fig. 2.1   Outline map of the Archaean craton of the North Atlantic region flanked in Greenland by the Nagssugtoqidian and Ketilidian mobile belts of Proterozoic age (after Bridgwater, Watson and Windley, 1973)

rador (Collerson *et al.*, 1976) and the Limpopo belt (Robertson and du Toit, 1981).

Now let us look at the nature of the most important of these interlayered rocks and some of the problems connected with their origin.

Quartzo-feldspathic Gneisses

The most common type is a quartzo-feldspathic gneiss containing biotite and/or hornblende, hypersthene with or without diopside, and plagioclase, potash feldspar and quartz in various proportions; such rocks may constitute as much as 80–90% of these regions. Plagioclase-rich trondhjemitic and tonalitic types with high Na/K ratios are particularly common (Collerson and Bridgwater, 1979; Tarney *et al.*, 1979).

The gneisses are typically foliated and may have felsic and mafic bands up to about 1 cm

wide. Detailed mapping may reveal an intercalation of gneiss types based on variations in composition, alternation of banded or more homogeneous types and the presence or absence of inclusions of a particular type such as amphibolites or anorthosites.

The gneisses contain a great variety of other rocks. Most prominent are thick layers of amphibolite, mica schist, quartzite and meta-anorthosite; rare, but locally thick, marbles and banded iron formations are a surprising feature in such old rocks; in places there are amphibolite dykes and inclusions of meta-ultramafics. Pegmatites tend to be common in amphibolite facies but rare in granulite facies areas.

Whilst it is possible to work out satisfactorily the origin of the amphibolites as volcanic (with pillows), or as dykes (with apophyses and discordances), the anorthosites as igneous (with   textures,   chromitite-layering, geochemistry) and the quartzites and marbles as sedimentary, it is more difficult to find unequivocal evidence of the precussor material of the gneisses, largely because they have suffered so much strain that they contain no field relations of any diagnostic value.

In the first edition I emphasized the then-prevailing problem of the origin of these gneisses in terms of source material, viz. arkosic-greywacke sediments, calc-alkaline volcanics and calc-alkaline plutonics such as tonalites. However, it is now clear that although some gneisses may be meta-sediments (e.g. Dougan, 1976; Guyana Shield), they appear to be a volumetrically small component of many well-described gneiss terrains, and that no firm evidence has been adduced to demonstrate an origin from volcanic rocks. The prevalent current opinion is that the bulk of the gneisses are metamorphosed and deformed calc-alkaline plutonic rocks, such as tonalites and granodiorites. However, we should be cautious since this *might* reflect the sampling area of the best described rocks which are largely from the North Atlantic Craton, the Limpopo belt and S India because meta-sedimentary gneisses are considered to be predominant in NE China (Sun and Wu, 1981; Cheng *et al.*, 1982) and in eastern USSR (Shuldiner, 1982).

We owe much to McGregor (1973) who established without any reasonable doubt that many Amîtsoq and Nûk gneisses in West Greenland were derived by deformation of homogeneous granites. The leucocratic bands of the gneisses were formed partly by elongation of feldspar megacrysts, and partly by segregation during metamorphic differentiation of quartzo-feldspathic material during shearing. McGregor's results were fundamental and impressive as they demonstrated for the first time exactly how a high-grade granitic gneiss of widespread extent is derived. In a similar way Collerson *et al.* (1976) demonstrated that some Uivak gneisses in Labrador are derived by deformation of porphyritic granites. In so far as some Amîtsoq, Nûk and Uivak precursors are homogeneous, feldspar megacryst 'granitic' rocks, they have the appearance of typical post-tectonic granites. Both contain remnants of earlier meta-sediments and meta-volcanics, and thus cannot be regarded as the remains of some primordial granitic crust.

The only unequivocal way to solve the problem of the origin of the gneisses from field observations is to map them out structurally into high- and low-strain zones, and then to look for the precursor material in the latter. Fig. 2.2 shows an area in the Scourian of NW Scotland where lens-shaped, low-strain zones contain homogeneous undeformed tonalite and are surrounded by ductile, foliated, highly strained tonalitic gneisses. This deformation pattern is also interesting because it shows how the Scourie basic dykes were intruded preferentially into the foliated gneisses rather than the low-strain lenses. Also, where the dykes were intruded into lower strain zones they are thick and fewer in number, whereas in the foliated gneisses there are numerous thin dykes—an example of structural control of dyke emplacement.

In all gneissic terrains the deformation has been characteristically inhomogeneous so that high- and low-strain zones can be mapped. In my experience I have never observed a low-strain zone in gneisses on any continent in which the precursor material is obviously a sediment or an acid-to-intermediate volcanic rock. At present the onus is on those who envisage a meta-sedimentary or meta-volcanic origin for the widespread gneisses to demonstrate that such material is indeed present in low-strain zones.

Because most gneisses have a general calc-alkaline chemistry and because there are serious arguments against a simple sedimentary or volcanic derivation, an origin by deformation of tonalitic-granodioritic rocks for many gneisses is most likely (e.g. Rollinson and Windley, 1980a). The TiO_2–SiO_2 distribution of Archaean gneisses is certainly comparable to that of the British Caledonian continental margin calc-alkaline igneous trend (Tarney, 1976). The Amîtsoq gneisses have a lanthanide distribution pattern remarkably similar to that of the Mesozoic

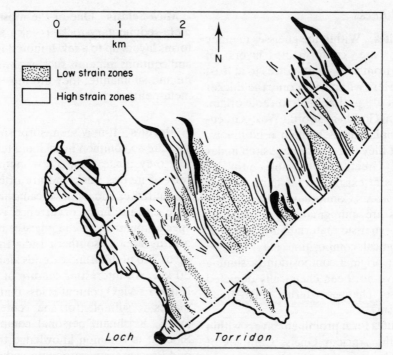

Fig. 2.2 High and low strain zones in the tonalitic Scourian gneisses of the Torridon area of NW Scotland. The low strain zones are remnant lenses of undeformed homogeneous tonalitic rocks. NW-trending basic dykes (in black) were intruded preferentially into the highly strained foliated gneisses (after Davies and Windley (in preparation) quoted in Windley (1982), Igneous rocks of the Lewisian complex, in D. S. Sutherland (Ed.), *Igneous Rocks of the British Isles*, Wiley, London, Fig. 2.4, p. 14)

Bonsall tonalite from the Peninsular Range batholith in southern California and their feldspar-related mineralogy and chemistry is comparable to that in calc-alkaline igneous suites (Lambert and Holland, 1976). Holland and Lambert (1975) and Tarney and Windley (1977) suggested that calc-alkaline magmas were intruded and underthrust into deep levels of the continental Archaean crust and this culminated in granulite-facies metamorphism.

Two principal hypotheses have been proposed to explain the genesis of meta-igneous Archaean gneisses. (1) They were derived by partial melting of pre-existing continental/sialic crust (Fyfe, 1973; Collerson and Bridgwater, 1979), or (2) they were derived by partial melting of mafic/basaltic crust and/or fractional crystallization of basic magma (Compton, 1978; Drury, 1978; McGregor, 1979; Moorbath, 1980; Rollinson and Windley, 1980a; Anhaeusser and Wilson, 1981). Many of these papers use trace and REE data to constrain these models, and currently more geochemists ascribe to the second proposal. This problem is related to such questions as whether younger Archaean gneisses were produced by remelting of older gneisses (see later in this chapter for isotopic constraints), whether Archaean granulites are residues of crust depleted in H_2O and lithophile elements (Fyfe, 1973), or whether they were created by the influx of mantle-derived, CO_2-rich fluids during stabilization of continental crust (Newton *et al.*, 1980), and whether the precursor calc-alkaline magmas could have been derived by partial melting of subducted ocean-type crust.

Supracrustal Rocks

Amphibolites Within the gneisses in many areas there are conformable layers of amphibolite from a few centimetres to at least 1 km thick. Here we consider only the thicker layers, which are probably of volcanic origin. Rarely, as in Godthaabsfjord, West Greenland, the amphibolities contain relic pillows showing that they were lavas deposited under water; this Greenland example provides proof that water existed on the earth at least 3000 Ma ago. Commonly, the thicker amphibolites are unmigmatized by penetrative granitic–gneissic material and thus their present chemical composition may well be close to their original composition (assuming that they have not been chemically depleted or otherwise changed during high-grade metamorphism).

Amphibolites form prominent layers within gneisses in the Ancient Gneiss Complex of Swaziland (Hunter *et al*., 1978), the Guyana Shield (Dougan, 1976), the Aldan Shield (Fedorouskij and Lejtes, 1980; Moralev, 1981), the Sargur supracrustals of S India (Janardhan *et al*., 1978), NE China (Sun and Wu, 1981), and the Limpopo belt (Key, 1977).

More has recently been published on the geochemistry and possible origin of amphibolites from W Greenland than elsewhere. Amphibolites make up 25–40% of the 3370 Ma Isua supracrustal belt (Gill *et al*., 1981) and are of two types, formed either from tholeiitic or high-MgAl basaltic magmas. Amphibolites from the Malene supracrustal belt in Buksefjorden have high Fe, Cr and Ni, reflecting a mid-Archaean mantle source enriched in these elements, and their volcanic parents may have been fed by the extensive Ameralik dyke swarm (Chadwick, 1981). Friend *et al*. (1981) found that two widely separated belts of amphibolite were distinct in composition, one having an island arc and the other an oceanic character. Trace element chemistry, especially chondritic La/Ta ratios of amphibolites in the Fiskenaesset region, suggest an ocean-floor origin which would be compatible with their pillow structure and very wide lateral extent (Weaver *et al*., 1982).

Mica Schists One of the most prominent meta-sedimentary rocks is mica schist which forms layers up to a few hundred metres thick and contains minerals such as garnet, cordierite and sillimanite; they are most reasonably meta-pelites.

Marbles It may seem surprising that marbles are so common in such old terrains. Usually they range from a few metres to a few tens of metres thick and are often bordered by quartzites and meta-volcanic amphibolites. One extraordinary occurrence is worthy of special mention. In the gneisses of the Sankaridrug area of southern India the marble is up to 250 m thick and extends along strike for 30 km. An interesting feature of this marble is that its MgO content is less than 4%, which makes it suitable for the cement industry (V. S. Krishnan, personal communication). Since it is common knowledge that early to mid-Precambrian carbonate sediments were mostly dolomites, and that limestones did not evolve in abundance until later times, this occurrence of an extremely old dolomite-free marble is unique.

Other high-grade marbles occur in the Limpopo belt of southern Africa (Robertson and du Toit, 1981), the Aldan Shield of Siberia (Moralev, 1981; Kazansky and Moralev, 1981), NE China (Sun and Wu, 1981), and S India (Janardhan *et al*., 1978). There is considerable scope for research into the geochemistry of these marbles.

Quartzites Quartzites are of special importance in many high-grade regions. In the Limpopo belt they reach 3 km thick and are closely interbedded with the marbles. In the Aldan Shield there is an aggregate thickness of several km. They form thin layers in the Lofoten Islands of NW Norway, the Androyan and Graphite Sequences of Malagasy, the Scourian and southern India. Some fuchsite quartzites in the Sargar Supracrustals of S India contain a heavy mineral suite of chromite, magnetite, tourmaline, rutile and zircon, an assemblage similar to that of modern beach placers (Ramiengar *et al*., 1978), whilst others contain beds of barytes

(Radhakrishna and Sreenivasaiah, 1974). Some quartzites, which are enriched in magnetite, grade into the quartzitic facies of iron formation. It is possible, but has yet to be demonstrated, that some quartzites within these high-grade terrains, are meta-cherts.

Iron Formations The Archaean Granulite-gneiss belts contain banded iron formations (BIF) which are a weakly banded quartz-magnetite facies associated with the above supracrustals. They are well represented in southern India, in Hebei and Liaoning Provinces in NE China, the Kola Peninsula, Stanovoy Range and the Ukraine in the USSR, the Kambui schists of Sierra Leone, the Nimba Series of Liberia, the Limpopo belt of southern Africa, the Imataca Complex of Venezuela and the Isua area in W Greenland (for details and references see Prasad et al., 1982). They are typically only a few metres thick, but reach 30 m in southern India, and in NE China are up to 80 m in thickness and several kilometres in length and thus form the foundation of the steel industry of NE China at Anshan (Sun and Wu, 1981). These Chinese BIF are hardly known in the English language literature and yet are so abundant that they are unique in the Archaean. For example, several mines are so large that they employ between 10 000 and 30 000 workers. These BIF are described by Cheng Yu-Chi et al. (1978), Wang Liankui et al. (1979), Zhang et al. (1981) and Li Shunguang (1982). Except for Isua, BIF are very thin and rare in the Archaean gneisses of the North Atlantic region. The Chinese BIF illustrate the fact that we are getting a biased picture of Archaean high-grade regions by over-emphasis of the North Atlantic region.

One iron formation is particularly important. In the Isua area of West Greenland a gneiss dome is mantled by a greenschist- to amphibolite-grade sequence with volcanics, calcareous quartzites and an economic magnetite-banded iron formation (Fig. 2.3) (Perry et al., 1978; Appel, 1979, 1980). The BIF has yielded a Pb/Pb age of 3770 Ma. There are no other comparable BIF in West Greenland, nor are similar rocks of this age known at present from any other continent. Discovery of BIF as old as this places new constraints on the model of Cloud (1968a)—that BIF are acceptors of oxygen produced by photosynthesizing organisms. Either the model is incorrect or blue-green algae flourished 3800 Ma ago.

Finally, what about the environment in which these sediments were laid down? They were deposited in a quartzite–pelite–carbonate association in an epicontinental platform/shelf environment under shallow water and stable tectonic conditions. They represent a proto-Superior-type of BIF formed at a time when stable cratons were not long lasting. They are distinct from the Algoma-type BIF in greenstone belts. It is important to emphasize that these sediments are very different from those in the Archaean greenstone belts (clastic – greywacke – flysch – conglomerate – shale association) which accumulated in unstable turbidite-type eugeosynclinal environments. The marked difference in these sediment types makes it impossible for the high-grade Archaean terrains to be simply highly metamorphosed greenstone belt sequences (see further in Chapter 4).

Layered Igneous Complexes

In many high-grade Archaean regions there are tectonic lenses and layers of a variety of deformed and metamorphosed layered igneous rocks (for general review see Windley et al., 1981). Two types can be distinguished.

Anorthosite–Leucogabbro Recent studies of these plagioclase-rich complexes include:

Fiskenaesset, W. Greenland (Fig. 2.4) (Bishop et al., 1980; Weaver et al., 1981)
Labrador (Wiener, 1981)
S India (Ramakrishnan et al., 1978; Janardhan et al., 1978)
Limpopo belt, S India (Barton et al., 1979)
Aldan Shield, USSR (Fedorovskij and Lejtes, 1980)
Kapuskasing belt, S Canada (Simmons et al., 1980)

14

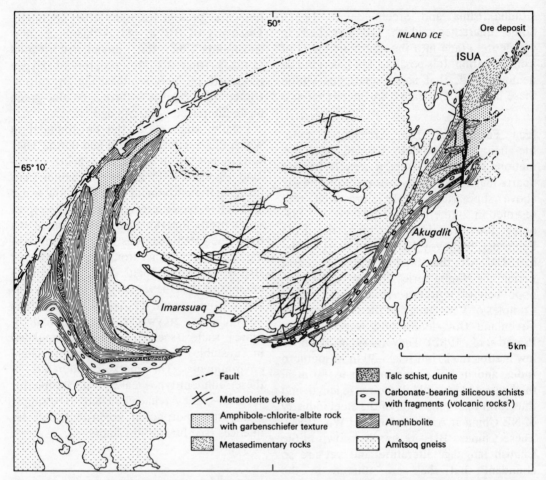

Fig. 2.3   Map of the Isua supracrustal belt, W Greenland; for location see Fig. 2.1 (after Allaart,1976; reproduced by permission of The Geological Survey of Greenland)

Pikwitonei Province, S Canada (Hubregtse, 1980)
Central Australia (Katz, 1981).

The complexes range up to about 1 km in thickness but often, due to tectonic disruption and thinning, they are only tens of metres to 100 m or so thick, and where extensively migmatized (invasion by late gneisses) they may be represented by only a few metre-sized pods.

The leucogabbros, which could also be termed gabbro(ic) anorthosites, typically have a prominent cumulate igneous texture marked by subhedral plagioclase megacrysts in a hornblendic matrix. The presence of the cumulate textures in leucogabbros, chemi-

cally graded layers and, in key complexes (Fiskenæsset, Limpopo and Sittampundi, S India), chromitite seams demonstrates that these rocks belong to layered igneous complexes. Gabbros form a minor component of some bodies (Fiskenæsset, Sittampundi) but ultramafic rocks are rare; eclogitic types occur at Sittampundi. Sapphirine-bearing rocks are associated with many complexes (Fiskenæsset, Limpopo, Central Australia, Sittampundi and Sakeny, Malagasy).

The complexes are commonly bordered by supracrustal rocks, either meta-volcanic amphibolities or a shelf-type assemblage of quartzites, marbles and sillimanite mica schists (meta-K pelites), and they are often intruded by the nearby tonalitic gneisses.

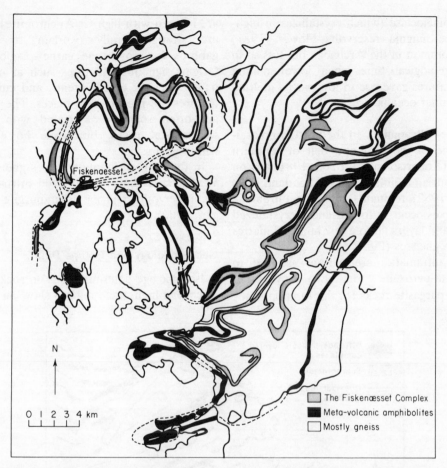

Fig. 2.4 Map showing the anorthositic Fiskenæsset Complex, West Greenland (see Fig. 2.1) bordered by metavolcanic amphibolites within a high-grade gneiss terrain (redrawn from J. Myers, Plate 1, Rapp. Grønlands geol. Unders., **73**, 1976; reproduced by permission of The Geological Survey of Greenland)

The crystallization pattern of these bodies is dominated by plagioclase-hornblende; the plagioclase is rather calcic with an An range of 80–100 where best preserved and much lower where retrogressively overprinted, and the hornblende is subsilicic tschermakite–hastingsite–pargasite–edenite. Chromites have a distinctive high FeAl composition (Cr : Fe = 1 : 1), different from that in ophiolites and layered intrusions such as Bushveld. The original crystallization pattern of the Fiskenæsset Complex has been only weakly modified by metamorphic re-equilibration as igneous cryptic chemical variations are preserved in some mineral species (Bishop *et al.*, 1980). The Fiskenæsset com-plex formed by crystal fractionation of a trace element-impoverished tholeiitic aluminous magma generated by hydrous fusion of previously depleted mantle. Part of this liquid escaped at an early stage to give rise to trace element-impoverished pillow-bearing basalts for which an ocean-floor environment is not implausible (Weaver *et al.*, 1981). The remaining liquid was emplaced into the basalt pile to form the Fiskenæsset complex. Sleep and Windley (1982) calculated from thermal data that the Archaean oceanic crust could have been much thicker (>20 km) than modern crust and it could have had a significant component of anorthositic cumulates. This may be the reason why anorthosites with calcic plagioclase (which crystallizes in high-pressure magma reservoirs–Flower, 1981) were common in the Archaean but less so in later geological time when lower mantle temperatures gave rise to less partial melting and thinner oceanic crust.

Mafic–Ultramafic   In the northern marginal zone of the Limpopo belt of southern Africa (Robertson and du Toit, 1981) and on the Scourian mainland of NW Scotland (Sills *et al*., 1982) remnants of mafic–ultramafic complexes occur as folded and recrystallized lenses and layers up to a few hundred metres thick in gneisses (Fig. 2.5).

The ultramafics are variably altered to tremolite-pyroxene or two pyroxene-olivine-pargasite rocks (in the Limpopo they are layered with high Cr/Al chromite) which are stratigraphically overlain by meta-gabbros (two pyroxene garnet-plagioclase). Original igneous structures such as opaque oxide layering, cryptic trends and cumulate textures are preserved in places. The bodies are often spatially associated with meta-supracrustal rocks, but some lie entirely within gneiss.

In the Yilgarn granulite facies gneisses of W Australia there is a layered intrusion of meta-lherzolite and meta-harzburgite (Morgan, 1982).

Geochronology and Isotope Data

Isotopic age determinations on rocks from Archaean high-grade regions show an evolu-

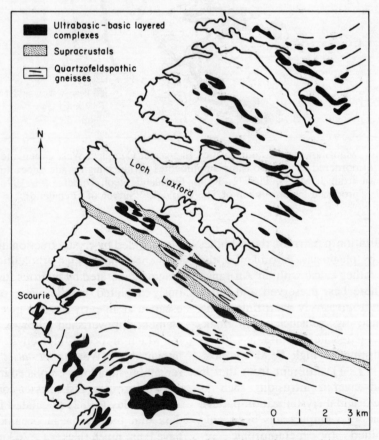

Fig. 2.5   Map showing remnants of ultrabasic–basic layered igneous complexes bordered by supracrustal mica schists in gneisses of the Scourie–Laxford area, NW Scotland (redrawn from F. B. Davies, personal communication)

tion from *c.* 3800 to 2500 Ma, about a third of geological time.

Only a few years ago there were no significant ages older than 3000 Ma and it was generally thought that typical Archaean high-grade regions had been through a single evolutionary stage. It is now realized that several regions have had a multistage history; in particular this refers to West Greenland, Labrador and the Limpopo mobile belt, the relative chronologies of which are shown in Table 2.1. Isotopic ages in most other high-grade regions in the world fall in the period 2700–3100 Ma (e.g. Gwenora in Zimbabwe, Uganda, Wyoming and Wisconsin in the USA, S India, N Sweden, NW Scotland, the Lofoten Islands in N Norway, E Greenland, British Columbia, Quebec, Pikwitonei in Manitoba, the English River gneisses in Ontario in Canada and the Ukrainian Shield in the USSR). Interesting older ages have recently been reported from the Minnesota River Valley, USA (3230 Ma, U–Pb: Michard-Vitrac *et al.*, 1977), Michigan, USA (3410 Ma, U–Pb: Peterman *et al.*, 1980; 3600 Ma, Sm–Nd: McCulloch and Wasserburg, 1980), Imataca in Venezuela (3670 ± 100 Ma, U–Pb: Montgomery, 1979), the Yengra gneisses, Aldan Shield (3300 ± 200 Ma, Pb–Pb: Moralev, 1981), the Omolon block, Siberia (3400 ± 150, U–Pb: Bibikova *et al.*, 1978), the Stanovoy Range, Siberia (3250 ± 100 Ma, U–Pb: Neymark *et al.*, 1981), Swaziland (3555 ± 111 Ma, Rb–Sr: Barton *et al.*, 1980), S India (3358 ± 66 Ma, Rb–Sr: Beckinsale *et al.*, 1980), and the Yilgarn block of W Australia (3348 ± 43, Rb–Sr: De Laeter *et al*, 1981). It is probable, indeed predictable, that older ages will sooner or later be revealed in several of these regions. This applies especially to S India which has similar tectono-stratigraphic units to those in West Greenland, Labrador and the Limpopo belt (i.e. gneisses and meta-supracrustals associated with layered ultramafic–mafic–anorthositic complexes) and comparable structural and metamorphic histories.

It is widely thought that the radiometric dates obtained from these high-grade gneissic regions are a measure of the age, in general terms, of a regional metamorphism or of a late reworking or volatile fluxing event (*cf.* Bridgwater and Collerson, 1976; Chadwick and Coe, 1976; Collerson and Fryer, 1978). However Moorbath (1978, 1980) has cogently pointed out that the isotopic data place severe constraints on our ideas about the evolution of these high-grade gneisses. In its simplest form the conclusion is that, rather than indicating any of the above overprint events, Rb/Sr whole-rock isochrons simply tell us the date when the precursors of the orthogneisses were added to the continental crust. Lead isotope studies may also be used to distinguish between the 'reworking' and the 'accretion' models (Taylor *et al.*, 1980). Moorbath's suggestion has such serious implications for the evolution of early sialic material that it is worth following briefly the basis of his argument.

The $^{87}Sr/^{86}Sr$ ratio of any given rock (or mineral) increases progressively with time because ^{87}Rb is decaying radioactively to ^{87}Sr with a half life of 50 000 Ma and the growth rate of $^{87}Sr/^{86}Sr$ for a given period of time is proportional to the Rb/Sr ratio in the sample. Now the Amîtsoq gneisses, for example, have an average Rb/Sr ratio of *c.* 0.3 and they had an initial $^{87}Sr/^{86}Sr$ ratio 3750 Ma ago close to 0.701. Because the decay rate of ^{87}Rb to ^{87}Sr is known, these Amîtsoq gneisses must have had an average $^{87}Sr/^{86}Sr$ ratio of 0.715 by 3000 Ma, which is the approximate age of the Nûk gneisses in the same area. If these Nûk gneisses had been derived by remobilization of the Amîtsoq gneisses they should have an initial strontium isotope ratio of 0.715, but in fact they have a ratio of about 0.702. This means that the younger gneisses could not possibly be reworked older 'basement' gneisses and their low initial strontium isotope ratio indicates that their immediate precursors, tonalitic intrusions, were juvenile additions to the continental crust at, or close to, their measured age of about 3000 Ma. In other words, the low initial $^{87}Sr/^{86}Sr$ ratio is inherited directly from the upper mantle or low Rb/Sr source region. Because large volumes of granitic gneisses formed in the period 2800–3000 in, for example, Scotland, Greenland and Zimbabwe where they have low initial $^{87}Sr/^{86}Sr$ ratios of approximately 0.701–0.702, a major implication of Moorbath's model is that continental growth on a major scale took place during this period.

Although the measured age of a high-grade gneiss could be related to the time of regional metamorphism, Moorbath demonstrates that this takes place not more than 50–100 Ma after the separation of the gneiss precursor from its source region and its emplacement into the sialic crust. It is interesting to note that this is about the same time span as between the formation of modern oceanic crust at a plate accretion boundary and the formation of a new island arc or continental margin at a plate-consuming boundary.

Metamorphism

Most of the Archaean regions concerned went through a period of high-grade metamorphism, commonly in the granulite facies sometime in the period 2800–3100 Ma soon after massive tonalite injection, tectonic intercalation and consequent crustal thickening, which, according to the Greenland and Labrador evidence, was late in a long sequence of events.

With one minor exception (Griffin et al., 1980), all the very old (>3600 Ma) gneisses in the world have reached the amphibolite rather than the granulite facies. In contrast, late Archaean gneisses (c. 3000–2600 Ma) commonly reached the granulite facies (e.g. S India: Harris et al., 1982; Anabar Shield in the USSR: Bogomolov, 1981). In the North Atlantic craton this could be ascribed to the fact that the metamorphism represents the thermal culmination of thrust stacking and tonalite injection that led to crustal thickening (tectonic and magmatic accretion, Wells, 1980, 1981).

The mineral chemistry of the granulites demonstrates reaction over a pressure range of at least 7–12 kb corresponding to a depth range of 25–40 km before uplift (Newton 1978). Because these granulites today are still underlain by some 30–35 km of continental crust, it is commonly concluded that, in the absence of sialic underplating for which no evidence has ever been adduced, the orogenic belts that gave rise to these granulites must have reached a crustal thickness of some 60–75 km by the late Archaean, comparable to that of the modern Andes and Himalayas (Dewey and Windley, 1981). This would be consistent with the fact that these granulites formed in the range of medium-pressure gradients of 18–35°C km^{-1} (Tarney and Windley, 1977), which is also close to that of Mesozoic–Cenozoic gneissic belts.

The granulites typically are depleted, relative to upper crustal gneisses, in large ion lithophile, heat-producing elements like K, Rb, Cs, U and Th (Tarney and Windley, 1977), although such depletion was, in places, selective (Rollinson and Windley, 1980b). This depletion probably took place via a fluid phase with a high CO_2/H_2O ratio which left the granulites with common CO_2-rich fluid inclusions (Newton et al., 1980; Harris et al., 1982). The elements were transported to upper crustal levels, which were extensively eroded during the enormous uplift that took place at the end of the Archaean as a result of the earlier major crustal thickening. Transport of this released uranium enabled the concentration of the first substantial deposits of this type in early Proterozoic clastic basins (see Chapter 6).

Deformation Patterns

The intercalated rock units have usually been folded several times with the result that the typical deformation pattern in these high-grade regions is a complex interference structure in which domes and basins are accompanied by refolded isoclines. Such patterns occur on a metre scale in many medium- to high-grade gneisses, but they can also be seen on a regional scale particularly well in the Archaean craton of West Greenland, the Limpopo belt of southern Africa and the Aldan Shield of Siberia.

Fig. 2.4 illustrates at 500 km^2 area in West Greenland where the individual folds are up to a few kilometres across. The formation of such a fold pattern is dependent on the varying geometrical relationships between

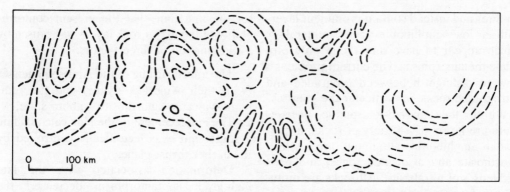

Fig. 2.6 The deformation pattern of the Archaean of the Aldan Shield, USSR (redrawn from Salop and Scheinmann, 1969; reproduced by permission of L. J. Salop)

the structures of the early and late fold sets. Most of the folds formed in association with high-grade regional metamorphism late in the history of the regions (e.g. V. R. McGregor, 1973). The first major isoclines in these regions are best displayed by the layered igneous anorthositic complexes (e.g. Windley *et al.*, 1981). They may well have formed as nappe-like structures in response to a dominant horizontal tectonic regime (Bridgwater, McGregor and Myers, 1974a; Coward, Lintern and Wright, 1976; Myers, 1976).

A somewhat different structural pattern is seen in Fig. 2.6. This half a million square kilometre area of the Aldan Shield is dominated by dome-shaped ovals up to 300 km across in granitic gneisses (Salop and Scheinmann, 1969). There is a close relationship between this style of folding and granitization (partial melting), and the formation of the domes was commonly accompanied by high-grade regional metamorphism and the diapiric ascent of granitic rocks in the dome cores. Salop and Scheinmann emphasize that this type of structural style is characteristic of, and exclusive to, the deep levels of the Archaean crust and they thus proposed the term 'permobile' to describe this early stage in earth history.

Some Key Regions

It is convenient at this stage to review briefly what is known about the nature, origin

and build-up of some classic high-grade Archaean regions. Evidence of what happened in early Precambrian times is not limited to these regions, but they do provide a useful fund of data as they have generally been studied in some detail.

West Greenland

The stratigraphy, metamorphic grade and tectonic pattern of the rocks in West Greenland are little different from those in other high-grade regions. Because of these similarities the region has been grouped with Labrador, East Greenland and the Scourian of Scotland to form the Archaean craton of the North Atlantic region (Bridgwater *et al.*, 1978).

Several large hornblende granulite-grade areas occur within a widespread amphibolite-grade terrain. There is commonly evidence that some amphibolite-grade areas were formerly at a higher grade. In the granulite–amphibolite border zones hypersthenes are partially altered to hornblende, in the amphibolite facies gneisses there are relict pods of hypersthene gneisses and in the bordering amphibolite-grade areas many of the rocks less susceptible to chemical change, such as amphibolites, retain their granulite facies mineral assemblages. On the other hand, some amphibolite facies areas, such as central Godthaabsfjord and South Fiskenæsset, appear never to have been at a higher grade; they have none of the above features and instead contain prominent layers with a low amphibolite facies mineralogy which appear to have undergone only prograde metamorphism. The current problem is how to distinguish between prograded and retrograded rocks, in particular the gneisses.

Within the quartzo-feldspathic gneisses there are layers of a variety of rocks, in particular amphibolite, aluminous mica schist, anorthosite and associated rocks, and pods and lenses of ultrabasics. All rocks are mutually conformable and their relative age relationships, especially between the older and younger gneisses, are best known in the Godthaabsfjord region (Fig. 2.7) where V. R. McGregor (1973) erected a sequence of events, most of which have since been dated isotopically, in particular by the Oxford Geochronology Group (for reviews see Moorbath 1977, 1978, 1980). The chronology for West Greenland in Table 2.1 is taken from McGregor (1973), who showed

for the first time that the present conformability of the major rock layers was caused by a combination of three factors:

1. The interthrusting of the Amîtsoq gneisses (with their Ameralik dykes) and the sediment–volcanic–anorthosite–gabbro suite.
2. The emplacement of the Nûk calc-alkaline granitic suite was tectonically controlled by the earlier thrust planes.
3. Deformation associated with the late high-grade metamorphism decreased the angle of many earlier tectonic or intrusive discordances.

It was the presence of the amphibolite dykes in the Godthaab region that enabled McGregor to break through the conformability barrier; they are present in the layers of older Amîtsoq gneisses but they are absent in the younger rock units. Where the amphibolite dykes are absent, as in most of the Archaean of West Greenland, it is difficult to

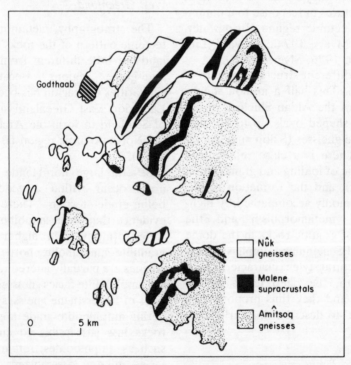

Fig. 2.7 Map of the Godthaab area, West Greenland, showing the distribution of Malene supracrustals and the younger Nûk and older Amîtsoq gneisses (redrawn from McGregor, 1973; reproduced by permission of The Royal Society)

Table 2.1 Major events in the Archaean craton of the North Atlantic region (modified from Bridgwater *et al*., 1978; Barton and Key, 1981; Robertson and du Toit, 1981)

Labrador	Ma	Greenland	Ma	Limpopo Belt	Ma[a]
Post- to late-tectonic potassic granites	2600 ± 200	Intrusion of post-tectonic potassic Qôrqut granite	2530–2580 (Moorbath *et al*., 1981)	Intrusion of post-tectonic potassic granites	2200
Granulite and amphibolite facies metamorphism. Major folding	2800 ± 200	Granulite and amphibolite facies metamorphism. Interference pattern folding	2800 ± 100	Emplacement of Bulai granite and granulite facies metamorphism	2700
Remobilization of earlier gneisses, granite intrusion, tectonic intercalation	3121 ± 160	Intrusion of Nûk gneisses. Tectonic intercalation	2890–3065 (Baadsgaard and McGregor, 1981)	Granulite facies metamorphism and interference pattern folding	3150
Intrusion of anorthosites		Intrusion of anorthosites	>3000	Intrusion of anorthosites (Messina Intrusion)	3270
Deposition of Upernivik Supracrustals		Deposition of Malene Supracrustals		Deposition of Limpopo Supracrustals	
Intrusion of Saglek dykes	3300 ± 300	Intrusion of Ameralik dykes		Intrusion of Causeway dykes	3570
Metamorphism and deformation. Intrusion of Uivak gneisses	3760 ± 150 (Wanless *et al*., 1979)	Intrusion and deformation of Amîtsoq gneisses	3600–3800		
Deposition of Nulliak volcanics and sediments	>3800	Deposition of Isua Supracrustals and 'pre-Amîtsoq' (Akilia) association	>3700	Deformation and metamorphism of Sand River gneisses	3790

[a]Ages recalculated with decay constants of Steiger and Jäger (1977).

demonstrate unequivocally the above sequence of events. However, the fact that the stratigraphic relationships elsewhere are similar to those in the Godthaabsfjord region suggests that the general chronology may be widely applicable.

Important aspects to come out of McGregor's work are:

1. It is necessary to map out an intercalation of rock units in these high-grade areas almost like stratigraphic mapping in Phanerozoic rocks. Subtle variations in gneiss types, inclusions of rocks such as amphibolite in some layers and not in others, and the intercalation of major rock layers may enable a sequence of events to be established. The present rock intercalation has only tectonic significance. Few original sedimentary-volcanic contacts are preserved. The older and younger gneisses can be distinguished on the basis of the absence and presence of amphibolite dykes.

2. The Amîtsoq and Nûk banded gneisses developed by deformation of homogeneous tonalites and granodiorites (McGregor, 1979). Because of this indubitable relationship geologists in Greenland are not inclined to interpret the quartzo-feldspathic Archaean gneisses as being recrystallized sediments or volcanics.

3. The 2800 ± 100 Ma isotopic age reflects a late thermal event that overprinted virtually a thousand million years of earth history. In many high-grade regions throughout the world the oldest isotopic age lies in the range 2700–3000 Ma; it is probable that many likewise have an older history.

The Isua Supracrustal belt (Fig. 2.3) contains the oldest known sediments and volcanics on earth, including quartzites, chert, carbonate-rich sediments, semi-pelites, iron formation, conglomerate, basic lavas and tuffs and cumulate ultramafics (Allaart, 1976; Appel, 1979). Boulders in the conglomerate have a U–Pb zircon age of 3.769 ± 8 Ma (Michard-Vitrac et al., 1977), whilst the sediments and volcanics have a Sm–Nd age of 3770 ± 42 (Hamilton et al., 1978) and a Rb–Sr whole-rock isochron of 3740 ± 60 Ma (Moorbath et al., 1977). Petrology of schists with kyanite, staurolite and garnet, etc. indicate a temperature of c. 550°C and a pressure of c. 5 kb during prograde metamorphism and a depth of burial of at least 15 km (Boak and Dymek, 1982). Yeast-like microfossils described by H. D. Pflug were re-interpreted by Bridgwater et al. (1981) as limonite-stained fluid inclusions, and re-interpreted by Roedder (1981) as limonite-stained dissolution cavities.

Amîtsoq gneisses from Isua have a Rb–Sr isochron age of 3720 ± 60 Ma (Moorbath et al., 1977) and, from near Godthåb, a U–Pb age of 3590 Ma (Gancarz and Wasserburg, 1977) and a Lu–Hf total-rock age of 3550 ± 220 Ma (Pettingill and Patchett, 1981). The Amîtsoq gneisses contain many inclusions of meta-basic-ultrabasic rocks with komatiitic or Fe-rich tholeiitic affinities, banded ironstones, meta-sedimentary detrital gneisses. McGregor and Mason (1977) group these inclusions under the term Akilia association and suggest that they may be fragments of a greenstone belt that was intruded by the precursors of the Amîtsoq gneisses.

The amphibolites of the Malene supracrustals and their equivalents have been discussed earlier in this chapter. They were intruded by anorthositic complexes. There appears to be more anorthositic material in this Archaean craton than in most others —a few layers reach 6–8 km in thickness but mostly they are up to 2 km thick, and some single layers can be followed along the strike for 100 km. These must be at least 1500 km strike length of anorthositic layers in the Archaean of West Greenland as measured on the present ground surface, of which 500 km make up the Fiskenæsset Complex (Fig. 2.4 Myers, 1981). The chemistry of all silicate, oxide and sulphide phases through the ultra-mafic-gabro-anorthosite cumulate sequence shows that igneous cryptic trends are preserved in spite of high-grade metamorphism (Bishop et al., 1980 and earlier monographs). The cumulate sequence/metavolcanic amphibolite association most likely represents an ocean-floor assemblage—see earlier this chapter for details (Weaver et al., 1981, 1982).

The tonalitic-granodioritic precursors of the Nûk gneisses were intruded into all earlier rock units in the period 2890–3065 Ma, commonly during the major thrusting and tectonic interleaving (Baadsgaard and McGregor, 1981). This involved some crustal contamination from the Amîtsoq gneisses (Taylor et al., 1980). Following regional metamorphism of amphibolite to granulite grade and the formation of complex fold interference patterns about 2600 Ma, the Archaean history was completed 2550 Ma by the post-tectonic intrusion of the potassic Qôrqut granite which involved lower crustal remelting (Brown et al., 1981; Moorbath et al., 1981).

Labrador

In Saglek fjord on the coast of Labrador in the Nain Province there are the remains of a crustal evolutionary sequence similar to that recorded in West Greenland (Bridgwater et al., 1978; Baadsgaard et al., 1979). The high-grade complex is made up of the following mostly conformable units:

1. Uivak gneisses (the oldest rocks). A varied group of highly deformed quartzofeldspathic gneisses derived from homogene-

ous prophyritic 'granitic' rocks and characterized by the presence of relics of amphibolite dykes (Saglek dykes). They are subdivided into two types: layered granodioritic gneisses (Uivak I) derived from trondhjemitic--granodioritic parents (Collerson and Bridgwater, 1979), and less extensive iron-rich porphyritic granodioritic and ferro-dioritic gneisses (Uivak II) (Bridgwater and Collerson, 1976). They account for at least 50% of the area and both types yield a whole-rock Rb/Sr isochron of 3622 ± 72 Ma with an initial ratio of 0.7014 ± 0.0008 (Hurst *et al*., 1975). Uivak II have a Pb–U zircon age of 3760 ± 150 Ma (Wanless *et al*., 1979). The Uivak gneisses contain lenses and layers up to 100 m wide of earlier Nulliak supracrustal rocks; viz meta-basalts, hornblendites, oxide iron formations, semi-pelites and carbonates. Late isotopic re-equilibrations are defined by Collerson (1983).

2. Saglek dykes. A suite of amphibolite or pyroxene granulite (depending on metamorphic grade) dykes that transect foliation and other structures in, and are confined to, the Uivak gneisses.

3. Upernavik supracrustals are a group of meta-sedimentary schists and gneisses (meta-pelites, quartzites and marbles) and meta-volcanic amphibolites with layered ultramafic and gabbro-anorthosite complexes (Wiener, 1981).

4. Young quartzo-feldspathic Iterungnek gneisses which do not contain Saglek dykes. They form two generations of deformed small sheets and veins that intrude the Uivak gneisses and Upernavik supracrustals but which rarely form mappable units. They have a Rb/Sr age of 3121 ± 160 Ma (Hurst *et al*., 1975).

5. Post-tectonic, K-rich granites with an age of 2600 ± 100 Ma.

The authors concerned conclude that the Uivak gneisses were intruded by a swarm of basic dykes, then tectonically interleaved with a cover sequence, intruded by granitic sheets, metamorphosed and deformed under amphibolite and granulite facies conditions, and finally intruded by post-tectonic granites

–a chronology essentially similar to that in West Greenland (Table 2.1). Further south at Hebron in Labrador there are comparable 3645 Ma gneisses (Collerson *et al*., 1982).

Scourian, Scotland

Fig. 2.5 shows the geology of the type Scourie area for the Archaean Scourian complex of NW Scotland. This area lies in the central region of the Lewisian complex which records structural and chronological evidence of the earliest (Scourian) granulite facies metamorphism. The main rocks are granulite facies gneisses which contain remnants of meta-sediments and recrystallized layered ultramafic-gabbro complexes and which are cut by a major swarm of dykes (the Scourie dykes) which have a Rb–Sr whole-rock age of 2390 ± 20 Ma (Chapman, 1979). Bowes (1978) gives a good review of the Lewisian complex.

Sm–Nd whole-rock measurements suggest that the calc-alkaline protoliths of the Scourian granulites separated from a previously undifferentiated chondritic mantle 2920 ± 50 Ma (Humphries and Cliff, 1982), about 240 ± 90 Ma before the Scourian metamorphism which has a Rb–Sr and Pb–Pb age of about 2680 Ma (Chapman and Moorbath, 1977; Hamilton *et al*., 1979a). There is evidence of two periods of granulite facies mineral growth at 12–15 kb and 1000°C and at 10–14 kb and 800–900°C (Savage and Sills, 1980). Granitic sheets emplaced before the metamorphism crystallized at about 1000°C (Rollinson, 1982a). Many layered ultramafic-gabbro bodies occur in the gneisses as tectonically thinned and broken up lenses and layers up to about 400 m thick and 12 km long. Trace element chemistry suggests that the ultramafics are cumulates and the gabbros are the derived liquids (Sills *et al*., 1982). The gneisses predominantly have a high Na/K tonalite composition (Tarney *et al*., 1979; Rollinson and Windley, 1980a) which formed by high pressure partial melting of a mafic source, in contrast to low pressure crystallization of basic (amphibolitic) gneisses and granulites (Weaver and Tar-

24

ney, 1980). The granulite facies metamorphism selectively depleted the gneisses in K, Rb,
Th, Cs and U in a fluid rather than a melt
phase (Rollinson and Windley, 1980b;
Weaver and Tarney, 1981b). The synmetamorphic deformation that affected the
intruded tonalites gave rise to complex, foliated, banded and intermixed gneisses with
disrupted relics of meta-sediments and
layered complexes (Bowes, 1978). The
Scourie dykes were intruded into this foliated
gneiss complex (Fig. 2.2).

Limpopo belt, southern Africa

This high-grade ENE-trending mobile belt
separates the Zimbabwean and Kaapvaal cratons of southern Africa. Recent geological
and isotopic studies on this belt have shown
that it has a long history not unlike that of

West Greenland and Labrador (Ermanovics
et al., 1977; Barton and Key, 1981; Robertson and du Toit, 1981). It is divided into
three zones. The North and South Marginal
Zones are metamorphosed equivalents of the
greenstone belt-granite terrains of the adjacent cratons; the granulite to amphibolite
facies metamorphisms have apparently obscured all isotopic ages older than 2870 to
2650 Ma. The mafic-ultramafic complexes
referred to earlier belong to the Northern
Marginal Zone.

The Central Zone has no evidence of a
greenstone belt precursor. It contains a relict
belt of 3790 Ma Sand River gneisses (Fig.
2.8) of granodioritic to quartz dioritic composition (Fripp, 1981), which are cut by five
generations of dykes. Most of the region consists of the Limpopo or Beitbridge meta-
supracrustals which are paragneisses contain-

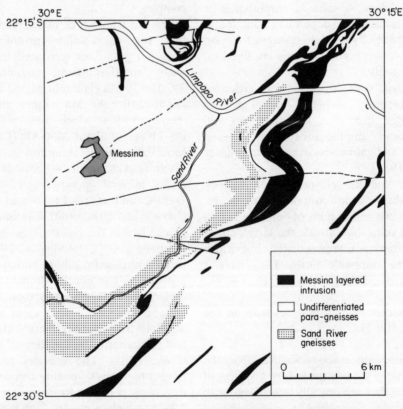

Fig. 2.8   Map of the Messina area of the Limpopo mobile belt showing the
occurrence of the 3790 Ma old Sand River gneisses within younger meta-
sedimentary gneisses together with 3270 Ma old Messina layered intrusion
(after Barton *et al.* (1977), *Geol. Surv. Botswana Bull.*, **12**, Fig. 2, p. 77)

ing relict amphibolites, quartzites and marbles (the 'old' Messina Formation) which Fripp (1981) and Barton and Key (1981) concluded were deposited on a continental basement of Sand River gneiss. The supracrustals were intruded by the Messina layered intrusion, 3270 Ma ago, which consists of recrystallized cumulate calcic anorthosites, gabbros and ultramafics with chromitites (Barton et al., 1979). The oldest set of basic dykes, the Causeway dykes, dated at 3570 Ma, cut the Sand River gneisses but not the Beitbridge gneisses or the Messina intrusion. The main period of granulite facies metamorphism and folding was at 3150 Ma which was followed by further metamorphism and granite emplacement (Table 2.1).

Southern India

Earlier controversies regarding the basic chronology of the Archaean of southern India have gradually been resolved. Today the Sargur Group and the Peninsular Gneisses are usually considered to be older than the Dharwar Group of greenstone belts, although the precise ages of these groups are not yet known with confidence (Radhakrishna and Vasudev, 1977; Chadwick et al., 1978; Janardhan et al., 1978; Sarkar, 1980; Swami Nath and Ramakrishnan, 1981). In contrast, Pichamuthu (1982) still holds to the view that the Dharwar belts are older than the Peninsular gneisses. A much quoted critical age relationship is in the Holenarasipur belt where Dharwar rocks rest unconformably on Sargur rocks (Naqvi et al., 1978).

The age of the Sargur Group is not known, except that it is older than the Peninsular Gneiss which has a Rb/Sr whole-rock age of 3358 ± 66 Ma (Beckinsale et al., 1980). The Sargur Group contains para-gneisses, quartzites, mica schists, khondalites (garnet–sillimanite schists and gneisses), amphibolites, quartzites, marbles, banded iron formations (Prasad et al., 1982), and also remnants of anorthosite-gabbro-ultramafic complexes locally with chromitites (Janardhan et al., 1978; Ramakrishnan et al., 1978).

The tonalitic-granitic parents of the gneisses have invaded the Sargur supracrustals, rafting off large enclaves that are now surrounded by gneiss. This intrusion was contemporaneous with deformation and metamorphism of the supracrustals (Chadwick et al., 1981b). There is a southward transition from the greenstone-granite to the high-grade terrain (Condie et al., 1982). Metamorphism reached the granulite facies (Raith et al., 1982a) with PT data ranging from 5.0 ± 1.0 kb and 700 ± 20°C (Harris et al., 1982) to 9–10 kb and 780 ± 60°C (Weaver et al., 1978). During the Sargur metamorphism maximum crustal thicknesses were in excess of 45 km (Rollinson et al., 1981). Much of the Peninsular Gneiss has initial strontium isotope ratios of 0.701–0.704, suggesting a multi-stage derivation from the mantle (Saha, 1979).

A Comparative Review

In the foregoing pages of this chapter the rock units and chronological evolution of some classic Archaean regions have been summarized. Several marked similarities will be immediately obvious to the reader. Perhaps one of the most distinctive is the presence in the gneisses of all regions concerned of layers and lenses of meta-basic, meta-ultrabasic and meta-anorthositic rocks that are the remains of layered igneous complexes. Commonly the complexes are bordered by recrystallized shelf-type sediments that include mica schists, quartzites and marbles together with meta-volcanic amphibolites into which they (the complexes) were probably intruded. In West Greenland, Labrador and the Limpopo belt the complexes and supracrustals clearly postdate some gneisses and predate others. Similar age relations with respect to different gneiss generations should be looked for in southern India. Several regions have amphibolite dykes in the older but not the younger gneisses and most regions were subjected to high-grade metamorphism and deformation in the late Archaean, about 3000–2800 Ma, which transformed inter alia the intrusive calc-alkaline 'granitic' rocks into gneisses. It is not premature to suggest that these parts of the Archaean crust underwent a broadly similar chemical, tectonic and relative chronological evolution although, naturally, each region has its own distinctive features in terms of, for example, proportions of rock units and chemical differences. Archaean greenstone belts are remarkably similar the world over (see Chapter 3); the high-grade gneissic parts of the Archaean crust likewise share many similarities.

Mineralization

There are not many mineral deposits in Archaean high-grade regions. If the original sedimentary and igneous rocks did contain many mineral concentrations, these might have been destroyed by metamorphic/tectonic processes, or else these early rocks might have been relatively impoverished in ore deposits. In this respect these high-grade Archaean regions contrast markedly with the Archaean greenstone belt-granite terrains which are enriched in ores (Chapter 3). The following are the main types of mineral concentrations. Many are uneconomic or only subeconomic, but nevertheless they provide us with data on the types of mineralization that formed in the early stages of the earth's evolution.

Iron Formations

Most Archaean granulite–gneiss belts contain some lenses and layers of metamorphosed iron formations (BIF) which are usually only a few metres thick and typically are interbedded with magnetite quartzite in supracrustal sequences with mica schists, marbles, amphibolites and paragneisses. They are mostly of the banded quartz–magnetite type (Prasad *et al.*, 1982). The BIF are usually in the amphibolite to granulite facies and their percentage of Fe is <40. In S India the layers reach 20 m in thickness, but in NE China they are up to 80 m thick and form the basis of a major steel industry (see page 13 for references). The BIF were deposited in a quartzite-pelite-carbonate association under shallow water, possibly early shelf-type conditions. They are a proto-Superior type of iron formation; Archaean cratons were unstable and did not survive long and therefore we see only relics of these BIF.

The oldest BIF is at Isua, West Greenland (Allaart, 1976; Appel, 1980). This occurs in a greenschist-to amphibolite-grade sequence of calcareous quartzites and amphibolites (tuffs) mantling a gneiss dome. The BIF has oxide, sulphide, silicate and carbonate facies. Copper sulphide mineralization occurring in the last three has a submarine, exhalative origin (Appel, 1979).

Cr in Anorthosites and Ultramafics

A chromium metallogenic province exists in these high-grade regions because chromitite seams are common in the anorthosites and ultramafics.

One of the most prominent deposits occurs in the Fiskenæsset Complex, West Greenland (Ghisler and Windley, 1967) where the anorthosites at the top of the layered succession contain chromitite seams, that locally reach 20 m thick, throughout the 500 km strike length of the complex. The chromitives have been metamorphosed, deformed and faulted, and the aluminous chromites have Cr : Fe ratios of about 1 : 1.

There are similar metamorphosed and deformed chromitites in the following bodies:

1. In the Sittampundi and Coimbatore, South India complexes where chromitite seams are up to about 6 m thick and occur towards the top of the stratigraphy.
2. In ultrabasic layers associated with anorthosites in the Kondapalli complex, South India.
3. In hornblendite and serpentinite layers in the meta-anorthosites of the Limpopo belt, southern Africa there are chromitite seams up to about 60 cm thick.

Ni, Cu, Pb in Amphibolites

Amphibolitic layers in the gneisses commonly contain rusty sulphide-bearing zones but they are rarely economic. One exception is the Ni–Cu sulphide deposit at Selebi–Pikwe in the Limpopo belt, Botswana, which occurs in a 50 m wide amphibolite layer (Robertson and du Toit, 1981). (There is stratabound galena mineralization with sulphides of Ni, Sb, Cd, Ag in tuffaceous amphibolites in the Isua belt: Appel, 1978). For a general review of Cu–Ni sulphide mineralization see H. D. B. Wilson (1982) who describes examples from the English River gneissic belt in Ontario, Canada.

Corundum in paragneisses

In the Aldan Shield in the USSR there are lens-shaped corundum bodies (corundite) up to 25 m × 70 m in metasedimentary gneisses and schists. They are interpreted as recrystallized early Archaean alumina-rich weathering crusts, i.e. bauxites (Tenyakov *et al*., 1982).

Graphite in paragneisses

Graphite-bearing schists are common in paragneisses, often associated with sillimanite-kyanite schists, marbles etc., but they are rarely economic. In S Malagasy, however, there is an 800 km long belt of graphite schists, the weathered zone of which is mined (Besairie *et al*., 1951). Similar rocks are prominent in the Aldan Shield in the USSR (Sidorenko and Sozinov, 1982). A general review is given by Goosens (1982).

Chapter 3
Archaean Greenstone Belts

Archaean greenstone belts are the oldest major group of well-preserved volcano-sedimentary basins and so they give us much direct evidence of early crustal conditions. They occur in many shield areas, vary in age from c. 3500 Ma to 2300 Ma and range in size up to 250 km across. Their stratigraphy, general structure, volcanic geochemistry, types of sediments and ore deposits are remarkably similar and this uniformity allows them to be treated as a single integral group (Condie, 1981; Anhaeusser, 1982).

Important occurrences are the Swaziland Supergroup in the Barberton Mountain Land of South Africa, the Sebakwian and the Bulawayan–Shamvaian Groups of Zimbabwe, the Abitibi, Yellowknife and many other belts in the Superior and Slave Provinces of Canada, the Dharwar Supergroup in India and many belts in the Pilbara and Yilgarn blocks of W Australia, in Finland and adjoining USSR, in the Wyoming Province of the western USA, in the Liberian Province of W Africa and in the Central African Province.

The greenstone belts are bordered and intruded by 'granitic' plutons which are not open to dispute. However, the relationship of the belts to nearby high-grade quartzo-feldspathic gneisses is subject to various interpretations, the merits of which are reviewed in the next chapter.

Study of Archaean greenstone belts has reached an interesting stage of development as there is currently a proliferation of contrasting ideas about their mode of development. These vary from the fixist to the mobil-ist and include models such as the remains of primordial oceanic crust, downsagging basins, pinched downfolds between colliding micro-continents proto-oceanic ridge systems, terrestrial equivalents of lunar maria, and marginal basins formed in back-arc environments. In this chapter we shall consider the characteristic features of the greenstone belts, and in the next we shall look at how they and the Archaean high-grade regions may have evolved and how they may be interrelated.

General Form and Distribution

The shape of greenstone belts is easier to illustrate than to describe. At their simplest they have a linear plan and a basin-shaped cross-section, at their most complex a cuspate form with a triangular plan. The traditional idea of the structure of a typical greenstone belt, based on Barberton, envisages a basin-shaped infold or downfold in a sea of granitic material. Theoretical and gravity model shapes are shown in Fig. 3.1. Recent data from many belts on three continents show that early nappes have inverted the successions and that the structural evolution is more complicated than a simple downfold (see later under 'structure').

Geochronology

A review of isotopic data (Fig. 3.2) from Archaean greenstone belts the world over points to the following conclusions (Condie, 1981). There were major periods of magmatic activity in the periods 2600–2700 Ma and

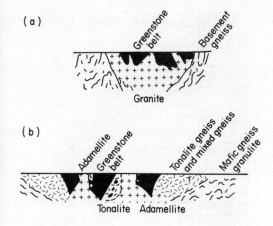

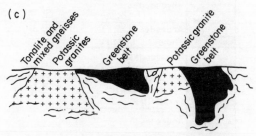

Bulawayan–Shamvaian Groups were forming in the Zimbabwe greenstone belts about 2700–2600 Ma ago, flat-lying cratonic, largely clastic, sequences with volcanics were being laid down in the Dominion Reef-Witwatersrand Systems not far to the north. This demonstrates an overlap between the formation of Archaean greenstone belts and cratonic cover sequences under more stable platform-type conditions typical of the early Proterozoic (Sutton, 1976).

Fig. 3.1 Some models for the structure of greenstone belts and their surrounding rocks. (a) A theoretical model of Windley (1973). (b) A gravity-based model for belts in the Keewatin–MacKenzie districts of Canada (Gibb and Halliday, 1974). (c) A gravity-based model for Dharwar belts in Southern India (Subrahmanyam and Verma, 1982)

2800–3000 Ma. Between 3100 and 3600 Ma there was a diachronous evolution at different places at different times; this was contemporaneous with maria basaltic magmatism on the moon. In the 3000–2800 Ma period there is no evidence of supracrustal deposition, only of plutonism, thus the period 2700–2600 Ma was by far the most important period of Archaean volcanism and plutonism. In the Rhodesian and Yilgarn Provinces there were two or three periods of greenstone–granite formation—the upper and lower greenstones of Glikson (1976a). Each individual greenstone belt period involving deposition, magmatism, deformation, metamorphism, uplift and erosion lasted about 50–100 Ma.

It is important to note that whilst the

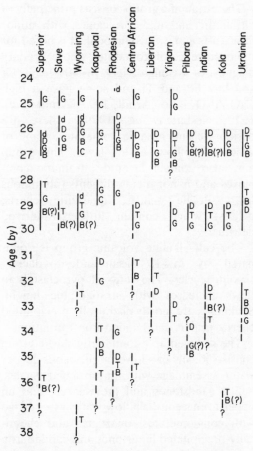

Fig. 3.2 Geochronologic summary of Archaean granite–greenstone provinces. B: greenstone belt development, T: intrusion of tonalite—trondhjemite, D: deformation and regional metamorphism, G: intrusion of granites and pegmatites, d: intrusion of mafic dykes, C: cratonic sedimentation ± volcanism (from Condie (1981), *Archaean Greenstone Belts*, Elsevier, Amsterdam, Fig. 1.20, p. 43)

30

Stratigraphy

General Points

The main subdivision common to many greenstone belts is into a lower, dominantly volcanic, and an upper sedimentary group. The lower group in most belts is further divisible into a lower, primarily ultramafic, and an upper volcanic group in which calc-alkaline, mafic-to-felsic rocks predominate. The typical subunits of this threefold division are shown in Fig. 3.3. Cyclicity is common; individual belts may contain 5–10 major volcanic cycles.

The ultramafic group consists principally of ultramafic and mafic volcanics with minor felsic tuffs and is thus bimodal. It is noted for the occurrence of komatiites whose primary diagnostics are high MgO (8–40%), Ni, Cr and low Fe and Ti and some have a high CaO/Al_2O_3 ratio (Arndt, 1977; Arndt *et al.*, 1979; Nesbitt *et al.*, 1979). There are peridotitic and basaltic komatiites, some of which are pillow-bearing lavas. This group may also contain layered ultramafic complexes and minor metasediments (aluminous quartz-sericite schists, quartzitic cherts, pelites) which contain little sial-derived detritus.

The calc-alkaline volcanic group is dominated by low-K basalt–andesite–dacite–rhyolite cycles. The ratio of pyroclastics to flows increases with stratigraphic height. Sediments are mostly chemically precipitated cherts, jaspers and banded iron formations.

The sedimentary, dominantly clastic group consists ideally of a lower argillaceous deeper water assemblage with, in particular, shales, pelitic sandstones and greywackes, and an upper arenaceous, shallow-water assemblage with conglomerates, quartzites and chemically-precipitated limestones and banded iron formations that tend to occupy the tops of cyclic units (Lowe, 1980a).

Notwithstanding the overall uniformity amongst the world's greenstone belts, many have their distinctive features.

Belts in southern Africa and Australia are typically enriched in ultramafics and mafics, whereas those in North America have a

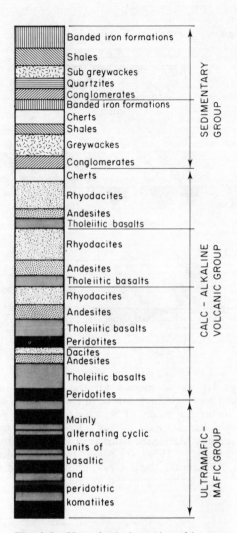

Fig. 3.3 Hypothetical stratigraphic succession for an Archaean greenstone belt based on the Barberton model (modified after Anhaeusser, 1971b; reproduced by permission of The Geological Society of Australia)

higher proportion of calc-alkaline rocks. But these variations depend on the difference in age of these belts. Lowe (1982) provides a useful comparative study of these older and younger belts. Sedimentary layers in the 3400–3500 Ma Onverwacht and Warrowoona Groups of the Barberton Mountain Land and the Pilbara Province, respectively, are made up of silicified volcaniclastic sediments, banded and carbonaceous cherts, silicified evaporites and, locally, stromatolites deposited under shallow water anorogenic conditions. The volcanic sequences in these belts essentially lack terrigenous debris derived by weathering and erosion, Algoma-type iron formations, volcanogenic massive sulphide deposits, hyaloclastic breccias and evidence for the existence during their accumulation of older sialic basement. In contrast, sedimentary rocks in the 2600–2800 Ma belts of the Superior and Slave Provinces of Canada, and probably in belts of comparable age in the Rhodesian and Yilgarn cratons, include generally unsilicified volcanic-lastic units, mafic hyaloclastic breccias, terrigenous clastic layers, abundant Algoma-type iron formations and massive sulphide deposits. Sedimentation took place under relatively deep water conditions, local tectonic activity was widespread and older sialic basement was present. These younger volcanic sequences contain few shallow water deposits, such as carbonaceous cherts, evapo-

rites and stromatolites. Differences between the older and younger belts reflect the increasing abundance and influence of sialic crust from early to late Archaean time and an accompanying increase in ocean-water depth over the volcanic platforms. With these considerations in mind we shall now consider the stratigraphic make-up of the major belts on different continents.

Swaziland Supergroup, Barberton, South Africa

The Barberton greenstone belt is well reviewed by Anhaeusser (1971a, 1978, 1981), Anhaeusser and Wilson (1981) and Viljoen and Viljoen (1969) and its main features are summarized in Fig. 3.4. The lower volcanic succession (the Onverwacht Group) is overlain by the sedimentary Fig Tree and Moodies Groups.

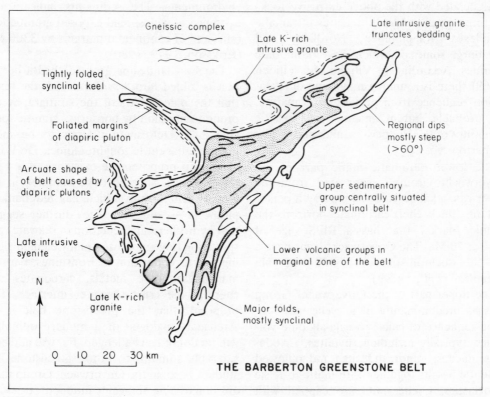

Fig. 3.4 The characteristic features of the Barberton greenstone belt, South Africa (modified after Anhaeusser, 1971a; reproduced by permission of The Geological Society of Australia)

The lower part of the Onverwacht Group consists largely of ultramafic and mafic flows and intrusive complexes, divisible into three formations: the Sandspruit, Theespruit and Komati, in upward sequence (Viljoen and Viljoen, 1969). The Sandspruit Formation occurs as relict inclusions of serpentinized rocks in bordering intrusive tonalitic gneisses. The Theespruit Formation, comprising principally metamorphosed ultramafic and mafic lavas, is noted for primitive fossil microorganisms in carbonaceous cherts (see later this chapter). The Komati Formation contains peridotitic komatiites with pillow structures and quench textures which indicate that a mobile ultramafic magma was subaqueously extruded (Viljoen and Viljoen, 1969). In this and the underlying two formations are pillow-bearing basaltic komatiites, recrystallized to amphibolites and divisible into three chemically distinctive types based on their MgO contents: Barberton ($\pm 10\%$), Badplass ($\pm 15\%$) and Guluk ($\pm 20\%$).

Associated with the above extrusive rocks are three types of layered igneous ultramafic complexes, Kaapmiuden, Nordkaap and Stolzburg, some of which have komatiitic affinities. According to Viljoen and Viljoen (1969) there is a unique magmatic differentiation sequence from peridotitic komatiite, the probable parent magma, to basaltic komatiites of the Badplass, Guluk and finally Barberton types.

The lower ultramafic–mafic part of the Onverwacht Group is separated from the upper calc-alkaline volcanic part by a persistent 6m thick chert-carbonate horizon, the Middle Marker that has a Rb/Sr age of 3280 ± 70 Ma. The Onverwacht lavas have a Sm–Nd isochron date of 3450 ± 30 Ma (Hamilton et al., 1979b).

The upper part of the Onverwacht Group consists predominantly of a cyclic mafic-to-felsic sequence of calcic to calc-alkaline volcanics, typically including rhyolites, rhyodacites, dacites, minor andesites and pillowed tholeiitic basalts together with mafic-to-felsic pyroclastics. Cyclic units are capped with chert and calcareous and ferruginous shales. The upward increase in K_2O and calc-alkaline character of the volcanics in the Onverwacht Group as a whole may reflect the progressive evolution of a thickening sialic crust or an emerging volcanic arc (Condie, 1981, 1982a).

The Onverwacht Group is unconformably overlain by the argillaceous Fig Tree Group which passes upwards into the arenaceous Moodies Group. The Fig Tree rocks include rhythmically alternating shale-greywacke sequences, with some chert and banded ironstones and trachytic tuffs. The Moodies Group contains polymictic conglomerates, quartzites and feldspathic sandstones—a molasse-type sequence, with some chert and banded ironstones and trachytic tuffs.

Sedimentary studies of the Onverwacht Group indicate that it developed in an anorogenic submarine basin unaccompanied by major tectonic uplift or subsidence (Lowe and Knauth, 1977). The Fig Tree and Moodies Groups developed in evolving back-arc or passive continental margins from early rifts to continental shelf and shelf-rise environments. The sediments indicate the existence of widespread exposed granitic terrains and of continental margins by 3300 Ma (Eriksson, 1979, 1980).

The Swaziland Supergroup of the Barberton belt is folded into a synform with the result that the oldest parts of the stratigraphy are prone to invasion by bordering granitic rocks (Fig. 3.4). Deformation tends to be more intense adjacent to tonalite domes. De Wit et al. (1980) proposed the existence of a 1 km thick vertical sheeted dyke complex in the Onverwacht Group which has fed flanking pillow lavas and sills. They further suggest that hydrothermal circulation of seawater at a spreading ridge mobilized many elements giving rise to deposition of ironstones, barytes, manganese, base metals, carbonates and chert at the crust–seawater interface. This proposal, that the greenstone belt is an Archaean analogue of a modern ophiolite, will no doubt be challenged, but will add considerable stimulus to future greenstone belt studies. Also, in the Onverwacht Group there are remarkable abiogenic mudpool structures formed by fluid emission in mineralized hydrothermal vent systems (De Wit et al., 1982).

The close association of these with stromato-lites and possible microfossils supports the idea of Corliss *et al.* (1981) that hydrothermal vents may have been important environments for prolific biological activity in the Archaean, as they are today around deep ocean hot springs along the East Pacific Rise and Galapagos Ridge (Corliss *et al.*, 1979).

Zimbabwean Belts

In the Zimbabwean Archaean craton there are a great many greenstone, or schist, belts (Fig. 3.5) with arcuate, cuspate and synfor-mal shape (Anhaeusser and Wilson, 1981; Nisbet *et al.*, 1981).

There are at least two ages of greenstone belts (the upper and lower greenstones of Glikson, 1976). 3500 Ma tonalitic-trond-hjemitic gneisses contain infolded remnants of the Sebakwian (Group) belts (Wilson *et al.*, 1978). Some tonalitic gneisses underlie the Selukwe belt and are pre-Sebakwian in age (Stowe, 1974). The 3450 Ma Mushandike

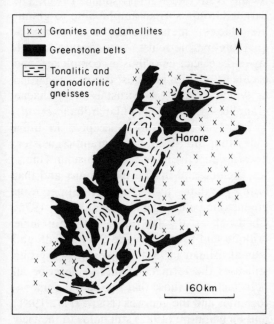

Fig. 3.5 Map of the greenstone belts of Zimbabwe together with gneisses and granites (redrawn from Viljoen and Viljoen, 1969; reproduced by permission of The Geological Society of South Africa)

granite cuts the gneisses and the 3350 Ma Mont d'Or granite intrudes the Selukwe greenstone belt (Moorbath *et al.*, 1976). The Bulawayan-Shamvaian Groups belong to greenstone belts formed in the period 2700–2600 Ma (Hawkesworth *et al.*, 1975; Hamilton *et al.*, 1977). They overlie uncon-formably both the gneisses (Bickle *et al.*, 1975) and the Sebakwian Group (Nisbet *et al.*, 1977). The Great Dyke transected all these rocks 2460 Ma ago (Hamilton, 1977).

The Sebakwian Group includes serpentin-ized ultramafics and recrystallized mafics, such as tremolite-chlorite schists, talc schists, actinolite-chlorite schists and hornblende and chlorite schists. There are also arenaceous sediments and banded iron formations. Most of these rocks occur as inclusions within intruded granitic bodies, such as the Rhodes-dale batholith.

The Bulawayan Group locally lies uncon-formably with a basal conglomerate on Sebakwian rocks and contains metamor-phosed basaltic pillow lavas (some are low-K tholeiites), andesites, dacites and calc-alkaline tholeiites; there is a progressive upward increase in the proportion of felsic volcanics and there are conspicuous mafic-to-felsic volcanic cycles, tuffs and agglomer-ates. Stromatolites occur in Bulawayan limestones (see later in this chapter).

The Shamvaian Group consists largely of poorly sorted, clastic meta-sediments such as polymictic conglomerates (with pebbles of gneiss and granite), greywackes, quartz mica schists and phyllites, together with limestones and banded iron formations.

Western Australia

Greenstone belts occur in the Yilgarn and Pilbara blocks.
(a) The Yilgarn block in SW Australia con-sists of granites and gneisses that enclose a network of northerly striking greenstone belts in three tectonic provinces (Gee *et al.*, 1981; Hallberg and Glikson, 1981). There are three volcano-sedimentary sequences, each sepa-rated by an unconformity and typically con-

sisting of the following successions (in ascending stratigraphic order): pillowed and massive ultramafics and basalts; basalts, andesites, greywackes, slates, argillites and phyllites; conglomerates and greywackes. These belts are characterized by a large proportion of clastic sediment. Some of the lowermost ultramafic extrusive flows contain spinifex quench textures and have komatiitic chemistry and there are some differentiated layered ultramafic–mafic intrusions.

The oldest rocks in the Yilgarn block are tonalitic gneisses with a Rb–Sr age of 3348 ± 43 Ma (De Laeter *et al.*, 1981). Some gneisses formed a sialic basement to greenstone belts (Archibald and Bettenay, 1977). The greenstone belts formed in the period 2800 and 2600 Ma, and there is some suggestion of older and younger belts within that time span (Condie, 1981). The older group at Kambalda has a Sm–Nd age of 2790 ± 30 Ma (McCulloch and Compston, 1981) and at Agnew a Rb–Sr age of 2720 Ma (Cooper *et al.*, 1978). The younger belts have a minimum age of 2640 Ma on granitic intrusives (Cooper *et al.*, 1978).

(b) The Pilbara block, NW Australia, contains two ages of greenstone belts (Hallberg and Glikson, 1981; Hickman, 1981). The greenstone belts are much older than those in the Yilgarn block. Deposition of the Pilbara Supergroup began about 3600–3550 Ma when tholeiitic basalts were extruded over a postulated sialic crust of reworked granitic, volcanic and sedimentary material. Calc-alkaline volcanism with a U/Pb age of 3450 Ma (Pidgeon, 1978), a Sm–Nd age of 3450 Ma (McCulloch and Compston, 1981), and a Rb–Sr and Sm–Nd age of 3560–3570 Ma (Jahn *et al.*, 1981) was contemporaneous with intrusion of trondhjemites and granodiorite and was followed by widespread deposition of chert and carbonate sediments, and extrusion of pillowed tholeiite, high-Mg basalt and local peridotitic komatiite. Then synclines and domes developed from this blanket of flat-lying supracrustal rocks. Erosion resulted in deposition of sandstone, conglomerate and turbidite. There was extensive deformation and intrusion of granodiorites and adamellites at 2950 Ma. Good exposure and interlinking of greenstone belts precludes their deposition in separate, isolated basins. Some greenstone belts were tectonically interthrusted with high-grade gneisses (Bickle *et al.*, 1980). One extraordinary feature of Pilbara sedimentation is the evidence in cherty meta-sediments of silicified and baritized gypsiferous evaporites up to 15 m thick and at least 25 km in extent (Barley *et al.*, 1979).

Dharwar Supergroup, southern India

The Dharwar greenstone belts of southern India occur as well-defined northerly striking strips up to 60 km across, separated by gneisses and granites. The gneisses contain older strips of meta-supracrustal rocks belonging to the Sargur Group. Controversy has long raged in India about the age relations of these rock units. Discovery of unconformities in the Chitradurga, Bababudan and Holenarsipur belts demonstrates that the Dharwars are younger than the Sargurs (Naqvi *et al.*, 1978; Swami Nath and Ramakrishnan, 1981). The Sargur Group is in an amphibolite to granulite facies metamorphic state, has a quartzite-marble-schist association with layered, calcic, anorthosite complexes and occurs as relics in gneisses; thus it is regarded as typical of the supracrustals of Archaean granulite–gneiss belts (Janardhan *et al.*, 1978). It is increasingly accepted in India today that the tonalitic and granitic gneisses, known collectively as the Peninsular Gneisses, include and invade the Sargurs and that both pre-date the Dharway Supergroup greenstone belts (Ramakrishnan *et al.*, 1976; Chadwick *et al.*, 1981a,b); for a rationale critique and much data see Swami Nath and Ramakrishnan (1981). However, one group still uses the term Dharwars to describe all belts including those that pre-date the unconformities and the gneisses (Naqvi *et al.*, 1981) and Pichamuthu (1982) still holds to the view that the Sargurs are 'keels of greenstone belts' in the high-grade terrain and that the Dharwar belts are older than the Peninsular gneisses.

The Dharwar Supergroup greenstone belts are divided into the lower Bababudan and the overlying Chitradurga Groups, separated by an unconformity. The basal oligomictic conglomerate of the Bababudan Group lies unconformably on a basement of Sargur schists and gneisses. This is overlain by amydaloidal basalts, mafic to felsic volcanics and ultramafic rocks, shallow-water quartzites, chloritic schists, siliceous phyllites, agglomerates, tuffs and iron formations. The Chitradurga Group contains polymictic conglomerate, quartzite, phyllite, limestone, dolomite, ferruginous chert, greywacke-argillite and calc-alkaline mafic-to-felsic volcanic rocks.

The age of the Sargur Group is not know except that it is older than much of the Peninsular Gneiss, which has a Rb–Sr whole-rock isochron age between 3358 ± 66 Ma (Beckinsale *et al.*, 1980) and 3000 Ma (Rajagopalan *et al.*, 1980). On the basis of limited isotopic age data on granitic and volcanic rocks and unconformable relations, the age of the Dharwar Supergroup is probably between 3000 Ma and 2500 Ma (Venkatasubramaniam, 1974; Chadwick *et al.*, 1981b).

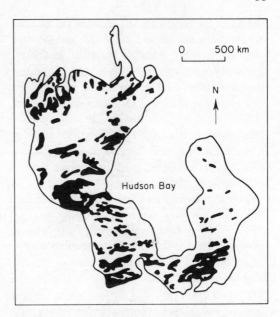

Fig. 3.6 The supracrustal rocks, mostly within Archaean greenstone belts, in part of the Canadian Shield (redrawn after Goodwin, *Geol. Ass. Can. Proc.*, **19**, 1968, 2; by permission of The Geological Association of Canada)

Superior and Slave Provinces, Canada

In the Canadian Shield, greenstone belts form a prominent part of the Archaean Superior and Slave Provinces (Fig. 3.6) but they are also present in the Churchill Province where they have been overprinted by Proterozoic (Hudsonian) plutonic activity.

The typical volcano-sedimentary pile in a greenstone belt developed sequentially in three main stages:

1. Construction of a thick broad mafic platform by widespread effusion of predominantly tholeiitic basalt.
2. Increasingly felsic pyroclastic eruption leading to erection of high-rising piles upon the mafic platform.
3. Partial denudation of the volcanic piles and construction of volcaniclastic blankets.

In the Superior Province basalts account for 50–60% of total volcanic rocks, 'andesites' about 20–30% and more felsic rocks about 10–15%. There are also rare late alkaline shoshonitic rocks. A typical volcanic pile is 10–17 km thick (Goodwin, 1977a). Orthogneisses bordering some belts may have formed a basement (Park and Ermanovics, 1978; Clark *et al.*, 1981).

Of the multitude of Canadian belts the two most well known (the Abitibi and the Yellowknife) will be discussed here.

The Abitibi belt (Dimroth *et al.*, 1982; Goodwin, 1982a) is 800 km long and 200 km wide and thus contrasts markedly with the comparatively small Barberton belt. It is truncated to the east and west by the younger Grenville and Kapuskasing high-grade belts, and so it was originally even longer; it is the largest single continuous Archaean greenstone belt in the world. It contains 58% volcanic, 10% sedimentary and 32% granitic rocks in batholiths (Goodwin, 1977b).

The distribution of the volcanic rocks is dominated by the presence of eleven elliptical volcanic complexes (Fig. 3.7), each with

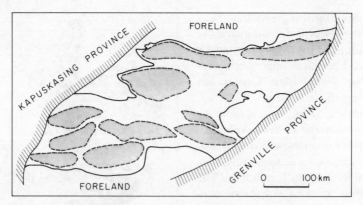

Fig. 3.7 Eleven elliptical volcanic complexes (stippled) in the Abitibi greenstone belt, Canada (redrawn from Goodwin and Ridler, 1970, by permission of The Geological Survey of Canada)

mafic-to-felsic extrusive rocks and coeval intrusions and sediments. These complexes lie close to the northern and southern borders (forelands) of the belts, the intervening median part being occupied by uniform tholeiitic basalts, fine-grained clastics and major granitic batholiths. This pattern probably reflects the original linear distribution of a series of strato-volcanoes.

The lowermost tholeiitic lavas contain local komatiites up to 2.5 km thick. Andesite flows and pyroclastics intercalated with basalts increase in amount upwards, and felsic rhyolites to dacites predominate in the upper levels. The mafic lavas commonly contain pillows, palagonite, variolites, amygdules and hyaloclastites, thereby indicating subaqueous accumulation; primary igneous textures are remarkably well preserved. In the overlying felsic volcanic rocks pyroclastic types are very common (Goodwin, 1982a).

Sediments in the Abitibi belt are particularly of the 'poured-in' turbidite type suggesting rapid accumulation in a tectonically unstable environment. Two principal facies are present:

1. Volcanogenic, comprising greywackes, shales, lithic sandstones, conglomerates and breccias, the constituent clasts being from recognizable volcanic rocks. They are associated with soft sediment deformation structures,

chaotic textures, polymictic unsorted materials, graded bedding and abrupt facies changes.
2. Flyschoid, comprising rhythmically bedded greywacke–argillite sequences of uniform composition and construction lacking in lateral facies changes.

The Canadian volcano-sedimentary assemblages contain shallow to deep water transitions which represent remnants of the original basins and trenches (Goodwin, 1977b). A typical transition is marked by:

1. Off-shore thickening of the total assemblage (i.e. all stratigraphic units tend to be thicker nearer the axes of the belts). This fact lends support to the idea that the sedimentation kept pace with progressive opening of the shore lands.
2. A corresponding thick-to-thin and coarse-to-fine clastic transition in the direction of deeper water.
3. In banded iron formations there is a basinward transition from predominantly cherty-oxide to sulphide facies (Fig. 3.11).

The stratigraphy and palaeogeography of part of the Abitibi belt were analysed by Dimroth et al. (1982) who pointed out that the linearity, volcanic and sedimentary sequences, and presence of voluminous granitoid plutons associated with the later stages of volcanism, were all shared by modern volcanic island arcs. Abitibi tholeiitic basalts closely resemble modern mid-ocean ridge basalts (which presumably form the foundation of intra-oceanic arcs), and the calc-alkaline andesites and alkali rocks, are very similar to modern oceanic island arc andesites and to some modern volcanic-arc high-K rocks, respectively (Capdevila *et al.*, 1982). In a comparison of the morphology of Abitibi pillow lavas with that of oceanic pillows from the mid-Atlantic Ridge, Wells *et al.* (1979) concluded that, while the stratigraphy of the Archaean and modern oceanic crust is similar, there are more pillows in the latter, that Archaean flows are longer and thicker than those of today and Archaean pillows are more flattened and welded, suggesting that the rate of eruption in the Archaean was greater than on the mid-Atlantic Ridge. Archaean pillow morphology, vesicle size and distribution indicate a shallow depth of eruption of less than 1 km.

In the Slave structural province the Yellowknife Supergroup occurs in a number of discontinuous greenstone belts (Henderson, 1975, 1981; Jenner *et al.*, 1981). The typical stratigraphy is divisible into: (a) a lower volcanic sequence of massive and pillowed metabasaltic to meta-andesitic flows, intermediate-to-acidic lavas, and tuffaceous rocks including dacites, latites and quartz latites, the acidic rocks occuring in the upper parts of the volcanic piles; and (b) an upper predominantly sedimentary sequence of immature greywackes, shales and mudstones. Sediments make up 80% of the Supergroup.

Geochemistry of the Volcanic Rocks

In the preceding sections it became clear that whilst greenstone belt successions on different continents resemble each other there is substantial vertical variation within them. In particular, confining one's attention to the predominantly volcanic section, the lower part is often composed of ultramafic and mafic rocks, whereas the upper part is characterized by a calc-alkaline assemblage including mafic-felsic suites and basalt–andesite–rhyolite cycles (Goodwin, 1977a; Condie, 1981).

1. Komatiitic extrusives and intrusives occur in the lower parts of many belts, e.g. in Yilgarn and Pilbara blocks, the Barberton belt, Belingwe and Que Que in Zimbabwe and the Abitibi belt (Sun and Nesbitt, 1978; Nesbitt *et al.*, 1979, 1982; Arndt. *et al.*, 1979; Arndt and Nesbitt, 1982). They have high Mg, Ni, Cr and low Fe and Ti and high CaO/Al_2O_3 and Al_2O_3/TiO_2. Peridotitic komatiites require a very high degree of melting (50–80%, Arndt, 1977) of the mantle and basaltic komatiites extensive (40–60%) melting. This fact may be explained by shallow depths of melting which may be consistent with expected high rates of heat flow and steep geothermal gradients in the Archaean.

2. Meta-basalts in the lower ultramafic–mafic parts of the belts have K, Na/K, Sr, Zr and Fe^{3+}/Fe^{2+} similar to, Al and Ti lower than and Mn, Ni, Cr, Co, Rb and Fe/Fe + Mg (total iron) higher than, modern oceanic tholeiites. The MgO-Ni ratios in meta-basalts, etc. are similar to those of modern oceanic ridge basalts, etc., implying similar olivine compositions and Mg–Ni ratios in parental mantle material. Chondrite-normalized REE patterns typically show near-flat curves and no La and Ce depletion; such patterns are comparable with those of modern oceanic tholeiites (Condie, 1981). Zr–Y–Ti ratios fall in the field of oceanic tholeiites. Archaean K_2O contents are typically in the range 0.14–0.26% comparable with the range of 0.16–0.22% of oceanic tholeiites. The Archaean oceanic crust should contain more K_2O than present-day oceanic basalts because the low K_2O content of the modern rocks is caused by the fact that the liquids were derived from a mantle source that had previously been depleted. In the Archaean one would expect the mantle to have been less depleted than that of today because less had been extracted from it. Marked regional variations in major, trace and REE data suggest that the upper mantle was already distinctly heterogeneous by the late Archaean (Sun and Nesbitt, 1977; Jahn *et al.*, 1980).

3. The volcanic rocks in the central part of the stratigraphy of greenstone belts typically

consist of a calc-alkaline basalt–andesite–dacite–rhyolite association that becomes increasingly felsic upwards (Condie, 1982b; Goodwin, 1982a). The abundance of pyroclastics and andesites also increases with stratigraphic height. Andesites make up 30% of Canadian greenstone belts (Goodwin, 1977a), but a far smaller proportion of belts in southern Africa and Australia. In Canadian belts there is a progressive increase in Al, K, Sr and Ba with height in the stratigraphic pile. Alkaline to shoshonitic volcanics are rare in greenstone belts, being only recorded from Canada (Brooks *et al.*, 1982). All the features described above are very similar to those in modern immature to mature island arcs. In both cases there is evidence of a progressive upward chemical trend in the evolution of the volcanoes. Goodwin (1977a) concluded that the calc-alkaline suite of belts in the Superior Province of Canada corresponds to that of Quaternary island arcs of thin to intermediate-type crust approximately 15 to 25 km thick. It seems most likely that greenstone belts with a large proportion of calc-alkaline rocks formed either in an arc-type or back-arc environment (Condie, 1981). Much current trace element and isotope geochemistry is devoted to the comparative study of Archaean and modern volcanic rocks (Gill, 1979; Jahn and Sun, 1979).

Structure

In this section we shall look at the structural make-up of the greenstone belts, leaving the more theoretical models for the next chapter.

The classical picture of the structure of all the greenstone belts is one of synforms formed largely by vertical downsinking of volcano-sedimentary rocks associated with the diapiric uprise of granite bodies. The essential theme running through these ideas, defined or implicit, is that the formation of the basin-shaped structure was largely controlled be vertical movements.

More recent detailed field work and rethinking have revealed that not all belts have a simple synformal structure and that large-scale lateral displacements have played a significant role in their formation. Ramsay (1963) was probably the first person to appreciate the importance of the horizontal strain component in the formation of major structures in the belts, a fact corroborated by Coward and James (1974). D. S. Wood (1973) calculated that the dimensional changes associated with the formation of some Zimbabwean belts involved a shortening across the subvertical fold axial surfaces of 75% and subvertical extension in the axial surfaces of 300%. The first evidence that horizontal nappe-type movements played a major part in the evolution of greenstone belts came from the Rhodesian craton (Stowe, 1974) and such folding has since been tentatively proposed for belts in Botswana (Key *et al.*, 1976). Coward *et al.* (1976) demonstrate that similar large-scale crustal shortening was responsible for the early development of belts in southern Africa and that some belts are not synclinal as they have downward-facing sedimentary structures; they are late folds formed on the inverted limbs of early nappes. Similar nappe structures and inverted successions are now known in belts in W Australia (Archibald *et al.*, 1978), Ontario, Canada (Poulsen *et al.*, 1980), in the Barberton belt (Williams and Furnell, 1979; De Wit, 1982) and in the Dharwar belts of India (Drury and Holt, 1980). From a theoretical standpoint Burke *et al.* (1976a) suggest that there may be many more thrust and slides than are so far recognized in greenstone belt successions (implying that the enormous stratigraphic thicknesses of the order of 10–20 km usually listed are an illusion), and that horizontal shortening expressed by vertical cleavage and nappe structures has been an important element in the evolution of many belts. De Wit (1982) provided evidence of major gliding and overthrust nappe tectonics in the Barberton belt, and Fripp *et al.* (1980) showed that there are shear zones and mylonitic fabrics both in the greenstone belt and along its base, and interpreted the base as a major low-angle sole thrust which transported the whole greenstone belt as an allochthonous tectonic unit onto the gneisses. Drury (1977) demonstrated that early isoclinal folds in the Yellowknife belt in Canada cascaded off rising diapiric granitic plutons before the main synformal keel or basin developed. All the results quoted above show that evidence is gradually accumulating to indicate that many greenstone belts are not simple down-sagged basins.

Metamorphism

All the sedimentary and volcanic rocks of the world's greenstone belts have suffered some degree of recrystallization, varying from the zeolite to the granulite facies. The most common type of metamorphic imprint has given rise to greenschist facies assemblages—hence the term *greenstone* belt. There are two significant departures from this grade.

1. In many belts there is an increase in grade, mostly to the amphibolite facies, firstly from the centre to the margins and secondly towards intrusive granitic plutons.
2. Some belts have been regionally metamorphosed to a higher grade—to the granulite facies in the Yilgarn block of SW Australia (Gale and Klein, 1981), and to the amphibolite facies at Hongtoüshan in NE China (author's observation).

Ultramafics are usually serpentinized with varieties such as talc schists and tremolite–actinolite schists. Basic volcanics commonly have assemblages like actinolite–albite–chlorite–epidote or, at higher grade, hornblende–andesine–epidote–quartz. Metasediments are typically schists containing actinolite–chlorite–quartz, sericite–quartz and biotite–chlorite–quartz. The metamorphic climax in belts in Sierra Leone was at 565–595°C and 4.9–5.5 kb (Rollinson, 1982b).

For all the metamorphic variation in greenstone belts, one thing is certain: the high pressure facies types with glaucophane and eclogite are absent.

Mineralization

In this section the mineral deposits that occur in Archaean greenstone belts will be summarized, particular regard being paid to the stratigraphic controls on their mode of occurrence (Boyle, 1976; Condie, 1981; Lambert and Groves, 1981; Kazansky, 1982). The types of mineral deposit in different shield regions are remarkably similar and any particular ore is confined to a limited environment. A metallogenic framework thus emerges in which the formation of the mineral deposits can be related to the development of the greenstone belts and their granites, e.g. in the Dharwar craton (Radhakrishna, 1976).

The greenstone belts contain a multitide of economic minerals; they are one of the main depositories in the world of elements such as Au, Ag, Cr, Ni, Cu and Zn. In the past they were termed gold belts and since the gold rush days, which took place in most of the continents concerned around the turn of the last century, they have been the subject of extremely detailed mapping by mining companies and surveys.

The mineral deposits can be related to the major rock groups that make up the greenstone belt/granite terrains as follows:

1. Ultramafic flows and intrusions: chromite, nickel, asbestos, magnesite and talc.
2. Mafic-to-felsic volcanics: gold, silver, copper and zinc.
3. Sediments: iron ore, manganese and barytes.
4. Granites and pegmatites: lithium, tantalum, beryllium, tin, molybdenum and bismuth.

Fig. 3.8 shows the distribution of the main mineral deposits within the volcanic and sedimentary rock groups of the Abitibi belt.

Many elements (e.g. Cr, Ni, Au, Ag, Cu, Zn) can be related to the magmatic differentiation of the mantle-derived ultramafic–mafic–felsic volcanic rocks. Sedimentary (e.g. Fe, Mn), ultramafic cumulate and volcanic exhalative ores were stratigraphically controlled but the final formation

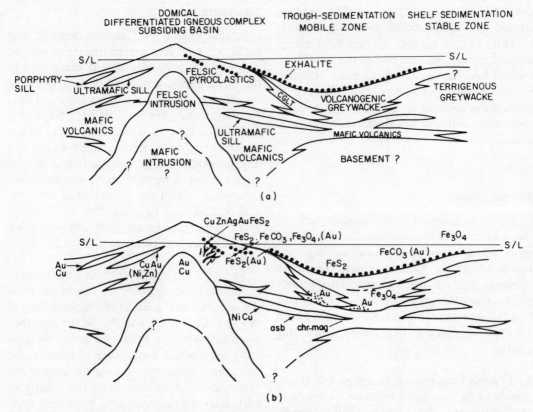

Fig. 3.8 A diagrammatic cross-section of the Abitibi belt. (a) Shows the tectonic–stratigraphic relations with the volcanic–sedimentary complex and (b) The main mineral deposits. Full width of section 50 miles, maximum vertical thickness 10 miles. (After Hutchinson *et al*., 1971; reproduced by permission of *Trans. Can. Inst. Metall.*)

of the majority of the deposits was structurally controlled. The first Sn deposits to form in the evolving continents occur in pegmatites at Bikita, Zimbabwe; other minor mineralization in greenstone belts includes barytes, talc, manganese, asbestos and tungsten.

Chromite

Deposits of chromite are not common in greenstone belts. The most important are in the Sebakwian ultramafics in Zimbabwe but there are minor occurrences in the Abitibi belt.

The Selukwe chromite occurs in serpentines and talc–carbonate rocks that are weakly metamorphosed ultramafics belonging to differentiated sill-like lenses conformably intruded into Sebakwian schists. This is one of the largest occurrences of high-grade chrome ore in the world and far exceeds any other Archaean greenstone belt chromite deposit.

Nickel

Surprisingly there are no economic nickel deposits in the Barberton greenstone belt, particularly since the lower ultramafic group is so well developed there.

The most important Archaean Ni deposits occur in SW Australia (Kalgoorlie belt), S Canada (Abitibi belt) and Zimbabwe (north-west extension of the Selukwe belt). In these greenstone belts the Ni typically occurs in magmatic segregations at, or close to, the base of mafic to ultramafic sill-like bodies.

There are a number of Ni deposits in the

Yilgarn block but these are overshadowed by the Kambalda deposit. Most of the nickel sulphide mineralization at Kambalda occurs at the base of a meta-ultramafic sill intrusion in contact with meta-basalt, but some ore occurs as lenses within the ultramafic rocks.

It is ironic that in the Canadian greenstone belts, originally thought by Anhaeusser *et al.* (1969) to have no lower ultramafic group, there are so many nickel deposits. Of the 16 economic deposits the majority occur at the base of differentiated peridotitic bodies.

Gold–Silver

Gold is the most important economic mineral in Archaean greenstone belts throughout the world; for example, there are 197 known gold occurrences in the Superior Province of Canada, almost all of which are situated in the volcanic-rich greenstone belts. The gold distribution pattern has been traditionally related to regional fractures connected with, for example, granitic stocks, lava flows and unconformities, but this structural disturbance has partly obscured the more fundamental stratigraphic control of the gold concentration. Fig. 3.9 shows the gold distribution in Zimbabwean greenstone belts, most of which are in the igneous Bulawayan and Sebekwian units.

The gold occurs in the following types of stratotectonic environment:

1. At felsic-mafic volcanic contacts in the upper parts of the central volcanic group, in particular in the Canadian belts and in the Barberton Mountain Land. In the Abitibi belt most gold deposits lie in intermediate to felsic volcanic rocks or their intrusive equivalents.
2. The gold is sometimes located in the carbonate facies iron formations, e.g. in the Abitibi belt, but more typically in the oxide facies as in southern Africa. Anhaeusser (1971b, 1976) points out that the iron formations tend to terminate the sedimentary cycles by overlying the final stage volcanics and pyroclasts; this type of gold mineralization is thus connected with the most highly differentiated exhalative volcanic phases. In reviewing the gold metallogeny of Zimbabwe, Fripp

(1976) suggests that the gold in Sebakwian banded iron formations closely related to aquagene tuffs was precipitated from subaqueous volcanic exhalations of thermal brines in the temperature range of 300–400°C (Fig. 3.10). (See also Kerrich and Fyfe, 1981).
3. Basic igneous rocks are commonly the dominant host rocks: in SW Australia the Golden Mile dolerite and the underlying tholeiitic Paringa basalt and in Zimbabwe the Bulawayan greenstone volcanics.
4. In the Abitibi, gold occurs in association with felsic alkaline intrusions in the form of subvolcanic sills, discordant plugs or stocks, and with their flow and pyroclastic equivalents (Goodwin and Ridler, 1970).

The most important gold mineralization lies in the marginal zone of the greenstone belts near the bordering granitic plutons and this mineralization typically decreases progressively away from the granitic contacts. Thus, although the ultimate source of the gold and associated sulphide mineralization is commonly reckoned to be the vast pile of mafic and ultramafic volcanics, the elements concerned appear to have been mobilized by the action of thermal gradients set up by the invading granitic bodies (Anhaeusser *et al.*, 1969).

Silver commonly occurs with gold in the greenstone belts. In the Abitibi belt it is found typically with Au, Cu and Zn at felsic–mafic volcanic contacts (Goodwin and Ridler, 1970).

These granites also have prophyry-type Cu–Mo mineralization comparable to that associated with modern active plate boundaries (Elbers, 1976; Davies and Luhta, 1978; Ayres *et al.*, 1982; Barley, 1982).

Copper–Zinc

Chalcopyrite–sphalerite mineralization typically occurs in the volcanics and pyroclastics in the upper parts of the volcanic cycles of greenstone belts, but the world distribution is variable.

In the Superior Province belts there are 185 copper–zinc occurrences (40% of all mineral occurrences) but only seven in each

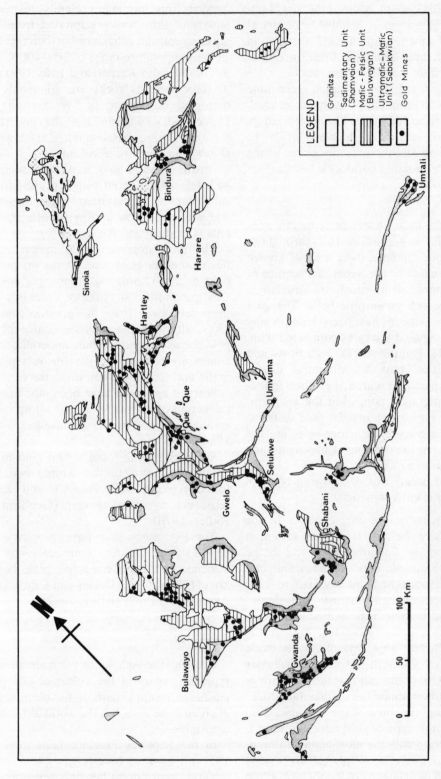

Fig. 3.9 Map showing the distribution of the more important gold mines in relation to the stratigraphy of the greenstone belts in Zimbabwe (redrawn after Anhaeusser, 1976; reproduced by permission of *Minerals, Science and Engineering*—discontinued publication)

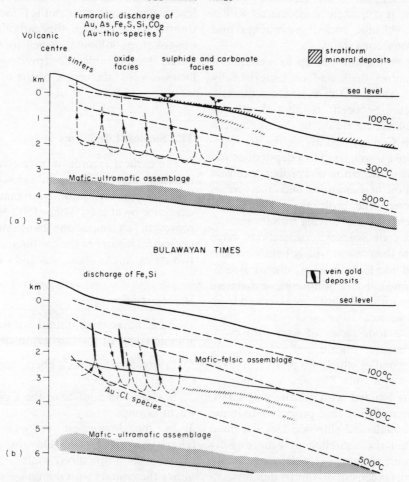

Fig. 3.10 A model for the evolution of gold deposits in the Archaean. (a) Stratiform deposits in Sebakwian iron formations formed by convecting brines leaching the gold from the volcanic pile. (b) Vein gold deposits formed under a lower geothermal gradient in Bulawayan times (after Fripp, 1976; reproduced by permission of J. Wiley)

of the Slave and Churchill Provinces (MacGeehan and MacLean, 1980a,b). Lead-bearing sulphide (galena) occurrences are rare in rocks of Archaean age—there are a few in the Abitibi belt (none elsewhere in Canada), the lead usually being associated with Cu–Zn but subordinate to both (Large, 1977).

Tantalum

The largest known concentration of pollucite and the world's largest source of tantalum occurs in the Tanco pegmatite in the Bird River greenstone belt in Manitoba, Canada (Crouse *et al.*, 1979).

Iron

Banded iron formations are common in Archaean greenstone belts but this ore type reached its peak of development in early Proterozoic basins about 2000 Ma ago and so will be discussed in more detail in Chapter 5.

Of the four types of iron formation, Algoma, Superior, Clinton and Minette the first, characteristic of the Archaean greenstone belts, is consistently associated with a variety of volcanic rocks, greywackes and black carbonaceous slates.

The iron formations are up to a couple of hundred metres thick and characteristically form short lenses in the volcanic sequence. Some occur between rhyolite–dacite or pyroclastic-andesite flows. In the Michipicoten area in Canada underlying dacite–rhyolite flows are carbonatized to a depth of nearly 200 m. This carbonation was probably caused by the action of fumaroles and hot springs that affected the volcanics during the development of the overlying iron formation. These types of volcanic associations have given rise to the present widely held view that the banded iron formations are chemical sediments formed by a fumarolic–exhalative process, the iron and silica being derived from a volcanic source.

There are four facies of iron formation, oxide, carbonate, silicate and sulphide, the first and second of which are predominant in the Algoma-type ores. The iron ore bands are usually interbanded with ferruginous chert. H. L. James (1954) first proposed that the oxide, carbonate and sulphide facies formed further from the shoreline in progressively deeper water. Goodwin (1973) made a considerable advancement on this by demonstrating that the three facies defined the original shelf–basin bathymetry (Fig. 3.11). The shallow water oxide (magnetite) facies grades transitionally through the carbonate (siderite–ankerite–dolomite) facies to the deeper water, basinal sulphide (pyrite–pyrrhotite) facies towards the centre of the orogen.

The Earliest Life Forms

The earliest records of terrestrial life occur in sedimentary rocks in the greenstone belts. These organisms demonstrate that life was in existence by at least 3400–3500 Ma and thus represent an important benchmark in the geological history of the continents. There are two types, micro-organisms and tromatolites.

Micro-organisms

These have been found in three stratigraphic levels of the Barberton succession.

1. In organic-rich black cherts and shales of the Fig Tree Group.
2. In chert and argillite of the Upper Onverwacht Group.
3. In the lowermost sedimentary rocks (cherts) of the Lower Onverwacht Group (Theespruit Formation) only 350–600 m above the contact with the underlying granite gneisses.

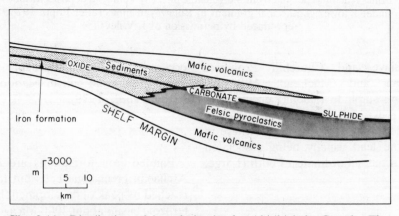

Fig. 3.11 Distribution of iron facies in the Abitibi belt, Canada. The oxide–carbonate–sulphide facies transitions delineate the original shelf-to-basin slope of the 'orogen' (redrawn after Goodwin, 1973b; reproduced by permission of *Economic Geology*)

In the Swaziland Supergroup there are three types of microstructures: sphaeroidal unicell-like structures, rod-shaped bacterium-like bodies and filamentous thread-like structures (Schopf, 1975).

The micro-fossils in the Lower Onverwacht are half the size of those higher up the succession—spheroids are 10 mm in diameter and filaments about 7 mm long. Some of the carbonaceous spheroids of the Fig Tree Group resemble algae and other cysts of flagellates; the filamentous forms look like blue–green algae (Schopf, 1974). Some sphaeroids from cherts in the Swartkoppie Formation contain evidence of binary cell division (Knoll and Barghoorn, 1977). Carbonaceous sphaeroids, similar to those in the Onverwacht Group, are reported in the 3500 Ma Warrawoona Group in the Pilbara Province of NW Australia (Dunlop et al., 1978); these are the oldest micro-fossils so far discovered. In the 2600 Ma Pietersburg greenstone belt of South Africa there are 0.0–1.0 mm spherical carbonaceous aggregates that might be vegetative diaspores of primitive columnar plants, if comparison with similar structures in the Witwatersrand Basin are correct (Saager and Muff, 1980). One must bear in mind that some authorities still question the biogenic origin of these micro-structures (Schopf, 1975), but the statistical studies of size distributions and distinct populations (Muir and Grant, 1976), and the discovery of evidence of cell division (Knoll and Barghoorn, 1977) suggest that some of these micro-structures are indeed micro-fossils.

It is worth mentioning here that the Fig Tree cherts contain much organic material arranged in laminations parallel to the bedding, indicating its original sedimentary deposition. The algal bodies have coatings that mainly consist of compounds of Cu, Fe, Ni and Ca, showing that the organisms were able to precipitate metal salts from water by action of their body processes, whilst the sulphur in the coatings, bound in the metal sulphides, originated from the organic matter of the algae.

The carbon isotope data and the occurrence of micro-fossils from the Barberton suc-cession are shown in Fig. 3.12. The insoluble organic carbons from the Lower Onverwacht have $\delta^{13}C$ values ranging from -14.3 to -18.9%, whilst the kerogens in the younger beds range from -26.1 to -33.0%. The high values in the Lower Onverwacht have received much discussion—they might be due to secondary metamorphic effects (Sylvester-Bradley, 1975). The ratios in the upper beds are consistent with fractionation by photosynthesis, in which case conditions must have existed soon after 3355 Ma permitting the growth of photosynthetic plants.

The relationships described in this section may suggest that there was a significant evolutionary burst during the period when the Barberton succession was laid down. Although it has been claimed that this was the time when photosynthesis began and 'proto-life' evolved into 'life', it now seems more likely that these events took place much earlier in time (more than 3700 Ma), and evidence of them might be better sought in Archaean terrains like West Greenland with a sedimentary history back to more than 3760 Ma.

Stromatolites

Stromatolites are unequivocal macro-fossils. 2700–2600 Ma stromatolites are known from the Belingwe greenstone belt in Zimbabwe (Bickle et al., 1975; Martin et al., 1980), the Slave Province in Canada (Henderson, 1975), and in the Steep Rock Lake belt, S Canada (Joliffe, 1966). 3000 Ma examples occur in the Pongola Supergroup (not a greenstone belt) of South Africa (Mason and von Brunn, 1977). Recently, 3400–3500 Ma stromatolites have been reported from the Warrawoona Group in the Pilbara Province of NW Australia (Lowe, 1980b; Walter et al., 1980) and in the Sebak-wian Group in Zimbabwe (Orpen and Wilson, 1981). Thus we have indubitable evidence of the existence of life as far back in time as the oldest known greenstone belts.

Stromatolites result from accretion of detrital and precipitated minerals on successive sheet-like mats formed by communities of

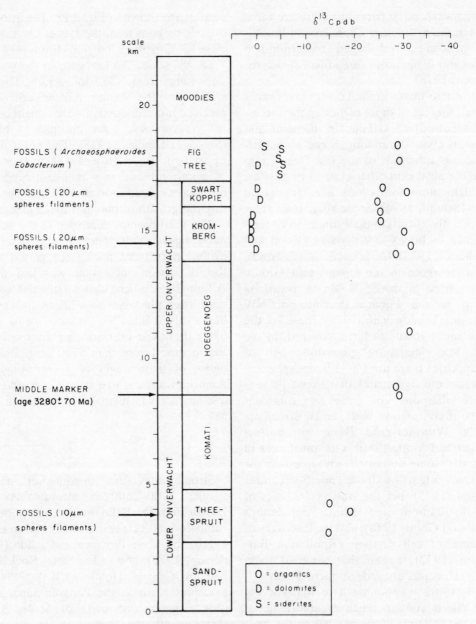

Fig. 3.12   The occurrence of microfossils and the distribution of carbon isotope data in the Swaziland Supergroup of the Barberton belt, South Africa (after Sylvester-Bradley, 1975; data from several sources)

micro-organisms predominated by filamentous blue–green algae. The existence of stromatolites, about 3500 Ma, places a minimum age on the time of origin of algal photosynthesis producing gaseous oxygen as a by-product, and of integrated biological communities of primitive micro-organisms presumably including producers (blue–green algae), reducers (aerobic and anaerobic bacteria) and consumers (bacteria, predatory by absorption).

One of the main obstacles to the evolution of life in Archaean times was, according to Cloud (1968a), the lack of an ozone screen in the upper atmosphere. The present ozone layer prevents the penetration of high-energy ultra-violet radiation in the 2400–2900 (especially the 2600) angstrom wavelength range which would inactivate the life-inducing DNA molecule. The lack of an Archaean ozone screen allowed the damaging radiation to penetrate the surface layers of water, preventing the healthy growth of early life forms which were necessarily restricted to local shielded habitats. It is not surprising, therefore, that primitive life forms are only found in a few isolated localities in Archaean rocks.

There are two fundamentally different types of primitive organism (Cloud, 1968a, 1976a,b).

1. Procaryotes, consisting of procaryotic cells which lacked a wall around the nucleus and were incapable of cell division. The nucleus is that part of the cell in which the DNA or genetic coding material is concentrated.

2. Eucaryotes, consisting of eucaryotic cells which had a nucleus enclosed within a membrane and were capable of cell division by which the genetic coding material (DNA) was successively parcelled out among the different cells and descendants of the organism.

The procaryotes were relatively more resistant to ultra-violet radiation than the eucaryotes and so were less likely to be adversely affected by radiation-induced mutations and thus were the first organisms to develop in the early Archaean, when the atmosphere lacked a radiation-protective ozone screen, as typified by the bacteria and algae of the Fig Tree Group. They were the predominant organism up until about 1300 Ma. The eukaryotes are mainly oxygen-using organisms whose appearance was triggered by the development of a relatively oxygenic atmosphere during the mid-late Proterozoic.

Stromatolites are restricted to shallow, stable, shelf, marine environments. The fact that early cratons were unstable and thus were rarely preserved in the permobile tectonic regime of the Archaean means that there were fewer niches for stromatolite growth compared with the early Proterozoic when stable cratons were well developed.

Chapter 4

Crustal Evolution in the Archaean

In Chapters 2 and 3 the main features of the Archaean high- and low-grade terrains were reviewed. Here we are concerned with the tectonic evolution of these terrains and so come against the problem of the spatial and temporal relationships between them. There is so much current interest in the early history of the earth that there is no lack of thought about evolutionary models and these will be considered below. All models, of course, reflect, if not the preference and prejudice, the degree of experience of the proposer and this limitation often hinders the viability of the model. A major problem has been that there are insufficient constraints to limit speculation, but the current tendency to integrate structural, stratigraphic and geochemical data should give rise to more viable hypotheses.

Greenstone Belts

There are as many hypotheses to explain the formation of greenstone belts as once there were for 'granites and granites'. Although much mapping of these belts took place in the early decades of this century, it was not until 1968 that the recent spate of tectonic modelling began.

Classical Models

1. Anhaeusser *et al.* (1969) and Viljoen and Viljoen (1969) envisaged that greenstone belts formed either along fundamental fractures in downwarps at the interface between thin continental crust and oceanic crust or in parallel fault-bounded troughs in an unstable thin primitive sialic crust. The basins developed as *in situ* depositories by progressive downsagging of the heavy volcanic pile; this is the beginning of the downsagging basin model that was the foundation of the later models of Anhaeusser and Glikson (see below). During the late stages the underlying granitic crust was thickened by some form of underplating (cause unspecified) and the greenstone belts were invaded internally and marginally by diapiric granites that caused compression of the belts (giving rise to their arcuate form). These authors were unlucky in that they produced their models before the plate tectonic revolution had an impact on interpretations of continental development.

2. Anhaeusser (1971a) produced a further development of the downsagging basin model (Fig. 4.1). A limitation of this and the earlier models is that the authors did not entertain the possibility that the nearby high-grade gneisses could represent the basement.

3. Glikson (1972, 1976a, 1979 and earlier papers) and Glikson and Lambert (1976) produced a set of variations on a basic theme. Important aspects of these models are:

(a) They followed previous authors in not accepting evidence for an early gneissic basement. Note, however, that the models were an attempt to explain the development not just of greenstone belts but also of Archaean shields and cratons. The high-grade 'mobile belts' were not considered in the early papers; they appeared as a 'postscript' in 1972 (Fig. 4.2), having formed after the greenstone belts, and in

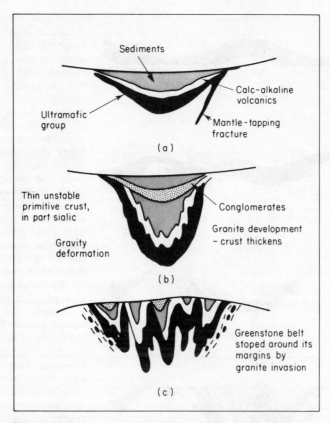

Fig. 4.1 Diagrammatic model showing the evolutionary development of a greenstone belt by progressive down-sagging of a volcano–sedimentary basin into a thin unstable primitive crust (redrawn after Anhaeusser, 1971a; reproduced by permission of The Geological Society of Australia)

later papers they are considered to represent coeval infracrustal roots of the greenstone belts.

(b) They emphasize the primitive chemical nature of the lower ultramafic and mafic volcanics (often komatiitic) and so, in the absence of a basement, the formation of a widespread primitive oceanic crust is envisaged as the first stage in proto-continental evolution. In other words, the ultramafic–mafic rocks at the base of greenstone belts are interpreted as relics of 'the primordial crust' (1972) or 'a once extensive simatic crust' (Glikson and Lambert, 1976).

However, seen in historical perspective these suggestions are no longer viable. Viljoen and Viljoen (1969) correctly iden-

tified the primitive chemical character of the Barberton komatiites and it is likely that they, and associated lower tholeiites, did form in part of some oceanic crust (although some komatiites lie unconformably on older sialic basement—Bickle *et al.*, 1975). But the conclusion that this is a remnant of the earth's most primitive or primordial oceanic crust is not acceptable.

(c) It is envisaged that parts of the oceanic crust subsided (by downsagging or rifting) and underwent partial melting to give rise to sodic acid magmas which rose upwards to form island nuclei, the greenstone belts forming between them in the intervening troughs. The key to the formation of the sodic granite magmas was the experimental

50

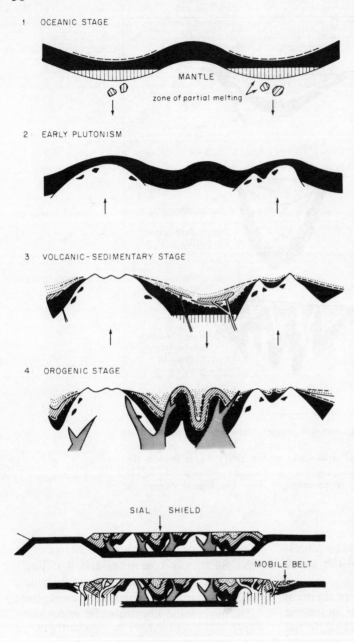

1: OCEANIC STAGE

MANTLE

zone of partial melting

2: EARLY PLUTONISM

3: VOLCANIC-SEDIMENTARY STAGE

4: OROGENIC STAGE

SIAL   SHIELD

MOBILE BELT

Fig. 4.2 A model of evolution of Archaean shields. 1: Oceanic stage; mega-rippling of the oceanic crust, minor sedimentation of chert, ferruginous chert, banded iron formations and pelitic sediments derived through erosion of the oceanic crust at structural highs. Partial melting of subsiding eclogite and/or amphibolite segments of the crust. 2: Early plutonism; low degrees of partial melting of the oceanic crust gives rise to sodic acid magmas which rise as diapiric oval-shaped or elongated batholiths. 3: Volcanic–sedimentary stage (greenstone belts); the isostatic rise of the early batholiths associated with the subsidence of intervening tracts of the ocean crust. Further partial melting of the oceanic crust below these troughs gives rise to calc-alkaline volcanism. The erosion of the batholiths as well as of the intrabasinal volcanics results in the accumulation of detrital sediments. 4: Orogenic stage; further subsidence of the volcanic–sedimentary troughs between the granitic nuclei into warmer regions of the crust results in folding and low-grade metamorphism. Low degrees of partial melting at the base of the thickening crust give rise to younger (potassic) granites. Shield formation stage cooling of the volcanic–sedimentary troughs results in the aggregation of the granitic nuclei into shields. The boundaries of these shields with the surrounding oceanic crust constitute the loci of development of high-grade 'mobile belts', possibly above subduction zones. (After Glikson, 1972; reproduced by permission of The Geological Society of America)

work of T. H. Green and Ringwood (1968) which showed that sodic acid liquids can be produced by low degrees of partial melting of amphibolite or eclogite under wet conditions. In the 1970s Green and Ringwood applied their earlier data to subduction zone environments, but the subsidence model continues today (Goodwin and Smith, 1980; Glikson, 1981, 1982; Goodwin 1981).

(d) It is well known that the lowest ultramafic-mafic rocks of some greenstone belts, especially in Zimbabwe (the Sebakwian) and at Barberton (the Lower Onverwacht), occur as inclusions within bordering late intrusive tonalites—this is not under dispute. But high-grade gneisses typically contain lots of inclusions and since Glikson and Lambert (1976) and Glikson (1976a) do not believe that the gneisses belong to any form of basement they suggest that they formed late in the greenstone belt cycle and so equate the real greenstone belt inclusions in the tonalites with the inclusions in the gneisses (Fig. 4.3). In this way they conclude that Archaean gneiss–granulite terrains are the late coeval roots of early greenstone belts that are remnants of some, once extensive, simatic crust. Goodwin (1977b) concludes that greenstone belts in the Superior Province of Canada become progressively more gneissic in composition with depth, and that the crystalline gneissic craton is the infrastructure to the extrusive–rich greenstone basins which have been variably removed by erosion. However, the detailed field, structural and geochronological data of a great number of specialists suggest to them a quite different relationship between greenstone belts and nearby high-grade granulites and gneisses (Bickle *et al.*, 1975; Hawkesworth *et al.*, 1975; Hawkesworth *et al.*, 1975; Coward, Lintern and Wright, 1976; Pavlovskiy, 1980; Chadwick *et al.*, 1981a,b; Swami Nath and Ramakrishnan, 1981). For further details on these interrelations see later in this chapter.

(e) Glikson and Lambert then consider that the calc-alkaline volcanic part of greenstone belts are 'upper or secondary greenstones' (Fig. 4.3) formed (with the overlying sediments) within linear troughs; these make up the bulk of late Archaean greenstone belts. The reason for giving the lower and upper parts of greenstone belts significantly different ages and origins is the fact that locally they are separated by an unconformity and the intrusion of some Na-rich granites. Thus whilst the primary greenstones 'evolved in environments showing no evidence of pre-existing or proximal sialic crust', the secondary greenstones 'rest on post-granite unconformities and/or include granite-derived arenites and conglomerates' (Glikson, 1976). Lowe (1982) puts these relationships into a further perspective (see later this chapter).

4. The classical model in a modern context: a still popular idea is that greenstone belts formed in rifts. This may be a palaeo-aulacogen (Pavlovskiy, 1980). Grachev and Fedorovsky (1981) emphasized the rarity of andesites in early Archaean (>3000 Ma) belts compared with their abundance (60%)

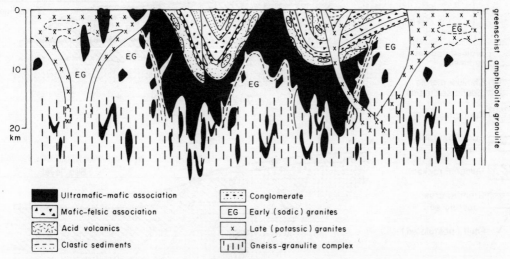

Fig. 4.3 An interpretation of the field relationships between major Archaean rock units in western Australia, South Africa–Zimbabwe and India. The figure is meant to portray the general concept of the relations between greenstone belts and gneiss–granulite belts (modified after Glikson, 1976a; reproduced by permission of J. Wiley)

52

in modern island arcs, and the typical occurrence of a bimodal basalt–rhyolite association which occurs today in oceanic islands. They therefore ascribed to the view that this earlier Archaean association formed in central-type volcanoes on proto-oceanic crust.

Greenstone belts with thick calc-alkaline volcanic successions (30% andesite and 10% felsic volcanics in Canada) largely formed in the period 2700–2600 Ma (Condie, 1981). To explain the formation of this calc-alkaline material Goodwin and Smith (1980) suggested that the great thicknesses of earlier mantle-derived volcanics led to downsagging, and that partial melting of their lower parts produced the felsic magmas; the process was termed sag-subduction (Goodwin and Smith, 1980) or rift-and-sag (Goodwin, 1981). This was described as an Archaean plate tectonic process. Glikson (1982) also envisaged derivation of these magmas by 'depression and partial melting of the volcanic repositories'.

If I were in favour of the rift model, I would emphasize the ideas of McKenzie et al. (1980) and Bickle and Erikkson (1982)

based on McKenzie (1978). Greenstone belts underwent two-stage subsidence comparable to that of modern sedimentary basins and rifts. Initial fault-controlled subsidence was due to thinning or loading of the crust and led to extrusion of volcanic rocks in the lower parts of the basins. Later thermal subsidence took place as the lithosphere relaxed to equilibrium thickness and gave rise to fault-free downsagged sedimentary sequences Fig. 4.4).

Trace element studies show that the Abitibi lower tholeiitic basalts closely resemble modern mid-ocean ridge basalts derived from a depleted mantle source. Consequently Capdevila et al. (1982) propose an origin by pulsating migrating mantle diapirism, a type of 'hot-spot' tectonics in a plume-controlled rift (Fyfe, 1978).

Plate Tectonic Models

The following plate tectonic models have been applied to the possible evolution of greenstone belts.

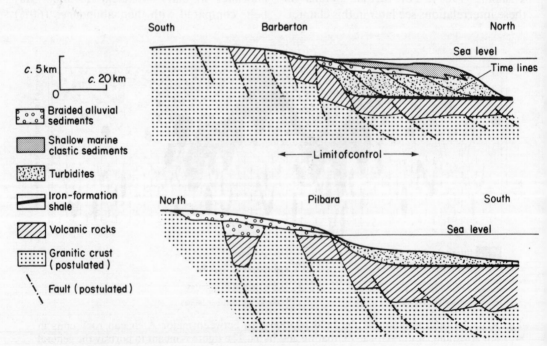

Fig. 4.4 Schematic palaeographic reconstructions in Barberton and Pilbara greenstone belts (from Bickle and Erikkson, 1982, *Phil. Trans. R. Soc. Lond.*, **A305**, 225–247)

1. Goodwin and Ridler (1970) deduce that the Abitibi belt formed intracratonically upon a thin mafic crust between adjoining forelands of predominantly sialic composition. There are three main tectonic units: the primitive sialic cratons, comparatively unstable marginal sedimentary basins and troughs and orogenically active off-shore volcanogenic belts. The distribution pattern of sedimentation and volcanism is what would be expected from progressive tectonic spreading or opening of the sialic forelands as a result of incipient ocean floor spreading.

In order to better demonstrate the relationship between greenstone belts and many high-grade (basement) gneisses and in order to take account of the 'oceanic' character of the lower tholeiites (Glikson, 1972, 1976) Windley (1973) proposed a model according to which the belts formed in proto-oceanic ridge systems involving only small amounts of (continental) crustal separation. It was envisaged that some belts may have formed in ensialic graben-like rift zones and others, like perhaps the Abitibi belt, in narrow, linear oceanic basins caused by relatively small amounts of opening. A somewhat similar tectonic model based on an ascending mantle plume was applied to the belts of the Kaapvaal craton by Hunter (1974a) and to the Barberton belt by Condie and Hunter (1976) (Fig. 4.5).

2. Anhaeusser (1973) first proposed a subduction-controlled origin for greenstone belts and the island arc model has since proved popular. Important comparable features include a base of low-K tholeiitic volcanics overlain by calc-alkaline flows and pyroclastics, often with up to 30% andesite in late Archaean belts, an increase upwards in the proportion of felsic volcanics, and the presence of chemically similar tonalitic plu-

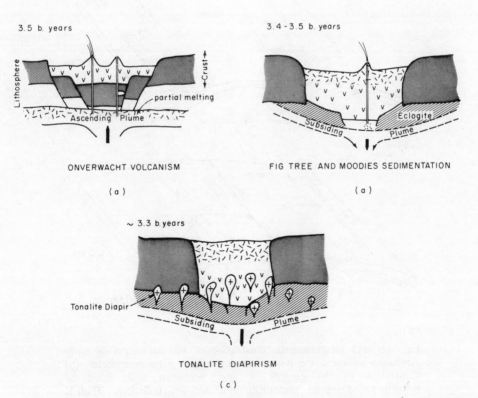

Fig. 4.5 A mantle plume model to explain the evolution of the Barberton belt (modified after Condie and Hunter, 1976; reproduced by permission of Elsevier Scientific Publishing Co.)

54

tons (Maaloe, 1982). From a consideration of stratigraphy and palaeogeography, Dimroth *et al*. (1982) concluded that the Abitibi belt was an Archaean analogue of an immature oceanic island arc, whilst Gélinas *et al*. (1977), from a study of some of the volcanics, suggested that a continental island arc environment like that of the Taupo province of New Zealand was most likely. Dimroth *et al*. (1982) also laid emphasis on the volcanic and sedimentary asymmetry of the Abitibi belt, on the 20–50 km spacing of volcanic complexes, on caldera collapse and volcano-tectonic subsidence, all features shared by Cenozoic oceanic island arcs. The greenstone belts of Finland are also similar to modern ensimatic island arcs (Blais *et al*., 19 7). The parallel alternation in the Superior Province of greenstone belts and paragneiss belts was

regarded by Langford and Morin (1976) as comparable to that of the major alternating greenstone–granite gneiss belts of the Canadian Cordillera which developed coevally and sequentially as part of a continuing process of merging arcs and intervening sedimentary basins in the Mesozoic–Cenozoic. Blackburn (1980) pointed out that none of the data in these belts of Ontario is incompatible with plate tectonic theory and produced a stimulating rational model which deserves serious consideration. After a study of all the features of Archaean greenstone belts, Condie (1981) concluded that the low-K tholeiites were formed at plume-generated rift systems, the andesites and tonalites at arc systems at convergent plate boundaries and the growth of this part of the Archaean crust by arc–arc collisions (Fig. 4.6).

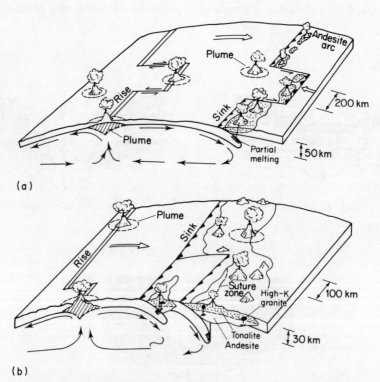

(a)

(b)

Fig. 4.6 (a) Diagrammatic illustration of the formation of early Archaean andesitic arc systems at convergent plate boundaries. (b) Illustration of the growth of late Archaean sialic crust by tonalite–trondhjemite plutonism and arc–arc collisions. High-K granitic plutons are produced by partial melting of andesitic rocks in the lower crust (from Condie, 1981, *Archean Greenstone Belts*, Elsevier, Amsterdam, p. 52)

3. There is a current, popular attempt to explain greenstone belts as fossil marginal basins. Reasons for suggesting that the greenstone belts formed in back-arc spreading centres rather than in main oceanic or arc environments are:

(a) In a mobile system the main oceanic lithosphere soon 'self-destructs' by subduction and cannot be expected to be preserved to any extent.

(b) Even without eventual continent–continent collision (where the uplift process is accentuated) volcanic arcs tend to be uplifted and eroded so that only their plutonic roots are exposed. Therefore the greenstone belts with their thick volcanic successions are most likely not to be remnant arcs, but rather fossil marginal back-arc basins which in contrast tend to be well preserved.

Burke *et al.* (1976a) expound a stimulating rationale for the back-arc basin concept and suggest that the Palaeozoic Round Pond area of Newfoundland and a Mesozoic area in the northwestern Sierra Nevada of California bear close resemblance to greenstone belts.

Tarney *et al.* (1976) present a detailed and cogent argument for regarding the Rocas Verdes marginal basin in South Chile as an actualistic counterpart of Archaean greenstone belts (Fig. 4. 7). The synclinal form, dimensions, greenschist metamorphic grade, structural style, volcano-sedimentary rock associations and stratigraphy, volcanic geochemistry, relationship with older basement and younger intrusive tonalite–granodiorite plutons of the Rocas Verdes are all remarkably similar to their equivalents in greenstone belts. The main difference is that the younger rock belt, in contrast to the older, contains a mafic dyke complex (100 % dykes) typical of ophiolites: but the presence or absence of such a dyke complex is dependent on many factors, such as rate of spreading and supply of magma.

The Rocas Verdes complex has pillow-bearing basalts, pillow breccias and tuffs, shales and greywackes with a high proportion of volcanogenic detritus, and cherts and jasper. Marginal basin basalts have the high Cr and Ni values typical of ridge basalts (but unlike arc tholeiites) and the higher lithophile elements (K, Rb and Ba) of arc tholeiites (unlike ridge basalts) and are slightly light-REE enriched: in all three respects comparable with Archaean greenstone belt basalts. Andesites are not prominent as in most early greenstone belts. The range of bulk compositions and trace element abundaces of the plutons and batholiths that engulf the marginal basin rocks (often leaving them as mega-xenoliths in a granitic terrain) are similar to the diapiric plutons that invade Archaean belts.

From their experimental studies Toksöz and Bird (1977) concluded that a necessary consequence of the subduction of oceanic lithosphere is an induced convective circulation in the wedge above the slab and this gives rise to back-arc basins. If the higher rate of heat production in the Archaean gave rise to increased plate accretion and subduction, then more marginal basins would have formed than today (Sleep and Windley, 1982). Toksöz and Bird (1977) also emphasized that marginal basins vary from the undeveloped to the active spreading, mature and the inactive. There are important variations in the degree of crustal extension along the length of the marginal basins in S Chile (De Wit and Stern, 1981). In the south, high extension enables mafic magmas to be emplaced in oceanic-type spreading centres, but less extension to the north caused the mafic melts to be intruded into continental crust and so here the back-arc basin developed on a sialic basement. This relationship is particularly relevant for early Archaean greenstone belts, such as Barberton and Pilbara which contain no evidence of significant continental basement, and for late Archaean belts which in places are unconformable on granitic gneisses and which contain abundant clastic sediments derived from uplifted continental margins (Lowe, 1982). Much of the diversity that is present in greenstone belts of various ages may be explicable in terms of the infilling of different types of rifts in progressive stages of evolution in inter-arc and back-arc settings. A marginal basin model was pro-

56

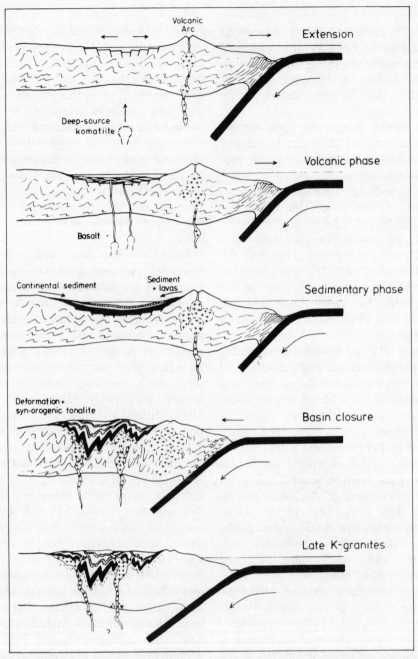

Fig. 4.7 Suggested development of an Archaean greenstone belt in a back-arc marginal basin position on comparison with the 'Rocas Verdes' complex in S Chile. Back-arc extension in an Archaean ductile lithosphere produces a greater component of crustal thinning rather than crustal rifting, as today. Magma production exceeds that required for simple extension. Sediment source is mixed volcanigenic/continental and the sequence may include calc-alkaline andesitic and salic lavas from the adjacent volcanic arc. Later movement of arc towards continent produces deformation and synclinal form of greenstone belt. Andean-type tonalitic-to-granitic plutons (derived from mantle by at least two-stage fractionation process) may be syn- to post-tectonic with compositions dependent on depth of melting of subducted oceanic crust (after Tarney *et al.*, 1976; reproduced by permission of J. Wiley)

posed for greenstone belts in S Canada (Goodwin, 1974; Vogel, 1978), Zimbabwe (Condie and Harrison, 1976), South Africa (Erikkson, 1980), S India (Drury, 1983) and the Aldan Shield (Moralev, 1981). However, Groves *et al*. (1978) and Rutland (1982) argue that the marginal basin model is inappropriate for the coeval and non-linear greenstone belts of W Australia.

Granulite–Gneiss Belts

In contrast to the greenstone belts there have been very few attempts to postulate a tectonic model for Archaean high-grade regions. It is clear that as long as the rock units in these regions were thought to have developed as a conformable group little insight would emerge, worthy of realistic consideration, of possible tectonic environments. A major step was made by V. R. McGregor (1973) in breaking through the conformability barrier of part of West Greenland and this led to the development of an isotopically confirmed chronology (e.g. Moorbath, 1977, 1978, 1980; Bridgwater *et al*., 1978). A somewhat similar chronology of events has been established in Labrador (Bridgwater *et al*., 1978; Hurst *et al*., 1975) and the Limpopo belt of southern Africa (Barton and Key, 1981; Robertson and du Toit, 1981). But it is necessary to go beyond the erection of local chronologies in order to understand the tectonic environment in which these kinds of rocks and relations evolved.

Now let us look at several ideas related to the tectonic environment of these high-grade terrains.

Recrystallized Supracrustal Rocks

Some Archaean granulite–gneiss belts are interpreted as a pile of recrystallized supracrustal rocks. Such an idea has been applied to the rocks in the Limpopo belt of southern Africa (Barton and Key, 1981), the Wyoming Province of the USA (Condie, 1976), the Ancient Gneiss complex of Swaziland and many other examples (Barker and Peterman, 1974), Western Australia (Glikson and Lam-

bert, 1976), the USSR (Moralev, 1981) and NE China (Sun and Wu, 1981). In the USSR much emphasis is placed on the fact that the oldest (>3500 Ma) cratons consist largely of recrystallized supracrustal rocks such as quartzite, high alumina gneiss, garnet–biotite gneiss, paraschist, basic gneiss, amphibolite and enderbite (Salop, 1983). The oldest rocks in the world according to Salop are quartzites at the bottom of the Iyengran Seies of the Aldan Shield.

The natural corollary of this approach might be to regard the granulite–gneiss belts as just highly metamorphosed greenstone belts (e.g. the Hebei Province of NE China, Li *et al*., 1982); but the lithological differences between the two terrains are so great that this conclusion is hardly likely. Take the sediments, for example: a shelf-type association of quartzites, marbles and pelites in the high-grade and a greywacke–turbidite association in the low-grade regions. A traverse from the Limpopo mobile belt into the Rhodesian or Kaapvaal cratons is sufficient to bring out the very great differences between these two types of Archaean crust (Coward, *et al*., 1976). Tarney (1976) and Lambert *et al*. (1976) conclude from a study of chemical data that gneissic complexes differ significantly from the greenstone–granitic pluton association. It is important to note, however, that the Northern and Southern Marginal Zones of the Limpopo belt are made up of former greenstone belts that were metamorphosed to granulite and high amphibolite facies (Robertson and du Toit, 1981).

Oceanic Crust

There are probably remnants of oceanic crust in some Archaean granulite–gneiss belts, and the best candidate to date is in the Fiskenaesset region of W Greenland where, within the gneisses, there are many tens of kilometres (strike length) of basaltic amphibolites which reach 1–2 km in thickness and locally contain well-preserved pillows (Escher and Myers, 1975). Clearly, it would be fruitless to use major elements to assign a particular type of tectonic environment, but it may be possible to use trace elements to discriminate between an ocean floor as against an island arc affinity. Weaver *et al.* (1982) show that the Fiskenaesset amphibolites have low La/Ta ratios close to 15, comparable to that of modern mid-oceanic ridge basalts, in contrast to that of destructive plate margin magmas which are greater than 30.

The Fiskenaesset anorthosite complex (Myers, 1981) is consistently intercalated with the amphibolites. The geochemistry of the basaltic amphibolites is dominantly controlled by fractional crystallization processes and the required plagioclase-dominated crystal extract is very similar to the primary mineralogy of the cumulates of the Fiskenaesset complex; there is also a marked similarity between the observed trace element chemistry of the amphibolites and that of the calculated successive liquids of the complex (Weaver *et al.*, 1981). These results strongly suggest a genetic relationship between these rock suites. The first batch of liquid gave rise to the basaltic volcanics and the second residual liquid to the Fiskenaesset complex, and this sequence is likely to be an Archaean analogue of oceanic crust.

Sleep and Windley (1982), from a consideration of thermal constraints, proposed that the Archaean oceanic crust must have been more that 20 km thick, in contrast to modern oceanic crust (*c.* 5 km), and that it had a significant volume of cumulates. Shallow subduction of young, hot, oceanic lithosphere and buoyancy, caused by the proportion of plagioclase-rich cumulates, would assist in flaking off and introducing the anorthosite–amphibolite association into the deep crustal levels of active continental margins (Windley and Smith, 1976). Similar anorthositic layered complexes often bordered by amphibolites are common in many Archaean granulite–gneiss belts (Windley *et al.*, 1981). An oceanic environment was proposed in general for such complexes by Moores (1973) and Burke *et al.* (1976a) and, specifically for Labrador, by Wiener (1981). The anorthositic complexes would form at different depths in such a thick basaltic volcanic pile, the composition of the cumulate plagioclase being

dependent on the ambient load pressure as known from experimental work and with reference to modern ocean-ridge magmas (Flower, 1981). The primary assemblage in the Fiskenaesset complex of anorthite–clinopyroxene–orthopyroxene–spinel indicates crystallization of 7 kb at 23 km depth (Windley *et al.*, 1973) in the volcanic pile into which the magma was emplaced.

In the Limpopo mobile belt of southern Africa there is an ultramafic suite of dunitic granulites and serpentinites with a tholeiitic differentiation trend, probably derived from gabbroic or plagioclase-bearing ultramafic rocks, and these are locally associated with thick metabasalts. The whole assemblage is regarded by Light (1982) as relict oceanic crust-mantle within the framework of his stimulating and enlightened plate tectonic model for the Limpopo belt (Fig. 4.8).

The metavolcanic amphibolites are commonly associated with a quartzite–thin marble–mica schist association that may be regarded as typical of an Atlantic-type trailing continental margin marine environment. This implies a close relation between oceanic crust and continent-derived clastics. However, it seems likely that some of the quartzites and marbles are recrystallized cherts and exhalative carbonates.

Active Continental Margins

The predominent rock type in most Archaean granulite–gneiss belts is a tonalitic gneiss which has been recrystallized under granulite or amphibolite facies conditions. In NW Scotland these two gneiss types have the same REE and immobile trace element compositions, except that the granulites are depleted in the mobile radioactive heat-producing elements K, Rb, U and Th (Weaver and Tarney, 1980; Tarney *et al.*, 1982). The removal of trace elements took place not by magmatic processes or by the action of a melt phase, but of a CO_2-rich fluid phase during the granulite facies metamorphism (Touret, 1974; Newton *et al.*, 1980). Thus the granulites are not residues from intracrustal partial melting, but metamorphically depleted tonal-

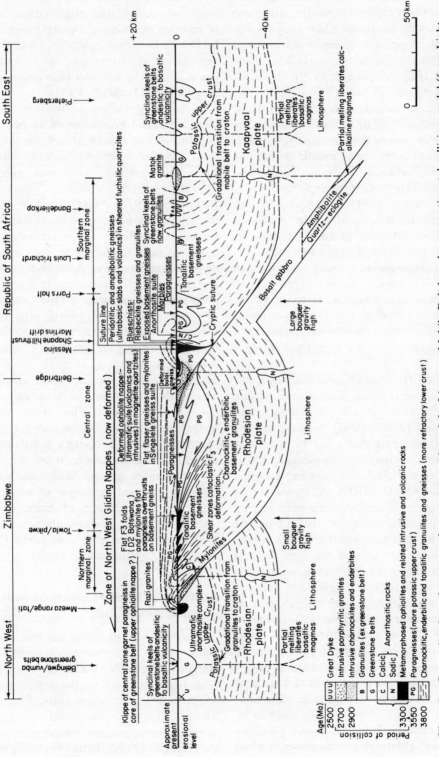

Fig. 4.8 Generalized northwest–southeast cross-section of the Limpopo Mobile Belt based on a continental collision model (from Light, 1982, *Tectonics*, **1**, 325–342. Copyright by the American Geophysical Union)

ites. The tonalitic precursor magmas of the amphibolite- and granulite-grade gneisses were most likely injected by underplating into the deep continental crust where they crystallized with a primary gneissic texture and metamorphic mineralogy (Holland and Lambert, 1975; McGregor, 1979); this is the under-accretion of Wells (1980). The geochemistry of radiogenic isotopes (^{87}Sr, ^{143}Nd, ^{206}Pb and ^{208}Pb) of these calc-alkaline orthogneisses shows that the parental liquids were derived from upper mantle or oceanic lithosphere sources during 'continental accretion-differentiation superevents' (Moorbath and Taylor, 1981). The peak of continental growth was in the late Archaean by means of intrusion of such tonalites, as well as by formation of greenstone belts.

The predominent rock type in Mesozoic–Cenozoic Andean-type calc-alkaline batholiths is tonalite (Pitcher, 1978). Archaean and Andean tonalites have similar major and element chemistry, including a marked depletion in the high field strength elements Nb and Ta relative to both the LREE and LIL elements, and the major difference between them lies in the strong HREE depletion in the Archaean rocks (Weaver and Tarney, 1980). As these authors point out, assuming that some form of subduction or downwelling (McGregor, 1979) operated in the Archaean, the most logical explanation for this HREE difference is that it represents an increased component derived through partial melting of the subducting slab, leaving residual garnet and such a process would be favoured in the high thermal regime in the Archaean. Tectonic models which place Archaean high-grade terrains above sites of convective upwellings with no plate subduction (e.g. Young, 1978; Condie, 1981, pp. 374–6) are inadequate because they do not allow sufficient generation of tonalitic magmas. Only a subduction model provides a constantly replenished source to be fed to the site of magma generation and, moreover, is able to dispose of the residues by recycling them back to the mantle (Weaver and Tarney, 1980). A subduction-controlled mechanism was ascribed to the gneisses in the

Lewisian complex of NW Scotland (Bowes, 1978) and in the Baltic Shield (Bowes, 1976). Such subduction was disguised as 'a zone of mantle downwelling' by McGregor (1979) for the W Greenland gneisses. Crustal evolution in the Archaean and later times took place largely by arc and continental-margin magmatism (Hamilton, 1981).

Archaean granulite–gneiss belts typically have a bimodal association of acidic tonalitic gneisses on the one hand and, on the other, layers and lenses of amphibolites, ultrabasic rocks and layered anorthositic gabbro or ultramafic-gabbro complexes (Barker and Peterman, 1974). The tonalitic gneisses formed by high-pressure fractionation processes (residual garnet ± hornblende), whereas the amphibolites and associated layered complexes, as described in the last section, formed by low-pressure fractionation mechanisms in an oceanic-type environment (Weaver and Tarney, 1980). Any tectonic model must satisfactorily account for the tectonic interleaving of these rock units. The downsagging *in situ* basin model is obviously again inadequate, but an active continental margin does provide a realistic and reasonable environment. The best modern analogues are the Cordilleran margins of southern Chile in S America and British Columbia in Canada where there are deeply eroded tonalitic gneissic batholiths with tectonic screens and inclusions of amphibolitic rocks, thrust-nappe and structures (in B.C.) and intercalated meta-sediments with a quartzite–schist–marble association (Tarney *et al.*, 1982; Hutchinson, 1982)

Relationship between Greenstone Belts and Granulite–Gneiss Belts

It is important to appreciate that there are many types of greenstone and granulite–gneiss belts. No doubt each belt has its own signature and its particular mode of origin. Some greenstone belts (Pilbara and Barberton) are very early and have little evidence of older continental basement, whereas the majority of such belts formed in the late Archaean, many with an unconformable relationship to either earlier greenstone belts or gneiss belts. Some greenstone belts have been metamorphosed to an amphibolite (Hong-tushan belt, NE China) or granulite grade (N Marginal Zone of the Limpopo belt). The Akilia association represents relics of green-stone belt material within the Amîtsoq gneis-ses in W Greenland. The Dharwar belts of S India are very late Archaean in age and begin to share characteristics with early Proterozoic basins. Many granulite–gneiss belts are domi-nated by tonalitic orthogneisses, but others have a higher proportion of paragneisses. Some such belts have prominent layered anorthosite complexes, others have none. We must be careful not to over-generalize about the origin of these high and low belts; we should begin to consider the origin of each individually and separately. At the simplest there may be just a fault between a high-level greenstone and a deep-level gneiss belt. The parallel greenstone belts of Ontario in Canada are separated by narrow linear belts of paragneisses interpreted as sediment-filled basins by Langford and Morin (1976). Increasingly, more complicated relationships are being discovered. In the Pilbara block of NW Australia gneiss and greenstone belts have been complexly inter-thrusted (Bickle *et al*., 1980).

A Modern Interpretation for Archaean Crustal Evolution

At this point I must admit to a bias in favour of some early variant of the plate tec-tonic scheme of events to explain Archaean crustal evolution, and for very good reasons. Only this model is capable of continously replenishing over hundreds of millions of years the sites of continental growth with new, mantle-derived, basaltic and associated material which had to be partially melted to create prodigious quantities of calc-alkaline magmas which enabled 50–60% (Moorbath and Taylor, 1981), 85% (Dewey and Wind-ley, 1981) or near 100% (Armstrong, 1981) of the present mass of continental crust to be created by the end of the Archaean, 2500 Ma ago. Partial melting of downsagging basins or

rise of magmas up fractures are obviously inadequate models. The advantage of a proto-plate tectonic model is that we know that it can explain the growth of both the oceanic and the continental crust, but no other *one* mechanism can satisfactorily explain the formation of *both*. Also we know that the modern type of plate tectonic mechanism is exceedingly efficient at generat-ing oceanic and continental materials which have chemical and other characteristics which are not grossly different from Archaean ma-terials. But it would be naive to think too closely in uniformitarian terms, because it is increasingly realized that modern-style plate tectonic processes began in the early Pro-terozoic when large stable cratons had formed, against which trailing margins, Andean-type margins and collisional belts could develop. We know that nowadays con-tinental growth takes place predominantly at continental margins where oceanic litho-sphere is subducted, and we know that the continental growth was several times higher in the Archaean because of the higher ther-mal output of the earth (Brown, 1977; Bickle, 1978; Burke and Kidd, 1978; Davies, 1979; Sleep and Langran, 1981). As Moor-bath and Taylor (1981) pointed out, no one yet has produced a *compelling* reason to sup-pose that this state of affairs has changed fun-damentally since the formation of the earliest continental crust, which was obviously in the Archaean. In fact it would be reasonable to infer that the higher thermal regime in the Archaean gave rise to more, rather than less, plate accretion, destruction, partial melting, crustal growth and collision, than today. And that is surely what we observe in the permobile/high-growth record of the Archaean. Therefore, rather than seek to erect some new tectonic scheme of events that cannot be constrained, it is preferable at this stage of development of the earth sci-ences to test rigorously the possibility of a variant of the plate tectonic scheme of events, modified to take into account likely dif-ferences in factors such as rates of accretion and subduction, the higher rate of breakdown of radioactive material, the higher mantle temperature, the thickness of oceanic and continental crust, the angle of dip of subduction zones, the age and temperature of subducted plates and the chemistry of mantle-derived and slab-melted magmas (Sleep and Windley, 1982). If these differences were seriously taken into account, together with likely secular trends that have taken place throughout earth history (see Chapter 22), then it should be possible to erect a viable model for early crustal growth. Those who accept that some variation of a modern process is most likely for the Archaean include Burke *et al.*, 1976; Langford and Morin, 1976; Bickle, 1978; Bowes, 1978; Moores, 1979; Blackburn, 1980; Tarney and Windley, 1980; Dewey and Windley, 1981; Dickinson, 1981; Hamilton, 1981; Moorbath and Taylor, 1981; Tarney and Windley, 1981; and Light, 1982. However, others favour more conservative models based on hot spots (Fyfe, 1978), upwelling and downsagging (Young, 1978), downsagging (Glikson, 1982), sag-subduction (Goodwin and Smith, 1980; Kröner, 1981b), rift-and-sag (Goodwin, 1981) and fixism–mobilism (Salop, 1983). But in discounting the potential of a modern tectonic process, these authors disregard, or have not fully taken account of, information from the most relevant deep-seated sections of modern Andean-type and Himalayan-type orogenic belts. A major problem lies in the fact that very little is known about the roots of Mesozoic–Cenozoic orogenic belts because of lack of sufficient uplift; therefore, how can we expect to be able to interpret the roots of deeply eroded early Precambrian belts in a comparable manner? Let us consider two relevant sections.

1. Deep sections of the Andean belt. Hutchinson (1982) described the central gneiss complex that forms the deep root/core of the British Columbia batholith which has been so rapidly uplifted and eroded that Tertiary granulites are exposed (Hollister, 1975, 1979) (for details see Chapter 19). The high-strain tonalitic gneisses and migmatites and the associated subhorizontal intrusion of tonalitic magmas in nappe complexes provide a modern analogue of the late Archaean gneiss terrain seen at its best in West Greenland. Recent investigations by Tarney *et al.* (1982) of the deeply eroded section of the Mesozoic Patagonian batholith in S Chile have shown that tonalite makes up more than 70% of the deep Andean crust, that tonalite, diorite and mafic inclusions are complexly intermixed, highly foliated and commonly converted to banded gneisses and that these foliated gneisses contain tectonically incorporated inclusions and strips of quartzite, pelite and marble. Anorthosite–gabbro complexes form small bodies in the batholiths in both Chile and S California (Windley and Smith, 1976). Thus all the main components, deformation and intrusive structures of the Archaean gneiss complexes are present in modern Andean-type belts, if one looks in the comparable tectonic level of the crust. Moreover, granulitic xenoliths brought up by alkali basalts in southern Chile contain a CO_2-rich fluid phase that attended granulite facies metamorphism (Selverstone, 1982), just as in Archaean granulites (Newton *et al.*, 1980).

2. Deep sections of the Himalayan fold belt. East of Nanga Parbat the uplift of the Himalayan belt has been so little that the Tertiary sediments and volcanics on both sides of the suture are well preserved (see Chapter 21). Dewey and Burke (1973) used the Tibetan section as a modern example of Precambrian basement reactivation but this high-level section has little relevance to deeply eroded Archaean belts. In contrast, west of Nanga Parbat, in Pakistan, the uplift has been so high (Zeitler *et al.*, 1982a) and the deformation and metamorphism so high (Coward *et al.*, 1982) that the Tethyan sediments on the Indian plate have been removed to expose the gneissic basement, and the Kohistan arc has been turned on end to expose its amphibolite–granulite facies base (see Chapter 21). In essense the Cretaceous Kohistan island arc consists of an upward succession of the Chilas layered complex, arc-type calc-alkaline volcanics and intrusive tonalitic plu-

tons (Bard *et al*., 1980a). The granulite-grade stratiform Chilas complex is comparable to the greenschist-grade stratiform Dore Lake complex of the Chibougamau greenstone belt of S Canada (Allard, 1970). Both stratiform complexes have been isoclinally folded, both arc volcanics are greenschists and both belts were deformed to a vertical position before tonalitic plutons were intruded. The Indian continental margin consists of widespread orthogneisses and paragneisses with prominent ductile shear belts and tectonic strips of marble, pelite and quartzite. The Indus Suture is a conformable boundary between amphibolite and gneiss—all Precambrian geologists should walk across this boundary to see how unremarkable a suture can be. Thus, there are extraordinarily close similarities between the western Himalayas and Precambrian gneiss–greenstone belts, if one looks at the comparable tectonic level of the crust.

Turning to Archaean crustal evolution, the most viable model basically combines the back-arc marginal basin idea for greenstone belts and the main arc (plutonic batholith) interpetation of the high-grade gneissic complexes. These two models are mutually complementary.

According to the combined model (Fig. 4.9) the main arc would form slightly before the back-arc. Amîtsoq/Uivak-type gneisses formed the 'basement' in small sialic plates. The extrusion of tholeiitic lavas to form oceanic crust was accompanied by emplacement of complementary anorthositic complexes and deposition of shallow water sediments close to a continental margin. Intensive subduction activity, probably at a shallow angle, flaked off anorthosite-buoyant slices of oceanic crust and emplaced them into deep levels of the growing tonalitic crust, which formed by the intrusion of batholithic-proportions of slab- and mantle-derived tonalitic magmas. This intrusion was associated with thrusting and nappe formation during a widespread horizontal tectonic regime which produced a tectono-igneous stratigraphic pile that contributed to a major thick-

ening of the crust in late Archaean times. The intrusions probably occurred through a wide depth of crust, from high levels where they engulfed the earlier volcanics to deeper levels where they may have passed directly into foliated gneisses or granulites. These events were synchronous with, or immediately followed by, high temperature metamorphism that transformed all rocks into an amphibolite or granulite grade, whilst deformation converted the 'granitic' rock into gneisses.

Extension in the back-arc area began before the completion of the above events. Early ultramafic–mafic volcanics may have been extruded onto thinned sialic crust or may have formed in a rift-type oceanic environment; their composition was characterized by a low-K content together with komatiites which resulted from partial melting of hot Archaean mantle. Later volcanics with a calc-alkaline character and much pyroclastic material formed during the emergence of arc-type volcanoes and later sediments derived much clastic material from the erosion of gneissic basement in the adjacent uplifted arc areas. The closure of the back-arc basins, which gave rise to the tpyical greenstone belt structure, may have taken place contemporaneously with the last deformation and metamorphism in the arc area.

At best, the combined model suggests that the major period of growth of the continents in the late Archaean took place in a manner broadly similar to the accretion taking place at the leading edges of modern continental plates. This implies some form of ridge and subduction activity in the Archaean which, in turn, means that the kinds of rock association, stratigraphy and rock chemistry (bulk and trace element) in the late Archaean and Mesozoic equivalents should be largely similar, which they are. At worst, the combined model provides a basis for comparing future Archaean geochemical results with those from a modern environment whose mode of evolution is better understood.

In the Mesozoic continental growth was limited to relatively narrow linear fold belts whereas in the permobile regime of the Archaean it was widespread. This was prob-

64

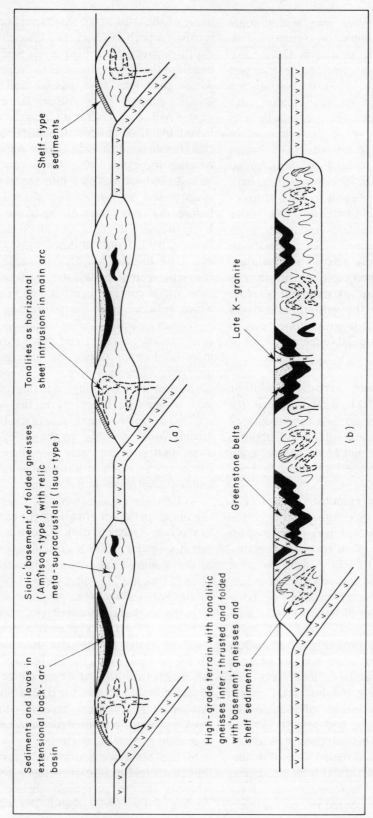

Fig. 4.9 A plate tectonic model to explain the growth of continents in the Archaean. (a) Widespread lateral movement of many early Archaean mini-continental plates with shelf-type quartzites, carbonates, and K-pelites and with mantle-derived tonalites in batholithic proportions in proto Andean-type arcs and of volcanics in back-arc environments. (b) Aggregation of mini-continents gives rise to extensive continental plate by the end of the Archaean consisting of greenstone belts, and granulite–gneiss belts with older and younger gneissic components. Amphibolite- to granulite-grade metamorphism (heat flow) and deformation of tonalites to give rise to tonalitic gneisses takes place in the roots of the main arc

ably because plates were small in the Archaean and so subduction activity affected the whole plate in contrast to Phanerozoic times when plates were so large that orogenic activity was confined to their leading edges. Deformation was not necessarily more intensive in the Archaean than at present—it was merely distributed more widely throughout smaller plates.

The proposals made here are consistent with the interpretation of Archaean whole-rock isochron dates of Moorbath (1977, 1978, 1980) as giving the age of major accretion–differentiation events. On the basis of their low initial strontium isotope ratios the gneisses, greenstone belt volcanics and granitic plutons are predominantly juvenile additions from the mantle to the continental crust, at or close to the measured age, implying continental growth on a major scale, especially in the mid-late Archaean. This 'early continental accretion episode' took place in a period 'not exceeding 50–100 Ma in duration, and this time interval is of the same order as that found at present between the most recently formed complementary oceanic ridges, island arcs and continental margins'.

As illustrated in Fig. 4.9, the accretion was

related to the aggregation of mini-plates in the late Archaean, to give rise to larger more stable plates by the early Proterozoic.

Postscript

The ideas on Archaean crustal development reviewed in this chapter are more than a documentary list. They are of educational value to the student for two reasons:

1. They show how ideas in this field have evolved in the last decade. It is sometimes easy to criticize early models in the light of the experience of hindsight, but they are the building blocks with which newer ones are invariably created. The historical approach also serves to demonstrate that even the last and/or most widely accepted concept is only a working hypothesis.

2. It is interesting to see how two contrasting uniformitarian (plate tectonic) and non-uniformitarian models can be erected to explain one set of data. Thus, rather than have just a set of data to learn, the student can exercise his judgement in making a critique from a certain standpoint or in making a balanced appraisal of opposing ideas.

Chapter 5

Early to Mid-Proterozoic Basic–Ultrabasic Intrusions

Following the permobile tectonic regime in the Archaean there evolved a new stage in earth history when major cratons stabilized, allowing, for example, the deposition of thick supracrustal sequences (as described in Chapter 6) and also the formation of mega-fracture systems which might be regarded as initial attempts to fragment these first-formed extensive continental masses. Magmas were intruded into many of these fractures giving rise to transcontinental dyke swarms and many layered complexes (the subject of this chapter). Subsequently, most regions remained stable for the remainder of geological time and the bodies are therefore still well preserved in Zimbabwe, South Africa, East Africa, West Australia, West Greenland, western USA and North Canada. There is, of course, little or no evidence of the bodies in those regions that underwent later tectonic activity, and in others they were covered by later supracrustal deposits. The development of the early Proterozoic bodies, and the basins and belts described in Chapter 6, was clearly diachronous; thus, whilst the Great Dyke was intruded in Zimbabwe, the Witwatersrand Supergroup was being deposited in South Africa.

The intrusions are divisible into three types: giant dyke-like layered bodies such as the Great Dyke (Zimbabwe) and the Widgiemooltha Dyke Suite (West Australia), major swarms of dolerite/diabase dykes, and layered stratiform igneous complexes. The first were intruded into the consolidated greenstone belts and high-level granites of Gondwanaland, the dyke swarms cut high-grade Archaean gneisses as well as the greenstone belts, and the layered complexes were intruded into older basement or unconformable early Proterozoic cover sediments.

The Stillwater Complex, Montana

The Stillwater Complex is a 2700 Ma layered intrusion of basic and ultrabasic rocks that has not been metamorphosed; but it has been tilted into a steeply-dipping position and heavily faulted (Page, 1977). It was originally intruded as a sub-horizontal sheet into early Precambrian schists and gneisses of the Beartooth Mountains, Montana, but only about 60% of the initial thickness is now visible. The lower part was invaded by a 1530–1580 Ma granite, the upper part was eroded and the remainder tilted in late Precambrian times, and Middle Cambrian sediments were deposited unconformably on the eroded surface of steeply-dipping rocks. The present exposed strike length is about 48 km (faulted at each end) and the maximum stratigraphic thickness about 6000 m. The contact aureole has a Rb/Sr whole rock age of 2750 ± 45 Ma (Mueller and Wooden, 1976), whilst the gabbros of the complex have a Sm/Nd internal isochron age of 2701 ± 8 Ma (DePaolo and Wasserburg, 1979), and the chilled margin has a zircon U/Pb age of 2713 ± 3 Ma (Nunes, 1981).

The Great Dyke, Zimbabwe

This is a well-known layered ultrabasic–basic intrusion with a Rb–Sr age of 2514 Ma (Hamilton, 1977).

The dyke has a length of 480 km and an average width of 5.8 km (Fig. 5.1). It is a remarkable and unusual body as it contains four layered lopolithic subcomplexes, Musengezi, Hartley, Selukwe and Wedza, in which the layering dips inwards at a shallower angle than the steeply-inclined dyke contacts (Worst, 1960). According to Bichan (1970) each complex consists of cyclic sequences of ultrabasic rocks (A. H. Wilson, 1982) which reach a maximum exposed thickness of 2100 m in the Hartley Complex and are overlain by a 900 m thick gabbroic capping. Ideally each ultrabasic cycle has a basal chromite seam, forming a sharp footwall with underlying pyroxenites, followed upwards by peridotites and then pyroxenites; in the Hartley Complex there are 11 chromite seams. The gabbroic capping has a sharp contact with the underlying ultramafic rocks and consists of a lower zone of anorthositic gabbro followed upwards by gabbros and norites which, at the very top, give way to quartz gabbros. Bichan (1970) concluded that each complex formed by pulsatory injection of magma derived from a parent source. According to Hughes (1976) the parental magma of the Great Dyke was akin to a high magnesian basalt.

Worst (1960) concluded that three major events took place in the evolution of the Great Dyke:

1. Formation of a linear zone of weakness: successive heaves of magma were injected through fissures developed at four positions

Fig. 5.1 Igneous complexes and dolerite dyke swarms in southeastern Africa

along this line until they met a horizontal plane of weakness in the earth's crust where they spread out laterally and differentiated as individual units.

2. Subsidence of the floor and formation of a graben with layers sagging into their present synclinal structure and concomitant shearing of the dyke's contacts.

3. Erosion down to the present level with removal of any lateral extension of the layers beyond the present dyke margins.

Bichan (1970) considered that the dyke was located over the position of a thermal updraft in a mantle convection cell and the waning of the heat flow pattern resulted in slumping of the dyke into its graben.

The Widgiemooltha Dyke Suite, West Australia

This suite includes the Coronation, Jimberlana and Binneringie dykes; the first two have a Rb/Sr age of 2420 ± 30 Ma. These are giant northeast-trending ultrabasic–basic dykes that traverse the greenstone belt-granite terrain of West Australia.

The Binneringie Dyke has no lopolithic shape, but vertical layering throughout; it is 320 km long and 3.2 km wide (McCall and Peers, 1971). Marginal bronzite gabbros pass inwards to more ferroan augite-pigeonite gabbros and there are minor intermediate and acid phases with granophyric and devitrified acid glass in the form of segregation patches, dykes and dykelets representing infillings by late magmatic phases of shrinkage cracks in the gabbro. In spite of the fact that the layering is vertical there is much rhythmic and cryptic layering as well as graded and cross bedding. In order to explain these unusual features McCall and Peers considered that the crystallization took place within vertically moving convection currents and that all the rhythmic and cryptic effects were caused by extensive marginal heat loss, whilst the bedding effects were caused by pressure and supersaturation rhythms. The average composition of the chilled bronzite gabbro is dissimilar to that of the Great Dyke

and its satellites and more akin to that of other major layered intrusions.

The Jimberlana Dyke is at least 180 km long and 2.5 km wide and is a small analogue of the Great Dyke of Rhodesia (Campbell, 1977; McClay and Campbell, 1976). It has a very steep, V-shaped cross section and many internal canoe-shaped sub-complexes along its length, just like the Great Dyke. The intrusion largely consists of cumulate bronzitites and norite gabbros with phase, rhythmic and cryptic layering. It formed by the emplacement of several pulses of magma that gave rise to two magmatic unconformities.

The Sudbury Irruptive, Canada

The Sudbury Irruptive is a basin-shaped body with an area of 58 × 26 km, intruded partly into Huronian rocks. Its Rb/Sr whole-rock isochron age ranges from 1956 ± 98 Ma (Gibbins and McNutt, 1975) to 1883 ± 136 Ma (Hurst and Farhat, 1977). Two high precision U–Pb zircon ages on norite are 1849.6 Ma (Krogh et al., 1982). Dietz (1964) suggested an origin by meteorite impact (overturned rim beds, breccias and shatter cones), supported by evidence of shock metamorphism (French, 1972) and basin reconstruction (Brocoum and Dalziel, 1974).

The Bushveld Complex, South Africa

The intrusion was emplaced between 2095 ± 24 Ma (mafics—Hamilton, 1977) and 1920 ± 40 Ma (late granites—Coertze, et al., 1978). It is the world's largest igneous body covering approximately 66 000 km^2, it has a vertical thickness of up to 9 km, and it is the largest repository of magmatic ore deposits in the world (see later in this chapter) (Hunter and Hamilton, 1978).

The Complex occupies an elliptical area (Fig. 5.1) with a central part underlain by granite, microgranite, felsite and granophyre, and two marginal lobate belts, the western and eastern, made up of rocks ranging from dunite to norite, anorthosite and ferrodiorite.

Following on the classical work of Hall, Daly, Wagner and many other geologists in the early part of this century, recent research has demonstrated two important facts:

1. The Complex is not a lopolith: the mafic and ultramafic rocks do not extend under the central granites.

2. The Bushveld granites are younger than the mafic and ultramafic rocks, and did not form *in situ* from the magma that yielded these rocks.

The Complex was intruded into a structural depression (Hunter and Hamilton, 1978) into the Transvaal Supergroup under 4 Kb crystallization pressure (Cawthorn, 1977) by a process of continuous basin subsidence (Sharpe and Snyman, 1980).

Basic Dyke Swarms

Distribution and Extent

Vast swarms of dolerite (diabase) dykes were intruded into the Archaean terrains (both high- and low-grade) and early Proterozoic cover rocks after 2700 Ma and particularly in the period between 2500 and 1500 Ma. Table 5.1 shows prominent examples (see Figs. 5.1 and 5.2). For most references see first edition. For Canada see Baragar (1977).

Particularly impressive is the scale of dyke intrusion at this time in earth history. The Scourie dyke swarm is not less than 250 km across its strike direction, the Kangamiut swarm is about 12.5 km across and 240 km long and is one of the world's densest dyke swarms (Escher *et al.*, 1975); the whole of the Archaean craton of West Greenland is crisscrossed by numerous intersecting dyke swarms, the Labrador–Slave swarm in Canada can be followed intermittently along strike for at least 2500 km and probably continues in West Greenland; the Sudbury–Mackenzie swarm is 3000 km long and more than 500 km wide; the Waterberg swarm is at least 200 km long; and the Umkondo and Mashonaland dykes occur throughout an area about 400 km by 300 km in NE Zimbabwe (Fig. 5.1). The intrusion of these enormous swarms represents a new

Table 5.1 Examples of early to mid-Proterozoic dyke swarms

Dyke Swarm	Age (Ma.)
Scourie, NW Scotland	2390 ± 20
E. Antarctica	2350 ± 48
Kangamiut, W Greenland	1950 ± 60
Lofoten, NW Norway	1795 ± 20
Molson, Canada	1800–2000
Kaminak, Canada	2370 ± 200
Matachewan–NS, Superior Province	2500
Slave–Superior Province–NW	2150 ± 2165
Abitibi–NE, Superior Province	2100
Nipissing Diabases, Canada	2155 ± 80– 2162 ± 27
Mackenzie	1200
Sudbury–NW swarm	1250 ± 50
Granite Mountains, Wyoming	1600
Bighorn Mountains, Wyoming	$\begin{cases} 2826 ± 58 \\ 2200 ± 35 \end{cases}$
Ivory Coast, W Africa	1740 ± 170
Waterberg, southern Africa	1750–1950
Mashonaland, Zimbabwe	$\begin{cases} 1850 ± 20 \\ 1910 ± 280 \end{cases}$
Umkondo, Zimbabwe	>1785
Pilansberg dykes, S Africa	c. 1450
Soriname, S America	1600–1750
Roraima, Guyana	1500
Hart, Australia	1800 ± 25
Chopan Dykes, India	2370 ± 460
Kopinang, British Guiana	2000
Outukompo, Finland	2250–2150

type of igneous and tectonic activity in the earth, totally different from that of the Archaean; they demonstrate a major change in crustal conditions and serve to separate the permobile late Archaean from the early–mid Proterozoic when extensive rigid or semirigid plates had formed.

Two features of the dykes are worth considering in some detail: their structural relationships and their chemical composition.

Structural Relationships

Many of the dykes under discussion are particularly interesting because they tell us of the conditions operating within, or on the margins of, the continental plates in early–mid-Proterozoic times (Halls, 1982).

Escher *et al.* (1976) suggest that there is a genetic link between the major dyke swarms,

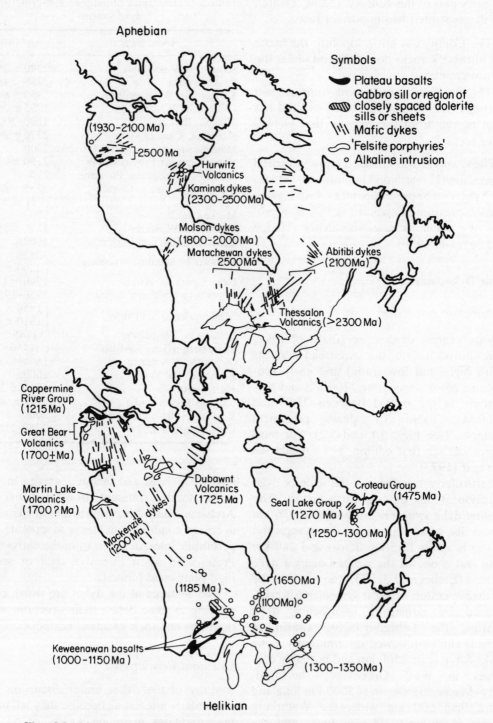

Aphebian

Symbols

⬤ Plateau basalts

▨ Gabbro sill or region of closely spaced dolerite sills or sheets

|\| Mafic dykes

◠ 'Felsite porphyries'

○ Alkaline intrusion

(1930–2100 Ma)

2500 Ma

Hurwitz Volcanics

Kaminak dykes (2300–2500 Ma)

Molson dykes (1800–2000 Ma)

Matachewan dykes 2500 Ma

Abitibi dykes (2100 Ma)

Thessalon Volcanics (>2300 Ma)

Coppermine River Group (1215 Ma)

Great Bear Volcanics (1700 ± Ma)

Martin Lake Volcanics (1700 ? Ma)

Dubawnt Volcanics (1725 Ma)

Croteau Group (1475 Ma)

Seal Lake Group (1270 Ma)

(1250–1300 Ma)

Mackenzie dykes (1200 Ma)

(1650 Ma)

(1185 Ma)

(1100 Ma)

Keweenawan basalts (1000–1150 Ma)

(1300–1350 Ma)

Helikian

Fig. 5.2 Maps of the Canadian Shield showing the distribution of Aphebian (2500–1800 Ma) and Helikian (1800–1000 Ma) volcanic rocks and dykes, etc. (from Barager, 1977, *Geol. Ass. Canada*)

with ages from 1950 Ma to 2390 Ma, in Scotland, East and West Greenland (Fig. 5.3) and Labrador. There are two sets, trending roughly NE–SW and E–W. Their close association with ductile shear zones in the gneissic wall rocks shows that the dykes represent a conjugate swarm along shear fractures rather than along tensional openings. Watson (1980) showed that the Matachewan dyke swarm in Southern Canada was intruded 2690 Ma in connection with important sinis-

tral transcurrent movement on the Kapuskasing lineament.

The basic dykes in the Lewisian can be used to separate the Scourian and Laxfordian tectonic events (Chapman, 1979). These Scourie dykes show a great variation in structure and petrology throughout the Lewisian (Weaver and Tarney, 1981a)—many of these variations are considered to date from the period of intrusion and reflect diversity in their conditions of emplacement at slightly

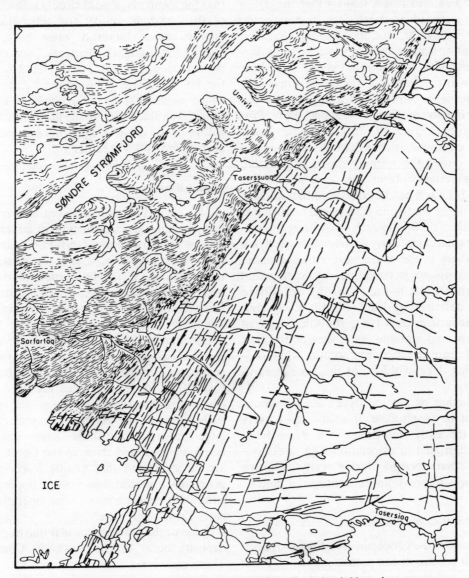

Fig. 5.3   Part of the Kangamiut dyke swarm in West Greenland. Note the two generations of dykes and the more highly deformed zone to the north (after Escher *et al.*, 1975; reproduced by permission of *Canadian Journal of Earth Sciences*)

different times and in different parts of the complex. The dykes were generally intruded at depth into crust that was still hot; temperature and pressure estimates range from 5–6 kb corresponding to a depth of 15–20 km at 500°C to more than 8.5 Kb (>30 km depth). In places early members of the suite show evidence of synkinematic intrusion, elsewhere they are regarded as just synmetamorphic, whilst later members were intruded under cooler, more brittle conditions. Features which suggest that the dykes were intruded at depth into highly ductile crust include: oblique and sigmoidal foliation indicating simple shear of the dyke walls during intrusion, a close relationship with shear zones in the wall rocks, a coarse grain size, and presence of primary hornblende and garnet. Finally in East Antarctica early Proterozoic (2350 ± 48 Ma) tholeiitic dykes were emplaced at considerable depths in the crust during the warning stages of granulite facies metamorphism (Sheraton and Black, 1981). Thus it can be concluded that in these regions the continental plates had not become entirely rigid by the early–mid-Proterozoic.

To what extent early Proterozoic basic dykes elsewhere in the world were intruded under such synplutonic conditions at depth is not known because not many of the dykes have been described in detail. But the Kangamiut and Scourie dykes serve to show that depth of erosion and variable conditions of the host rocks must be taken into account when considering dyke evolution at this time in earth history.

Whether or not the dykes are intracontinental or synplutonic, it seems most likely that the early–mid-Proterozoic basic dykes were intruded in association with stress systems that formed during early abortive attempts to break up the continental plates.

Chemical Composition

Study of the geochemistry of these dykes is useful because it can potentially tell us about the composition and evolution of the subcontinental mantle from which the dyke magmas were derived.

Surprisingly few modern chemical studies have been made of these dykes. The 2390 Ma Scourie dykes include bronzite–picrite, olivine–gabbro, norite and quartz–tholeiite types (Weaver and Tarney, 1981a). Most show enrichment in light REE and large ion lithophile elements, but not an equivalent enrichment in other incompatible high-field strength ions such as Nb and Ta. Petrogenetic modelling by Weaver and Tarney indicates a high degree of mantle melting which suggests that the elements present closely reflect those of their mantle source; the subcontinental mantle source therefore must have been enriched in lithophile elements. Such enrichment reasonably took place 2920 Ma when the host Lewisian gneisses were generated, which means that the sub-continental mantle source was able to retain the geochemical signature of the crust-forming process that took place some 500 Ma earlier.

Mineralization

The early Proterozoic dykes and layered complexes have a similar type of mineralization belonging to the Cr–Ni–Pt–Cu association, which occurs in basic-to-ultrabasic host rocks (Table 5.2). These rocks and their metals represent a significant influx of material from the mantle into the continental crust at this time.

Chromite occurs in the Great Dyke as the dominant phase in chromite seams and as disseminated crystals in all olivine-bearing rocks. The ultrabasic cycles ideally have a basal chromite seam and there are eleven such seams in the Hartley Complex. Many large layered intrusions have chromite accumulations, and those in the Great Dyke are an indication that mantle fractionation processes responsible for appreciable chromium concentrations were operating by early Proterozoic time.

Some of the most spectacular magmatic ore deposits occur in the Bushveld Complex. These include:

1. Tin and fluorspar in the granite.
2. Vanadiferous and titaniferous magnetite

Table 5.2 Some early–mid-Proterozoic layered complexes and their types of mineralization

Intrusion	Host rock	Mineralization	Age (Ma.)
Bushveld, S Africa	(various)	Cr Fe Ti V Pt Ni	2095
Lynn Lake, Manitoba	gabbro	Ni Cu	1700
Sudbury, Ontario	norite–micropegmatite	Cu Ni Co Au Pt	1850
Sarqâ, South Greenland	hornblende peridotite	Pl Au Ag Cr Cu Ni	c. 1800?
Thompson-Moak Lake, Manitoba	meta-sediments associated with serpentinized peridotite	Ni Cu Co Au Pt	c. 1800
Great Dyke, Zimbabwe	ultramafics	Cr	2460
Stillwater, USA	ultramafics	Cr	2701
Usushwana, S Africa	gabbro	Cu Ni	2870

in the upper layered sequence (anorthosite).
3. Platinum and nickel in the Merensky Reef and in bronzitite pipes.
4. Chromite in the lower layered sequence (pyroxenites, norites and anorthosites).

The Bushveld Granite contains economic tin deposits, largely occurring as disseminated cassiterite and in pipe-like bodies and fissure fillings in the late granites (Hunter and Hamilton, 1978). These are important because the fractionation processes responsible for tin accumulations were rather ineffective and slow to operate in earth history. The Bushveld deposits indicate that the processes were operating 2100 Ma ago and it is probably no coincidence that they were associated with the intensive magma fractionation that gave rise to the Bushveld Complex.

There are many early–mid-Proterozoic layered intrusions in Finland, many of which are mineralized (Mikkola, 1980) (Table 5.3).

Table 5.3 Mineralized early–mid-Proterozoic layered intrusions in Finland

Intrusion	Host Rock	Mineralization	Age (Ma)
Kemi	Mafics	Cr	2440
Porttivaara	Gabbro	VFe	2440
Koitelainen	Gabbro	Cr	2440
39 bodies Lake Ladoga	Ultramafics and gabbro	NiCu	1890
Oulujärvi	Amphibolite from gabbro	VFe	2060

Chapter 6

Early to Mid-Proterozoic Basins and Belts

On and against the eroded remnants of the Archaean greenstone belts and their granitic rocks and of the high-grade gneisses, there were deposited vast sequences of volcanics and sediments. These rocks herald a new major stage in Earth history as they, and their geotectonic environment, are different from comparable Archaean examples. For the first time asymmetrical fold belts were formed; in fact the first Andean-type orogenic belts and aulacogens are recognizable from this period. These are particularly exciting rocks to study as the time boundary between the Archaean and Proterozoic represents the most dramatic and fundamental period of change in the evolution of the continents—far more important than that separating the Precambrian and the Phanerozoic.

Basins in Southern Africa

Archaean crustal development in the Kaapvaal craton of southern Africa was completed by 3000 Ma, thus allowing the accumulation of vast thicknesses of unconformable flat-lying sediments and volcanics in broad basins at a time when greenstone belts were still forming in Zimbabwe, India, Australia and Canada. Approximately 43 km of sediments and volcanics belonging to the Pongola, Dominion Reef, Witwatersrand, Ventersdorp, Transvaal and Waterberg–Matsap Supergroups (shown in Fig. 6.1) were laid down between about 3000 Ma and 1800 Ma; for a detailed description, see Button *et al.* (1981) and Tankard *et al.* (1982). The rates

of vertical movement decreased progressively with time from 0.27 (in the Archaean) to 0.023 mm yr^{-1} (Table 6.1).

The Pongola Supergroup was deposited 3000 Ma ago. It consists largely of sandstones with prominent sedimentary structures similar to those in modern tidal flat environments; the palaeotidal range is inferred to be 12–25 m, much in excess of that today, lending support to the hypothesis of an early Precambrian origin for the earth–moon systems (von Brunn and Hobday, 1976). The Supergroup was intruded 2870 ± 30 Ma ago by the Usushwana Complex which comprises a layered suite of pyroxenites, gabbros and granophyres in the form of two dyke-like bodies.

The Dominion Reef, Witwatersrand and Ventersdorp Supergroups constitute the 'Witwatersrand Triad'. The major part of the

Table 6.1 Estimated rates of vertical crustal movement during deposition of sedimentary–volcanic systems in southern Africa (after Hunter, 1974b; reproduced by permission of Elsevier Scientific Publ.)

Supergroup	Maximum thickness (km)	Time-span (Ma)	Rate (km/Ma or mm/yr)
Swaziland	21.3	80	0.27
Pongola	10.6	70	0.15
Witwatersrand/ Ventersdorp	16.7	300	0.05
Transvaal	9	250	0.036
Waterberg	6.5	160	0.023

Pongola, Dominion Reef and Ventersdorp Supergroups consists of extrusive igneous rocks whereas those of the Witwatersrand and overlying *Transvaal and Waterberg Supergroups* are predominantly sedimentary. In decreasing order of abundance, the volcanic rocks include andesites, rhyolites, trachytes and tholeiitic basalts, and prominent pyroclastic deposits of agglomerate, tuff and tuff breccia. The sedimentary rocks, on the other hand, include conglomerates, orthoquartzites, arkoses, sandstones, shales, dolomites and limestones, cherts and banded iron formations.

Anhaeusser (1973) has shown how the progressive stabilization of the continental crust can be followed indirectly by examining the thickness and areal extent of these basins; their areas of outcrop, estimated depositional boundaries and basin axes are depicted in Fig. 6.1. From the earliest supracrustal rocks in South Africa (the Swaziland Supergroup), through the Pongola to the Waterberg–Matsup Supergroup, Anhaeusser (1973) demonstrated a progressive increase in size of the depositional basins. There is also a noticeable change in the proportion of volcanic to sedimentary rocks, the older Supergroups having relatively greater volcanic components than the later predominantly sedimentary accumulations. The north-westerly migrating basin axes (see also Hunter, 1974b) might be a reflection of some form of continental accretion about an early Archaean nucleus.

Anhaeusser (1973) demonstrated that strong contrasts emerged between crustal growth in the Archaean and early Proterozoic. The structural characteristics of the deeply infolded synclinal greenstone belts mostly gave way to broad basins containing flat-lying or gently dipping strata, the main deformation features being mild warping, doming, epeirogenic subsidence, faulting and dyke invasion; metamorphism was unimportant. The associated sediments are marked by a high proportion of stable platform types, in particular conglomerates, orthoquartzites and sandstones. The volcanic rocks include thick extensive non-sequential varieties embracing continental flood basalts of tholeiitic type,

and a calc-alkaline sequence with predominant potash-enriched andesites and rhyolites and important pyroclastics. Bickle and Eriksson (1982) showed that the basins underwent subsidence as follows. An initial rapid volcanic stage due to thinning or loading of the crust was followed by a more protracted sedimentary stage due to thermal relaxation of the thinned lithosphere—a type of basin development common in the Phanerozoic (McKenzie, 1978). Van Biljon (1980) is the only person to interpret any of these basins in a uniformitarian way; he suggested that the Witwatersrand basin formed in an embayment along a subduction zone between two Archaean mini-continents.

The Svecokarelian of the Baltic Shield

The Svecokarelian orogeny produced much new continental crust in southern Finland and Sweden at 2200–1700 Ma. The Archaean basement in SE Finland (Fig. 6.2) is bordered on its south side by the 2200–2100 Ma Kalevian Group of flysch which have been thrusted eastwards over the craton. Along this belt are tholeiitic pillow lavas and copper sulphide deposits (Park and Bowes, 1981), large serpentinite bodies and cherty quartzites (the Outokumpu association) that all constitute an oceanic crust-mantle sequence in a major suture (Gaal, 1982). To the southwest there is a Jatulian belt of meta-turbidites followed by volcanic arc rocks (amphibolites) that evolved 1900–1850 Ma ago. The volcanics were folded and intruded by calc-alkaline 'granitic' batholiths in the period 1850–1750 Ma.

There are two plate tectonic models of the Svecokarelian orogeny based on different directions of subduction dip. With a dip to the northeast (Hietanen, 1975; Berthelsen, 1980; Gaal, 1982), the Kalevian sediments were laid down on the continental shelf and the turbidites in inter-arc and back-arc basins. If the Kalevian sediments formed in a back-arc basin, the Cu–Co deposits at Outokumpu in the oceanic crust rocks would be comparable to the Cyprus-type deposits of the Troodos ophiolite (Mäkelä, 1980). The vol-

76

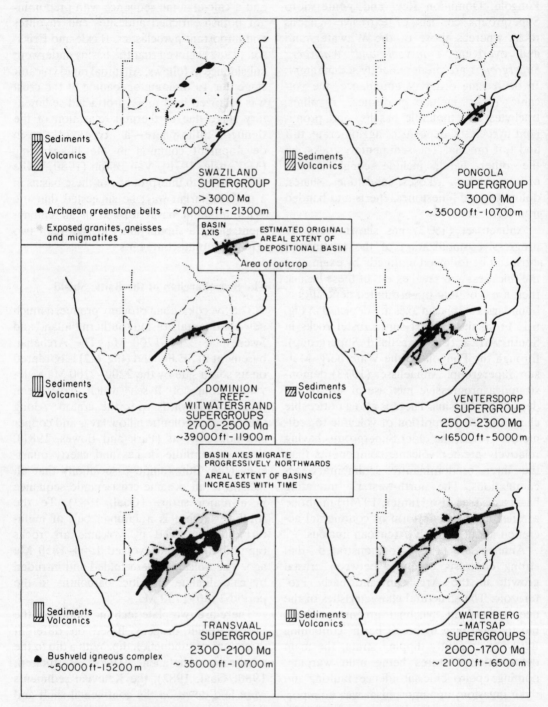

Fig. 6.1 Generalized geological maps depicting the progressive development of cratonic basins from the Archaean to the early Proterozoic in South Africa. Crustal thickening and stabilization results in increased areal extent of each successive stratigraphic sequence. The progressive increase with time in the sedimentary/volcanic ratio is indicated (after Anhaeusser, 1973; reproduced by permission of The Royal Society; ages from Condie, 1982d)

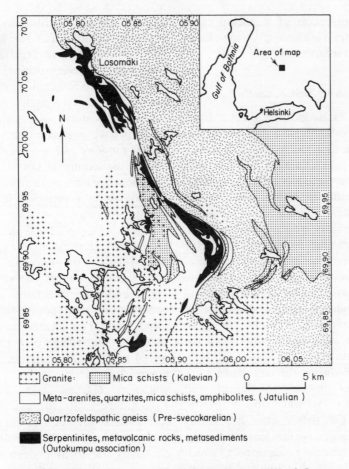

Fig. 6.2 Map of part of SE Finland, 50 km NW of Outokumpu, showing a Svecokarelian suture zone (the Outokumpu association) separating Presvecokarelian basement to
the NE from Svecokarelian mobile belt to the SW (after Park
and Bowes, 1981, *Bull. Geol. Soc. Finland*, with permission)

canic arc belt contains porphyry Cu–Mo
deposits.

With subduction towards the southwest, a
Tibetan-type model is created (Bowes, 1980).
The serpentinites represent oceanic crustal
relics thrust eastwards in ophiolitic nappes
close to the main NW-trending suture. The
Kalevian Group represents flysch-type deposits formed as the two continental masses
converged. The high-grade, low-pressure
meta-turbidites are deeply eroded rocks
uplifted as a result of crustal thickening in the
overriding plate (Campbell, 1980).

In southern Sweden there are altered
tholeiites of ocean-floor type, andesites and

basalts comparable with present-day arc volcanics (Loberg, 1980) and granitic batholiths
(M. R. Wilson, 1982). The geographical distribution, age relations and geochemistry of
these rocks are what would be expected of
progressive cratonization. In the period
1750–1620 Ma there was fracture-controlled
emplacement of granitic batholiths and voluminous ignimbrites associated with the cauldron subsidence (Nyström, 1982); this was a
high-level environment very similar to that of
the Coastal Batholith of Peru (Myers, 1975,
1976).

Although all the detailed relationships are
not yet clear, the Svecokarelian belt promises

to contain the essentials of a Wilson cycle, and many Scandinavian geologists are clearly keen on such an interpretation.

The Circum-Superior Belt, Canada

This fold belt is made up of the Labrador Trough, the Cape Smith Belt and the Belcher Fold Belt surrounding the Archaean Superior Province (Fig. 6.3) (Baragar and Scoates, 1981); it was interpreted as a suture between colliding continents by Gibb and Walcott (1971). The structure is internally asymmetrical, the sedimentary western zone towards the craton comprising much orthoquartzite, dolomite and iron formations, and the predominantly volcanic eastern zone away from the craton composed of mafic volcanics and intrusives and shales and greywackes. A medial geanticlinal ridge separates these two zones. Sediment input was mainly from the cratonic side during the early stages and from the orogenic side during the later history.

The volcanics include tholeiitic and komatiitic basalts (Schwarz and Fujiwara, 1977). The fact that they contain hardly any andesites, dacites or rhyolites contrasts markedly with the Archaean volcanic rocks which typically evolved towards acidic end fractionates. Frequency distribution curves for differentiation index and for K_2O, Na_2O and TiO_2 contents show that the Labrador Trough lavas have a compositional spread that is closer to modern oceanic tholeiites than to Archaean tholeiitic basalts.

On the western side Archaean basement gneisses have not been affected by the Hudsonian orogeny and are overlain unconformably by the sediments. On the eastern side the basement gneisses were reworked during the Hudsonian and intruded by granodioritic-granitic plutons.

Following Gibb and Walcott (1971) there are increasing attempts to explain the evolution of the Circum–Superior belt in terms of plate tectonic theory. The earliest sediments are commonly rift-related continental red beds locally with potash-rich basalts. The Cape Smith belt clearly records large-scale 'Atlantic-type' rifting of a continental plate with 4 km of craton-derived, quartz-rich clastics and continental-type, Ti-rich tholeiitic volcanics (Hynes and Francis, 1982). Komatiitic basalts formed at the onset of spreading at the edge of the narrow oceanic basin and more evolved tholeiites with trace chemistry similar to modern MORB are representatives of the early Proterozoic oceanic crust. Wardle (1981) recognized a shelf–slope–basin transition in the Labrador Trough developed as a passive continental margin on the craton to the west. To the east there are thick accumulations of flysch and submarine mafic volcanics sheeted by comagmatic basaltic sill complexes; Wardle suggested these rocks formed in a narrow oceanic rift system. The Richmond Gulf graben is failed-rift aulacogen that extends at a high angle from the craton into the Belcher belt (Chandler and Schwarz, 1980). The palaeomagnetic data of Irving and McGlynn (1981) indicate that the bordering cratons behaved as a single entity during the Hudsonian Orogeny (1750–1600 Ma); probably there was an incipient oceanic rift up to 1000 km wide which allowed the extrusion of oceanic-type tholeiites. Barager and Scoates (1981) suggested that subduction was not involved in the closure of the Labrador

Fig. 6.3 Map showing the distribution of the Huronian and Animikie Supergroups and the Circum-Ungava fold belt in SE Canada; A: Richmond Gulf graben (compiled from Dimroth *et al.*, 1970 and Card *et al.*, 1972; reproduced by permission of The Geological Survey of Canada)

Trough to Belcher belt, and Dimroth (1981) that continental delamination (Bird, 1979) was responsible for the fracturing and closure of the lithospheric slabs. In my opinion both these models suffer from the fact that they are based solely on the Circum–Superior 'geosynclinal' belt and do not take account of the contemporaneous high-grade, highly deformed belt with calc-alkaline granitic plutons that lies to the east of the Labrador Trough and north of the Cape Smith belt; the evolution of the Himalayas would never be understood if one over-concentrated on the sedimentary-volcanic rocks. In contrast, Thomas and Kearey (1980) interpret the gravity profile of the suture (*sensu strictu*),

which extends eastwards from the Labrador Trough, in terms of an Andean model and thus account for the linear belt of granitic plutons centred 70–150 km to the east of the Trough. Thomas and Gibb (1977) used the 350 km long, linear, positive gravity anomaly of the Cape Smith belt to define a northerly dipping suture belt.

The Wopmay Orogen, Canada

In the NW Canadian Shield there are the well-preserved remnants of a northerly trending orogenic belt, developed in the period 2100–1800 Ma, termed the Wopmay Orogen; the earlier term Coronation

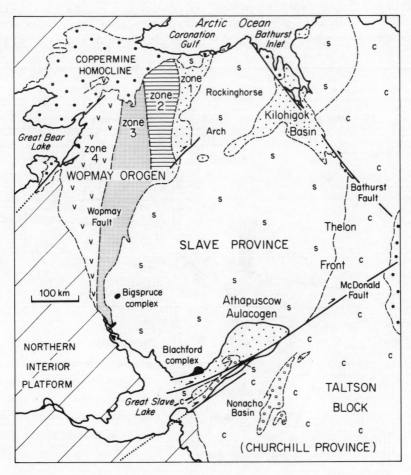

Fig. 6.4   Major tectonic elements in the northwest corner of the Canadian Shield. North is to the top of the page (from Hoffman, 1980, *Geol. Ass. Canada*)

80

Geosyncline is now obsolete (Hoffman, 1980, 1981). This is a unique belt because it contains the record of a complete Wilson Cycle. It provides the best evidence for the operation of modern-style plate tectonics in the early Proterozoic; it is, therefore, worth looking at in some detail.

The orogen is situated on the west side of the Archaean Slave Province (the craton). Fig. 6.4 shows how it is divisible into four zones:

1. Autochthonous continental platform sediments (600 m) overlain by 800 m of distal flysch, and above that molasse.
2. Allochthonous continental shelf dolomites (2–4 km) overlain by 80 m of hemipelagic shale deposited during foundering of the dolomite bank, in turn overlain by flysch of granitic-metamorphic provenance. This zone has been thrusted eastwards onto the autochthon.
3. Continental rise sediments intruded by the S-type Hepburn and Wentzel batholiths and metamorphosed to sillimanite-orthoclase grade at 10–12 km depth. No volcanic equivalents are known.
4. Weakly metamorphosed calc-alkaline volcanic rocks, ignimbrites (with continental sediments) intruded by I-type epizonal granites of the Great Bear Batholith (Hoffman and McGylnn, 1977).

Important tectonic events in the evolution of the orogen are as follows.

1. The formation of the Athapuscow Aulacogen and the Kilohigok Aulacogen (Basin) (Fig. 6.4) which are failed traverse rifts along which craton-derived sediment was channeled towards the evolving continental

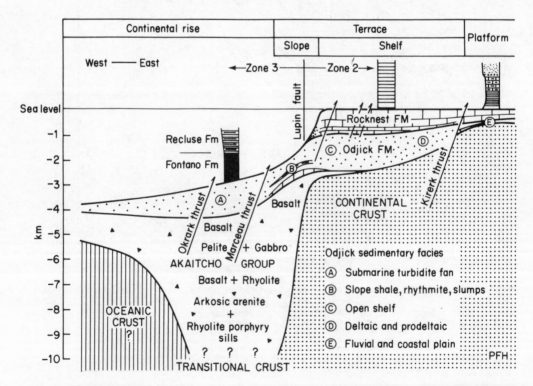

Fig. 6.5 Pre-orogenic reconstruction of the mature Wopmay continental margin. The depth of the continental rise is drawn the same as in modern oceans (from Hoffman, 1980, *Geol. Ass. Canada*)

margin. These rifts are associated with three alkaline-peralkaline complexes (e.g. the Blachford Complex), including carbonatite and nepheline syenites, and major dyke swarms which parallel the aulacogen for 240 km within and outside it. There is evidence of early rifting of the Wopmay continental margin under zone 3 in the form of major bimodal (basalt–rhyolite) volcanic rocks (Fig. 6.5). It is interesting to note that the rift succession underlies the inner continental rise rather than the outer shelf as in the modern North American Atlantic margin.

2. The first event leading to the destruction of the continental margin was the foundering of the continental terrace and deposition of the deeper water hemipelagic shale in association with growth faults (as in the Timor Trough). The overlying flysch was deposited by turbidity currents that flowed southwards along the trench axis.

3. Swarms of gabbro sills were intruded largely in the hemipelagic shale. Such mafic magmatism released into a trench by subducting lithosphere is unparalleled in the Cenozoic.

4. Deformation of the continental margin involved eastward thrust translation relative to the craton, similar to that in Cenozoic collision zones. Thrusts and folds without basement involvement formed in the flysch detached from the subducting plate before collision (as in the Makran), and in the terrace sediments underlying the flysch during the early stages of collision (as in the Zagros). Flat thrusts became folded when plate convergence continued after collision, and continued convergence led to crustal underthrusting in the foreland (as in the Himalayas).

5. During westward subduction (Fig. 6.6A,B) the peraluminous hornblende–poor S-type plutons of the Hepburn–Wentzel batholiths were generated by crustal thickening and anatexis caused by thrust stacking and imbrication and were intruded in zone 3 as tabular sheets with hot-side-up (inverted) and hot-side-down (normal) metamorphic aureoles (St-Onge, 1981). Continental andesitic volcanic rocks, post-tectonic, I-type,

granitic plutons and the ignimbrites of zone 4 were emplaced in two stages, as in the Turkish-Iranian plateau. First was the Andean-type calc-alkaline activity before collision, and second was the ignimbritic extrusion generated after collision by deep crustal anatexis. This westward subduction led to collision of the Slave Plate with the Bear Plate (a microcontinent) to the west, leaving an oceanic plate further to the west.

6. Eastward subduction (Fig. 6.6C,D) of this oceanic plate beneath the Bear Plate led to closure of the ocean and a second terminal collision with an incoming continental (Mackenzie) plate from the west. This caused the intrusion of calc-alkaline plutons further east in the Athapuscow aulacogen and late conjugate transcurrent faulting on the foreland to the east, comparable to the slip-line indentation faults in Asia north of the Himalayas.

The transcurrent faulting also extended along, and dislocated, the Athapuscow aulacogen and the Kilohigok Basin, and this indentation may have cracked the Churchill lithosphere giving rise to the 1800 Ma alkaline volcanism of the Dubawnt Group and related E–W lamprophyre dykes (Fig. 6.7).

In contrast, Gibb (1978) interpreted the master faults bordering the Slave Province not as aulacogen rifts, but as slip-line wedge faults caused by the eastward indentation of the Slave Province (wedge) into the Churchill Plate. The nearby N–S Thelon Front is marked by paired gravity—anomalies between crustal blocks of different mean density and thickness, and this density discontinuity may represent the cryptic suture between the collided Slave and Churchill Provinces (Gibb and Thomas, 1977).

The Wopmay Orogen has been brilliantly analysed by Paul Hoffman from some two decades of mapping; it provides detailed documentation of a complete Wilson Cycle of early Proterozoic age.

The Wollaston Fold Belt, Canada

One of the most exciting recent developments in early Proterozoic geology has come

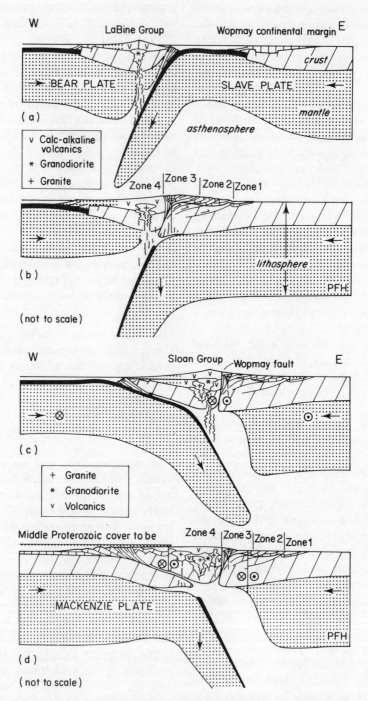

Fig. 6.6 A—B: First collision of the Wopmay Orogen, resulting in accretion of the Great Bear arc-bearing microcontinent. C—D: Renewed subduction against the accreted microcontinent leading to terminal continental collision. The collision suture is possibly marked by the Keith Arm gravity high—Fig. 6.7 (from Hoffman, 1980, *Geol. Ass. Canada*)

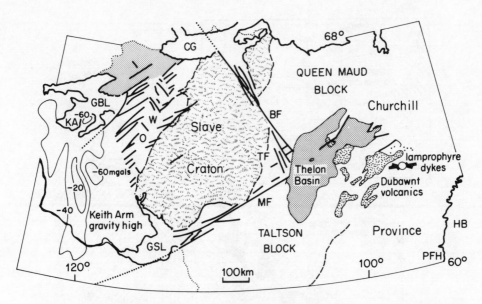

Fig. 6.7 Conjugate transcurrent faults (heavy lines) caused by east–west compression related to a terminal continental collision suture possibly located along the Keith Arm gravity high. The relatively rigid Slave craton has indented and cracked the Churchill Province lithosphere, giving rise to the Dubawnt volcanics and lamprophyres. Faults: BF: Bathurst Fault, MF: McDonald Fault. Locations: CG: Coronation Gulf, GBL: Great Bear Lake, KA: Keith Arm, GSL: Great Slave Lake, HB: Hudson Bay, TF: Thelon Front (from Hoffmann, 1980, *Geol. Ass. Canada*; dyke orientation data courtesy of A. N. Le Cheminant; gravity data from Earth Physics Branch (1974))

from the western Churchill Province of Canada, where the main parts of a Wilson Cycle have been well defined (Lewry and Sibbald, 1980; Ray and Wanless, 1980; Lewry *et al.*, 1981; Lewry, 1981).

The major tectonic belts are as follows (Fig. 6.8):

1. The Cree Lake Zone consists of Archaean gneisses overlain unconformably by the Aphebian Wollaston Group which comprises immature arkoses, polymictic conglomerates and mafic lavas interpreted as an early continental rift succession. This, in turn, is overlain by mature quartzites, calc-silicates and pelites which formed part of a stable platform on the edge of the continental margin.

2. The Rottenstone–La Ronge magmatic belt. On the southeast side there is an arc-type succession of pillowed basaltic flows, volcanic breccias and greywackes

intruded by quartz–diorite–tonalite-granodiorite plutons. Further to the northwest, across a 'quartz–diorite' line similar to that in the western Cordillera of North America, is the plutonic core of an Andean-type arc (the Wathaman batholith) comprised largely of quartz monzonite–granodiorite–granite (Lewry *et al.*, 1981).

3. The eastern La Ronge domain mostly consists of gneissic meta-psammites and meta-semipelites with relict pebbly psammites and polymictic conglomerates. These are interpreted as volcanogenic proximal greywackes, volcaniclastic-epiclastic sediments derived from the Rottenstone-La Ronge arc and deposited in a fore-arc basin. Thrust stacking and isoclinal folding have severely telescoped this fore-arc.

4. The Glennie Lake domain in composed of orthogneisses and later granites with

84

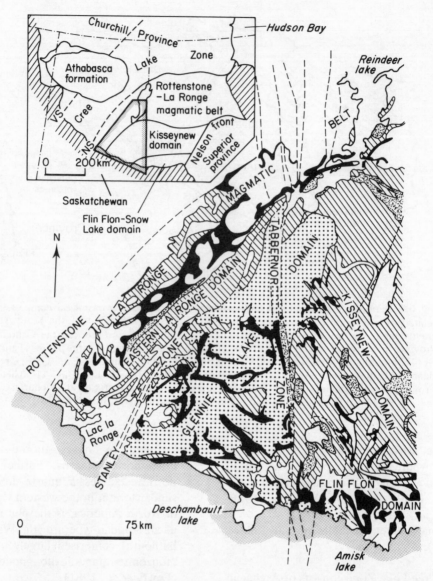

Fig. 6.8 Geology of the southeastern part of the Churchill Province in northern Saskatchewan (location indicated by heavy, hatched lines in inset). Solid black area, undifferentiated metavolcanics; ▨, mainly volcanogenic meta-sediments; ▥, meta-arkoses; $\left|\begin{smallmatrix}++\\++\end{smallmatrix}\right|$, undifferentiated granitoids of Glennie Lake domain and Hanson Lake area; unshaded area, quartz dioritic to granitic plutons of other domains; ▨, Phanerozoic cover (oblique line in inset). Dashed lines indicate fault and shear zones. VS (inset): Virgin River Shear Zone; NS (inset): Needle Falls Shear Zone (Reprinted by permission from Lewry, *Nature*, **294**, 69–72. Copyright © 1981 Macmillan Journals Limited.)

narrow supracrustal belts with a distinctive structural character which is different from that of adjacent domains. It is interpreted as an Archaean microcontinent.

5. The Kisseynew domain mostly comprises

migmatized volcanogenic metagreywackes interpreted as detritus deposited in a remnant inter-arc basin and derived from the La Ronge and Flin Flon arcs.

6. The Flin Flon arc, once thought to be an

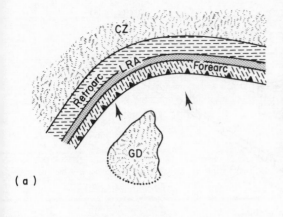

(a)

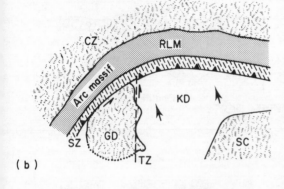

(b)

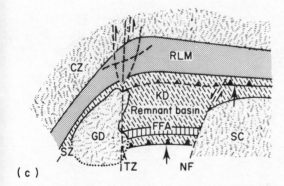

(c)

Fig. 6.9 Possible tectonic evolution of the south-east Churchill Province in Saskatchewan and Manitoba. (a) (~1950–1900 Ma?): early Hudsonian development of subduction zone and La Ronge–Lynn Lake volcanic arc marginal to Cree Lake continental margin. The Glennie Lake domain is an 'outboard' microcontinent within a predominantly oceanic subducting plate. (b) (~1900–1870 Ma?): collision of Glennie Lake domain with eastern La Ronge forearc and initial telescoping of the forearc prism. Uncoupling occurs between the impeded Glennie Lake microplate and still subducting Kisseynew microplate with initiation of the southern Tabbernor Zone. The arc is uplifted and stripped to form a major Cordilleran-type arc massif, the Rottenstone–La Ronge Magmatic Belt. The Flin Flon–Snow Lake arc may be initiated by migration of subduction site southwards between the Tabbernor and Thompson 'transforms' or may already have been active in the outboard part of the southern plate. (c) (~1860 Ma?): the Glennie Lake domain is sutured to the arc massif with immense telescoping of the fore-arc and possible transcurrent reactivation along the suture. Subduction north of the Kisseynew domain has ceased and the suture buried beneath sediment influx. The Kisseynew subplate becomes an inter-arc remnant basin receiving sediments both from north and south. Differential movements propagate the Tabbernor Zone northwards as a splaying wrench system. The Thompson Belt forms another transcurrent/transform junction, to the east of which incoming of the Superior continent leads to terminal collision along the northern part of the Nelson Front, followed by general compression, late folding, continuing propagation and reactivation of major fractures. Assigned dates are speculative, based on limited radiometric data. CZ: Cree Lake Zone, GD: Glennie Lake Domain, SC: Superior Craton, LRA: La Ronge–Lynn Lake Arc, FFA: Flin Flon–Snow Lake Arc, RLM: Rottenstone–La Ronge Magmatic Belt, KD: Kisseynew Domain, SZ: Stanley Zone, TZ: Tabbernor Zone, NF: Nelson Front (Thompson Belt) (Reprinted with permission from Lewry, *Nature*, **294**, 69–72. Copyright © 1981 Macmillan Journals Limited.)

Archaean greenstone belt, is widely regarded today as an early Proterozoic island arc formed by northward subduction.

7. The Superior Craton is the well-known Archaean craton.

8. The Stanley and Tabbernor Zones are straight belts or shear zones of mylonitic gneisses interpreted as a suture zone

and a transform fault, respectively. The Thompson Belt (or Nelson Front) forms another transform junction oriented at a high angle to the main orogenic belt.

Figs. 6.9 and 6.10 illustrate a possible tectonic evolution of the Wollaston Fold Belt. Early rifting was followed by deposition of a platformal sequence on the southern edge of the

86

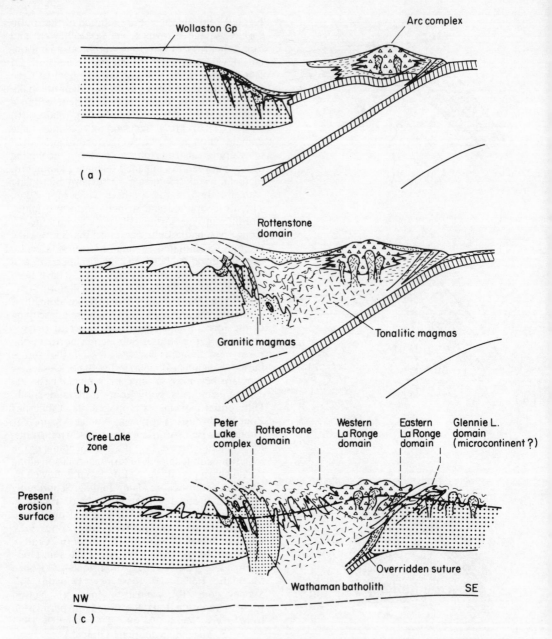

Fig. 6.10 Schematic crustal sections showing possible plate tectonic evolution of the Rottenstone–La Ronge magmatic belt (from Lewry *et al.*, *Precambrian Research*, **14**, 277–313; Fig. 10, by permission of Elsevier, Amsterdam).

(a) Situation in late Aphebian (circa 1900 Ma?): establishment of La Ronge–Lynn Lake volcanic arc following a long interval of Atlantic-type continental margin sedimentation. The Rottenstone domain now becomes the site of a back-arc basin with early continental rise sediments being overlain by proximal to distal arc-derived clastics and volcaniclastics.

(b) Early Hudsonian: initiation of Cordilleran-type orogeny, magmatism and tectonic 'rolling-up', with general fusion of continental margin and continental rise/arc clastics in the Rottenstone domain. The indicated upthrusting of basement slices and early mafic–ultramafic plutons as an early stage of development of the Peter Lake complex is speculative but might account for the observed general lack of pervasive tectonic fabric and low metamorphic grade in this region.

(c) 'Mid'-Hudsonian (circa 1850–1800 Ma): emplacement and consolidation of the Wathaman batho-

Archaean continental margin. Northward subduction of an oceanic plate created the La Ronge island arc, the back-arc basin on its north side becoming the site of intrusion of the main Andean-type batholith whilst a fore-arc basin was created on the south side. The Glennie Lake microplate collided northwards with the fore-arc, telescoping it severely. Uncoupling between the impeded Glennie Lake microplate and the still subducting Kisseynew oceanic microplate initiated the transform Tabbernor Zone. The Flin Flon arc started in the south, and the Kisseynew subplate became an inter-arc remnant basin receiving sediments from north and south. The Superior Craton moved northwards against the Thompson Belt transform junction leading to terminal collision with the arc to the north.

This sequence of events provides a viable explanation in plate tectonic terms for this Hudsonian orogenic belt. The geological relationships are exactly what one would expect to find in a deeply eroded and uplifted section of the Himalayas.

The Wollaston Fold Belt is defined by a positive geomagnetic anomaly and its continuation through Saskatchewan to Wyoming is marked by a narrow belt of very high electrical conductivity (Camfield and Gough, 1977). In Wyoming this is coincident with a geological boundary which Hills *et al.* (1975) suggested was an early Proterozoic plate margin.

The First Aulacogens

Aulacogens are large, long-lived, graben-like trenches first described in the USSR (Milanovsky, 1981). They extend from the early Proterozoic mobile belts into the Archaean cratons, often with a radial disposition. They are rifts that failed to open into oceans.

The formation of aulacogens requires the existence of stable continental platforms, therefore they could not form in the unstable Archaean terrains; they first appear in early Proterozoic times and constitute an excellent tectonic marker of the progressive cratonization of the continents. They are common in Proterozoic and Phanerozoic platforms and some have formed recently, opening not into a mobile belt but into an ocean, e.g. the Benue Trough (Burke and Dewey, 1973a) (see further in Chapter 16).

Detailed work by Hoffman (1973) in an aulacogen bordering the Wopmay Orogen has revealed the following characteristics and evolution. The aulacogen is a deeply subsiding trough in which sedimentary rocks, much thicker than on the adjacent platform, accumulated during every phase of the orogenic cycle. The sediment transport was longitudinal along the length of the trough, the sediments increasing in thickness longitudinally towards the platform margin where basic volcanics occur at five horizons and where tonalite-to-granodiorite laccoliths were intruded, attesting to the great depth of the bounding faults.

During the shelf stage in the orogen the aulacogen was in an incipient rifting stage (Fig. 6.11) during which the lips of the fault–trough stood high and shed thin quartzitic sediments into the trough, and subalkalic basalts were extruded, as might be expected in an active rift with thinned crust. As the continental shelf foundered and was buried by the clastic wedge, the aulacogen passed

lith is now complete, as is intrusion of the Peter Lake complex by related granitic magma. Major mylonitization along the Needle Falls shear system has occurred and late- to post-kinematic intrusion in other areas is in progress or is completed. The Glennie Lake domain is postulated to be a microcontinental block: collision with the La Ronge arc causes obliteration and overriding of the suture by southeastward-moving thrust slices and fold sheets of the arc–trench gap accretionary prism and microcontinental basement to produce the intensely strained zone of the eastern La Ronge domain. Later stages of compression produce refolding of these fold–thrust sheets. Further or contemporaneous arc development is envisaged to the southeast (in the Flin Flon–Snow Lake domain) followed by eventual closure at the Superior Province boundary in late Hudsonian times (? circa 1800–1750 Ma)

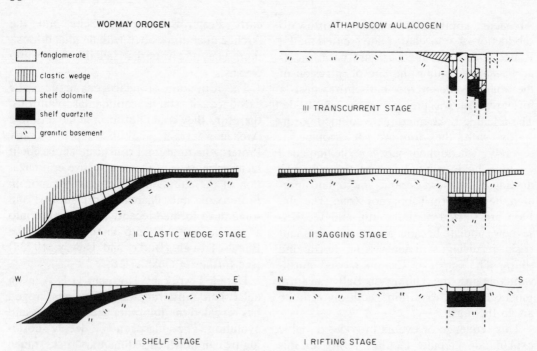

WOPMAY OROGEN   ATHAPUSCOW AULACOGEN

☐ fanglomerate
▥ clastic wedge
☐ shelf dolomite
■ shelf quartzite
꙳ granitic basement

III TRANSCURRENT STAGE

II CLASTIC WEDGE STAGE   II SAGGING STAGE

I SHELF STAGE   I RIFTING STAGE

Fig. 6.11   Three-stage evolution of an aulacogen and its relation to the evolution of the Wopmay Orogen (after Hoffman, 1973; reproduced by permission of The Royal Society)

into a sagging stage during which time the lips were depressed and covered with thick mudstones, greywacke turbidites and redbeds in a broad downwarp during the flysch to molasse phases of the geosyncline. Alkalic basalts were extruded, as might be expected in an area of thickened crust. Finally, the aulacogen was the site of regional transcurrent faulting giving rise to scattered downdropped basins in which thick alluvial sediments were deposited. Badham (1978) disputed the aulacogen model for the Athapuscow graben, suggesting that it was initiated as a strike–slip system in the late Archaean along the border of the Slave and Churchill Provinces.

Tillites

The Gowganda Formation in Ontario of Huronian age has been recognized as a glacial deposit since the beginning of this century, but it is only recently that further examples have been demonstrated, indicating a more

extensive early Proterozoic glaciation (Fig. 8.4).

Gowganda Formation, Ontario	Headquarters Schists, Wyoming
Chibougamau Series, Quebec	Pretoria Series, Transvaal, S Africa
Fern Creek and Enchantment Lake Formations, Michigan	Griquatown Series, Cape Province, S Africa
Reany Creek Formation, Michigan	Witwatersrand Supergroup, S Africa
Padlei Formation, Hurwitz Group, NW Territories	

Most of the North American occurrences lie in a belt nearly 3000 km long from Chibougamau to Wyoming (Young, 1970), but the most detailed account of the glacial rocks and structures is from the Gowganda Formation.

The Gowganda paraconglomerates are 2288 ± 87 Ma old, 15–200 m thick, and occupy an area of about 20 000 km^2. The primary evidence in support of a glacial origin for these beds is:

1. Massive polymict conglomerates with exo-

tic clasts up to boulder size, mainly of pink granite, variably distributed and concentrated.

2. Finely laminated argillites (varves) commonly with dropped clasts of plutonic rocks.

3. Presence of a striated basement.

The Animikie tillites of Michigan lie directly on the Archaean basement, in contrast to the Huronian Gowganda which overlies 4000 m of older Huronian sedimentary formations. The Reany Creek and Fern Creek Formations are correlated with the Gowganda. In the Medicine Bow Mountain region of SE Wyoming there is an 8 km thick sedimentary succession that includes polymictic conglomerates and laminated dropstone argillites (the Headquarters Schist), correlated with those of the Gowganda Formation. The glacial deposits in both regions are overlain by four very similar sedimentary formations.

From chemical studies Nesbitt and Young (1982) concluded that early Huronian lutites were deposited under tropical conditions. Later, climatic deterioration culminated in deposition of the Gowganda Formation and was followed by climatic amelioration. On a uniformitarian basis the depositional basin must have passed through about 60° latitude between 2500 and 2100 Ma. At present day spreading rates this would have taken between 100 Ma and 350 Ma. Palaeomagnetic data indicate that early Huronian volcanic rocks were extruded at a palaeolatitude of 32° and that the Gowganda Formation was deposited at 60° of latitude.

Little detail is known about the South African glacial sediments. Those in the Griquatown Series are less than 30 m thick but extend through an area of 20 000 km², and there are two or more tillites, 180 m apart stratigraphically, in the Government Reef Series of the Witwatersrand Supergroup.

Sedimentation Patterns

Early to mid-Proterozoic supracrustal rocks can be classified into three lithological assemblages (Table 22.1) (Condie, 1982a,d).

1. The quartzite–carbonate–shale assemblage comprises at least 60% of known Proterozoic sequences and commonly reaches 10 km in thickness. Quartzites are typically massive and cross-bedded, and sandstones indicate granite–gneiss source areas. Red beds, chert and banded iron formations are present in some successions (Animikie Group in central USA, and Mont Bruce Supergroup in Australia). Other examples are the Transvaal Supergroup in South Africa (Button, 1976b) and the Snowy Range Supergroup in Wyoming, USA. Comparison with Phanerozoic equivalents suggests this assemblage formed in three stable tectonic settings: rifted continental margins, cratonic margins of back-arc basins, and intra-cratonic basins.

2. The bimodal volcanic–quartzite–arkose assemblage accounts for 20% of successions, most of which are from 5 to 10 km thick. The bimodal rhyolite–basalt association is typical in some sequences but minor in others. Sediments are immature clastics, such as arkose, feldspathic quartzite and conglomerate derived from rapidly uplifted granitic source terrains. Iron formations, red beds, shales and carbonates occur in some sequences. The assemblage is characterized by rapid facies changes over short distances with mixed subaqueous and subaerial volcanics and sediments, and it occurs in cratonic rifts and aulacogens. Examples are the East Arim graben in the North West Territories of Canada, the Mount Isa aulacogen in Australia, the Dewaras rift and the Waterberg succession in southern Africa.

3. The calc-alkaline volcanic-greywacke assemblage amounts to more than 20% of successions and is characterized by calc-alkaline volcanics with pillow basalts, andesites, dacites and volcaniclastic rocks. Sediments are greywackes and argillites of turbidite origin. Examples are the Flin Flon–Snow Lake, Yavapai and Birrimian successions which can be regarded as Proterozoic greenstone belts (see Chapter 10) and are comparable in general respects with Archaean greenstone belt successions. As Condie (1982a) pointed out, today this assemblage occurs in back-arc, intra-arc and fore-arc basins.

Banded Iron Formations (BIF)

Early Precambrian banded iron formations are about the same age on many continents. They reached their peak of development in early Proterozoic basins or geosynclines situated near the boundaries of the Archaean cratons—the greatest development occurred between 2600 and 1800 Ma (Goldich, 1973). BIF account for 15% of the total thickness of early Proterozoic sedimentary rocks.

Important examples occur in:

Labrador Trough,	Mauritania, W Africa
Canada	Minas Gerais District,
Animikie Basin, USA	Brazil
Hamersley Group,	Krivoyrog, Ukraine
W Australia	India
Transvaal System,	Baltic Shield
S Africa	
Griquatown Formation,	
Cape Province,	
S Africa	

For a general review of the character and possible modes of evolution of these BIF see Dimroth (1977), Kimberley (1978), Goodwin (1982b), Mel'nik (1982) and Gross (1980).

The Proterozoic BIF belong to the Superior type which are thinly laminated rocks, mostly belonging to the oxide, silicate, sulphide and carbonate facies. They rarely contain clastic material and are stratigraphically associated with chert, dolomite, quartzite, black-carbon-bearing shale, argillite and volcanic rocks. The sequence dolomite, quartzite, red and black ferruginous shale, iron formation, black shale and argillite, in order from bottom to top, is common with local variations on all continents; Trendall (1968), however, pointed out that in the Hamersley, Animikie and Transvaal basins cherts locally underlie the iron formations, but otherwise the BIF 'do not consistently follow or precede any other sediment type'. They are not associated with as many volcanic rocks as are the Archaean iron formations, but there are normally volcanics somewhere in the succession, although their relative positions in the stratigraphy are highly variable. For example, the Lower Griquatown BIF is overlain by the 1200 m thick Ongeluk vol-

canics, the Fortescue Group is interbedded with the 500 m thick Woongarra acid volcanics, and the Ironwood BIF, East Gogebic Range, Minnesota is interbedded with volcanic rocks (Trendall, 1968).

Trendall (1968) quoted the following data on the thickness of some BIF:

	m
Animikie	
Gunflint Range	170
Cuyuna Range	0–150
Gogebic Range	210
Menominee Range	210
Marquette Range (Negaunee BIF)	>667
Hamersley	
Brockman BIF	>667
Total aggregate thickness	1000
Lower Griquatown	c. 500
Upper Griquatown	167
Transvaal Dolomite Series	233–267

The BIF typically extend for many hundreds of kilometres (cf. the Archaean BIF) outlining the former extent of the sedimentary basins.

The distinctive character of the internal subdivisions of the BIF enables them to be recognized and correlated over considerable distances. Individual parts of the main Dales Gorge Member of the Hamersley Brockman BIF can be correlated at the 2.5 cm scale over about 32 000 km^2 and correlations of varves within chert bands can be made on a microscopic scale over 296 km (Trendall, 1968). A 1.67 m thick algae horizon in the Biwabik BIF of the Mesabi Range can be followed over 80 km and Trendall (1968) stated that in the Griquatown BIF stilpnomelane bands, a few inches thick, can be identified in boreholes 64 km apart. These examples serve to demonstrate the fact that early Proterozoic sediments are incredibly well preserved and can be intercorrelated, not only on a large, but even on a small scale over enormous distances.

Blue-green algae and fungi have been identified in the non-ferruginous cherts of the Gunflint iron formation of Ontario (Awramik and Barghoorn, 1977) and some of these fossilized structures resemble modern-day iron-precipitating bacteria such as Sphaerotilus, Gallionella and Metallogenium. It is known that these bacteria are able to grow and precipitate ferric hydroxide; they live in ferruginous water, particularly where bog-iron ore is forming. The present day reducing environment is a result of bacterial activity and this may have been so in early Proterozoic BIF sediments (Cloud, 1973; Barghoorn *et al*., 1977; Awramik and Barghoorn, 1977; Klemm, 1979). It was during the early Archaean more than 3700 Ma ago that algal photosynthesis began and oxygen became available for the oxidation of ferrous iron in the oceans. The concentration of atmospheric oxygen was extremely low—Berkner and Marshall (1967) calculated that it remained at between 0.001 and 0.0001% of the present atmospheric level for 2000 Ma after algal photosynthesis began—but modern iron bacteria are able to oxidize ferrous iron at such low levels of concentration. Primitive iron bacteria therefore may well have accelerated the precipitation of iron deposits at a time when the prevailing low oxygen levels probably restricted the rate of purely chemical oxidation. But the fact that red arkoses and

conglomerates containing andesite pebbles with oxidized weathering crusts, and red pelites underlie the Sokoman BIF in the Labrador Trough means that oxidizing conditions must have been reasonably high during deposition and diagenesis.

It is increasingly thought that the Superior-type BIF were deposited on continental shelves (Fig. 6.12) (Kimberley, 1978; Gross, 1980; Goodwin, 1982b), in evaporitic barred basins (Button, 1976a), on flat prograding coastlines such as the Persian Gulf (Dimroth, 1977), or in intracratonic epeiric basins (Eriksson and Truswell, 1978). Many early Proterozoic BIF are closely associated with thick carbonate successions, e.g. Hamersley, Griquatown, and in Brazil and India. Eirksson *et al*. (1976) suggested that the ferruginous sediments in the Transvaal Supergroup are a proximal lagoonal facies of barrier bar, intertidal and subtidal limestones and dolomites. The Sokoman BIF in the Labrador Trough has limestone-like textures pointing to an origin of iron formation by limestone replacement. Dimroth (1977)

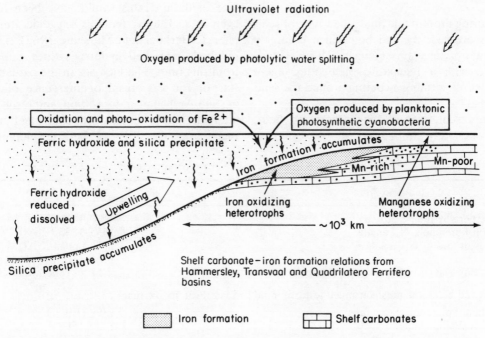

Fig. 6.12   Model for the deposition of Superior-type banded iron formations on a continental shelf in association with carbonates (from Goodwin, 1982b, *Rev. Bras. Geoc.*)

believes that the BIF were deposited as aragonite which was replaced by silica and iron compounds during diagenesis. There was evidently a high P_{CO_2} control on sedimentary processes in these early Proterozoic basins. Carbonates of this age are extensively dolomitized; ferruginous dolomites formed from a refluxing brine generated in a lagoon where the BIF were forming (Eriksson et al., 1976).

In contrast, Klein and Bricker (1977) point out that several modern sedimentary environments have highly reducing, iron-rich mineral assemblages a few centimetres below a highly oxygenated, aqueous medium (e.g. Chesapeake Bay, Santa Barbara Basin and Gulf of California in the USA, and the Black Sea. Early Proterozoic BIF contain little or no detritus, whereas modern sediments contain large amounts of detrital material. Today the precipitation of iron silicates is unlikely because of the abundance of detrital aluminium which causes the precipitation of aluminous silicate clays instead of pure Fe-silicates.

Red Beds

During Proterozoic time a new type of sediment evolved: the red bed which is largely absent in Archaean sequences. This is a sandstone with a haematite–pigmenting agent commonly forming a coating around the sand grains.

Representatives of the earliest red beds, approximately 2000–1700 Ma old, occur on several continents (Table 6.2) (for references see 1st edition):

The red beds are molasse-type immature sediments deposited in shallow water. They are particularly well developed in northern Canada where they have a wide time range from late Aphebian (c. 1800–1640 Ma) to Neohelikian (at least 1200 Ma). They were deposited in fault-controlled intra-cratonic basins, which implies the existence of a broad stable platform throughout most of Helikian time (1640–880 Ma). In other parts of the world they are likewise thought to have formed under stable cratonic conditions.

The last great episode of banded iron formations, about 2000 Ma ago, marks the excess accumulation of O_2 in the oceans. About this time O_2 began to escape from the hydrosphere and to invade the atmosphere. According to Cloud (1968b), in the absence of an ozone screen at this time, ultraviolet light in the range of 2000–2900 Å impinged on the Earth's surface and converted some of the escaping O_2 to O and O_3 (ozone). Since the reaction rates with respect to surface materials of both these products are many orders of magnitude greater than those of O_2, surface oxidation rates would have been high even in such early tenuous oxygenic atmospheres (Berkner and Marshall, 1967). Thus red beds, enriched in ferric oxides, appear from this time. The increase in the oxidation state of iron was a result of emergence following late Archaean cratonization, and resultant interaction of sediments with oxidizing fluids. Incipient stages of red bed development

Table 6.2 Some early to late Proterozoic red beds

System and Area	Age
Waterberg System, S Africa	>1790 (U/Pb)
Martin Formation, N Canada	1635–1835 (K/Ar)
Dubawnt Group, N Canada	1716 (mean K/Ar)
	1732 ± 9 (Rb/Sr isochron)
Echo Bay–Cameron Bay Groups, N Canada	c. 1700–1800 Ma (K/Ar)
Thick red bed sequences continued to form in mid to late Proterozoic time:	
Roraima Formation, Suriname, S America	1599 (Rb/Sr)
Jotnian, Baltic Shield	1300–1500
Torridonian, NW Scotland	935–751 (Rb/Sr)
Bathurst region, N Canada	c. 1200 (K/Ar)

occur in late Archaean arenites in Canada (Shegelski, 1980). The first indication of early Proterozoic iron-oxidation and enrichment is in 2300 Ma palaeosols underlying the Huronian Supergroup (Gay and Grandstaff, 1980).

Phosphorites

These are phosphorus-rich sedimentary rocks that did not form in the Archaean. Minor phosphorites began to appear in the early Proterozoic, but major deposits not until the late Proterozoic. Examples about 2000 Ma old occur in Finland, at Rum Jungle in N Australia, in the Marquette Range of Michigan (phosphoritic pebbles in conglomerates), the Hamersley Group of W Australia and at Broken Hill in SE Australia (for references, see Cook and McElhinny, 1979). There are wave-brecciated, stromatolite-bearing phosphorites (indicating the presence of tides) in the early Proterozoic Aravalli Group in NW India (Chauhan, 1979), and U-bearing 1900–2000 Ma phosphatic sediments in Finland (Vaasjoki *et al.*, 1980).

Major Phanerozoic phosphorites form at low latitudes, especially between 10° and 20° N and S, in areas of oceanic upwelling on trailing shallow continental margins near the shelf-rise break. It was probably the lack of such margins in the Archaean that prevented the deposition of phosphorites. The formation of the first extensive continental plates by the early Proterozoic allowed the first deposition of phosphorites.

Evaporites

There is increasing evidence of the deposition of sulphate evaporites in the early to mid-Proterozoic. In the Great Slave Supergroup (c. 2100–1800 Ma) in N Canada there are halite and gypsum casts, dolomite nodules after anhydrite associated with stromatolitic dolomites, pelletoid ironstones and mudstones with wave ripple marks and shrinkage cracks. These features point to deposition by evaporite-derived brines in a supratidal sabkha environment (Badham and Stanworth, 1977). Similar pseudomorphs and

relics of sulphate minerals occur in the McArthur Group (1600–1400 Ma) in N Australia (Walker *et al.*, 1977), in silica-dolomites (1600–1500 Ma) at the Mount Isa Cu–Pb–Zn mine in N Australia (McClay and Carlile, 1978; Neudert and Russell, 1981), and in the Belcher Group (>1798 Ma) in Canada (Bell and Jackson, 1974). Highly metamorphosed evaporites occur at the pre-2000 Ma Caraiba Cu deposit in Bahia, Brazil (diopside-anhydrite rocks) (Leake *et al.*, 1979), and in the Churchill Province (pre-1750 Ma) on Baffin Island in Canada (calcite, lazurite, diopside, nepheline, scapolite) (Hogarth and Griffin, 1978). These occurrences negate the suggestion by Cloud (1976b) that bedded sulphates did not appear until 1000 Ma.

Mineralization

In this section we shall consider some of the principal early-mid-Proterozoic mineral accumulations. These are important, not only because of their economic value, but also because they tell us a great deal about the depositional environments and the conditions operating in the atmosphere and oceans in this critical period in the evolution of the crust—critical because they represent a dramatic change from those operating in Archaean times.

Gold and Uranium in Conglomerates

The Huronian Supergroup in Canada contains major uranium-bearing conglomerates (Robinson and Spooner, 1982) and the Jacobina Series in Brazil has conglomerates enriched in gold and uranium. All the systems of the Witwatersrand Triad (Rundle and Snelling, 1977; Pretorius, 1976) and the Transvaal in South Africa have conglomerates enriched in both elements. Other examples of one or other element are in the following Supergroups: the Krivoi Rog and Kursk (USSR), the Tarkwajan (Ghana), Karelian (Finland), Hurwitz (Canada), Fortescue (Australia) and the Pongola (South Africa) (Pretorius, 1981). These deposits represent an important, even unique, metallogenic event in the early Proterozoic as this type of metal concentration is rare in the Archaean (e.g. the Bababudan conglomerates of India).

The conglomerates contain a great variety of detrital minerals within the rock matrix, in particular pyrite, gold and uraninite. A widely recognized view is that the detrital minerals were derived by erosion of the earlier greenstone terrains and concentrated by fluviatile and deltaic processes in a shallow-water high-energy environment; minor later modifications gave rise to small ore-bearing veinlets. Deposition took place along the interface between a fluvial system that brought the sediments and heavy minerals from an elevated source area and a lacustrine littoral system that reworked the material and redistributed the finer sediments along the shoreline. The goldfields were formed as fluvial fans around the periphery of an intermontane, intra-cratonic lake or shallow-water inland sea (Pretorius, 1976). Fig. 6.13 shows how the Witwatersrand placer gold province is located within the Kaapvaal craton, the entry points of fan conglomerate deposits being related to a provenance area containing uplifted gold-bearing Archaean greenstone belts.

Finally, these rocks and ores should give us information about the early Proterozoic oxygen balance, but there is a current controversy. The presence of easily weatherable pyrite and uraninite has commonly been used to indicate deposition of near-source placers under anoxic atmospheric conditions (Stanton, 1972). This model has been challenged by Dimroth (1979), Clemmey (1981) and Simpson and Bowles (1981) who ascribe to deposition under an oxygen-rich atmosphere involving weathering, erosion, redeposition and oxidative diagenesis. Comparison with modern rivers suggests transport of 10–30 km (Hallbauer and Utter, 1977).

Manganese in Sediments

There were appreciable accumulations of manganese particularly in carbonate sediments (limestones and dolomites), in the period 2000–2300 Ma. In the following

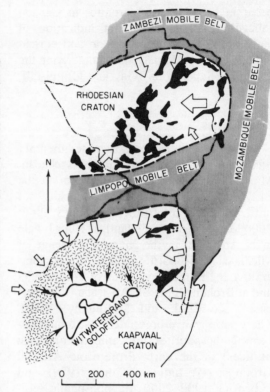

Known entry points of sediments and gold

Provenance area for Witwatersrand gold

Gold-poor zones. High-grade metamorphic belts

Gold quantity increases towards craton centre

Archaean greenstone belts (gold belts)

Fig. 6.13 Schematic map of southern Africa showing the provenance area of the Witwatersrand gold field in relation to the gold-bearing Archaean greenstone belts (after Anhaeusser, 1976; reproduced by permission of *Minerals, Science and Engineering*—discontinued publication)

account it is interesting to see what different geological processes they later suffered.

In South Africa there are large manganese deposits associated with the 2300–2100 Ma Transvaal Supergroup and the Loskop System. The manganese concentrations have two main modes of occurrence and host rock. First there are the basal ferruginous shaly beds of the Gamagara Formation of the Loskop System at the contact with the Dolomite Series of the Transvaal Supergroup; the second host is a siliceous (cherty) or banded ironstone cataclastic thrust breccia, up to 30 m thick, in close proximity to the Dolomite but formed as a result of post-Waterburg low-angle thrusting. In both cases the origin of the manganese is thought to lie in the Dolomite Series where it was deposited under supergene conditions by replacement of the banded ironstone, shale and breccias. However, the manganese ore, although occurring in tectonic breccias, is itself undeformed; thus it is concluded that the dolomite was decomposed and the ore-bearing solutions deposited in the breccias during the Carboniferous ice-age and in post-Karroo times.

In South America manganese deposits occur in:

1. The Serra do Navio Schists of the Amapá Series in the Guiana Shield (Brazil), which underwent a high-grade metamorphic event at least 2000 Ma ago. The main ores are lenses of carbonate or garnetiferous mangano-schist (with spessartine–rhodonite and Mn-rich olivine) within graphitic quartz–biotite–garnet schists. Manganiferous clayey limestones were apparently metamorphosed in the sillimanite–almandine subfacies of the almandine amphibolite facies.
2. In the Minas Gerais district of Brazil where the ores occur in lenses and bands intercalated with 2400 Ma old schists, gneisses and amphibolites.

Syngenetic manganese deposits occur in the Aravalli Group in India which is dated at 2000–2500 Ma. The ores occur as folded beds with phyllites, quartzites and dolomitic limestones, regionally metamorphosed in the chlorite and biotite zones and also, within impure limestones, thermally metamorphosed by a biotite granite. The manganiferous beds are little more than a metre thick but, like the South African and South American ores, they indicate an appreciable accumulation of manganese in a sedimentary environment not long before 2000 Ma ago.

Lead–Zinc in Carbonates

By 1700 Ma continental margins similar to those of today were well developed. Andean types had shelf and rise sequences with algal carbonate reefs and platforms. The carbon dioxide content in the oceans had increased to such an extent that thick dolomite sequences were deposited and biological activity was well advanced, caused by the oxygen increase produced by the proliferation of blue–green algae which probably entrapped carbonates. Thus we find the relatively sudden appearance of algal reefs and stromatolitic dolomites, especially in the shallow-water platforms. These features combined to give rise to the first favourable environment for major lead–zinc sulphide accumulations.

The minor concentrations of sphalerite–galena in the 2300 Ma Dolomite Series of the Transvaal Supergroup, South Africa, are some of the oldest ores of this type in the world. They occur in gold-bearing veins and as replacement deposits below impervious layers of chert and shale and Pb and Zn are enriched in the Malmani dolomite (Button, 1976b). The lead–zinc–dolomite association did not reach maturity, however, until about 1600–1400 Ma in the Mount Isa Geosyncline, the McArthur Basin in Australia, and the Black Angel deposit in West Greenland by 1700 Ma.

The McArthur Basin contains at least eight important stratiform base metal deposits including McArthur River (Pb–Zn–Ag) and Mount Isa (Pb–Zn–Ag and Cu). The McArthur River ore member is a black carbonaceous dolomitic pyritic shale which overlies vitric tuffs (Lambert, 1976; Oehler and Logan, 1977). The local sequence contains numerous algal stromatolite biostromes and algal reef dolomites, diagnostic of a shallow marine or intertidal environment, whilst salt crystal pseudomorphs and mud-crack impressions indicate complete marine withdrawal.

The Mount Isa ores occur in the Mount Isa geosyncline (Finlow-Bates and Stumpf, 1979). There are two types of ore in a shale–dolomite–siltstone sequence with cross-bedding, convolute bedding and slump breccias: copper in a chalcopyrite-rich silica-dolomite which is an algal reef and reef breccia, and lead–zinc in the famous galena–sphalerite-rich black carbonaceous Urquhart shale which interdigitates with the reef deposits and is regarded as an organic product of off-reef sedimentation. Williams (1969) considered that the ores were deposited in a broadly volcanic environment, the lower part containing the Eastern Creek Volcanics whilst acid tuffs are dispersed throughout the ore-bearing strata. He suggested that the metals are of exhalative-sedimentary origin and were introduced into a basin of deposition by submarine fumaroles. The ores show a close spatial relationship with a major triple junction and continental rifting (Sawkins, 1976b).

The Black Angel lead-zinc deposit occurs within the 1.3 km thick metamorphosed tremolite-bearing dolomitic marble of the Marmoralik Formation (which has a minimum age of 1700 Ma) in the Umanaq area of West Greenland (Pederson, 1981). In places the marble has a dark colour due to the presence of finely disseminated graphite. There are no volcanic rocks associated with the ore body or indeed within the Formation, but there are meta-volcanic amphibolite formations below and above it.

In the Lower Purcell Sediments of the Canadian Cordillera (reported in Chapter 8) there is a major Pb–Zn ore body within an argillite at Sullivan in British Columbia. The Sullivan lead is estimated to have an age of 1250 Ma or 1340 Ma; this agrees with the fact that the Purcell Sediments began to accumulate about 1450 Ma ago (Harrison 1972). On the basis of deep seismic reflections, Kanasevich (1968) proposed that a major mid-Proterozoic rift valley existed beneath southern Alberta and British Columbia. He then proposed a very interesting model: the Sullivan Pb–Zn ore body lies just within the edge of the rift and therefore the ore-bearing solutions may have been emplaced in the rift structure, i.e. in a tectonic environment similar to that of the modern hot metalliferous brines in the Red Sea. See Sawkins (1976a,b) for further discussion of relation of Sullivan-type Pb–Zn deposits to rifts.

Uranium–Vanadium–Copper in Clastics

Stratabound deposits of U, V and Cu occur in conglomerates, sandstones and siltstones, derived by erosion of continental material and deposited under fluviatile-deltaic conditions. The formation and emergence of thick continental plates by the early Proterozoic allowed the formation of the first deposits of this type (Windley, 1980). Salop (1977, p. 183) stated that Cu deposits are characteristic of 2600–1900 Ma red beds, and quoted examples in the Udokan Group of E Siberia, the Olekma River Basin (USSR), the Upper Huronian of Canada, and the Onega Group of N Karelia (USSR). There is a U–V deposit in 1740 Ma sandstone in Gabon (Cesbron and Bariand, 1975), and U mineralization in clastic beds in the 2100–1750 Ma Athapuscow aulacogen of the Wopmay Oregon (Hoffman, 1980).

Micro-fossils

The term Proterozoic is derived from the Greek for 'early life'. Although a few microfossils and stromatolites developed locally in Archaean sediments, they began to proliferate in early Proterozoic times. The more complex multicellular Metazoa, which require an oxygenous atmosphere for their growth, should also occur in Proterozoic rocks since the rapid increase in the oxygen content in the atmosphere is indicated by the transition from banded iron formations to red beds. However, unequivocal examples only appear in abundance in late Precambrian rocks; the problem of the precise time of their first appearance remains unsolved.

The best known indubitable early Proterozoic micro-fossils are the abundant algal flora from the black stromatolitic cherts of the Gunflint Iron Formation of the Huronian in Ontario (Barghoorn et al., 1977) which formed about 2000 Ma ago. These organisms include 30 types of filamentous, spore-like and anomalous forms only a few microns in size; some are morphologically comparable to living blue–green algae, others resemble living bacteria although they are smaller (Awramik and Barghoorn, 1977).

In the Witwatersrand goldreefs there are remarkable fossilized remains of microorganisms that include bacteria, algae, fungi and lichen-like plants. Hallbauer (1975) envisages a carpet-like colony of columnar carbonaceous individuals, each about 0.3–0.55 mm in diameter and up to 7 mm in length. Gold and uranium were extracted from the environment by the organisms in a way similar to modern fungi and lichens.

There are beautifully preserved organic remnants in the well-exposed early Proterozoic rocks in the Graenseland area of southwest Greenland (Bondesen *et al*., 1967). The Ketilidian sediments here, about 2000 Ma old, have suffered extremely little deformation and metamorphism with the result that primary stratigraphy and sedimentary structures are intact. Besides stromatolite-like macro-structures, there are abundant remnants of bacteria-like structures, filaments with an irregular cellular structure, irregular lumps of organic material, parts of threads which occasionally branch and, in particular, spore-like spheres of 0.5–1.5 mm diameter which are described as *Vallenia erlingi* (Bondesen, 1970). But it is not just the existence, but the widespread extent, of the organic remains that is impressive here. Some remnants occur in almost all sedimentary formations and rock types—dolomites, dolomitic shales and cherts are the most favourable hosts, but they even occur in greywackes and quartzites, and finally there is a 1–3 m thick coal–graphite bed. Moreover, the remains have an extensive horizontal distribution, the 2 m thick Vallenia-bearing dolomite having been found at four localities 25 km along strike and it is thought possible to trace it for more than 50 km.

Pedersen and Lam (1968) found that the Vallenia-rich dolomite still contained alkanes, aromatic hydrocarbons and methyl esters, whilst in their 1970 paper they reported that the coal layer contained alkanes, alkyl benzynes, naphthalene, single alkyl naphthalenes, single monoterpenoids and esters of fatty acids—all original organic compounds, indicating extensive biological activity 2000 Ma ago. Isotopic studies by Bondesen *et al.* (1967) suggested that Vallenia may have been photosynthetic and that the host dolomite was precipitated in oxygen-poor water. Also the composition of the coal indicates that it represents marine organic material accumulated under extremely reducing, euxinic conditions. The widespread distribution of the organic remnants suggests that organic material was present under nearly all sedimentary conditions, and that they may have had an influence on processes such as the formation of chert, dolomite and carbonaceous shale. If the organisms were photosynthetic, they would have contributed considerably to the development of an oxygenic atmosphere (Bondesen, 1970).

Black cherts associated with the 1600–1400 Ma McArthur Pb–Zn–Ag deposit contain diverse micro-fossils. Oehler and Logan (1977) suggested that bacteria consumed oxygen during their decomposition of detrital algal organic matter and thereby maintained the anoxic bottom conditions necessary for the accumulation and preservation of the sulphide minerals.

These early Proterozoic micro-organisms were largely of the procaryotic type, incapable of cell division and relatively resistant to DNA-damaging ultraviolet radiation. It was not until after about 1800 Ma that an effective ozone layer formed in the atmosphere which prevented radiation penetration and thus allowed the eucaryotes to develop (Fig. 22.3).

The evolution of stromatolites during the whole Proterozoic period will be reviewed in Chapter 8.

Chapter 7

Mid-Proterozoic Anorogenic Magmatism and Abortive Rifting

One result of the combined effects of the Hudsonian orogeny (1800–1700 Ma) in Canada, the Ketilidian (1900–1800 Ma) in S Greenland, the Laxfordian (1800–1600 Ma) in NW Scotland, and the Svecokarelian (1850–1900 Ma) in the Baltic Shield was to create an extensive consolidated landmass or plate across this region. During the next few hundred million years, until about 1200–1000 Ma, we see little or no evidence of orogenic cycles (sedimentation, volcanism, plutonism, deformation, metamorphism) throughout this region. Instead there was important and distinctive anorogenic magmatism giving rise, in particular, to andesine-labradorite anorthosites, rapakivi granites, quartz porphyry lavas, plateau basalts, alkaline complexes, carbonatites and basic dykes (see Bridgwater and Windley, 1973, for an early description). Many of these appeared as prominent rock groups for the first time in earth history and therefore they represent an important landmark in the evolution of the continents.

Chronology and Correlations

Fig. 7.1 shows that the magmatic suites extend in a belt from California to the Ukraine. A similar, but lesser known, belt with prominent anorthosites extends across what is now the southern hemisphere from S. America via Africa and India to SW Australia. Note from Fig. 7.1 that the intrusive-extrusive suite developed not in the old consolidated Archaean cratons, but in the recently formed early Proterozoic mobile belts; whether there is any syn- to post-orogenic relationship is not known.

Correlation of the main rock groups along the belt is made difficult by the fact that they were not contemporaneous (Table 7.1). At present it appears that there was a diachronous development: the anorthosites and rapakivi granites have ages of 1400–1500 Ma in North America (Emslie, 1978a,b), but 1550–1700 Ma in Finland (Simonen, 1980). As we shall see in this chapter there are reasons for believing that some anorthosites and granites are genetically related, that some rapakivi granites are associated with porphyry lavas, being extrusive equivalents of the granitic magma. Vertical block movements controlled the uprise of these magmas, the deposition of later red continental sandstones and the extrusion of plateau basalts. This was followed by the intrusion of major swarms of basic dykes of transcontinental scale and several large layered igneous intrusions. Late events include the intrusion of alkaline complexes and carbonatites, the extrusion of tholeiitic lavas in graben and the deposition of more clastic sediments. There is thus considerable evidence for a close relationship between the major rock groups. Continental rifting was initiated soon after 1400 Ma, and this rifting controlled the formation of the later rocks. Table 7.1 gives the simplified chronology of events in four main regions (for general reviews of this 'Elsonian' magmatism in North America, see Emslie, 1978a, b).

Anorthosites

Anorthositic plutons are distributed along two linear belts in the northern and southern hemispheres when plotted on a pre-Permian continental drift reconstruction. Most crystallized in the period 1700–1200 Ma with a peak at 1400 Ma (Morse, 1982). They were commonly intruded into rocks that had been through high-grade regional metamorphism, some 200–300 Ma earlier. There are many bodies within the Grenville Province in Canada (Fig. 7.2); only 10 years or so ago they were thought to be causally related to the 1000 Ma Grenvillian orogenesis, but it is now established that several still record their 1400 Ma igneous cooling ages (Emslie, 1978b) and have been only partially overprinted by the Grenvillian event.

Individual bodies vary considerably in size, although most fall within the range 100–10 000 km², and account for some 20% of the surface area of the Grenville Province. Gravity data indicate that the Adirondack massif in the USA is a 4–4.5 km thick subhorizontal slab, the Morin anorthosite in Canada a 2–3 km thick sheet (Emslie, 1978a), and that the anorthositic bodies in the Ukraine are sheet-like with thicknesses up to 4 km (Moshkin and Dagelaiskaya, 1972). Some bodies were probably intruded along cover-basement interfaces (e.g. the Michikamau, Morin and Adirondack bodies) and the sheet-like form implies a lack of horizontal tectonic stress. The Laramie body in Wyoming was intruded into a major shear zone interpreted as a suture separating older continental crust from younger supracrustal rocks (Hills and Armstrong, 1974). Many bodies were built up by multiple intrusions. Examples include the Rogaland complex, SW Norway (Demaiffe and Javoy, 1980); the Nain complex, Labrador (Berg, 1977); the Michikamau intrusion, Labrador (Emslie, 1970).

The anorthosites consist of more than 90% plagioclase, and associated gabbroic, noritic and troctolitic anorthosites (deep- and shallow-level intrusions, respectively: Simmons and Hanson, 1978) have 78–90% plagioclase. Plagioclase composition mostly lies in the range An_{45-55}. Due to their common lack of layering and their homogeneous textures they are often referred to as massif-type anorthosites (e.g. Philpotts, 1981). The magmatic textures are taken to indicate crystallization in a relatively calm tectonic environment (Woussen et al., 1981). The anorthositic rocks are typically associated with a suite of granitic rocks (see below) suggesting that they are products of bimodal magmatic processes.

Most anorthositic complexes were intruded at moderate depth where it is not easy to see a relationship with extensional tectonics. However, Berg (1977) proposed that the Nain Complex was intruded at 3.7–6.6 kb pressure into a graben defined independently by gravity studies. An intracontinental rift was also favoured by Wiebe (1980) for this complex. From their geochemical study of several complexes Simmons and Hanson (1978) concluded that magma crystallization took place at 50–60 km depth, consistent with derivation from an orogenically thickened crust as under the present Tibetan Plateau. There is some dispute about the type of magma responsible for these types of rocks: ideas range from high-Al basalt (Emslie, 1978a) to a highly feldspathic magma (Wiebe, 1980). Whichever is correct, these magmas are distinct from the tholeiitic, calc-alkaline and alkaline series, and represent a unique type of anorogenic magmatism (Morse, 1982) that was new to continental evolution in the mid-Proterozoic.

In the USSR there are many Proterozoic anorthosites in the Kola peninsular, the Ukraine, the Aldan Shield, the Anabar Shield and the Mongolo-Okhotsk region (Moshkin and Dagelaiskaya, 1972; Bogatikov, 1974). Dudenko and Zhdanov (1979) and Dagelaiskaya and Moshkin (1979) concluded that the bodies in the Kola and Aldan regions, respectively, formed by a process of anorthositization in which the earlier rocks were altered in such a way that the composition and fabric

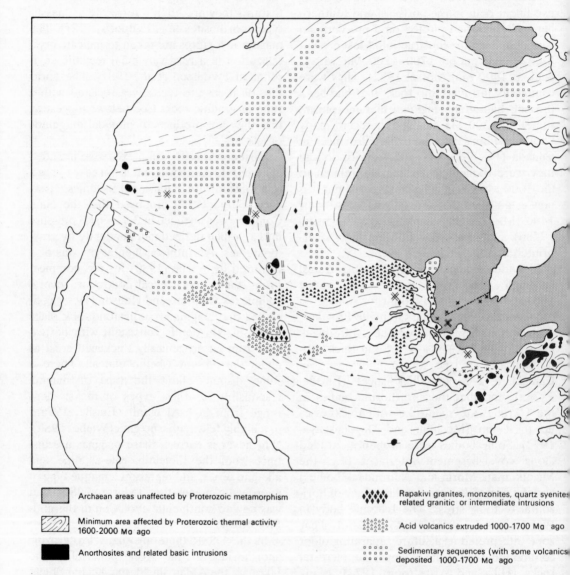

	Archaean areas unaffected by Proterozoic metamorphism		Rapakivi granites, monzonites, quartz syenites related granitic or intermediate intrusions
	Minimum area affected by Proterozoic thermal activity 1600-2000 Ma ago		Acid volcanics extruded 1000-1700 Ma ago
	Anorthosites and related basic intrusions		Sedimentary sequences (with some volcanics) deposited 1000-1700 Ma ago

Fig. 7.1   The distribution of the principal rock groups in the North Atlantic Shield belt formed during the mid-Proterozoic (modified after Bridgwater and Windley, 1973)

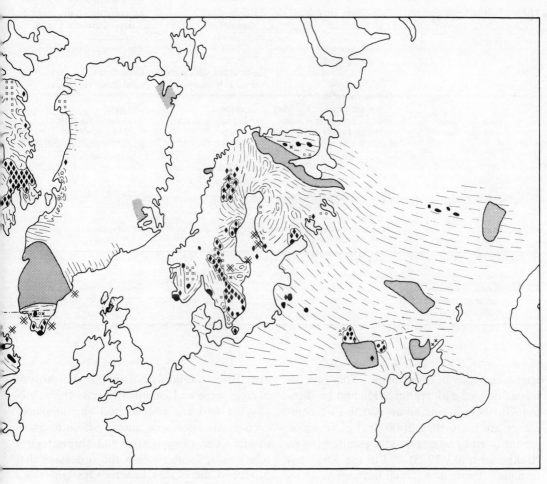

Basic extrusives and associated sediments deposited
1000-1200 Ma ago

Alkali intrusives emplaced 1000-1700 Ma ago

⬡ Dykes

═ Cataclasis

╌ Faults

Table 7.1 Chronologies of events that occurred between 1700 and 1100 Ma in four main regions of the North Atlantic Shield (modified after Bridgwater and Windley, 1973)

Ma BP	Ukraine	Baltic Shield	S Greenland and E Canada	S Canada and Mid-continental USA
		Rogaland anorthosite		Rapakivi granites (Nevada) Acid volcanics (Colorado)
1100	Post-tectonic granites	Peralkaline intrusions (S Sweden)	Basaltic–alkali magmatism Basic dykes	Keweenawan basic magmatism in graben. Clastic sediments.
		E–W dykes with *anorthosite* fragments		*Anorthosite* fragments in gabbros
1200			Basic dykes with *anorthosite* fragments.	Basic dykes Acid–basic volcanics
		Dykes controlled by	Alkali intrusions, dykes	
1300		graben faults (1270 Ma)	in graben	Granites
1400	Basaltic–alkali magmatism	Jotnian basalts and sandstones (1300–1400 Ma)	Sandstones in graben	Acid–basic volcanics (Wisconsin) Rapakivi granites and anorthosites
			Adamellite–*anorthosite* complexes. Rapakivi	Acid volcanics (Missouri)
1500		Major dyke swarms Egersund *anorthosite* and granites	granites	Granites (Nebraska, Dakota) Rhyolites
1600		Acid volcanics, porphyries, *anorthosites* and rapakivi granites (1550–1700 Ma) Basic dykes		Basalts, rhyolites, granites (Arizona)
1700	Koresten rapakivi granite–*anorthosite*–alkaline complex			Granites (Dakota, Colorado)

approach those of anorthosite. This process was associated with granitization and basification. In the Ukraine some anorthosite complexes are more than 2000 km^2 in area and are intruded by large rapakivi granite plutons (Bukharev *et al.*, 1973) and in the Koroston Complex there is a zonal arrangement of anorthosites, rapakivi granites and alkaline rocks (see p. 112) (Semenenko *et al.*, 1960; Gorokhov, 1964).

Adamellites, Rapakivi Granites and Acid Volcanics

A variety of granitic rocks are constantly associated with the anorthosites. Adamellites and granites typically form separate intrusive plutons. Mangerites and charnockites have the bulk composition of monzonite and granite, respectively, and contain orthopyrox-ene. They commonly border the anorthosites in an aureole and are intrusive into them; locally the two are interlayered and in places even grade via norite anorthosite into anorthosite. The close spatial and chronological relationships suggest that the processes that produced the earlier anorthosites generated granitic magmas at a more advanced stage. This bimodal anorthosite–adamellite association is characteristic of anorogenic magmatism (Emslie, 1978a). Malm and Ormaasen (1978) concluded that the 1900–1700 Ma mangerite–charnockite intrusions in Lofoten, N Norway, were emplaced at 11 kb pressure (35–45 km) under anorogenic conditions into the root zone of a developing rift.

Rapakivi granites are characterized by megacrysts of potash feldspar mantled with a rim of oligoclase. They occur widely in the North Atlantic belt, particularly in SE Fennoscandia (Fig. 7.3).

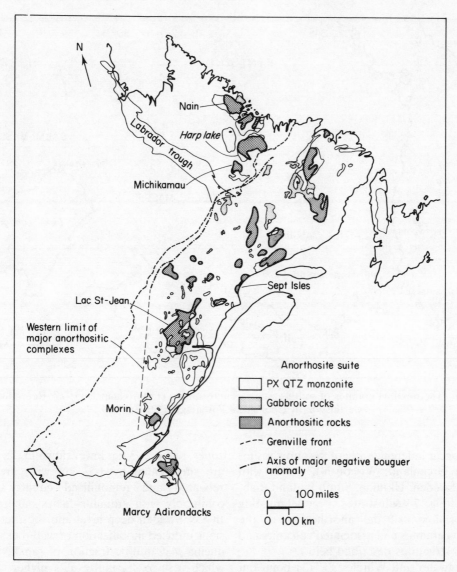

Fig. 7.2 Distribution of anorthositic and related rocks in the Eastern Canadian Shield. The Grenville Structural Province is outlined with a stippled pattern (modified after Emslie, 1975, *Geoscience Canada*)

Examples, together with ages (Ma), are:

1040	Colorado	Barker and coworkers (1975)
1050	Nevada	Volborth (1962)
1550	Sweden	Kornfält (1976); Wilson (1982)
1500	Wisconsin	Anderson (1980)
1570–1600	Berdyaush, Urals	Salop (1977)

1700	S Finland	Vorma (1976)
1720 ± 70	Koreston, Ukraine	Gorokhov (1964)
1500	Missouri	Bickford and Mose (1975)
1750	S Greenland	Gulson and Krogh (1975)
1610	Irel complex, Siberia	Salop (1977)
1600–1750	E European platform, USSR	Salop (1977); Klevtsova (1979)

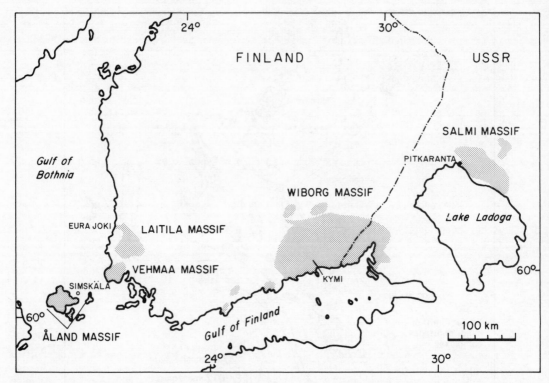

Fig. 7.3   The rapakivi granites of south-eastern Fennoscandia. (From Haapala, 1977. Reproduced by permission of Geologiska Föreningen, Stockholm)

Some occur independently of the anorthosites (South Greenland, Wisconsin), others with them (Sweden, Ukraine, South Finland, Labrador). Fig. 7.4 illustrates a model relating the formation of the anorthosites to the rapakivi granites and associated volcanics and graben structures described below.

Bridgwater and Windley (1973) point out that the thermal effect of the rapakivi granites (and also the anorthosites) on the surrounding country rocks varies considerably from place to place, reflecting differences in present erosion, in the original depth at which the bodies were intruded and, perhaps most important, in the temperature of the country rocks into which these bodies were emplaced. In Finland, where the upper parts of rapakivi granites are exposed, the adjacent country rocks show little effect of thermal metamorphism. On the other hand, in South Greenland, where there is considerable relief, the upper parts of the granites show little effect on their wall rocks, while their root

zones, some 2–3 km lower down in the crust, are surrounded by 1–2 km wide areas of reheated and remobilized country rocks commonly with granulite facies mineralogy. In SW Sweden deep level anorthositic intrusions induced mobilization of wall rocks, producing a granulitic fraction of rapakivi-type which rose to crystallize at a higher crustal level (Hubbard and Whitley, 1978).

Chemically, the rapakivi granites are rich in potassium, iron and fluorine, expressed by the presence of potash feldspar, fluorite and iron-rich mafic minerals such as fayalite (Barker et al., 1975; Vorma, 1976; Anderson, 1980). Many are close to quartz monzonite in composition and are thus similar to the mangerites; in fact in the Wiborg area of Finland rapakivi granite grades into mangerite or quartz monzonite.

Extrusive acid volcanic rocks in the North Atlantic Shield were extruded at different times and in different places in the period 1700–1000 Ma. There is a close similarity in

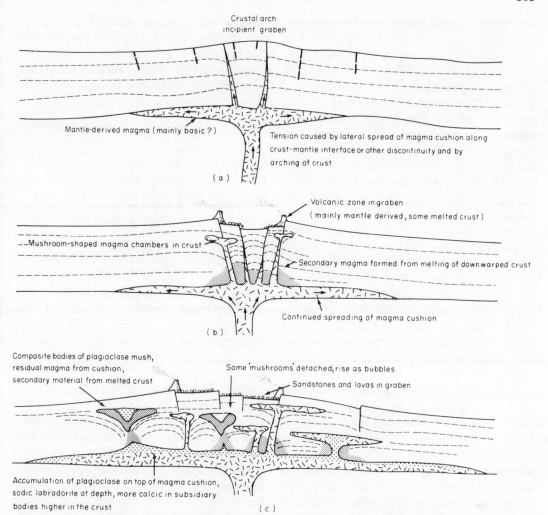

Fig. 7.4 A model that relates anorthosites, rapakivi granites, acid and basic volcanics and graben structures formed in continental crust (after Bridgwater, Sutton and Watterson, 1974). (a) Formation of crustal arch above magma cushion. (b) Continued spread of magma cushion, formation of graben and intrusion of mantle material into the crust along graben fractures. Some volcanicity. Downwarping of crustal material and melting near the base of the crust. (c) Continued spreading of magma cushion and further formation of graben structures. Accumulation of plagioclase on top of magma cushion at depth and in smaller chambers in the crust. Rise of plagioclase cumulates through the crust in a manner similar to salt domes, partly lubricated by residual material from magma cushion and partly by remobilized crustal material. Basic and acid volcanicity on surface, deposition of intermontane sediments

age and petrochemistry between these volcanic sequences and the rapakivi granites. In the Wiborg area of Finland there are labradorite porphyry lavas, chemically similar to the rapakivi granites, in association with rhyolite flows. But the classic area for seeing these interrelationships is in the Loos–Hamra region of Finland. Here there is a spectacular sequence of acidic volcanic rocks grading from syenites to rhyolites which occur together with a rapakivi granite of identical composition. Apparently the granitic rocks were intruded almost contemporaneously with the volcanics and thus there are gradations between them. And, on Suursaari island, the rapakivi magma extruded as a lava and formed quartz porphyry and agglomerates (Simonen, 1980). In the 900 km^2 anorogenic 1500 Ma rhyolitic field in the St. Francois terrain of Missouri in the midcontinent region of the USA (Bickford and Mose, 1975) geological and geochemical relationships indicate that biotite granites and associated granophyres intruded their own volcanic products and are comagmatic with alkali rhyolites (Kisvarsanyi, 1980). These are subvolcanic granitic plutons that were intruded along ring fractures in cauldron subsidence structures.

Clearly the rapakivi granites concerned crystallized from magma at high crustal levels, as suggested by Simonen (1980). One can conclude that the adamellites, the rapakivi granites, labradorite porphyry lavas and rhyolites were derived from a similar magma and that this was related to the magma that gave rise to the anorthosite–mangerite suite (see also Barker *et al*., 1975).

A further interesting fact is that some of the earliest ignimbrites formed at this time in Earth history. In Wisconsin, USA, there is evidence of 1760 Ma granitic-rhyolitic magmatism that represents the first appearance of anorogenic, probably rift-related, igneous activity of the North American region. Aluminous biotite granites were intruded at depths of less than 6 km with comagmatic melts giving rise to major volumes of ignimbrite (Anderson *et al*., 1980). Ignimbrites are also known in the 1850–1900 Ma Svecofennid meta-volcanics (hälleflintas) of Sweden and in the 1685 Ma silicic porphyry lavas of Finland (Simonen, 1980).

Continental Sandstones and Basalts

In the North Atlantic Shield towards the end of the mid-Proterozoic there was extensive faulting which controlled the location of sedimentation and volcanism as well as of later alkaline intrusions in graben-like structures. This was a period of high-level tectonics associated with surface deposition and postectonic igneous events which generally occurred after the emplacement of the anorthosites and rapakivi granites (Bridgwater and Windley, 1973). The formation of these structures implies the existence of wide platforms on stabilized late Precambrian cratons within which epicratonic troughs were sited (Stewart, 1976). Examples are:

1000–1300 Ma	Belt–Purcell Supergroup, western USA and Canada
1000–1100	Keweenawan Trough, mid-USA
1250–1300	Seal Lake, Labrador
1100–1400	Apache and Grand Canyon Series, Arizona
1300–1400	Jotnian, Sweden and Finland
1300	Gardar Province, South Greenland
1600–1700	Lake Onega beds, Baltic Shield
1600–1700	Palaeo-Helikian basins, NW Canadian Shield

The sediments are mostly coarse, clastic, red continental sandstones of molasse type, comparable with modern, shallow-water, piedmont facies flood plain deposits, and the volcanics mostly vary from tholeiitic to alkalic basalts.

Many of the deposits were laid down in elongate basins or troughs controlled by major faults which were active for a considerable time. For example, Laurén (1970) considered that the rise of the rapakivi granites in southern Sweden and the subsequent deposition of the Jotnian sandstones were controlled by the same fault lineaments. In Finland the rapakivi granites are generally pre-Jotnian, but some in Sweden are post-Jotnian. P. Eskola in 1963 suggested that once the rapakivi magmas formed they remained in the liquid state for sufficiently long, encapsuled in their deep-seated reservoirs, for them to be intruded diachronously, some just before, some immediately after the deposition of the Jotnian sediments and volcanics. In the Gardar Province of South Greenland (Upton and Blundell, 1978) the east-west fault system, that controlled the deposition of the first sandstones and lava extrusion about 1400 Ma, were active long beforehand and continued to be intermittently so, localizing the intrusion of the major alkaline complexes until 1000 Ma. Although the main movements on these east–west faults were horizontal, there were vertical displacement components of as much as 2 km.

Soon after 1700 Ma great thicknesses of red arkosic sandstones and conglomerates were deposited in northwest Canada in fault-controlled basins (e.g. Martin Formation, Echo Bay and Cameron Bay Groups, etc.). The extreme compositional and textural immaturity of these sediments reflects a continuously rising rugged source terrain. The basins are partly filled with acidic-to-basic lavas, the extrusion of which was probably triggered by faulting. This fault–basin cycle was followed by deposition of thick ortho-quartzites of the Athabasca, Thelon and Tinney Cove Formations and of the lower Hornby Bay Group; for relevant discussion, see Condie (1982d).

There are close similarities in lithology and environment between some successions in different parts of the North Atlantic Shield. The Jotnian basalts in Scandinavia overlie sandstones comparable to the Hornby Bay Group sandstones of northwest Canada, and these in turn overlie unconformably pre-Jotnian intermediate-to-acid porphyries similar to those of the Echo Bay and Cameron Bay Groups; in both places the last groups are intruded by porphyritic (rapakivi-type) granites. Moreover, the sub-Jotnian and Jotnian, taken as one unit, is comparable with the Keweenawan succession of the U.S.A.

However, the continental deposits differ considerably in age. For example, they range from the Lake Onega sediments of the Baltic Shield and the Red Beds of the northwest Canadian Shield, deposited in the period 1600–1700 Ma (Kratz et al., 1968; Salop, 1977), to the sediments and lavas of the Keweenawan Trough in the mid-USA with an age of 1000–1100 (Annells, 1974). The widespread deposition of such fault-controlled, epicratonic coarse clastic sandstones in many parts of the North Atlantic Shield during the later part of mid-Proterozoic time suggests that the high-level conditions which controlled their formation were broadly similar over a major portion of the earth's crust (J. H. Stewart, 1976).

A remarkable structure that is worth considering in a little more detail is the Keweenawan Trough. In the Lake Superior region continental red clastics and basic amygdaloidal lavas of the Keweenawan Group (1000–1100) rest unconformably on early Proterozoic and Archaean rocks. The most prominent gravity high in North America suggests that the 20 km thick succession continues within a major graben or rift zone, nearly 160 km wide, beneath Phanerozoic cover rocks for over 2000 km from Lake Superior to Kansas (Halls, 1978; Green, 1983) (Fig. 7.5). The Trough has the character of an intra-continental aulacogen infilled with basic volcanics, (Annells, 1974; Green, 1977; Baragar, 1977) marginal fan-conglomerates, mainly alluvial sediments and an absence of underlying granite crust.

It is important to note that the Keweenawan aulacogen, plateau basalts at Seal Lake and in SW Greenland, basic dyke swarms and a variety of alkali complexes are all approximately contemporaneous with the early stages of the Grenville Orogenic belt, and some of them are marginal to the belt (Fig. 7.5). Thus it is possible that these rocks formed in rift structures, particularly in the continental foreland related to the early opening stages of the orogenic belt. This idea should be borne in mind when considering the relevant basic dykes and alkaline complexes in the next few pages.

Intrusions into Cratons

Basic Dyke Swarms

During the period from about 1200 to 1000 Ma vast numbers of basic dykes were introduced into the continents, mostly in intense swarms. Some may have been feeders to higher level basalts, such as those described above.

1000–1200 Ma	Baffin Island, Canada
1150	Arizona, USA
1150–1170	Logan sills, Lake Superior
1200	Mackenzie III, Canada
1200	Death Valley, Calif., USA
1220	Central Sweden
1250	Gardar Province, S Greenland
1250	Sudbury, Canada
1270	Finland

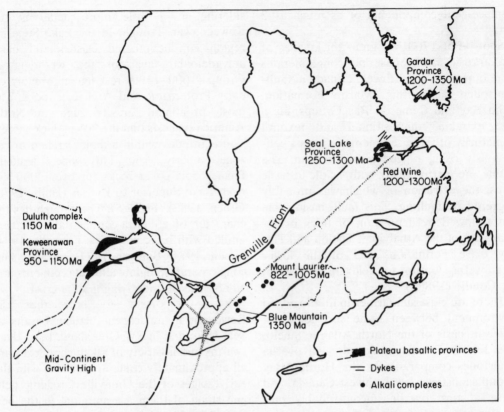

Fig. 7.5 Possible early Grenville rift zones with associated igneous rocks within and marginal to the Grenville orogenic belt. Greenland is restored to its pre-drift position relative to North America (from Barager, 1977, *Geol. Ass. Canada*)

Like the basic dyke swarms intruded in the period 2000–2500 Ma (Chapter 5) many of these late Precambrian swarms are of considerable size; indeed, some are transcontinental. In the Gardar Province of South Greenland (Emeleus and Upton, 1976) there are hundreds of basic dykes, most of which belong to a northeast-trending swarm (Fig. 7.6). Individual dykes are of considerable width, 50 m being not uncommon although the main range is probably from a few hundred to one hundred metres. In Finland there are many dykes in all the Jotnian sedimentary rocks, rapakivi granites and related mafic rocks, and were regarded by P. Eskola in 1963 as the feeders of the plateau basalt extrusions.

Compositionally, the dykes are mostly quartz dolerites and olivine dolerites (or diabases), but where they occur in the late alkaline provinces such as the Gardar they are closely associated with varieties such as trachydolerite, gabbro and gabbro-syenite. There is thus an overlap here with the dykes described in the later section in this chapter on alkaline complexes.

These late Precambrian dykes which are usually vertical are commonly faulted—a criterion which helps to distinguish them from unfaulted late Phanerozoic dykes in the same regions. They may be sheared and mylonitized especially along their margins, but the most obvious feature is that they may be displaced by faults for distances of up to a few kilometres.

It is possible that the dykes are comparable to the Mesozoic dykes bordering the North Atlantic ocean which are an early expression of the continental rifting and ocean floor spreading. As far as can be seen at present, major swarms situated well within continental masses cannot be so related to continental drift but may be more an expression of sub-continental convection cells operating within continents that were undergoing rifting elsewhere.

Layered Complexes

At about the same time as the basic dykes were intruded and the plateau basalts extruded several large igneous complexes were emplaced; the Muskox and Duluth bodies in Canada will be considered here.

The *Muskox Intrusion*, situated in the Canadian Northwest Territories, is a layered dyke-like body with a funnel-shaped cross-section, an exposed length of 118 km and a maximum outcrop width of 13 km. It has been dated at 1100–1200 Ma by Rb/Sr methods on micas and whole rocks; this is the same age as the nearby Coppermine basalts. The intrusion was emplaced just below the unconformity between basement gneisses of the Epworth Group and relatively undeformed quartzites of the Hornby Bay Group. The shape of the intrusion was probably controlled by the unconformity which dammed the magma causing it to spread out into a funnel shape. It is thought that the body was emplaced at least 3.3 km below the surface of the Earth, as this is the maximum thickness of the overlying Hornby Bay sediments. Aeromagnetic and Bouguer gravity anomalies show that the intrusion continues for at least 120 km beneath its roof rocks.

The intrusion consists of three units. A feeder dyke of bronzite gabbro with internal zones of picrite, two marginal zones that grade inwards (upwards) from bronzite gabbro to peridotite and a layered cumulate series, 2.8 km thick, ranging generally from dunite at the base, through peridotite and various pyroxenites and gabbros, to granophyric gabbro, and a local capping of granophyre at the top. The geology of the intrusion is described by Irvine and Baragar (1972).

The *Duluth Complex*, Minnesota, is dated as 1150 Ma and was intruded largely into Keweenawan lavas and sediments (Fig. 7.5).

It mainly consists of a 5 km thick sequence of gabbroic anorthosites, and not gabbro as it is usually termed (Phinney, 1970). There are several separate intrusions, the main rocks of which range from troctolite, gabbro and ferro-diorite to ferro-hedenbergite granophyre—a sequence displaying an iron enrichment trend (Weiblen and Morey, 1980).

Although the Duluth Complex is listed here as an intrusion emplaced towards the end of mid-Proterozoic time, it has obvious affinities with the somewhat earlier anorthosite bodies referred to earlier in this chapter. Because the complex contains inclusions of anorthosite up to 8 m across, and many other dykes, sills and flows in the area include anorthosite accumulations, it seems that a sizeable anorthosite mass exists at depth.

Alkaline Complexes

Following closely on the deposition of the continental deposits was the intrusion of alkaline complexes in many area of the North Atlantic Shield (Fig. 7.1), mostly near the end of the mid-Proterozoic. The complexes commonly occur in the same fault-controlled blocks and rift valleys (aulacogens) that contain the continental deposits (Fig. 7.7).

The type of magmatism varied considerably from alkaline and peralkaline to tholeiitic. This is expressed by a remarkable variation of rock types from alkaline gabbros and granites, quartz syenites, syenites, nepheline syenites and peralkaline rocks such as foyaites, naujaites, lujavrites and kakortokites, to carbonatites, lamprophyres, camptonites, gabbros and dolerites. 'As a general rule the alkaline magmatism appears to be restricted to distinct fault-controlled belts and to have persisted for a longer period of time than the tholeiitic magmatism, suggesting perhaps deeper, more localized control of magmatic activity later in the period' (Bridgwater and Windley, 1973).

Some of the main examples are:

1000 Ma	Haliburton–Bancroft, Ontario
1000–1100	Carbonatite complexes, Ontario and Quebec
1170–1300	Gardar Province, S Greenland

1050	New Mexico
1278	Seal Lake, Labrador
1345	Red Wine complex, Labrador
1300–1350	Blue Mountain, Ontario and St Hilaire, Quebec
1580	Norra Karr, Sweden
1500–1750	Oktyabryski and Tersyanski, USSR
1650–1750	Carbonatite complexes, Ontario and Quebec

This table shows that most of these bodies formed in the period 1300–1000 Ma. However, this so far gives too simplified a picture of the magmatic history of the period. In some regions there was a long and complex sequence of events extending over several hundred million years; nowhere is this better illustrated than in the Gardar igneous province of South Greenland which is extremely well exposed and documented.

A summary of the magmatic evolution of the Gardar Province is given by Upton (1974), Emeleus and Upton (1976) and Upton and Blundell (1978). Fig. 7.6 shows the major intrusions and the main dykes and faults, together with the sandstones and volcanics referred to in the last section, and Table 7.2 gives the Gardar chronology. The establishment of the main fault systems was followed by volcanism and sedimentation, and then by eight intrusive phases.

The intrusive rocks can generally be divided into dykes and plutonic complexes. There are a great number of dykes of several generations, the general emplacement order being lamprophyres and trachytes, dolerites and olivine dolerites (the most common type), granophyres, alkali microgranites and microsyenites, and finally trachytes and tinguaites. Two types of giant dykes are outstanding: composite dykes up to 500 m wide with marginal gabbro and central syenite that formed by a non-dilatational stoping mechanism and secondly troctolitic gabbro dykes up to 800 m wide.

Most of the plutonic complexes post-date the dykes. The predominant rocks are augite syenite, alkali granite and nepheline syenite, together with subordinate calc-alkaline granite and gabbro. Probably the most well known is the Ilimaussaq Intrusion. Generally the bodies in the west of the province are saturated and akaline, whereas those in the east are undersaturated or peralkaline. Many exhibit spectacular rhythmic igneous layering (e.g. Ilimaussaq), all are composite having been formed by two or more pulses of magma, and were emplaced by cauldron subsidence or stoping. Several major intrusions are located at the intersection of fault and dyke zones.

Many of the alkaline complexes are associ-

Table 7.2   Chronology of main events during the Gardar period in SW Greenland (dates after Upton and Blundell, 1978)

Gardar subdivision	Main events	Isotopic age (Ma)
Late	Camptonitic dykes Agpaitic intrusions Saturated and undersaturated syenite intrusions, gabbros and granites	1170
Middle	Major NE-trending basic dyke swarms (several generations) trachytes and syeno-gabbros Some major intrusive centres	1250
Early	ESE- and local ENE-trending troctolitic dykes and early syeno-gabbro dykes Lamprophyric dykes (generally NE trending) Nepheline syenite and carbonatite intrusions Basic and trachytic lavas and sandstones Faulting (ESE-, ENE- and NS-trending sets), continued intermittently throughout Gardar time	1300

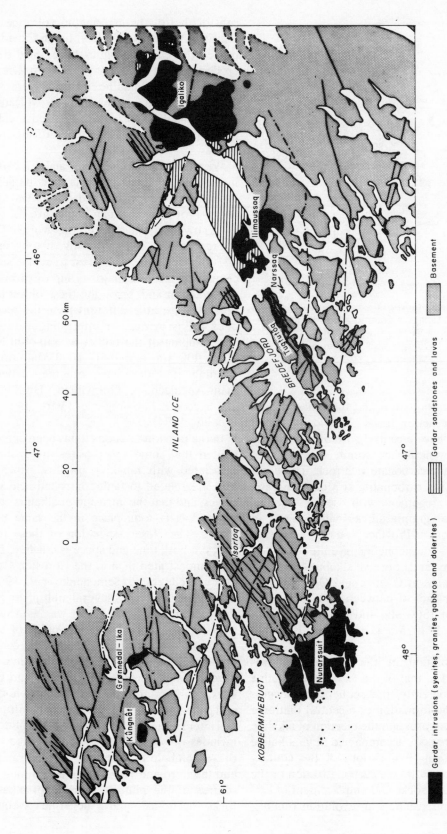

Fig. 7.6 Map of the Gardar Province of SW Greenland showing the main alkaline intrusions, dyke swarms and lavas and sandstones (after Watt; reproduced by permission of The Geological Survey of Greenland)

112

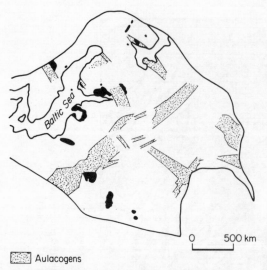

Aulacogens

Anorthosite, rapakivi granite and alkaline complexes

0   500 km

Fig. 7.7 Mid-Proterozoic aulacogens and anorthosite, rapakivi granite and alkaline complexes in the East European Craton (modified after Salop, 1977; see also Klevtsova, 1979 and Milanovsky, 1981)

ated with carbonatites and these likewise lie in or near major faults. Individual bodies commonly have a central core of carbonatite bounded by nepheline syenite and zones of pyroxene and carbonate-rich rock. There is a 800–1000 Ma carbonatite at Mountain Pass, California together with seven shonkinite–syenite–granite stocks with fenites; in the Gardar Province there is the Grønnedal–Ika nepheline syenite–carbonatite complex, and of twelve alkaline–carbonatite complexes in Quebec and Ontario eight have K/Ar ages of between 1005 and 1112 Ma, one of 1560 Ma, and three of between 1655 and 1740 Ma (Upton and Blundell, 1978).

The evolution of the carbonatites was closely tied to that of the associated alkaline rocks. Apparently, the sequence of events was first the formation of a central diatreme by explosive gases and the ejection of alkaline volcanic rocks, accompanied by alkaline metasomatism (fenitization) of the country rocks, followed by the carbonatization of the volcanic neck rocks and emplacement of carbonatite and syenite—an evolution reminis-

cent of that of the carbonatite volcanoes of the East African Rift Valley. In the Canadian bodies the early volcanics have since been eroded, but at Qagssiarssuk, South Greenland, minor carbonatite intrusions are still associated with early alkaline volcanics, amygdaloidal flows of carbonatized melilite rock, pyroclastic cones and tuffisite diatremes.

With the exception of the Palabora carbonite in South Africa (2060 Ma) and the Siilinjärvi carbonatite in Finland (2450 Ma), the 1650–1750 Ma Canadian bodies are the oldest known carbonatites in the world. Two interesting fact emerge: firstly, it took until 1750 Ma for the earth's crust to evolve sufficiently for the first major group of carbonatites to form and, secondly, these oldest carbonatites are little different from the recent ones in East Africa. An interesting feature is the grouping of the bodies at 350–400 Ma, 1000–900 Ma, and 1650–1800 Ma, corresponding to the age of the Caledonian–Appalachian, Grenvillian, Hudsonian and Svecokarelian orogenies (Vartiainen and Wooley, 1974).

In the previous sections it has been demonstrated that some anorthosites are spatially associated with rapakivi granites, which in turn are related to fault-controlled acid volcanics, and that the intrusion of alkaline complexes locally took place in the same fault blocks. The close association of these rock suites in both time and space is nowhere better demonstrated than in the Korosten Complex in the Ukraine (Semenenko *et al.*, 1960). The complex includes several multiphase plutons, the intrusion of which was associated with faulting, subsidence and uplift of separate blocks of the Shield. The plutons consist of three rock suites: the first, forming the border zones, includes labradorite anorthosites, gabbro–anorthosites, gabbro–norites and gabbro–monzonites; the second, intruded at a later stage in the upper part of the plutons, includes rapakivi granites, rapakivi-like biotite–amphibole granites and biotite granites; and the third and youngest suite, forming the centres of the plutons, consists of alkaline rocks such as quartz–aegerine syenites,

Whereas these rock suites formed at somewhat different times and often in different places in many parts of the North Atlantic Shield belt, they were telescoped together in the Ukraine to form this remarkable Korosten complex.

We thus arrive at a point in time, 1000 Ma ago, when the final stages took place of some 700–800 Ma of igneous activity in the North Atlantic anorogenic belt. In a way it was a culmination as it is significant that the last magma was chemically the most highly fractionated giving rise to, for example, the Ilimaussaq peralkaline intrusion with its 130+ minerals.

Mineralization

In the North Atlantic belt several distinctive types of mineralization were formed in association with the anorogenic volcanic and intrusive rocks.

Sn etc. in Granitic Rocks

There is uneconomic but significant mineralization of tin and/or related beryllium, tungsten, zinc and copper in all the five large rapakivi granite massifs in S Finland and adjoining USSR (Haapala, 1977). the mineralization only occurs in association with the youngest (c. 1600 Ma) intrusive phases of the massifs, and especially in their apical parts or at their contacts (particularly in leucocratic, topaz-bearing peraluminous granites with tin-rich protolithionite). The cassiterite and other ore minerals occur in disseminated grains, in pegmatite veins and pockets, in greisen and quartz veins and in contact skarns.

In the famous Black Hills pegmatites, associated with the 1620 Ma Harney Peak granite (non-rapakivi) stock in South Dakota, USA, there are economic concentrations of lithium minerals, beryl, mica, feldspar and tin.

In the St Francois terrain of Missouri, USA, tin mineralization occurs in pegmatite veins in late two-mica granites and this is associated with enrichments in W–U–Th–Nb and REE (Kisvarsanyi, 1980).

Ti, Fe, V in Anorthosites

The major anorthosite bodies commonly contain ilmenite–titaniferous magnetite–haematite deposits (Herz, 1976). These oxide minerals are disseminated through the plagioclase, often forming layered rocks or intrusive masses. Titaniferous magnetite is more prevalent in gabbroic anorthosite or noritic gabbro, and ilmenite–haematite in cores of anorthosite bodies. The gabbroic phases frequently contain more iron and titanium oxides than do the anorthosite phases. Iron-titanium deposits occur, for example, in anorthosites at:

Bergen, Egersund, Inner Sognefjord, South
  Norway
Lofoten, North Norway
St Urbain and Allard Lake, Quebec
Iron Mountain, Laramie Range, Wyoming
Duluth, Minnesota
Sanford Lake, Adirondacks, NY

The titaniferous magnetite in the gabbroic phases of the anorthosite intrusions in Canada is commonly enriched economically in vanadium (Rose, 1973).

Cu and U–Ag in Basalts

There are several distinctive mineral deposits in basic volcanics and tuffs extruded onto, or at the edge of, stable cratons in the period 1300–1000 Ma. We see examples here of the transfer of appreciable quantities of metals, presumably from the mantle, to the upper parts of the continental crust in this period (Watson, 1973).

The most prominent is copper which is concentrated in basalt flows in the Coppermine River Group (1100 Ma), NW Canada (Kindle, 1972), and in the Keweenawan basalts (1100 Ma) in Michigan, in 1250–1300 Ma basalts of the Seal Lake Group in Labrador, in 1000–1300 Ma basalts of the Volyn Group in the Russian plate (Salop, 1977, p. 255) and in the lavas of the slightly younger Bukoban System in East Africa, c. 800–1000 Ma). The ores consist of either native copper or chalcocite in amygdules within the tops of flows or in quartz–carbonate veins and breccias.

U, Th, Be, etc. in Alkaline Intrusions

Several 1300–1000 Ma alkaline complexes have a variety of element/mineral concentrations:

Ilimaussaq, South Greenland	U, Th, Nb, Be, Zr, Li, Rare Earths
Ivigtut, South Greenland	Cryolite, Pb, Ag
Blue Mountain, Ontario	Nepheline
Seal Lake, Labrador	Be, Nb, Th
Mountain Pass, California	Barytes, Rare Earths
Bancroft, Ontario	Nepheline, Corundum,
St Hillaire, Quebec	Uranium, Radium

One of the most interesting complexes is Ilimaussaq in which there is a progressive enrichment of the U, Th, Nb, Be, Zr, Li and rare earths, especially lanthanum and cerium. The radioactive and rare earth elements were concentrated in the residual fraction of the highly differentiated magma entering hydrothermal solutions in the last stage of solidification.

The Ivigtut granite (Rb/Sr age of 1248 Ma) pluton contains an impressive cryolite (Na_3AlF_6)–siderite (75–20%) deposit with marmatitic sphalerite and silver-rich galena.

An unusual body is the 800–1000 Ma carbonatite at Mountain Pass, California, which has remarkably large concentrations of barium and rare earths.

Summary and a Modern Analogue

So the picture emerges of a vast suite of igneous rocks (anorthosites and associated gabbros, rapakivi granites and other granitic rocks, acid and basic volcanics, basic dykes and alkaline rocks), many of which were closely related and which were intruded/extruded under anorogenic conditions within the broad time span of 1700–1000 Ma. They are characterized by a bimodal association of anorthosite and granite; the anorthosite is relatively sodic (in contrast to the calcic Archaean anorthosites), and has iron-titanium-vanadium mineralization; the granitic rocks are commonly peraluminous to peralkaline, contain two micas and fayalite, some have rapakivi textures, and they have associated tin mineralization. Many parts of the suite may be comagmatic and consanguineous. The erosion level along the belt varies considerably, in some places exposing anorthosites that formed at 35 km depth, and in others acid–basic volcanics and high level alkaline plutons. Some individual intrusions are related to local faults, but they can be best shown to be associated with graben or aulacogens in the East European Craton, the Gardar Province of South Greenland and the Keweenawan-Grenville Front belt in Canada. This anorogenic rock suite dominates mid-Proterozic crustal history, and this was the first time it developed in earth history. A precondition for its development was a stable continental crust and this was clearly absent in the Archaean. It took about 1200 Ma after the appearance (preservation) of the earliest rocks for this type of magmatism to generate. Such anorogenic mid-plate magmatism was not commonly developed in later earth history, but one excellent 'modern' analogue is the N–S belt of anorogenic Ordovician to late Jurassic intrusions in West Africa that is 200 km wide and 1600 km long, extending from the Hoggar in the Niger Republic southwards via the Air massif and the Jos Plateau to the margin of the Benue Trough in Nigeria (Husch and Moreau, 1982, for Palaeozoic; Bowden and Turner, 1974, for Mesozoic). The trend of this belt is on the extension of the African continental margin to the south, i.e. it is on an extension of the zone of rifting of Gondwanaland that developed into the Mesozoic opening of the South Atlantic (MacLeod et al., 1971). The Phanerozoic belt has not been as deeply eroded as its Proterozoic equivalent, but nevertheless remarkable similarities can be demonstrated. Typical is a bimodal association of peraluminous to peralkaline granites with two micas and fayalite, rapakivi types, and tin mineralization with Zn–U–Nb (Bowden and Kinnaird, 1978), and labradorite anorthosites with iron-titanium oxides, together with extrusive rhyolites and quartz porphyry dykes. This is a high-level anorogenic belt with ring complexes, extrusions, ring dykes, cone sheets and other evidence of cauldron subsidence. Barker et al. (1975) concluded that Pikes Peak batholith of Colorado, as now exposed, represents what would be found at 2–4 km below the present level of the Nigerian complexes. The St. Francois terrain of Missouri (Kisvarsanyi, 1980) has granitic ring complexes, rhyolites, dykes and cauldron subsidence structures and is more directly comparable in terms of crustal level with the Niger-Nigerian belt. We can conclude with the thought that there was a major but abortive attempt in the mid-Proterozoic to fragment the continental plates; the result was anorogenic magmatism. This has implications for Proterozic palaeomagnetism which will be discussed in Chapter 10.

Chapter 8

Mid–Late Proterozoic Basins, Dykes, Glaciations and Life Forms

The late Proterozoic is taken here to extend from 1000 Ma to the start of the Cambrian, about 600 Ma or 570 Ma. This period coincides with the Hadrynian Era in the Canadian time scale and the Epiproterozoic–Eocambrian of Salop (1977) but is not coincident with the late Precambrian subdivision used in Australia (the Adelaidean, 1400–600 Ma) (Brown *et al.*, 1968), or in the USA and Mexico where the late Proterozoic starts at 900 Ma (Harrison and Peterman, 1980). In the USSR the term Riphean is used for the period extending from 1650 ± 50 Ma to 680 ± 20 Ma, and Vendian for the 680 to 600 Ma period (Table 1.1).

There is some overlap of geological processes operating in the mid-late Proterozoic with those of the early Palaeozoic because the early stages of development of, for example, the Appalachian–Caledonian fold belt (continental rift, shelf and rise deposition, early sea-floor spreading) took place in the period 900–600 Ma. The deposition of the Torridonian, Moinian and early Dalradian sequences belonging to that fold belt will, consequently, be treated in Chapter 13.

This period is of special significance in the evolution of the continents for several reasons. A few rock groups, such as sedimentary copper deposits, appear for the first time, but more important is the fact that many rock groups, although they appeared in minor amounts in earlier periods, developed more abundantly in the late Proterozoic (e.g. red beds, tin and manganese deposits). The period is notable for the appearance of the eucaryotes and sexual cell structures and the rapid diversification of the first metaphytes and metazoans. During this time global glaciation occurred with the formation of widespread tillites. Himalayan-type orogenic belts reached a new peak of development and from this period our present pattern of cratons and plate tectonic regimes began to emerge (Windley, 1979).

Depositional Environment

Late Proterozoic sediments typically lie with a major unconformity with a 1000 Ma age on older sialic basement often on rifted continental margins notably in the belts surrounding the Proton-Atlantic Ocean—Marocanides, Avalonides, Cadomian, Iberides, Arvonian etc. (Choubert and Faure-Muret, 1980). Glacial tillites commonly occur in these sequences. Schermerhorn (1975) reviews the regular depositional environments at this (and succeeding Cambrian) time. A typical succession may be: shallow-water quartzites or sandstones, shallow platform carbonates, pre-flysch deep-water laminated fine-grained mudstones, flysch, molasse sandstones. Interpreted in plate tectonic terms this sequence is similar to that on the characteristic continental margins of Phanerozoic fold belts. The deposition of early terrigenous quartz sands is followed by the building of a carbonate bank on a continental shelf; with the onset of deformation the carbonate bank collapses and is covered by deep-water mud-stones starved of clastic material. Next a flysch sequence of greywacke turbidites is succeeded by lithic sandstones (molasse). The implication to be drawn from this particular type of sedimentary development is that the late Proterozoic sequences started to form at the trailing edge of stable continental margins which were converted by the onset of subduction, mostly by early Palaeozoic times, into tectonically active margins.

As Schermerhorn points out, individual basins have their peculiar characteristics; thus the Adelaide System lacks greywacke turbidites, flysch is absent in the Caledonian succession but present in the Appalachians, whilst the Brioverian of NW France has no carbonate stage. The platform deposits are usually thin compared with those in the mobile belts and do not extend for more than a few tens of kilometres from the margins of the late Precambrian belts. This distribution suggests that the shields stood above sea level at this time.

Let us now look at some mid–late Proterozoic sequences.

The Belt–Purcell Supergroup, western North America

The Belt–Purcell Supergroup (the term Belt is used in the USA and Purcell in Canada) extends from Idaho to British Columbia and is a 23 km thick sequence of remarkably homogeneous fine-grained clastics, dominantly siltstone and argillite with considerable interstitial carbonate; conglomerates and angular unconformities are rare. Radiometric data by Obradovich and Peterman (1968) on glauconite indicate that sedimentation took place intermittently from 1600 to 850 Ma (1450 to 850 Ma according to Harrison, 1972). Sedimentary rocks of Belt–Purcell age extend locally within the Cordilleran Structural Province to California and eastern Alaska (Wheeler and Gabrielse, 1972; Wheeler et al., 1974; Young, 1977).

Ross (1970) concluded that most Belt palaeoslopes were gentle, currents were slow, and the source terrain low with the result that thin marginal deposits covered large areas. Much of the fine-grained sediment was deposited on mud-flats with algal heads projecting above the water in a very shallow-water environment with an ill-defined shoreline, indicated by the lack of distinctive near-shore sediments and the presence of stromatolites and halite pseudomorphs. It is envisaged by Ross that the sequence was laid down in shallow salt water in a broad basin that was once essentially a large uninterrupted expanse of water and mud-flats. This environment required an extensive flat-lying stable platform of highly eroded basement rocks. From a consideration of its geometry and sedimention Harrison (1972) concluded that the Belt Basin was an epicratonic re-entrant of the sea to the west dominated tectonically by slow gentle downward warping.

In Canada, the Lower Purcell sequence consists of shallow-water siliceous clastics in the east and deeper water turbidites in the west, and the Upper Purcell of basaltic lavas and shallow water sandstones, shales and minor carbonates with stromatolites (Wheeler and Gabrielse, 1972; Wheeler et al., 1974).

Some of the Purcell rocks were affected by the East Kootenay Orogeny (maximum sillimanite grade) with a K/Ar thermal age of 750–850 Ma (corresponding to the low-grade Racklan orogeny in the northern Cordillera), possibly caused by the initiation of a short-lived easterly-dipping subduction zone (Fig. 8.1b).

The Windermere System of Canada was laid down unconformably on the deformed Belt–Purcell rocks. In Windermere time (800–600 Ma) there was an abrupt change to coarse, poorly bedded sediments suggesting that source areas of significant relief may have been caused by uplift by the preceding orogenies at the edge of the continent. The Windermere assemblage comprises in the lower part poorly sorted, rapidly deposited, quartzo-feldspathic grits, sandstones and shales, with thick tholeiitic volcanics (conglomerates may be glaciomarine tillites according to Aalto, 1971) and, in the upper part, pelites and distinctive carbonate units.

118

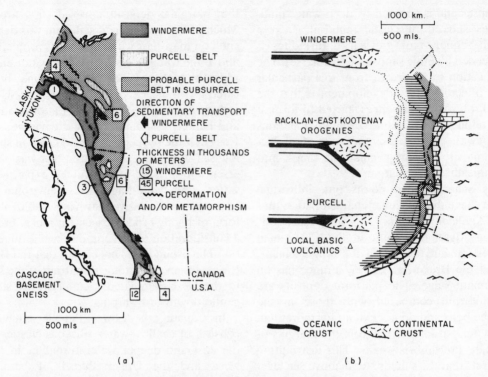

Fig. 8.1   (a) Map of the Belt–Purcell and Windermere groups in western North America. (b) A plate tectonic model to explain their occurrence according to Wheeler and Gabrielse, 1972 (reproduced by permission of The Geological Association of Canada)

Like those of the Purcell System, the sediments thicken appreciably and rapidly to the west (Wheeler *et al.*, 1974).

In plate tectonic terms Wheeler and Gabrielse (1972) proposed that the Purcell–Windermere 'Basin', with the similar succeeding Lower Palaeozoic strata, evolved between oceanic crust to the west and a North American craton to the east (Fig. 7.1b). Sedimentation took place during a miogeoclinal stage when westward prograding continental terrace wedges formed a continental shelf–slope–rise in which the sediments were derived consistently from the craton to the east. Features consistent with the continental terrace wedge are:

1. The constant polarity of sedimentary facies revealing a one-sided cratonal source.
2. A relatively uniform total thickness of stratigraphic column, presumably controlled by initial depth of water and maximum amount of downbuckling of oceanic crust caused by sedimentary loading.
3. The common occurrence of basic volcanic sills and flows in strata near the outer edge of the terrace—possibly the result of tension in an oceanward-tilted wedge.
4. Continued oceanward tilting of the sedimentary wedge as shown by unconformities that bevel towards the craton.

It is interesting to note that there are deposits of phosphorite in the Belt Series. Phosphate deposition typically occurs between latitudes 40°N and 40°S along the western sides of continents and is caused by the upwelling of cold polar currents as they move equatorwards along western continental margins (Cook and McElhinny, 1979). The occurrence of phosphorites in the Belt Series is corroborative evidence that these rocks were deposited on a westward-facing continental shelf.

Table 8.1   Stratigraphy of the Adelaide System showing the relative position of the two major late Precambrian tillites (after D. A. Brown *et al.*, 1968)

Cambrian	
Marinoan Interglacials:	The Pound Quartzite containing a medusoid–octocoral–annelid fauna Shales, siltstones, quartzites, dolomitic limestone with stromatolites Tillite boulder bed with sandstones. Upper glacials Siltstone, dolomite, muddy sandstone and local iron formation Silty shale with minor arkose
Sturtian	Tillite boulder bed with quartzites. Lower glacials
Torrensian	Argillite, dolomite, limestone, magnesite, quartzite, arkose Basal sandstone–conglomerate (1.95 km)
Willouran	Trachyte and minor andesite and rhyolite (600 m) Siltstone, slate, dolomite, quartz sandstone (3.6 km)

A refinement of this model was proposed by J. H. Stewart (1972) who concluded that the time of continental separation at the beginning of the Cordilleran geosyncline was marked by deposition of the unconformable Windermere Group on a miogeoclinal shelf along the trailing edge of the continent less than 850 Ma ago. The thick tholeiitic basalts near the bottom of the sedimentary sequence were a result of volcanic activity related to the thinning and rifting of the crust during early stages of separation. However, in an alternative model J. H. Stewart (1976) suggested that most of the Belt–Purcell sediments were deposited in epicratonic troughs and, consequently, that an ocean did not lie to their west.

The Adelaidean System, Australia

This system includes those beds deposited in the period 1400–600 Ma. The most well-known sequences occur in the Adelaide Basin, the Amadeus Basin and the Kimberley Basin; the first two are aulacogens (Rutland, 1973).

The Adelaide Basin   The sequence in this 700 km long basinal belt exceeds 15 km thickness, is divided into four time–rock series (Willouran, Torrensian, Sturtian and Marinoan) and is notable for its lack of turbi-

dites, its shallow-water marine clastics and carbonates, its few volcanics, its two tillite horizons which reach a maximum thickness of 2.1 km, and its famous Ediacara fauna. The main stratigraphy is shown in Table 8.1.

The Amadeus Basin   This 3.6 km thick, dominantly shallow-water sequence is situated in central Australia between high-grade granite gneiss complexes to the north and south (Arunta and Musgrave). The lower sedimentary quartzitic unit has been metamorphosed, part of the sequence has been recumbently folded, and there are prominent tillites associated with glacial erratics and striated blocks and correlated with the Sturtian glacials. The sequence is probably best known for the fact that a well-preserved, and biologically diverse, late Precambrian microflora occurs in the Bitter Springs Formation (Schopf, 1972). The major unconformity at the base of the overlying Cambrian, together with the deformation and metamorphism mentioned above, are taken to indicate a late Adelaidean 'orogeny'.

The Kimberley Basin   This basin contains the most complete Adelaidean sequence and comprises three groups: the lowest (550 m thick) is a shallow-water sequence of sandstone, subarkose with minor black shale and siltstone; the intermediate (1200 m) consists of siltstones, sandstones and subarkose with a prominent boulder bed (the Landrigan Tillite) containing polished and striated cobbles, pebbles and boulders; and the uppermost group (4000 m) commences with an arkose–siltstone–sandy limestone sequence, followed by a second tillite, which rests on a polished and striated pavement and succeeded by quartz-rich arenite, siltstone and shale.

Basic Dykes

The intrusion of swarms of dolerite (diabase) dykes and sills on several continents in the period 1000–600 Ma was a reflection of the continued widespread, relatively stable, tectonic conditions that prevailed at this time. Many of the dyke swarms may have formed in connection with early rifting along continental margins, like the early Mesozoic dykes bordering the present Atlantic ocean (May, 1971). So far these basic intrusions are best documented in northeast, east and southeast Africa and north Canada, but they are also noted in the Baltic Shield and England.

Surprisingly few of these dyke swarms have been well dated by the Rb/Sr isochron technique. We have to rely on K/Ar ages which will remain suspect until corroborated by more sophisticated methods, and the age of many dykes is guessed only by reference to dated older and/or younger formations.

The most abundant dykes occur in northeastern Africa. According to Vail (1970) the late Precambrian terrain bordering the Red Sea in eastern Egypt, Sinai peninsula, western Saudi Arabia and northeast Sudan, is one of the most intensely dyke-intruded areas of the world; Fig 8.2 shows that dykes occur throughout an area nearly 2000 km long and 500 km wide. In Sudan there were two periods of emplacement (K–Ar)—616 ± 18 Ma and 660 ± 19 Ma (Vail and Hughes, 1977). Some associated granitic bodies have Rb/Sr dates (quoted by Vail, 1970): Aswan Granite, Egypt (600 ± 20 Ma); Gattarian Granites, eastern desert, Egypt (500 Ma); Arabia (500–600 Ma). Some dykes are younger and some older than the granites,

some have different strike directions, and there are dykes of acid and basic composition. It is not clear whether some of the dykes are of Palaeozoic age, but they do predate the Mesozoic cover. The dykes in Sudan belong to a tholeiitic hornblende micrograbbro–microdioritic suite (Vail and Hughes, 1977).

In Zimbabwe some Umkondo dolerites have K/Ar ages of 650–1150 Ma and Waterberg dolerites likewise of 600–1130 Ma (Jones and McElhinny, 1966). This largely late Precambrian age is provisionally confirmed by the most recent palaeomagnetic results (J. D. A. Piper, 1973b).

The weakly metamorphosed and folded Bukoban System in Tanzania formed between 800 and 1000 Ma (upper basaltic lavas at 800 Ma and lower sandstones at 900–1000 Ma: J. D. A. Piper, 1973b). There are considerable numbers of dolerite dykes in this part of East Africa, some of which are associated with the Bukoban rocks whilst others are at a distance from them. Some swarms have impressive dimensions up to 300 km long and 200 km wide. With their associated basalts, many of the dykes constitute a Bukoban igneous subprovince. Although preliminary K/Ar dates on some dykes have yielded ages of c. 2500, 1900 and 900 Ma (Snelling and Hepworth, quoted in Vail, 1970), their significance is unknown or uncertain and so the dykes remain provisionally grouped with the Bukoban rocks.

About 675 Ma ago (Fahrig et al., 1971) a spectacular swarm of diabase dykes was intruded in northern Canada (Fahrig, Irving and Jackson, 1973). These 'Franklin diabases' crop out in a huge zone extending from northern Ungava Bay to Baffin Island where they are particularly concentrated (Fig. 8.3) Palaeomagnetic data suggest that the dykes were emplaced at low latitudes—a conclusion corroborated by the fact that late Proterozoic sediments of similar age contain gypsum, anhydrite and stromatolites which are indicative of a warm depositional environment. Also in the Bear Province diabase sheets and sills intruded into the Shaler Group on Victoria Island and the overlying Natkusiak lavas have a 'best estimated' K/Ar age of 650 Ma,

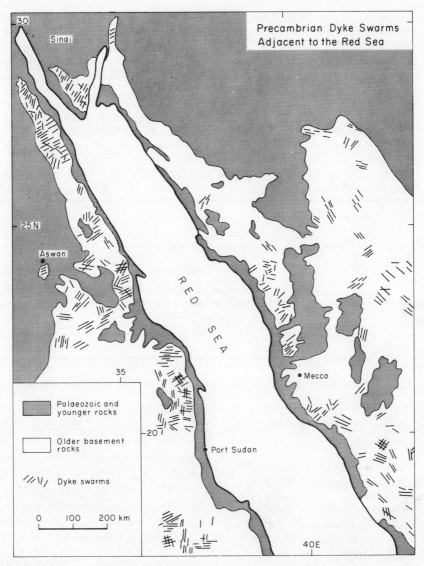

Fig. 8.2   Late Precambrian basic dyke swarms in the Red Sea region (after Vail, 1970; reproduced by permission of J. R. Vail)

and two groups of dykes near the Great Bear Lake have dates of 614–619 Ma and 630 Ma (Fraser *et al*., 1972). In the Canadian Slave Province shallow-dipping gabbro sheets that cut Goulburn and older rocks have radiometric ages of 650–700 Ma and in the Grand Canyon of Arizona the Cardenas lavas have a Rb/Sr age of 1090 ± 70 Ma.

Although their precise age is not known, there are many basic dykes of presumed late Precambrian age cutting mid-Precambrian rocks of southern Norway. According to

available maps they postdate 1000 Ma granitic rocks and the somewhat earlier anorthosite-norite suite of the Egersund Province.

The Precambrian inliers in central England, Wales and southeast Ireland contain undated small dyke swarms that most probably are equivalent in age. They occur in the Rosslare Complex in County Wexford, the Uriconian Group of Shropshire, the Stanner–Hanter Complex of Radnorshire, the Johnstone Series in Pembrokeshire and the

122

HADRYNIAN

Natkusiak basalts
(650 Ma)

Nauyat basalts

Symbols

〜 Plateau basalts

Gabbro sill or region of
closely spaced dolerite
sills or sheets

\!\ Mafic dykes

'Felsite porphyries'
Alkaline intrusion

Coronation sills
(650 Ma)

Franklin dykes
(675 Ma)

(570 Ma)

(640 Ma)

Grenville dykes
(570 Ma)

Fig. 8.3   Hadrynian age (1000–600 Ma) dykes, sills and volcanic rocks in
the Canadian Shield (from Baragar, 1977, *Geol. Ass. Canada*)

Malvern Complex and the Warren House Series of Worcestershire.

Tillites and Global Glaciation

The most extensive period of glaciation(s) in earth history took place in the period between 1000 Ma and 600 Ma and the effects were more widespread than those of the early Proterozoic, Permo-Carboniferous and Quaternary glaciations.

The main rock resulting from glacial action is *tillite* which is a consolidated till, whilst the term *tilloid* is used for a tillite-like rock of doubtful origin. The general term *diamictite* includes boulder beds, clays and sands, pebbly sandstones and mudstones, tilloids and tillites, and corresponds to *mixtite*. According to Chumakov (1981) only 25% of Precambrian 'tillites' can be accepted as glacial rocks.

Distribution

Most continents contain some evidence of late Precambrian glaciation, the main exception being Antarctica (for details of localities see Harland, 1964; Harland and Herod, 1975; Frakes, 1979; Chumakov, 1981; Hambrey and Harland, 1981). Fig. 8.4 illustrates the distribution of late Proterozoic tillites, the main type of glacial deposit.

The principal occurrences are as follows:

Europe: Scotland, Ireland, Norway and Spitsbergen, Sweden, Normandy (France), Czechoslovakia.

North America: many localities in the Western Cordillera from California to the Yukon and in the central Appalachians of the USA extending to Newfoundland in Canada; East Greenland.

South America: Brazil.

Australia: many localities extending in a broad arc from the southeast across central to northwest Australia. Two glaciations—Marinoan or Egan and Sturtian.

Africa: Equatorial, southern and southwest Africa are the main areas

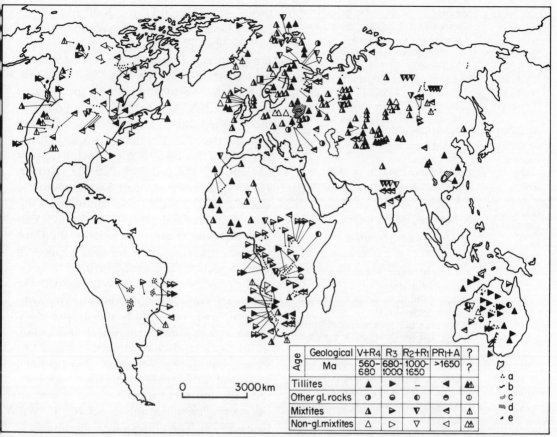

Fig. 8.4   A map of all Precambrian tillites, mixtites and other supposed glacial rocks. V: Vendian (56–650 Ma), R$_4$: Kudashian (650–680 Ma), R$_3$: Upper Riphean (680–1000 Ma), R$_2$ + R$_1$: Middle (1000–1350 Ma) and Lower (1350–1650 Ma) Riphean; PR$_1$ + A: Lower Proterozoic (1650–2600 Ma) and Archaean (>2600 Ma), ?: unknown age. a, b, c, d: outcrops or areas of distribution; e: glaciated bedrock (modified from Chemakov, 1981)

(Congo–Katanga–Angola, Zambia, Zaire, Southwest Africa, South Africa).

USSR: Belorussia, the Urals, Khazakhstan, the Tien-Shan, central and northeast Siberia, Lake Baikal, Bashkirian Highlands.

China: little detailed recent knowledge is available, but late Precambrian glacial deposits are quite common, especially in central and southern China (Sinian tillites), Sinkiang, Turkestan, west Shansi, Hunan and Anhui.

Age

It is not easy to date accurately a period of Precambrian glaciation; however, several tillites have been radiometrically dated and some have been given an approximate age within a small time range on the basis of stratigraphic relations with beds above and below which have been isotopically dated. Steiner and Grillmair (1973) suggested that there were three late Precambrian glacial episodes with mean ages of 616 ± 30 Ma (Eocambrian), 777 ± 40 Ma (Infracambrian I), c. 950 ± 50 Ma (Infracambrian II). The following occurrences are assigned to these periods:

1. Eocambrian

Ma	
650	Egan–Marinoan glaciation, Australia
668 ± 23 (660–680)	Varanger Ice Age, northern Norway
600	Mortensnes (≡ Varangian) tillite, Finnmark, Norway

600–640?	Sveanor tillite, Spitsbergen, Norway
600	Upper tillite. Eleonore Bay Formation, East Greenland
560–630	Granville tillite, Upper Brioverian, Normandy
570±10–675±25	Vendian period (tillites of Europe and Asiatic USSR)
570–600	Conception Group with tillite, Newfoundland, Canada
620–650	West Saharan tillite, Algeria
653±70 (max.)	Nama–Damara tillite, SW Africa
600+	Lavras tillite, Brazil
570	Cambrian–Eocambrian glaciation, China
600 (or 660–720)	Nantou glaciation, China
650	Wushsingsham glaciation, China

2. Infracambrian I

740–750 ±40	Moonlight Valley–Sturtian glaciation, Australia
715	Rybachy tillite of Kola Peninsula, European USSR
810	Volhynian tillite, Russian platform
747–810	Chingasan tillite, Siberia, USSR
750	Hsiho glaciation, China
750–850	Diamictites, West American Cordillera (California to the Yukon)
800±50	Glaciogen Toby conglomerate, Windermere System, British Columbia
<820	Southern Appalachian tillites, USA
750±50	Petit Conglomérat, Zaire
719±28	Bushmannslippe glacial sediments, SW Africa

3. Infracambrian II

950±50	Grand Conglomérat, Zaire, lower Congo and Katanga
950±50	Lower tillite of Tien-Shan near base of Late Riphean, USSR
950	Huishan glaciation, China

Tillites of probable late Precambrian age are found in Ghana, Togo, Dahomey, and the Anti-Atlas of Morocco (J. D. A. Piper, 1973a) and Sierra Leone and Guinea (Harland, 1964).

Stratigraphy

In Norway, Australia, west central and southwest Africa, East Greenland and the USSR there are two prominent tillite horizons separated by interglacial sediments. In Australia there is one tillite formation in the Amadeus Basin but two in the Adelaide Geosyncline (lower Sturtian, Upper Mari-noan—see Table 8.1), the Kimberley Basin (lower Landrigan and upper Walsh or Egan), and the Sturt Platform, equivalent to those of the Kimberley Basin (Brown et al., 1968). In west central Africa the lower and upper tilloids of the West Congo System are equated with the Grand and Petit Conglomérats (tillites) in the Katanga Supergroup of Zaire and with the two tillites in the Gariep and Damaran belts of southwest Africa which have ages of 910–870 Ma and 700–720 Ma (Kröner, 1976b). There were at least 5 late Proterozoic glacial events in the African continent (Hambrey and Harland, 1981). In central east Greenland there are two tillites in the Cape Oswald Formation at the top of the Eleonore Bay Formation (Harland, 1964b).

The principal sediments occurring with the tillites are sandstones (in places red with ripple drift and cross-stratification), quartzites, shales and dolomites (locally stromatolitic), whilst the interglacial sediments include siltstones, mudstones, dolomites, and proximal and distal turbidites in Finnmark. Evaporites, red beds and iron formations also characterize many glaciogenic sequences (G. E. Williams, 1975). The tillites vary considerably in thickness from, for example, 7 m (lower tillite of Finnmark) to 2.1 km (Sturtian tillite in Adelaide Geosyncline). According to Spencer (1975) the North Atlantic Varangian tillites range in thickness from 2 m to 1 km and they mostly lie from 50 m to 800 m beneath fossiliferous Lower Cambrian beds; the ice age probably lasted about 10–30 Ma (660–680 Ma). (See also Hambrey, 1983.)

Dunn et al. (1971) suggested that the late Precambrian tillites might eventually be adopted as stratigraphic time markers, but Crawford and Daily (1971) and Kröner (1976b) disputed this by emphasizing the non-synchroneity of the glaciations.

The Glacial Environment

The most sophisticated model explaining the complicated development of a late Proterozoic glaciation is by Spencer (1971) for the Port Askaig (Scotland) tillites.

Spencer showed that the formation of the

Port Askaig tillite involved seventeen successive ice advances and retreats (meltings). 'Many are recorded by the cycle: base (marine?) sediments deposited during a rise of sea-level, glacial mixtite, sub-aerial permafrost conditions (sandstone wedges), beach conditions (recording a transgression), marine sediments etc. (top).' (Fig. 8.5 summarizes the ideal glacial advance–retreat cycle.) The most useful time indicators in the sequence are the sandstone wedges which have a polygonal cross-section, interpreted as replacements of ice-wedge polygons formed under ice-free permafrost conditions, and which register as many as 27 periglacial periods. The tillite sequence, up to 750 m thick, contains granite and sedimentary fragments (the largest of which has the enormous size of 320 × 64 × 45 m) occurring within 47 mixtite beds ranging in thickness from 50–65 cm and separated by sedimentary interbeds from a few centimetres to 200 m in thickness. As Spencer pointed out, the tillite is considerably thicker than analogous Pleistocene sequences and 'represents a glacial period of comparable, or even greater magnitude than that of the Pleistocene Ice Age'. If this is true, think of the significance of the Sturtian tillite sequence in Australia which reaches 2.1 km in thickness.

Palaeomagnetism and Palaeolatitudes

Because the geomagnetic and rotational poles are closely associated, geomagnetic latitudes should be equivalent to palaeogeographical latitudes and therefore palaeomagnetic data may be used to indicate the position of continental masses with respect to former geographical poles and equator (Runcorn, 1961). The interesting result of much recent palaeomagnetic work on rocks associated with Eocambrian tillites is that they were formed near the palaeo-equator of the time; this contrasts with the Permo-Carboniferous and Pleistocene glaciations which were near-polar.

Harland (1964) reviewed his data with Bidgood and concluded that tillite horizons in southern Norway and eastern Greenland were deposited in near-equatorial latitudes. Likewise J. D. A. Piper (1973a) found that Africa lay in low latitudes from before 1000 Ma until post-lower Cambrian times and Tarling (1974) found that the Scottish tillites formed at a latitude of less than 10°S. Irving and Park (1972) produced an apparent polar wander curve (APWC) which indicated that the whole of North America was situated less than 20° from the palaeo-equator in the period 600–800 Ma and the APWC of McWilliams and McElhinny (1980) showed that Adelaidean Australian tillites formed in low palaeolatitudes.

As Harland and Rudwick (1964) pointed out, our usual concept of an ice age is so moulded by what we know of the Pleistocene glaciation that we initially find it difficult to envisage ice in the tropics; but since the late Precambrian ice was clearly far more extensive than the Pleistocene ice, we have to imagine a drastically different situation. Land ice was probably situated at high latitudes at the start and end of the glacial epoch, and extended to middle latitudes during the coldest periods; it is, however, unnecessary to postulate land ice at low latitudes as 'floating ice and drifting icebergs could have carried material far beyond the boundaries of the continental ice sheets, depositing glacial sediments of similar composition over wide areas of the sea floor. Such deposition would account for the existence of tillites in the Infra-Cambrian tropics' (Harland and Rudwick, 1964).

However, the above conclusion of an extensive synchronous worldwide glaciation essentially at low latitudes was disputed by Crawford and Daily (1971) and McElhinny *et al.* (1974). Their palaeomagnetic results indicated that very large polar shifts took place across almost all continents during the late Precambrian, from which they concluded that the poles migrated rapidly over different parts of the globe leaving records of glaciated regions as they passed over the various continents and that these glaciations took place at high latitudes.

G. E. Williams (1975) pointed out that the glaciogenic sequences contain substantial

126

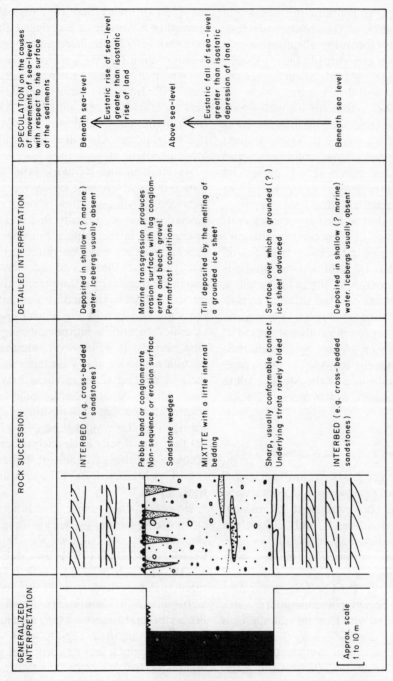

Fig. 8.5 The hypothetical glacial advance–retreat cycle characteristic of the Port Askaig tillite (after Spencer, 1971; reproduced by permission of The Geological Society of London)

evidence for a marked seasonal inequability of climate. To account for this he postulated a considerably increased obliquity of the ecliptic in the late Precambrian, the effect of which would be to weaken climatic zonation and so allow ice sheets and permafrost to form in low and middle latitudes. In his 1981 paper Williams calculated that in Australian 680 Ma old glacial varves conspicuous cycles reveal strong climatic periods of about 11, 22, 145 and 290 yr and a weaker period of near 90 yr. The 11, 22 and 90 yr cycles equate with sunspot periods, and the longer cycles with solar and climatic periods as indicated by tree-ring studies.

Mineralization

The following types of mineralization are found in sediments deposited during the late Proterozoic.

Copper in Sediments

There are anomalously high concentrations of copper in late Proterozoic sediments in many parts of the world. They are important from our point of view because, as Watson (1973) suggested, they represent the first major sedimentary accumulations in earth history. Their formation was dependent on a supply of copper eroded from earlier rocks, most probably basic volcanics. The host sedimentary sequences were laid down in aulacogens and along rifted continental margins.

The Katanga System in Zambia contains well-known economic sedimentary copper deposits. The main ore formation is a copper sulphide-enriched shale up to 70 m thick within the lower Roan Group which rests unconformably on a granite-schist basement (Mendelsohn, 1961). Sedimentary structures and lithologies indicate that near-shore shallow-water sedimentation gave rise to sandy pebble beds and carbonate deposits, largely of algal reef type.

The red beds of the Belt Basin in the northwestern USA contain very high amounts of copper (Morton *et al*., 1974). At least 100 ppm values occur in almost all formations throughout most of the thousands of square kilometres of the Belt terrain but there are no major economic deposits. The primary copper-bearing fluids may have been derived by exhalative activity associated with an aulacogenic structure, which apparently underlies the region (Kanasewich, 1968).

According to Watson (1973) the copper-bearing sediments of the Katanga System fringe the southern border of the large late Proterozoic craton on which the copper-bearing Bukoban volcanics were extruded about 800 Ma, and the cupriferous Belt sediments lie on the western edge of the Superior craton on the south side of which the copper-bearing Keweenawan lavas were extruded about 1000–1200 Ma. As the sediments were most likely derived by erosion of the cratons themselves, it is a reasonable assumption that the copper was supplied by the recently erupted lavas (see Chapter 7).

Further examples of mid–late Proterozoic sedimentary stratiform copper deposits are listed by Sawkins (1976a,b):

Keweenawan Trough, Lake Superior
Coppermine River, NWT, Canada
Seal Lake, Labrador
Southern Appalachians
Adelaide System (Rowlands *et al*., 1978)
Damara Belt, SW Africa

Manganese in Sediments

The first major manganese deposits formed in early Proterozoic basins where they are particularly associated with carbonate rocks. Conditions favourable for Mn deposition returned in late Proterozoic times when thick sedimentary sequences were laid down on, or at the border of, the broad cratons that had stabilized by this time.

Important concentrations occur in India and South West Africa. The Sausar Group was deformed and metamorphosed in the Sausar orogenic cycle that ended 869–996 Ma ago. The manganese ore belt lies in the States of Madhya Pradesh and Maharashtra in central India (Roy, 1966). The term 'gon-dite' is locally used for the spessartite–quartz ore derived from non-calcareous argillaceous and arenaceous manganiferous sediments originally in an oxidizing environment. The main ore horizons occur in the Mansar Formation, consisting largely of muscovite schists, whilst the remainder of the Sausar Group contains thick dolomites and marbles together with quartzites and quartz schists.

At Otjosundu in South West Africa the Damaran System contains manganese ore deposits in three major horizons, at the contacts of quartzite, iron formation, marble and biotite schists (Roper, 1956).

Tungsten in Limestones

There is a well-defined type of stratabound W(Mo)–Sb–Hg association that typically occurs in limestones (Maucher, 1968). In Argentina there are 580–620 Ma scheelite deposits in regionally metamorphosed limestones (marbles) interbedded with biotite schists and amphibolites (meta-lavas) (de Brodtkorp and Brodtkorp, 1977). The tungsten was derived from submarine volcanics and deposited in the marine carbonates.

Another occurrence of this metallogenic association is in Rogaland, SW Norway where stratabound wolfram–molybdenite mineralization occurs in 1478 ± 78 Ma graphite-bearing amphibolites in gneisses (Urban, 1971).

The Development of Stromatolites

The occurrence of the first stromatolites in Archaean greenstone belts was reported in Chapter 3, but these were only locally developed. The organisms responsible for these structures reached an advanced stage of development in Proterozoic times, being particularly abundant in the period 2250–600 Ma. Stromatolites are the commonest fossil in the Precambrian and are used as an intercontinental Proterozoic zonal fossil in the USSR (Semikhatov, 1976) and Australia (Preiss, 1977).

Stromatolites are layered stratiform, conical or columnar, biogenic sedimentary structures formed by sediment-binding algae.

Studies on modern stromatolites show that they grow in the intertidal and shallow subtidal zones where they are associated with cross-bedded sediments deposited in extensive tidal flats. Characteristically they occur in cherts, dolomites and limestones which are often associated with banded iron formations in early–mid-Precambrian environments.

Cloud (1968a) suggested that the binding of the micro-organisms in the stromatolites took place within originally gelatinous silica and mucilaginously entrapped calcium carbonate which provided a means by which the organisms could remain anchored below an ultra-violet shielding layer of sediment. The stromatolite-building blue–green algae are more resistant to DNA-damaging ultra-violet light than bacteria and other organized cells. This resistance, coupled with the shielding effect of the trapped sediments, was probably the reason why the blue–green algae adapted so well to the intertidal zones in the Precambrian; here they lived for the remainder of geological time, the effect of the ultra-violet radiation having been removed by the build-up of an ozone screen in late Proterozoic time.

The record that emerges indicates that stromatolite environments have scarcely changed during the last 1000–2250 Ma. They are commonly associated with interstitial breccias and ooliths, suggesting turbulent water, with contraction-cracked desiccated sediments and truncated, flat-topped ripple marks, implying exposure to the atmosphere (Cloud, 1968b), and with flat pebble conglomerates suggestive of a palaeo-beach zone (Vidal, 1972). According to Cloud (1968b) the maximum height (or amplitude) reached by many stromatolites at their maturity is a reflection of the tidal range. Compared with the modern examples at Shark Bay, Western Australia, which reach a maximum amplitude of 0.7 m, he pointed out that Precambrian stromatolites are often considerably larger, ranging from about 2.5 m to 6 m in amplitude; some of the largest occur in the late Proterozoic Belt Series in Montana, and the Otavi Series, South West Africa, and Cloud (1968b) inferred from this that the Precambrian tidal range was much greater in the past and hence that the lunar orbit was closer to earth; however, Walter (1970) challenged this as an incorrect extrapolation. He pointed out that early–middle Cambrian stromatolites near Lake Baikal are up to 15 m high and that Devonian stromatolites in western Australia grew in water as deep as 45 m, and he therefore suggested that many fossil stromatolites may have formed subtidally. This was supported by Awramik (1971) who concluded, from the lack of evidence indicating a periodically-exposed intertidal-to-supratidal environment, as well as the enormous size of some forms referred to above, that most columnar stromatolites grew in the subtidal zone.

An assessment of the number, widths and groupings of laminations in modern stromatolites compared with those in fossil forms provides an interesting means of using stromatolites as palaeontological clocks. Pannella (1972) demonstrated that the growth patterns may be of daily, tidal, monthly, seasonal and annual periodicity; he concluded that, because some 2200 Ma stromatolites in the Great Slave Supergroup have tidal bands, the earth–moon system must have been in existence since the early Proterozoic, although there is a tendency for the number of days to have decreased per tidal band.

Biostratigraphy of the Proterozoic

It was first realized by Russian biostratigraphers that stromatolites underwent sufficient evolutionary changes for them to be used as zone fossils, enabling the Riphean period (1650–680 Ma) to be subdivided into four major units; furthermore, the three main members of the late Riphean (1050–680 Ma) in the USSR are characterized by specific types of stromatolites that can also be recognized in Spitzbergen. The discovery that distinctive stromatolite assemblages are recognizable in their correct stratigraphic position in widely separated parts of the USSR was a remarkable breakthrough, providing a means of correlating late Proterozoic formations on an intercontinental basis.

Cloud and Semikhatov (1969) showed that many columnar stromatolites have intercontinental distributions (Karelia, Animikie Group, Labrador Geosyncline, Witwatersrand and Transvaal Systems) and occur over roughly the same stratigraphic ranges as defined in the USSR. The stromatolites contain three main types of microfossils: sphaeroidal unicells, cylindrical sheaths and cellular filaments, each of which increased in diameter and size range during the Proterozoic. Their taxonomic diversity increased especially after 1400 Ma (Schopf, 1977); thus Schopf supported the concept of stromatolite-based biostratigraphy.

The stratigraphic distribution of stromatolites in Proterozoic sequences in Australia is the most comparable to that in the USSR (Walter, 1972; Semikhatov, 1976; Preiss, 1977). They are distributed throughout eight different basins in sediments ranging from 2250–570 Ma. In particular, three groups of stromatolites have a distinctive distribution, occurring in early Proterozoic, late Riphean and Vendian sediments (Walter, 1972a).

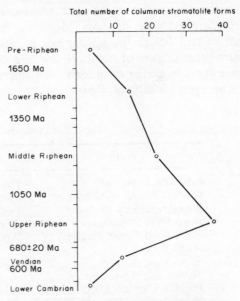

Fig. 8.6 Graph showing the diversity curve for Proterozoic columnar stromatolites (after S. M. Awramik, 1971, *Science*, **174**, 825–827; copyright 1971 by the American Association for the Advancement of Science)

The stromatolites were most prolific and reached their peak of diversity in the late Riphean, after which they declined appreciably in the Vendian and Palaeozoic. Awramik (1971) and others ascribed this rapid decline to the appearance of the Metozoa, similar to the Ediacara and Nama faunas in the Vendian; it is likely that these early animals not only ate the algae, but also destroyed the stromatolites by burrowing into them and inhibiting their growth by feeding on bottom deposits. Fig. 8.6 illustrates the sharp drop in the diversity of stromatolites in the Upper Riphean period just before the start of the Cambrian.

The First Metazoa

Metazoa are multicellular animals that require an oxygenous environment for their growth. They reached their peak of development in Phanerozoic time when many evolved protective shells and hard skeletons. During the last decade impressions of many soft-bodied animals have been found in late Precambrian rocks on several continents, providing evidence of early Metazoan life that existed before the appearance of hard parts; the origins of multicellular life lie somewhere in earlier Proterozoic sequences.

The type–locality of these late Precambrian metazoans is at Ediacara in the Flinders Range, 320 km north of Adelaide in south Australia. The fossiliferous beds lie in the current-bedded and ripple-marked Pound Quartzite, or Sandstone, the uppermost formation of the Marinoan Series of the Adelaide Geosyncline, 500 m below the unconformable basal Cambrian. The Pound Quartzite contains similar fauna at several localities up to 110 km from Ediacara. The age range of the Ediacara fauna is probably 590–700 Ma (Glaessner, 1971).

The fauna includes, in particular, jelly fish (Medusoids), worms (Annelids), sponges, soft corals related to the living sea pens, and several types of creatures unlike any known organisms. Glaessner (1971) lists the following assemblage:

Coelenterata (67% of specimens):
 Medusoids (3 species)
 Hydrozoa (3 species in 3 genera)
 Conulata (1 species)
 Scyphozoa (2 species in 2 genera)
 Anthozoa (Pennatulacea—4 species in 3 genera)
Annelida (25% of specimens):
 Polychaeta (5 species in 2 genera)
Anthropoda (5% of specimens):
 Trilobitomorpha (1 species)
 Crustacea (1 species)
Plus Tribrachidium of unknown affinities.

Table 8.2 Occurrence of the Ediacara fauna

Location	Rock unit	Age (Ma)
Ediacara, S Australia	Pound Quartzite	
Flinders Range, S Australia	Pound Quartzite	
Punkerri Hills, S Australia	Punkerri Sandstone, Officer Basin	
Deep Well, N Australia	Arumbera Sandstone	
Charnwood Forest, England	Woodhouse Beds	>684
South Wales		
South West Africa	Kuibis Quartzite, Nama Series	550–700
Torneträsk, N Sweden		c. 600
Podolia, Ukraine, USSR	Bernashov Sandstone	590
Yarensk, NE of Moscow, USSR	Gdov laminarites	c. 590
Olënek, N Siberia, USSR	Khatyspyt Formation	550–675
Rybatschii Peninsula, USSR		670–900
Central and E Russian Platform		650–675
SE Newfoundland, Canada	Conception Bay Group	620
N Carolina, USA	Carolina Slate Belt	620±20

Fossils similar to the Ediacara fauna occur in several parts of the world in rocks about 600–700 Ma old (Cloud, 1976a; Table 8.2):

The Appearance of Sex

One of the 'most consequential innovations to have occurred during the course of biological evolution' was the ability of microorganisms to change their mode of reproduction from asexual mitosis (simple cell division) to sexual meiosis (splitting of a body cell into two germ cells followed by their union, or fertilization). Schopf (1972) records evidence of this adaptation in fossil microflora in 900 Ma black carbonaceous cherts of the Bitter Springs Formation in the Amadeus Basin, central Australia. The population consists of 50 species including bacteria, eucaryotic algae and probably fungi, but it is dominated by blue–green algae very similar to extant forms. Schopf made the imaginative analogy that through much of geological time the blue–green algae have suffered from the 'Volkswagen syndrome—little or no evolution of external form concealing marked changes of internal machinery'. The critical evidence for sexual reproduction by this time lies in the fact that some eucaryotic algae have unicells in varying stages of mitotic cell division, whilst others have a tetrahedral arrangement of four spore-like cells, similar to the spore tetrads of living plants formed by sexual meiotic division. Thus by 900 Ma primitive micro-organisms had apparently evolved advanced techniques of sexual cell reproduction; in this respect the Bitter Springs microflora may provide an important benchmark in early biological evolution.

It is said that 'the rise of the eucaryotic cell from its procaryotic ancestors was the single greatest quantum step in evolutionary history'. This step involved the development of a sheath around the nucleus enabling cell division to take place and so for the DNA code to be passed on to the daughter organisms. It is usually assumed that the earliest evidence of possible eukaryotic remains is found in the 1300 Ma Beck Springs dolomite in southern California and that the oldest assured

eukaryotic organisms are in the Bitter Springs Formation. These occurrences give a minimum age of the evolution of the eucaryotic form of life. However this age has been challenged by Knoll and Barghoorn (1975) who found from experiments that partial degradation of certain modern algae simulated exactly the entire range of morphological variations including tetrahedral arrangements said to occur in the Bitter Springs Formation. They interpret the latter as just blebs of degraded protoplasm within undecomposed sheaths. If this conclusion is correct, it follows, firstly, that the eucaryotic cells may not have developed until the very end of the Precambrian, near the time of appearance of the Ediacara fauna and, secondly, that the tetrahedral structures referred to above are no more than pseudotetrahedral arrangements, implying that sexual reproduction evolved at a later date.

The Precambrian–Cambrian Boundary

Many estimates have been made of the age of the end of the Precambrian—see Chapter 1; we are concerned with the evolutionary changes that took place across this important time boundary.

Sedimentary structures suggest widespread shallow water conditions and the presence of glauconite indicates a marine environment. It has been suggested that the Precambrian sea water was too acid to allow growth of calcareous animal hard parts, too alkaline to allow sufficient concentration of Ca for precipitation of carbonate, and to have had too high a Mg/Ca ratio to allow precipitation of calcium carbonate.

Harland and Rudwick (1964) and Harland (1974) suggested that warming of the seas and melting of the ice after the last major period of Eocambrian glaciation, about 600 Ma ago, gave rise to flooding of peneplaned land and variable salinities with mixing melt waters, and this new environment provided ideal conditions for rapid biological evolution. The sudden increase in marine tides probably favoured protection from wave action or intertidal desiccation.

One of the main factors responsible for the evolutionary explosion was probably the increase in oxygen content in the atmosphere to 6.2% of the present level (Cloud, 1976a).

Although there are some unconformities at the base of the Cambrian and although there was locally tectonic activity at this time (e.g. the Katangan orogeny and the Cadoman phase of the Assyntian (Baikalian) orogeny at 650–680 Ma) the large number of successions that continue across the boundary without marked diastrophism suggests that the continental crust was not particularly mobile at this time (Harland, 1974). Thus there seems to have been little tectonic control, in the broadest sense, of biological evolution.

About 600 Ma ago the Metazoa evolved hard skeletons and shells (Stanley, 1976) in a process that concerned the formation of collagen, the main structural protein in Metazoan tissues. Its synthesis required molecular oxygen. Towe (1970) suggested that the organisms that were first capable of using oxygen for its synthesis were the most primitive (those that had the fewest organs, the least demanding muscles and epidermal respiration) and Cloud (1976a) that they obtained their oxygen solely by diffusion through their soft walls. These concepts are consistent with the fact that the earliest Ediacara Metazoans lacked hard parts, but presumably did contain collagenous tissue making their soft bodies more easily preservable. Thus the beginning of the Phanerozoic defines not the appearance of life but the rapid development of fossils with hard parts.

There is a remarkable end to this Precambrian story. In a few places late Precambrian organisms concentrated to such an extent that they led to reserves of indigenous gas and oil (Murray et al., 1980). Substantial gas reserves occur in a 5.5 m thick dolomite of the Pentatataka Formation just SE of Alice Springs in Australia, and there are light commercial oil fields in Lower Cambrian and Upper Proterozoic beds (Riphean–Vendian) in the Irkutsk basin of the Siberian platform, together with economic gas reserves, two-thirds of which are located in the Proterozoic beds.

Chapter 9

Late Proterozoic Mobile Belts

During the period extending from about 1300 Ma to the end of the Proterozoic (or shortly after) several mobile belts formed which are currently under much investigation. Firstly, the Grenville belt in North America which continues in southern Scandinavia as the Dalslandian or Sveconorwegian belt. Secondly, the Pan-African system of mobile belts which affected a large part of Africa and which extends westwards as the Braziliano belt in South America and is found in NW Europe as the Cadomian belt. There is considerable and increasing evidence from many different standpoints that these belts formed by modern-style plate tectonic processes and therefore we must consider them separately, and in some detail, because they document a very important stage in the evolution of the continents. The possible interrelationships between these belts in N America, Europe and N Africa were discussed by Young (1980).

The Grenville Belt

The Grenville is a high-grade, deeply eroded mobile belt which is well exposed in SE Canada and which continues southwestward to Mexico and Columbia, and northeastward to the Sveconorwegian province (which includes the Dalslandian) and to E Greenland (Fig. 9.1). Isotopic data from the Grenville show three peaks of activity at 1300–1250 Ma, 1100 ± 50 Ma and 950 ± 50 Ma (Baer, 1981a). At the time of writing the most recent detailed review of the Grenville is by Baer (1981b).

Important features of the belt (Fig. 9.2) are:

1. The presence within it of older rocks of four main groups:

(a) Remnants of Archaean and early Proterozoic (Aphebian) gneisses (Frith and Doig, 1975).

(b) Sediments of the Labrador Trough (*c*. 2000 Ma), including banded iron formations, extending well south of the Grenville belt where they are recrystallized to the

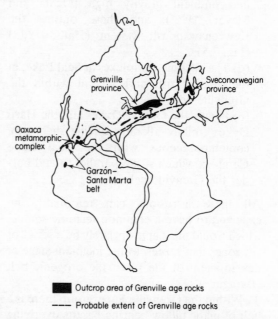

■ Outcrop area of Grenville age rocks

- - - Probable extent of Grenville age rocks

Fig. 9.1 Pre-drift reconstruction of North America, Africa and South America showing the outcrop areas of Grenville-age rocks (based on Sangster and Bourne, 1982, with additional data from Berthelsen, 1980, and Kroonenberg, 1982)

granulite facies by Grenvillian metamorphism.

(c) Several major anorthosite complexes (see Chapter 7) which have an age of 1500–1400 Ma.

(d) A Grenville Supergroup of metasediments (carbonates, pelites and sands) and meta-basalts which have a U/Pb zircon age of 1310 ± 5 Ma.

2. There is an important group of early basic to alkaline igneous rocks within and marginal to the belt (see Fig. 7.5). These include:

(a) Three plateau basalt fields, the Keweenawan, the Seal Lake and the Gardar of S Greenland, erupted between 1300 and 1100 Ma.

(b) The Mackenzie dyke swarm (1200 Ma) which links the Coppermine River basalts (1215 Ma) with the Grenville belt at a high angle (see Fig. 5.2) with a radiating structure as though focused on a mantle plume (Baragar, 1977).

(c) The Keweenawan basalts (Green, 1977) together with the Duluth Complex (Weiblen and Morey, 1980) centred on the mid-continent gravity high (Ocola and Meyer, 1973), and these outline the Keweenawan rift system (Halls, 1978) (Fig. 7.5).

(d) The alkaline complexes at Seal Lake, in the Gardar rift, and several within the Grenville belt.

(e) Finally the Gardar Dykes and the Harp Dykes (mildly alkaline olivine diabases), contemporaneous with the Seal Lake Complex, which strike parallel to and border the Grenville belt.

All these intrusions can reasonably be expected to have a common tectonic source, which would most appropriately be a series of rift zones that represent the incipient stage of development of the Grenville orogenic belt (Baragar, 1977).

3. Within the Grenville Supergroup there is a belt of mafic pillow-bearing basalts overlying a mafic-ultramafic meta-igneous complex; this has the character of early oceanic lithosphere (Chappel *et al.*, 1975). Sequentially there follows a 7 km thick section of 1310 ± 15 Ma mafic to intermediate volcanic rocks. Geochemical data suggest that the lower 4 km are low-K tholeiites similar to a modern, immature, emerging arc, whilst the upper 3 km are composed of andesite of modern mature arc type (Condie and Moore, 1977). These arc volcanics were intruded by several granodioritic to granitic calc-alkaline plutons and batholiths with lead isotope ages of 1250 ± 25 Ma (Pride and Moore, 1983). The volcanics were unconformably overlain by clastic and carbonate sediments of platformal type and, finally, the whole sequence underwent regional deformation and metamorphism up to amphibolite facies at 6 kb and 700°C, during arc-continental collision (Sethuraman and Moore, 1973; Brown *et al.*, 1975). The Grenville Supergroup (island arc) is bounded on the west by the Bancroft-Renfrew (BR) lineament along which there are 1280 Ma nepheline-bearing gneisses, which are most likely the remnants of alkaline complexes that formed in a rift prior to plate separation and which were converted to gneisses by the collision of the island arc with the older plate to the west, the gneisses of which abut discordantly into the lineament. On its eastern side the Grenville Supergroup is bounded by the Chibougamau-Gatineau (CG) lineament which has a marked aeromagnetic expression (Baer, 1976), and to the east of which there are gneisses, containing many anorthosite complexes, that strike parallel to the CG lineament. The gneisses to the east and west of the Grenville Supergroup are quite different in type. It seems that all these features add up to an allochthonous/suspect terrain of Cordilleran/Himalayan type, as implied by Brown *et al.* (1975). The island arc has been crunched between two continental plates of different age, type and structure. Furthermore, not far to the west, in the Parry Sound region, there are well-defined tectonic blocks separated by shear belts along which there are lenses of granulites, anorthosites, garnet gabbros and eclogites (Davidson *et al.*, 1982). These rocks seem very comparable with those along sutures in the Hercynian belt of NW Europe (Bard *et al.*, 1980a), the Pan-African belt of

the Hoggar (Caby *et al.*, 1981), and the NW Himalayas (Bard *et al.*, 1980c; Jan and Howie, 1981).

There is further evidence of early oceanic lithosphere in:

(a) The Llano Uplift of Texas where major serpentinites occur in an ophiolite (Garrison, 1981) which forms part of a 1200–1100 Ma province in SE New Mexico and W Texas. Condie's (1982c) model for the rocks in Texas involves the formation of a magmatic arc and a back-arc basin which was closed during an Andean-type orogeny.

(b) The granulite facies terrain of Quebec where there is an ophiolitic-type assemblage of gabbros, olivine pyroxenites, harzburgites and serpentinites (Rondot, 1978). Late deformation in the NE Grenville gave rise to a stack of thrust slices of gneisses which were driven over the foreland at about 950 Ma (Baer, 1981b). Metamorphism in the Grenville commonly reached amphibolite or granulite facies with pressures up to at least 6 kb indicating a depth of formation and an uplift of some 20 km.

4. It might be thought that palaeomagnetic data would provide a constraint on tectonic models of Proterozoic belts. However, from their review Nairn and Ressetar (1978) concluded that the apparent polar wander paths are ambiguous and from his re-evaluation of the subject Roy (1980) found that the average sampling for Precambrian poles is very poor with only one pole per 9 Ma, compared with 6 per Ma for the Carboniferous to the present. Also, palaeomagnetists over the last decade have had a poor record of working jointly with geochronologists in order to establish the isotopic rock age of the material on which they are employed. Many studies have been made of Grenvillian palaeomagnetism; about half favour a one-plate model (Morris and Roy, 1977; Berger *et al.*, 1979; Piper, 1980b), whilst the other half support the collisional or two-plate model (Irving *et al.*, 1974b; Burke *et al.*, 1976b; Irving and McGlynn, 1976; Patchett *et al.*, 1978). Dunlop *et al.* (1980) point out that the data could

support both models. However, McWilliams and Dunlop (1978) discovered that all Grenville palaeomagnetic poles are of postmetamorphic origin, and therefore they cannot be used to support or reject an ensialic or collisional model.

5. About 100 km within the Grenville belt there is a major negative Bouguer anomaly that extends for 1200 km parallel to the Grenville Front on the south side of the belt of dated Archaean rocks shown in Fig. 9.2. Seismic data indicate the crust is thickened by about 5 km beneath the anomaly. Thomas and Tanner (1975) and Gibb and Thomas (1976) have interpreted the geophysical anomaly as an edge-effect between juxtaposed crustal blocks of different mean density and thickness. This boundary possibly represents a cryptic suture. From their palaeomagnetic work, Irving and McGlynn (1976) independently suggested that the Grenville suture should lie along this same line.

At this point it will be convenient to outline some other models recently proposed to explain the tectonic development of the Grenville belt.

1. The classical model of Wynne-Edwards (1972). Firstly, there was deposition of up to 10 km of carbonates, quartzites and shales of platform type belonging to the Grenville Supergroup on an Archaean and early Proterozoic cratonic basement. This unconformable sequence was then deeply buried. Erosion has removed most of the supracrustal cover exposing the catazonal level of the orogenic belt.

2. The Dewey and Burke (1973) continental collision model. Between two colliding continental plates intervening oceanic lithosphere was largely destroyed, to be preserved only at a high level as ophiolites and blueschists. Extensive convergence was accommodated by thickening of the leading overriding plate causing geoisotherms to rise and partial melting to start in the lower part of the thickening plate.

The Grenville Front cannot be the site of the main Benioff Zone or intercontinental

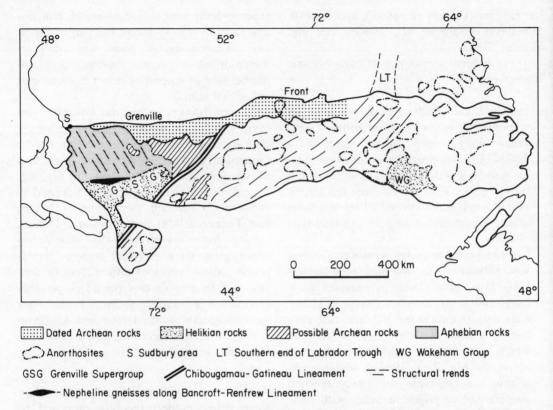

Fig. 9.2 Outline geological map of the Grenville mobile belt. The Grenville Supergroup is an island arc bounded by the Chibougaman–Gatineau (CG) lineament, and the Bancroft–Renfrew (BR) lineament along which there are nepheline gneisses. Anorthosites only occur to the east of the CG lineament. The Aphebian gneisses to the northeast of the Grenville Supergroup strike discordantly into the BR lineament, but the gneisses to the east strike parallel with CG lineament (compiled from Sangster and Bourne, 1982, and Baer, 1976, 1977)

suture because the distinctive iron formation of the Labrador Trough continues for at least 240 km south of the front into the Grenville Province. Dewey and Burke (1973) consequently suggested that the main suture must lie somewhere southeast of the present Grenville Belt.

3. Baer (1976) interpreted the geology of the Grenville Belt in terms of a plate tectonic model. Three evolutionary stages are envisaged:

(a) Emplacement of anorthositic rocks into pre-existing sialic basement in the period 1500–1400 Ma.

(b) About 1300 Ma ago a 'proto-Atlantic ocean' developed along NE-trending fractures to the southeast of the present belt. Associated rifting on the northwest continental plate gave rise to the Seal Lake graben, the Mackenzie dyke swarm and related Coppermine-Muskox volcanism (c. 1250 Ma), the Keweenawan rift filled with basic volcanic rocks (1200–1000 Ma—see Baragar in Baer *et al.*, 1974), an aulacogen near Bancroft–Renfrew in which the Grenville Group of marbles, clastics and volcanics was deposited (1310 ± 15–1250 ± 25 Ma) and, finally, a chain of nepheline-bearing alkaline complexes on one side of the aulacogen (1280 Ma).

(c) The ocean closed and the resulting continental collision gave rise to reactivation and deformation of the Grenville Province about 1100 Ma. According to this model the suture lies hidden somewhere in the Appalachians.

4. The millipede model of Wynne-Edwards (1976), according to which ductile spreading followed by contraction of the crust was the orogenic mechanism responsible for crustal reworking. The continental crust crept northwards like a millipede at a rate of about 10^{-1} cm yr^{-1} over an easterly trending spreading centre. The crust underwent thinning and extension as it became ductile over the zone of thermal upwelling, allowing the formation of an epicontinental sea and corresponding cover sediments. The ductile crust and cover then moved off the spreading system. Aligned trains of plutonic complexes provide relative movement vectors which track the motion of the plate as it passed across the spreading system.

5. Baer (1977) proposed that the whole Grenville belt was one dextral shear zone, 500 km wide, the only evidence being the postulated sigmoidal foliation trends. Baer (1981b) envisaged a model of intra-plate deformation, the necessary stresses for which were built up because of a 'plate-jam'. Baer's conclusion from this model was that 'characteristics that are typical of Phanerozoic belts formed according to present-day plate tectonics have not been found in the Grenville belt'. It is difficult to accept that conclusion when Baer so convincingly described many of those characteristics in his 1976 paper, and when one considers the data and ideas of many other specialists described earlier, based on the Grenville Supergroup.

Clearly none of the models completely satisfies everyone, but the presence of an early island arc is difficult to refute. This arc is comparable to the early Pan-African arcs in the Arabian–Nubian Shield (see later) and should be accepted as evidence of an evolving ocean-floor arc system which now occupies a mini-plate of its own, trapped by collision between the very different, earlier gneiss–granulite plates to the west and east. As a result of a conference on the Grenville, Baer *et al.* (1974) stated that 'most experts would agree that the Grenvillian orogeny may be explained in terms of plate tectonics'.

In the Columbian Andes there is a 1200–1400 Ma granulite belt (Garzón-Santa Marta) which formed at the western border of the Guiana Shield by continental collision according to Kroonenberg (1982).

The Dalslandian Belt

The Grenville continues eastwards via the early Moines of Scotland to the Dalslandian of southern Sweden and Norway (Max, 1979; Young, 1980). The Sveconorwegian orogeny occurred in the period 1200 Ma to 850 Ma. In southern Sweden there are a series of N–S trending thrusts, several hundred kilometres long and 50–100 km apart, which dip shallowly west (Zeck and Malling, 1976). Post-orogenic granites were intruded in belts parallel to the thrusts, which in turn separate slabs of crustal thickness that have been piled upon each other. Further west in the Telemarken area of S Norway, mylonitic thrust belts dip to the east, and Torsk (1977) and Falkum and Petersen (1980) have described a complete Andean-Himalayan igneous and tectonic sequence. Fig. 9.3 gives a map, section and model of the orogenic belt (Berthelsen, 1978, 1980).

Pan-African Belts

In the period 1200–450 Ma a series of mobile belts formed in several continents; the Pan-African extends throughout a large part of that continent (Fig. 9.4), reaching westwards to the Braziliano belt (Pedreira, 1979). The Cadomian is the equivalent in NW Europe (Cogné and Wright, 1980). Only a decade ago it was widely thought that many of these belts formed ensialically, but there has been considerable success in recent years in establishing geological relationships which point towards a plate tectonic mode of development. A major problem still remains, however, in the lack of isotopic age constraints with regard to the subdivision of the many belts that formed in this lengthy period.

138

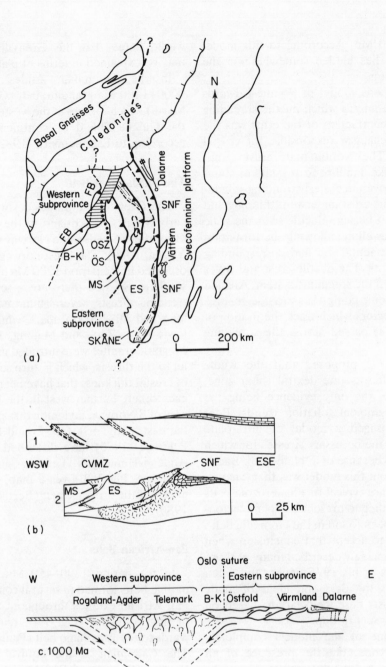

(a)

(b)

(c)

Fig. 9.3 Tectonic sketch of the Sveconorwegides of southern Scandinavia (after Berthelsen, 1980, BRGM, France). (a) SNF, Sveconorwegian front. ES: eastern segment, MS: median segment, ÖS: Östfold slab, OSZ: Oslo fjord shear zone, B-K: Bamle-Kongsberg segment, FB: friction breccia. (b) Diagram illustrating the development of the Sveconorwegian front (SNF) and the Central Värmland mylonite zone (CVMZ)—with thrusting along slightly age-different shears, MS: median segment, ES: eastern segment. (c) Tectonic cross-section of the Sveconorwegian belt. Reproduced with permission from Berthelsen (1980) in Cogné and Slansky, *Geology of Europe*, published by the Bureau de Recherches Géologiques et Minières

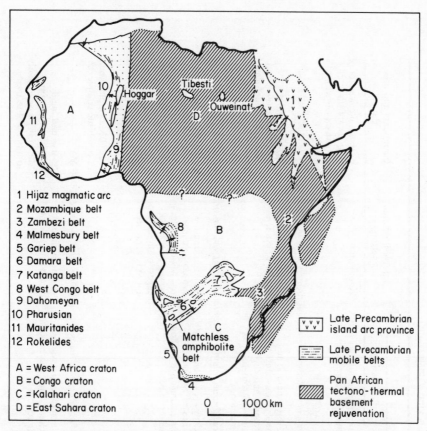

Fig. 9.4 Simplified tectonic map of Africa showing areas affected by the late Precambrian/early Palaeozoic Pan-African episode in relation to stable cratons. Note extensive tectono-thermal overprinting of East Sahara Craton (D) in northeast Africa. Arabia rotated to close Red Sea. Madagascar in its two possible pre-Phanerozoic positions (reproduced with permission from Kröner, 1979, *Geologische Rundschau*, **68**, 565–583)

There were probably several diachronous periods of development rather than one 'Pan-African event' (Jackson and Ramsay, 1980). These belts occupy an important place in time because the older ones overlap with those of Grenville age, and the younger with those of Caledonian–Appalachian age.

We shall now consider the major features that have been discovered in recent years that say something significant about the mode of development of Pan-African belts. The most exciting comprehensive results have come from two shields.

1. The Arabian–Nubian Shield (see many papers in vols. 1–3, of Al-Shanti, 1979). Gass (1981) provides the most up-to-date review.

The Upper Proterozoic crust in Saudi-Arabia, Egypt and Sudan was formed by the growth of several intra-oceanic island arcs, by later growth of Andean-type magmatic arcs and by coalescence (or cratonization) of these to form large continental masses, an idea first proposed by Greenwood *et al.* (1976) (see also Engel *et al.*, 1980). The evolutionary sequence is divisible into the Lower, Middle and Upper Pan-African, in the periods 1200–1000 Ma, 1000–600 Ma and 600–500 Ma (Table 9.1). The arc growth model is based on the absence of isotopic ages older than 1200 Ma, the generally low ($^{87}Sr/^{86}Sr$) ratios of 0.702–0.706, the suprabundance of volcanoclastic sequences and related cannibalistic sediments, the ubiquitous presence

Table 9.1 Rock types of the Arabian–Nubian Shield (from Gass, 1981; in A. Kröner (Ed.), *Precambrian Plate Tectonics*, reproduced by permission of Elsevier, Amsterdam)

Age (Ma)	Rock types			Inferred tectonic setting	Comments and other data
	Plutonic	Volcanic	Sedimentary		
Post Pan-African	Alkaline and peralkaline granites characterized by high Ti, Zr, Nb, U, Th	Alkaline and peralkaline trachytes and rhyolites	Terriginous arkoses and shallow water shales	Continental	Continental character of region established; all magmatism of 'within-plate' variety
500–600	———————————————— *Diachronous end of destructive margin process* ————————————————				
Upper Pan-African	Calc-alkaline granites and granodiorites with low Ti, Zr, U, Th and very low Nb	Rhyolites, dacites, trachytes and andesites	Conglomeratic and arenaceous units with granitic and rhyolitic clasts. Stromatolitic limestones	Continental with margins of Andean type	Regionally extensive, unmetamorphosed and structurally undeformed silicic volcanic and plutonic rocks
600–670	———————————— *Major break, regional unconformity, change in composition of magmatism* ————————————				
Middle Pan-African	Calc-alkaline diorites and granodiorites	Calc-alkaline andesites and basaltic andesites with subordinate rhyolitic and dacitic units	Greywackes and minor arkoses. Stromatolitic limestones and shallow water shales	Numerous mature intra-oceanic island arcs. Major stratigraphic and regional breaks suggest complex evolution of several arcs	c. 600 Ma emplacement of ophiolitic complexes c. 800 Ma emplacement of ophiolitic complexes Completely deformed and metamorphosed to greenschist facies c. 1000 Ma emplacement of ophiolitic complexes
c. 1000	———————————— *Distinct structural, compositional and metamorphic break: arc collision (orogenesis at c. 960 Ma)* ————————————				
Lower Pan-African	Gabbros diorites, granodiorites	Low-K basalts and basaltic andesites	Immature greywackes, cherts, shales, occasional limestones	Numerous immature intra-oceanic island arcs	Sparse and highly deformed outcrops. Metamorphosed mainly to amphibolite facies
1200?					

of calc-alkaline volcanic and plutonic products, and the identification of several linear belts of ophiolite complexes. About 10 ophiolites have so far been recognized in Saudi Arabia, 6–8 in Egypt, 2–3 in Sudan (Fig. 9.5) (Gass, 1981) and several in Ethiopia (Kazmin et al., 1978). There were three main emplacement ages at c. 1000, 800 and 600 Ma; some of these contain a complete ophiolite sequence according to the 1974 Penrose Conference definition, e.g. serpentinized peridotites, pyroxenites, gabbros, diorites, trondjhemites, sheeted dyke complexes, basalts and cherts (Bakor and Gass, 1976; Frisch and Al-Shanti, 1977; Al-Shanti and Roobol in Al-Shanti, 1979 vol. 1). There are olistostrome mélanges in Egypt in which the matrix of ophiolite blocks is a graphitic pelite (Shackleton et al., 1980; Ries et al., 1983), and tectonic mélanges in Saudi Arabia with a serpentinite matrix.

The arc-type volcanics are described by Engel et al., Fleck et al. in Al-Shanti (1979, vol. 3) in Saudi Arabia, Stern (1981) in Egypt and Shimron (1980) in Sinai. Roobol et al. (1983) demonstrate that three volcano-sedimentary sequences in central Arabia formed in progressively maturing volcanic arcs. Early-formed (> 900 Ma) immature island arcs were succeeded by 900–700 Ma chemically more mature island arcs and then by 700–570 Ma deposits formed in volcanic arcs transitional between island arcs and continental margins. The evolution did not reach the equivalent of an Andean continental margin. Acidic pyroclastics contain Pb, Zn, Cu mineralization (Delfour, 1976). Linear belts of calc-alkaline batholiths (Fig. 9.5) show a chemical polarity of K, Na, Rb/Sr and K/Rb (Brown in Al-Shanti, 1979, vol. 3; Rogers et al., 1980), and a zonation of Cu–Au to Ag–Pb–Zn mineralization (Al-Shanti and Roobol in Al-Shanti, 1979, vol. 1). Thrusting and crustal thickening during final stages of collision was followed by intrusion of 'within-plate' alkaline (Rogers et al., 1978) and peralkaline granites (Ries et al., 1983) enriched in Zr, Y and Nb (Radain et al., 1981), and with W, Sn, U, Nb mineraliz-

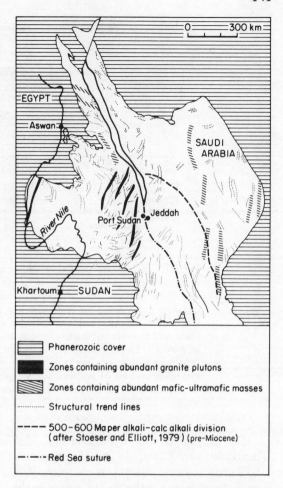

Fig. 9.5 Regional sketch map showing the disposition of mafic–ultramafic complexes (marking the approximate position of arc sutures), linear granitic zones (possible arc axes) and basement structural trends. The Red Sea has been closed to a Pre-Miocene position (from Gass, 1981, in A. Kröner (Ed.), Precambrian Plate Tectonics, Elsevier, Amsterdam, pp. 388–405)

ation (Sillitoe in Al-Shanti, 1979, vol. 1). Indentation-type wrench faults formed during post-collisional shortening (Fleck et al. in Al-Shanti, 1979, vol. 3). With such a sequence of rocks and events no significant reason remains to preclude acceptance of the modern-style ophiolite–arc-collision model for late Precambrian crustal growth in this Shield (Gass, 1981). Fig. 9.6 depicts four stages in such growth.

142

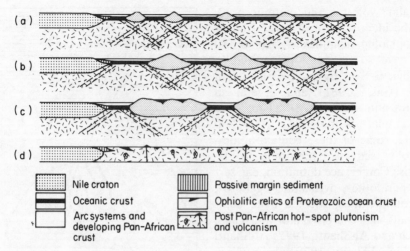

Nile craton

Oceanic crust

Arc systems and developing Pan-African crust

Passive margin sediment

Ophiolitic relics of Proterozoic ocean crust

Post Pan-African hot-spot plutonism and volcanism

Fig. 9.6 A cartoon depicting stages in the development of the Arabian–Nubian Shield. (a) Depicts the situation in the Lower Pan-African with many immature arc systems. By Middle Pan-African times (b) the arcs have matured and coalesced but have not attained continental dimensions. By Upper Pan-African times (c) the arcs have coalesced into continents but these still overlay subduction zones and magmatic activity had calc-alkaline affinity. (d) Depicts the post Pan-African (500–600 Ma) situation. When the continent was fully developed, subduction had ceased and magmatism was per alkaline and of within-plate affinity (from Gass, 1981, reproduced by permission of Elsevier, Amsterdam)

2. The Hoggar Shield of Algeria and Mali (Bertrand and Caby, 1978; Black *et al.*, 1979a,b; Caby *et al.*, 1981). A complete Wilson cycle is present in this shield, covering the time period 900–550 Ma. The major tectonic units in this N–S mobile belt are, from west to east (Fig. 9.7):

(a) The West African craton on the west, the eastern edge of which is a suture zone; the Gourma aulacogen extends from the craton to the suture.

(b) The Pharusian belt, with abundant volcano-detrital sediments up to 6 km thick and widespread calc-alkaline volcanics and batholiths with a U–Pb age on zircons of 831 ± 5 Ma. The belt records a transition from an island arc in the west to a Cordilleran-type continental margin on the east.

(c) A polycyclic central Hoggar-Air domain composed of ancient gneisses reactivated and injected by abundant Cordilleran-type plutons.

(d) The Hoggar-Ténéré domain contains

granodioritic batholiths and amphibolite facies metamorphics; it formed an early active continental margin of the Pan-African belt which was stabilized around 725 Ma.

Subsidiary key features of the mobile belt are also evident. The roots of an early palaeorift are represented by five generations of alkaline to peralkaline syenites and granites in dyke swarms and layered intrusions (Dostal *et al.*, 1979). An eastward-thickening sediment sequence developed on the passive continental margin of the West African craton; this comprises early clastics, carbonates with a platform–slope–rise transition, and late clastics. Further intrusions associated with continental fragmentation include garnet meta-gabbros, garnet amphiclasites and pyriclasites intruded into shelf sediments, and 810 ± 50 Ma layered amphibole–garnet–pyriclasites and two-pyroxene pyriclasites situated in the suture itself; all these rocks recrystallized deep in the crust in granulite-grade conditions and closely resemble

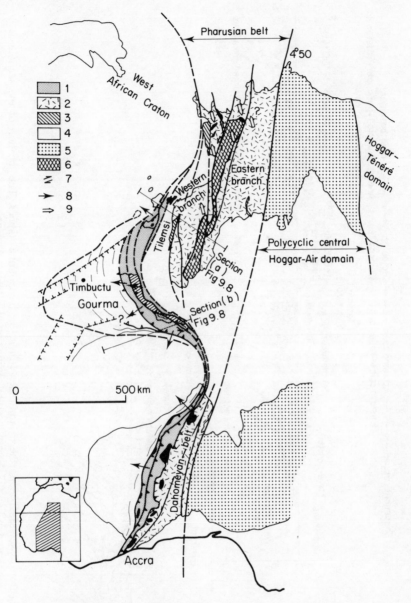

Fig. 9.7 Simplified sketch-map of the Pharusian–Dahomeyan belt after collision between the West African craton and the Touareg shield. *1*: Gourma and Buem–Atacora nappes; *2*: undifferentiated rocks mainly of Pan-African metamorphism; *3*: Greywackes and magmatic rocks of accretion zone; *4*: high-pressure–low-temperature metamorphism of the Gourma nappes; *5*: undifferentiated reactivated pre-Pan-African rocks in the central Hoggar and Nigeria; *6*: Eburnean granulites; in black: metabasic to ultrabasic rocks; *7*: strike-slip fault; *8*: late movements related to the collision with the West African craton; *9*: early collision in the Northern Iforas and western Hoggar (from Caby *et al.*, 1981, reproduced by permission of Elsevier, Amsterdam)

144

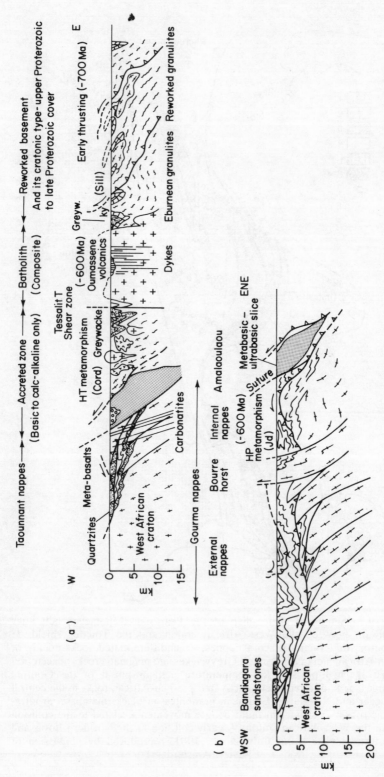

Fig. 9.8 Schematic geological cross-sections: a: northern Adrar des Iforas; b: the Gourma nappes. The location of the sections is indicated in Fig. 9.7. Reproduced by permission of Elsevier, Amsterdam

the Cretaceous garnet granulites of the Jijal Complex situated in the Indus Suture of N Pakistan (Jan and Howie, 1981). As a result of the eastward subduction and terminal collision, nappes of the passive margin sediments were thrusted westwards onto the West African craton (Fig. 9.8). Close to the suture these underwent high-pressure metamorphism with development of eclogitic schists with jadeitic pyroxene, and blue amphibole in metabasalts; low-pressure cordierite-bearing gneisses of the Cordilleran belt make up a paired metamorphic belt. During post-collisional shortening, caused by eastward indentation of the more rigid West African craton, conjugate strike–slip, brittle, fault systems formed in the less rigid, recently created Shield east of the suture (Ball, 1980). In conclusion, there can be no doubt of the existence of a Wilson cycle that ended by 550 Ma and which gave rise to the Hoggar mobile belt.

What about the Pan-African in other parts of Africa? At Bou Azzer in Morocco there is a 788 ± 9 Ma ophiolite that consists of serpentinized dunite and harzburgite, rodingite dykes, cumulate gabbro, diorite and trondjhemite, pillow basalts with copper mineralization, a sheeted dyke complex with 70% volume of dykes and a volcano-sedimentary sequence (Leblanc, 1981). The ophiolite was thrusted southwards onto sediments of the passive continental margin of the West African craton about 685 Ma ago, and it is overlain by calc-alkaline volcanics produced during associated southward subduction.

The Mozambiquian belt in East Africa largely consists of high-grade basement gneisses, and this has been thought to be a site of ensialic mobility because it did not easily fit into a conventional plate tectonic framework (Kröner, 1977b). However, Shackleton (in press) shows that this gneissic basement has a low-angle thrust structure exactly comparable to that on the northern margin of the Indian plate in the Himalayas south of the Zangbo suture. Shackelton concludes that the Mozambiquian belt formed by eastwards collision of the presently observed segment and that today we are looking at a westward-

directed post-collisional thrust belt at a deep erosional level (see also Chapter 10).

Following the suggestion of Burke and Dewey (1973b) that plate accretion and closure gave rise to the Dahomey suture in West Africa, McCurry and Wright (1977) described the calc-alkaline geochemistry of arc-type volcanics in Nigeria, which are consistent with the presence of a Pan-African suture. This is the southern extension of the Hoggar belt (Fig. 9.7).

Watters (1976) first identified the essentials of a Wilson cycle in the Damaran belt of Namibia in SW Africa. Martin and Porada (1977) interpreted this orogenic belt in terms of closed aulacogen, but Barnes and Sawyer (1980) gave a thoughtful critique of this ensialic idea and demonstrated that a model involving the formation of early alkali rocks, the development of oceanic crust, initiation of northwestward subduction and ocean closure terminating in continental collision can very adequately explain the major features. The presence of glaucophane in the Gariep belt meta-sediments is good evidence of subduction. This model was essentially corroborated by Downing and Coward (1981) who concluded that northward subduction of Damaran oceanic crust gave rise to the major belt of calc-alkaline granitic rocks, and southward overriding of the northern (Congo) over the southern (Kalahari) craton during terminal collision gave rise to low-angle, nappe-thrust stacks. Porada (1979) related the formation of the Damaran belt with the Ribeira belt of Brazil in terms of ocean opening and closure. The palaeomagnetic data of McWilliams and Kröner (1981) are consistent with the formation and closure of a narrow Damaran ocean.

In summary, we can say that the evidence for the operation of modern-style plate tectonic processes in Pan-African times is overwhelming. This conclusion, significantly, was reached by workers of many different nationalities in many different regions. Evidence for ensialic orogenesis is lacking according to most authorities (for a contrasting viewpoint, see Schermerhorn, 1981). The statement by Kröner (1979) that 'the Pan-

African event is representative of a transition from Precambrian ensialic plate tectonics to Phanerozoic Wilson cycle tectonics' is not supported by the great wealth of information now available from these belts and can be regarded as manifestly incorrect. Pan-African heat production would have been only 1.4–1.7 times present-day values (Brown in Al-Shanti, 1979, vol. 3), and arc dimensions would have been within 10% of those of today (Gass, 1981). The geological evidence suggests that an essentially uniformitarian approach can be applied to the Pan-African belts and that modern-style plate tectonic processes were going strong by, and during, the late Precambrian. Yet the palaeomagnetic data of Morel and Irving (1977) indicate tentatively that for the interval 1150 ± 950 Ma

the configuration of the continents precludes the existence of wide oceans. A maximum width of about 1000 km would be consistent with the spatial restrictions imposed by the palaeomagnetic data (Neary et al., 1976) and would enable the Wilson cycle to operate. Some geological evidence suggests that the ocean width was not great (Kazmin et al., 1978).

Taken as a whole, it is clear that in the last few years a major revolution has affected the interpretation of late Precambrian mobile belts, since, up until the mid-1970s in situ ensialic orogenies were considered more probable. The relationship of these late Precambrian mobile belts to the earlier stages of Proterozoic crustal development will be considered in the next chapter.

Chapter 10

Crustal Evolution in the Proterozoic

The subject of this chapter has reached an interesting and controversial stage of development, and the main problems now discussed are somewhat different to those apparent in 1976 when the final draft of the first edition of this book was written. The questions now being asked focus on the issue of when modern-style plate tectonics began: in the early Proterozoic or not until 1000 Ma ago? The chief new constraints bearing on this problem are: (1) new discoveries in early Proterozoic belts; (2) the realization that the Archaean types of tectonic belts continue through the Proterozoic; (3) new studies of deeply eroded Phanerozoic belts—in particular, of the Himalayan Range of Mesozoic–Cenozoic age. We can only learn about the possible application or non-applicability of plate tectonic processes in the Proterozoic by reference to modern orogenic belts, and in the past a major problem has always been that very little was known about the Himalayas and the deep sections of modern belts. New results in these fields thus have considerable bearing on the postulated mode of origin of Proterozoic belts. In this chapter we shall review those ideas and data which play a critical role in theories of continental evolution.

Proterozoic Greenstone Belts and Granulite–Gneiss Belts

Archaean terrains are made up of two fundamentally different types of tectonic unit: greenstone belts (GB) and granulite–gneiss belts (GGB). Conventional opinion has been that they, and in particular the GB, are a characteristic feature of just the Archaean and that they did not form in subsequent time. However, it is now apparent from recent isotopic age determinations and geological re-evaluation that GB and GGB continued to form throughout the Proterozoic, albeit in smaller number than during the Archaean.

The Greenstone Belts

Tarney and Windley (1981) list several Proterozoic GB and Table 10.1 gives some key examples (Windley, 1984). The main features are essentially identical to those of Archaean GB—only a decade ago the Amisk–Flin Flon belt in Canada was thought to be an Archaean GB. Two results of this re-assessment are that there are probably more Proterozoic komatiites than so far described (Wyman, 1980) and that the Algoma-type of banded iron formation, normally regarded as an Archaean sediment, continued to form throughout the Proterozoic (Windley, 1983a). Bimodal volcanics and greywackes (type 3 of Condie, 1982a,d) are characteristic of these belts and thus a rifted environment, comparable to a modern back-arc marginal basin, has commonly been proposed (Choudhuri, 1980; Bose and Chakraborti, 1981; Condie and Nuter, 1981; Tarney and Windley, 1981; Condie, 1982a,d); others are described as island arcs (Stauffer *et al.*, 1975; Condie and Moore, 1977; Moore, 1977; Engel *et al.*, 1980; Shimron, 1980; Lewry, 1981). The early and mid-Proterozoic

147

Table 10.1 Examples of Proterozoic greenstone belts (modified after Windley (1984))

Greenstone Belt	Age (Ma)	Reference
Trans-Amazonian Belts, Guiana Shield	2250	Choudhuri, 1980; Gibbs and Olszewski, 1982
Birrimian, W Africa	2300–1950	Sillitoe, 1979; Attoh, 1982
Lynn Lake, Manitoba, Canada	1700	Zwanzig et al., 1979
Amisk Group (Flin Flon-Snow Lake), Manitoba, Canada	1850–1650	Stauffer et al., 1975; Moore, 1977; Lewry, 1981; Bell et al., 1975
Yavapai Series, Jerome, Arizona	1770–1820	Anderson et al., 1971
Pecos, New Mexico	1800–1700	Wyman, 1980; Robertson, 1981
Dubois, Colorado, USA	1700	Condie and Nuter, 1981
Wyoming to Texas, USA	1720–1650	Condie, 1982c
Dalma, Singhbhum, NE India	1700–1600	Gupta et al., 1980; Bose and Chakraborti, 1981
Hastings area, Grenville Province	1300	Condie and Moore, 1977
Myole, N Kenya	1000–800	Kazmin et al., 1978
Nubian–Arabian Shield, Pan-African	900–600	Engel et al., 1980
Sinai, Pan-African	800–600	Shimron, 1980

GB provide a link between Archaean GB and the GB of the Pan-African which can reasonably be interpreted as ancient analogues of Phanerozoic arcs and marginal basins (Engel et al., 1980). This linkage provides an explanation of the remarkable similarity between the Kuroko massive sulphide deposits in Japan, which formed in a Miocene volcanic arc rift or marginal basin, and the massive sulphide deposits in Archaean GB (Cathles et al., 1980) and those in the early Proterozoic GB at Flin Flon and Snow Lake in Manitoba and at Jerome in Arizona. Condie (1982a) points out that there are four major differences between the calc-alkaline volcanic-greywacke assemblage (see Chapter 6) of Proterozoic and Archaean greenstone belts:

1. Ultramafic and komatiitic volcanic rocks are uncommon or absent in the Proterozoic successions.
2. Volcaniclastic rocks and, in particular, potassium-rich felsic volcanic rocks, are more abundant in the Proterozoic belts.
3. Greywacke is proportionally more important in most Proterozoic successions.
4. Chert and iron formations are less common in the Proterozoic belts.

Furthermore, he suggested that if we take account of the known decrease in the global thermal budget we can expect the Archaean greenstone belts to form in hot unstable rifts and basins, while after about 1000 Ma ago the calc-alkaline volcanic-greywacke assemblage is found in back-arc basins and other basins associated with convergent plate boundaries. In the early to mid-Proterozoic the assemblage may have formed in both environments (see further in Table 22.1, p. 332).

The granulite–gneiss Belts

The Proterozoic examples of GGB listed in Table 10.2 are broadly comparable with many Archaean GGB. One difference is the small proportion of intrusive tonalite in the Proterozoic belts. Typically they contain much reworked or reactivated basement gneiss, a fact used to support an ensialic origin for the belts (Kröner, 1977b), but wrongly so since the Himalayas contain a belt of reworked pre-Mesozoic gneisses (the Central Crystallines), up to 15 km wide and 2700 km long, and in the deeply eroded section of N Pakistan there are at least 15 000 km^2 of reworked gneisses occupying the northern edge of the Indian Plate (Coward et al., 1982). Examples of uniformitarian interpretations of tectonic evolution are the Telemarkian orogen of S. Norway as a remnant of a Cordilleran-type continental belt comparable

Table 10.2   Examples of Proterozoic granulite–gneiss belts (modified after Windley (1984))

Granulite–gneiss belts	Age (Ma)	Reference
Karelides, Finland	2200–1900	Raith *et al.*, 1982b
Namaqua–Natal, S Africa	2000–1000	Tankard *et al.*, 1982
Svecokarelian, Baltic Shield	1830–1700	Bowes, 1980; Nyström, 1982
Musgrave Range, Australia	1380	Davidson, 1973
Telemarkian, S Norway	1200–850	Torsk, 1977
Enderby Land, Antarctica	1100–800	Sheraton *et al.*, 1980
Grenville–Dalslandian	1000±200	Berthelsen, 1980; Baer, 1981
Mozambiquian, Pan-African	600±200	Kröner, 1980; Shackleton, in press

to the Columbian orogen of the Canadian Cordillera (Torsk, 1977), the Svecokarelian of the Baltic Shield, which evolved from an Andean-type batholith-dominated belt (Nyström, 1982) to a Himalayan-type belt (Bowes, 1980), the Karelides of Finland as a deep section of an arc–batholith–continental block belt (Raith *et al.*, 1982b), and the Musgrave Range of Australia as a Himalayan-type collisional belt (Davidson, 1973).

The main differences between the quartzite–carbonate–shale assemblage of Archaean and Proterozoic granulite–gneiss belts are:

1. There tends to be more associated volcanic amphibolite in the Archaean belts.
2. The associated anorthositic layered complexes did not continue to develop after the end of the Archaean.
3. Carbonate (marble) sediments are more common in the Proterozoic belts.

In terms of modern equivalents the quartzite–carbonate–shale assemblage in Archaean GGB formed on an unstable proto-shelf that was unable to survive the permobile tectonic conditions of that time. It was not until the early Proterozoic that this assemblage was able to survive in unconformable cover sequences on rifted stable continental margins.

In conclusion, the main point of emphasis is that the two principal types of Archaean orogenic belt continued to evolve, although with certain differences, throughout the Proterozoic. In this respect the Archaean–Proterozoic boundary should not be seen as such a major turning point in earth history as has

commonly been supposed; obviously, there were many other important differences across this time boundary and these are considered further in Chapter 22.

When did Modern-style Plate Tectonics Begin?

This is the great subject of debate today. I have quoted already, in the Proterozoic chapters of this book, over 60 authors who have concluded from their detailed studies that their early–mid-Proterozoic (>1000 Ma) rocks are so similar to modern equivalents that they proposed a plate tectonic scheme of events for their evolution. On the other hand, there is a small but vocal minority that asserts that modern-style plate tectonics did not begin until 1000 Ma (Piper, 1978; Baer, 1981a,b; Hutchinson, 1981; Kröner, 1981a,b; McCall, 1981). In this chapter we shall evaluate and discuss these issues.

Early Proterozoic Crustal Conditions

We know that massive thickening of the continental crust took place in the late Archaean (Chapter 4) and by the end of the Archaean continental lithospheric plates may have been up to 200 km thick (Davies, 1979). The fact that extensive, thick and rigid early Proterozoic plates were beginning to respond to deformation, deposition and intrusion in a mode comparable to that of today is indicated by the following features (Dewey and Windley, 1981):

1. Aulacogen re-entrants in stable continental margins (Wopmay Orogen: Hoffman, 1980; Richmond Gulf, Circum-Superior belt, Paradise Rift at Mount Isa) and continental rift successions (Wollaston Group: Lewry and Sibbald, 1980; Lake Superior region: Larue and Sloss, 1980) with alkali complexes (Wopmay Orogen: Hoffman, 1980).
2. Formation of oceanic crust (pillow lavas, serpentinites and cherty quartzites in the Svecokarelides; Park and Bowes, 1981; Gaal, 1982).
3. Shelf–slope–rise sequences on trailing continental margins (Labrador Trough: Wardle, 1981; Wollaston Group: Lewry, 1981).
4. Narrow linear/arcuate orogenic belts bordering stable continental plates (Circum-Superior belt, Wopmay Orogen).
5. The construction of island arcs (Arizona: Anderson, 1977; Flin Flon and Lynn Lake belts, Churchill Province: Lewry, 1981; Svecokarelides of Finland: Gaal, 1982; Raith *et al.*, 1982b; and Sweden: Loberg, 1980), of Andean-type batholith belts (Svecokarelian of Sweden: Nyström, 1982; Norway: Torsk, 1977; and Finland: Gaal, 1982; Wopmay Orogen: Hoffman, 1980; Wollaston belt; Lewry, 1981; in the Ketilidian belt of S Greenland) and of back-arc basins (Colorado: Condie and Nuter, 1981; Singhbhum region, NE India: Bose and Chakraborti, 1981; USA: Condie, 1982c).
6. Slip-line indentation fracture systems in cratonic forelands formed after terminal collision (deep-level shear zones in W Greenland: Watterson, 1978; Wopmay Orogen: Hoffman, 1980; Churchill Province-Slave Craton: Gibb, 1978).

Taken together these features indicate that early Proterozoic belts formed by processes that can be equated broadly with those of today.

The Depth-Basement Factor

A useful observation is that almost all of what we know about the development of orogenic belts by plate tectonic processes comes from study of high-level sections of weakly eroded Mesozoic–Cenozoic belts. In other words, we know very little about how the mid–lower continental crust in these belts behaves during the magmatic and collisional stages of the Wilson Cycle simply because there are very few modern belts that have been eroded to a deep level. Accordingly, it is not difficult to find similarities between high-level Proterozoic and Mesozoic–Cenozoic belts, but very difficult to understand how deeply eroded Proterozoic granulite–gneiss belts formed because there are no modern comparisons or equivalents. In my opinion herein lies the *raison d'être* for ideas on ensialic orogenesis applied to high-grade Proterozoic belts with much reactivated/reworked basement, such as the Grenville (e.g. Wynne-Edwards, 1972, 1976; Baer, 1981a,b) and many belts in Africa (Kröner, 1977b; Piper, 1978; Schermerhorn, 1981).

At least this was so until the recent discovery of the Coast Range batholith belt of British Columbia in Canada, and of the Kohistan arc section of the Karakoram Range of the Himalayas of North Pakistan. These are the two sections of modern orogenic belts that have been most highly uplifted with the result that they contain much reactivated gneissic 'basement'. More details of these can be found in Chapters 19 and 21, respectively, but the salient points are given here.

1. The Coast Range belt underwent an uplift prior to 52 Ma of up to 8 mm yr^{-1}, and 2–3 mm yr^{-1} since then (Hollister, 1979), with the result that the Central Gneiss Complex (Hutchinson, 1982) and early Tertiary granulites (Hollister, 1975) are exposed. Reactivation of the very complicated Palaeozoic(?) migmatites of the Central Gneiss Complex was caused by late-Cretaceous–early Tertiary nappe-thrust stacking, crustal thickening and associated deformation and metamorphism at *c*. 7 kb; this triggered the generation of granitic material which rose as parautochthonous plutons to higher levels. Hutchinson concludes that the lower crust was an essential component in the generation and evolution of the Mesozoic–Cenozoic batholith material. Thus the 'basement' of the Cordilleran

orogenic belt at 25 km depth consists of reactivated migmatites and gneisses.

2. In Kohistan the northern margin of the Indian plate has been uplifted in the Nanga Parbat region at a rate of nearly 1 cm yr^{-1} for the last 0.5 Ma (Zeitler *et al*., 1982a), and further west there was 5 km of uplift in the period 30–20 Ma (Zeitler *et al*., 1982b). Further east in the Himalayan range (Zanskar in NW India–S. Tibet) this margin of the Indian plate is covered with a well-preserved complete sequence of Palaeozoic and Mesozoic sediments (Ordovician to Cretaceous); however, in Kohistan these cover rocks have been largely eroded exposing at least 15 000 km^2 of basement of Precambrian sediments and Cambrian granites which have been extensively both gneissified and reactivated during Himalayan collisional and post-collisional deformation (Coward *et al*., 1982). Moreover, further east in India, Nepal and S Tibet the Palaeozoic sediments are underlain by a thick thrust-nappe stack of Precambrian gneisses which, in the lower crust, were reactivated in the Miocene to produce the well-known Miocene leucogranites. Thus when the 'basement' of the Himalayan Range is exposed we see that it consists of reactivated gneisses.

Because of the presence of so much reactivated gneissic basement, Baer (1981b), Kröner (1977a,b) and Schermerhorn (1981) could not conceive that the Grenville and African belts could have formed by modern-style plate tectonic processes and therefore must have involved intra-plate ensialic orogenesis. Kröner (1977a,b) emphasized the fact that African belts underwent more plate destruction by basement reactivation than plate accretion. But if our modern Cordilleran and Himalayan belts were eroded to 25 km depth (comparable to the 7 kb + pressure data from Proterozoic granulite-gneiss belts), we would likewise see more evidence of basement gneisses than plate accretion rocks like ophiolites, volcanics and high-level plutons which would all have been eroded. This idea is corroborated by the reappraisal of the Mozambiquian belt by Shackleton (in press) who shows that it is a deep section of a Himalayan type belt with post-collisional thrusts that have penetrated into the gneissic basement (Chapter 9).

The concept that Proterozoic mobile belts formed by intracontinental processes was based on the observation that they were built upon old continental crustal material. The observation is correct, but the concept is not because modern Cordilleran and Himalayan mobile belts are likewise built upon old continental crust. Thus the ensialic model for Proterozoic high-grade belts based on that argument can be regarded as obsolete.

Ophiolites and Glaucophane Schists

Another argument used to support the ensialic model for Proterozoic orogenesis is the fact that ophiolites and glaucophane schists are absent (Piper, 1978; Kröner, 1981b). The oldest glaucophane occurs in the late Proterozoic belts of the Mona Complex in Anglesey in Wales (Gibbons and Mann, 1983) and in Namibia (Kröner, 1980). Surprisingly it is absent elsewhere in the Pan-African belts where otherwise there is good evidence of the products of subduction. There are several reasons for this lack of glaucophane older than the late Proterozoic.

1. Trench sediments at an accretionary wedge underthrust and so uplift earlier members. Failure to uplift blueschist zones may reflect the small size of earlier oceans (Fyfe, 1981).

2. This reason relates to the shape of the PT trajectory of exhumed rocks. Due to the slow thermal compared with the pressure conductivity of rocks depressed to great depths, on initial uplift they tend to increase in temperature before reaching an equilibrium position with respect to PT conditions, after which they both decrease with further uplift. During this early T increase and P decrease, low-temperature glaucophane will be recrystallized to higher temperature hornblende and thus glaucophane schists will rapidly be transformed to greenschists. This process of partial conversion can be seen in many glaucophane schists and greenschists on sutures which con-

tain only relict glaucophane. Thus all Precambrian mobile belts that have been moderately to highly uplifted may not be expected to retain any glaucophane (England and Richardson, 1977).

3. Glaucophane schists that are uplifted to the surface may well be eroded rapidly and the glaucophane transferred to molasse sediments. Because the glaucophane schists only occur as small tectonic slices, they may be absent on sutures, and the only evidence of their former existence lies in detrital glaucophane. Not many Proterozoic molasse sediments have been searched for glaucophane.

4. Thermal constraints may well have caused ocean floor to subduct at a younger age than at present (Bickle, 1978 for the Archaean; Hynes, 1982 for the early Proterozoic). Subduction of young, warm, oceanic slabs would prevent the formation of low-T glaucophane schists.

Many presently exposed glaucophane schists occur at ocean-arc and ocean-continent plate boundaries, such as in the Fransiscan mélange in California. These cannot be used for comparison with possible Proterozoic equivalents because they have not yet suffered the complete Wilson Cycle. On the Indus-Yarlung Zangbo suture of the Himalayas there is remarkably little glaucophane–not more than 5 localities in a suture length of some 2700 km and each is less than a few kilometres long. Even where there is glaucophane in an ophiolite such as in the Yarlung Zangbo and Nagaland bodies, it occupies at the most one hundredth of the length of the ophiolite. Glaucophane rocks occur along not more than 0.3% of the Himalayan suture, and these few only as thin tectonic slices. With only a few kilometres more uplift there is little doubt that they would all be eroded. The same argument applies to the Himalayan ophiolites which occupy less than one hundredth of the length of the suture. They are also only thin tectonic slices and much uplift would certainly remove them by erosion.

The amount of deformation and meta-

morphism at these continental collisional sutures can be very high, with the result that an oceanic crustal section can easily be converted to a tectonic belt of amphibolites. Pillows are deformed so that their axial ratios reach more than 1 : 100 and they are converted to banded amphibolites. Gabbros are converted to foliated amphibolites. Basic dykes are rotated and become foliated amphibolites. Even with a few pods of serpentinite, this is no longer recognizable as an ophiolite. Significantly, the only young ophiolites that are well preserved are those that have been thrust out of the suture as flat slabs over platform sediments (Bay-of-Islands, Spongtang, etc.). Even small amounts of uplift would remove by erosion the highest levels of these platform sediments thus no well-preserved ophiolites can be expected in Proterozoic belts.

With the above considerations in mind one can hardly use the absence of ophiolites or glaucophane schists in Proterozoic mobile belts as an excuse to abandon a plate tectonic model and replace it with an ensialic one.

High-pressure granulites

Besides ophiolites and glaucophane schists there is a third rock group in Himalayan-type collisional sutures, exposed in deep crustal sections in particular. This comprises metamorphosed dunites, pyroxenites, gabbros, norites and anorthosites which are relics of layered igneous complexes. Common rocks are garnet-clinopyroxene granulites, eclogites, and garnet-bearing gabbros, amphibolites, pyroxenites and pyriclasites. Examples are the Tertiary suture of the Ivrea Zone of N Italy (Shervais, 1979; Rivalenti et al., 1981), the Cretaceous Jijal complex on the Indus suture of N Pakistan (Jan and Howie, 1981), the Hercynian suture of France and NW Spain (Bard et al., 1980a), and S Italy (Schenk, 1980), the 810 Ma Pan-African suture of the Hoggar in Algeria (Caby et al., 1981), and the 1300 Ma tectonic boundaries in the Grenville belt (Davidson et al., 1982). The last two examples provide evidence of Precambrian plate collisional processes.

Sutures

If modern plate tectonic processes were not in operation before 1000 Ma ago, then no sutures would exist; however, the following are good candidates:

1. The Grenville belt: the western side of the Grenville Supergroup is defined by the Bancroft-Renfrew lineament which separates arc rocks on the east from high-grade gneisses on the west, which strike discordantly into the lineament and along which there are several pre-tectonic nepheline syenite complexes (Baer, 1976). The lineament is well defined on both magnetic and gravity anomaly maps. Eclogites on tectonic boundaries in the Parry Sound region probably define palaeo-sutures (Davidson *et al.*, 1982).

2. In the Svecokarelides of Finland (Chapter 6) there are tholeiitic pillow lavas with large serpentinite bodies associated with cherty quartzites which are reasonably interpreted as remnants of oceanic crust (Gaal, 1982). The occurrence of a suture through the Outokumpu region is consistent with the conclusions of over a dozen recent authors drawn from the presence of oceanic crust with Cu sulphide deposits, arc-type volcanics, Andean-type granitic plutons, porphyry Cu–Mo mineralization, flysch deposits in a marginal basin, and a continent–continent collision zone (Bowes, 1980).

3. The Stanley shear belt in the Wollaston fold belt in the Churchill Province of Canada can well be argued, on the basis of much geological evidence, to be a suture (Lewry, 1981) (see Chapter 6).

4. The suture in the Wopmay orogen is located between zones 3 and 4 (see Chapter 6) in the tectonic position of the Wopmay Fault, along which there has been post-collisional strike-slip movement (Hoffman, 1980).

The documentation of these examples is good and the conclusions of many authors must be taken seriously. There is no longer any reason to doubt the existence of pre-1000 Ma suture zones.

Can knowledge of modern Himalayan geology help us to recognize sutures in Proterozoic mobile belts? There should not be too much difficulty in high-level sections of such belts. For a deep-level section every Precambrian geologist should be encouraged to walk across the Eocene Indus suture in N Pakistan (Coward *et al.*, 1982): one passes across the strike from vertical quartzofeldspathic gneisses to vertical amphibolites, the suture being along the conformable contact of these two formations which have the same upper amphibolite facies metamorphic grade. The gneisses are derived from a combination of late Precambrian sediments and Cambrian granites and the amphibolites from Cretaceous volcanic rocks. Even the deformation pattern within the gneisses and the amphibolites are similar. Although there are a few other geological constraints helping to define this boundary as a suture (see Chapter 21), taken on face value, the above account is true. Clearly it can be extremely difficult to recognize such a suture, whether it be Eocene or Proterozoic.

Palaeomagnetic Constraints

In the last chapter we considered briefly the inconclusive palaeomagnetic data that concern the Grenville orogeny. Here we must review the data pertinent to the question of a mid-Proterozoic supercontinent and which is related to the evidence for the anorogenic magmatism and abortive rifting documented in Chapter 7.

It must be pointed out that in all fairness most palaeomagnetists today treat Proterozoic data and their relevant conclusions with due caution in view of the lack of palaeo-longitudinal control, the poor data coverage, and the inadequacy of isotopic age control. With these limitations, a dispute is currently brewing over the conclusions of Piper (1978, 1980a,b, 1982) as to the existence of a Proterozoic supercontinent (Fig. 10.1) in the period from 2600 Ma to 570 Ma; he maintains that all the palaeopoles from the major shields conform to a single narrow path for the whole of that period and, therefore, that there were no periods of continental sep-

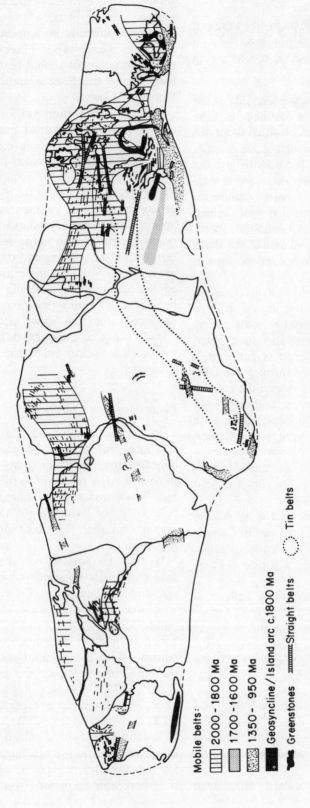

Fig. 10.1 The distribution of major tectonic and magmatic features applicable to the interval 3300–1000 Ma on the Proterozoic Supercontinent reconstruction. The tin belt (while probably not including all deposits of Proterozoic age) incorporates the belt of localities related to known, or probable, mid-Proterozoic plutonism and links with the deposits in the Hudsonian and 1600–1200 Ma terrains of the Laurentian Shield (from Piper, 1982, *Earth Planet Sci. Lett.*, reproduced by permission of Elsevier, Amsterdam)

aration, return and collision between the end of the Archaean and the beginning of the Cambrian. The discussion started when Herz (1969) and Bridgwater and Windley (1973) pointed out that there were belts of mid-Proterozoic anorogenic magmatic rocks aligned across both the northern and southern hemispheres. J. D. A. Piper extended the concept and produced palaeomagnetic data to include the whole of the remainder of the Proterozoic. This opinion is broadly supported by Embleton and Schmidt (1979) who concluded that: 'the cratons which comprise N America, Africa and Australia retained their relationship for much of Precambrian time and that the intervening mobile belts and major sutures did not result from convergence of previously widely separated microcontinents; their origin is ensialic'.

However, the following authorities disagree. Burke *et al.* (1976b) re-evaluated the palaeomagnetic data and concluded that the idea that plate tectonics did not operate in the period 2500–600 Ma is invalid. These authors make a very interesting point: in a

plate tectonic scenario most rocks that are preserved formed either when the ocean was young and small, or just before it closed when it was also small. Because of this, suture zone rocks are very similar, whether the ocean that has closed was 500 or 5000 km wide. This is relevant to the fact that palaeomagnetic errors are too large to detect relative motions less than about $15°$ in translation or rotation (Briden, 1976), and thus an ocean width of up to about 1000 km would be consistent with the spatial restrictions of the palaeomagnetic data (Neary *et al.*, 1976); this would be sufficient for the operation of the Wilson cycle. Irving and McGlynn (1979) critically reviewed the data pertinent to Piper's concept and concluded that: 'no substantial case can be made on palaeomagnetic grounds for the existence of a fixed Pangaea for the *whole* of the time period 2200–1300 Ma as Piper has argued' (author's italics) and that: 'the evidence favours the idea that the older continental nuclei have moved about relative to one another more or less continuously during the early Proterozoic perhaps

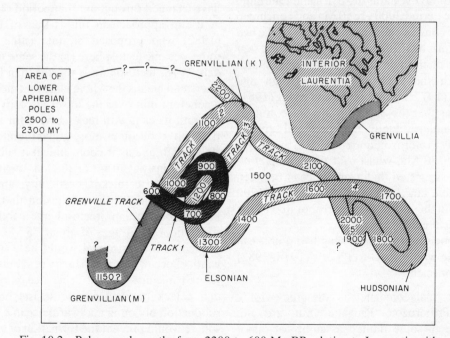

Fig. 10.2   Polar wander paths from 2200 to 600 Ma BP relative to Laurentia with the five main tracks indicated. Hairpins are named by the orogeny to which they are thought to correspond (after Irving and Lapointe, 1975; reproduced by permission of *Geoscience Canada*)

with the repeated assembly of Pangaeas', i.e. during the whole Proterozoic reassembly of most of the continents may have occurred several times. Irving and McGlynn (1981) concluded that: 'the idea of the operation of plate tectonic processes in the Proterozoic is not contradicted by the palaeomagnetic evidence from Laurentia'. Evidence does indicate that very large motions relative to the pole took place in the early Proterozoic, as would be expected from plate tectonics.

For North America the rate of change of palaeolatitude calculated from the APW path in the Proterozoic (2200Æ–600 Ma—Fig. 10.2) was 4–$5°$ Ma^{-1} compared with 1–$1.5°$ Ma^{-1} in the Phanerozoic (Irving, 1979). This probably reflects the more rapid mantle convection driven by higher heat flow in the Proterozoic. From this Irving points out that the long unimpeded drift trajectories in the Proterozoic would be difficult in an earth crowded with many continents, and therefore the continental crust at that time may have been grouped into fewer larger cratons than in the Phanerozoic. Morgan and Briden (1981) reassessed the APW paths of Africa and Laurentia over the period 2000–1850 Ma and concluded that the convergence and divergence of the two paths must be a reflection of collision and separation of the two continents. Within North America the evidence of Cavanaugh and Seyfert (1977) and McGlynn and Irving (1981) indicates that the Superior and Slave cratons moved independently in a horizontal sense in the early Proterozoic and collided in the period 1850–1750 Ma, whilst that of Dunlop *et al*. (1980) suggested that a small ocean between Grenvillia and Laurentia closed about 1200 Ma; both these results are corroborated by the geological evidence.

In conclusion, the following two points can be made with respect to Piper's (1978, 1982) supercontinent.

1. Most palaeomagnetists disagree with a whole-Proterozoic Pangaea, but are not adverse to some limited assembly of major continents at certain times and, in particular, during the mid-Proterozoic when abortive rifting and anorogenic magmatic activity were taking place.

2. The abundant geological evidence quoted in previous Proterozoic chapters is at total variance with a Pangaea in the early and late Proterozoic; supported by data from several disciplines, it seems more likely that mobile belts formed by plate interactions at these times.

So when did modern-style plate tectonics begin? By now it will be apparent where my own preference lies, but nevertheless I am confident that the majority of authorities that have recently published on this issue agree that it started in the early Proterozoic. But we know that conditions could not have been the same then as now, and we should not expect all geological features to be similar. We should be working out the variations that have taken place in deformational style, sedimentary facies, metamorphic conditions and geochemical patterns with a view to finding out in what ways and by how much plate tectonic processes have changed throughout that period of time. A useful approach in this is that of Hynes (1982) who proposed an interesting idea: today oceanic tectosphere (in the sense of Jordan, 1978, tectosphere is thicker than lithosphere and means the whole plate that moves as a coherent unit over the lower viscosity layer beneath it) can, with luck, survive for a long time; but it becomes more gravitationally unstable with age as it cools and so it subducts spontaneously at an age of 200 Ma. Evaluation of tectospheric thickness, stability, and age variations leads to the conclusion that early Proterozoic ocean tectosphere subducted spontaneously at ages of about 75 Ma. Consequences of this model would be a limited width of ocean basins, a paucity of subducted–related magmas and of relict oceanic crust, and a lack of glaucophane schists because subduction of young and warm oceanic tectosphere would prevent the formation of low-T/high-P metamorphic assemblages.

Chapter 11

Palaeomagnetism and Continental Drift

Palaeomagnetists are able to study the variation of the earth's magnetic field through geological time because permament magnetization takes place when igneous rocks and sediments form. It is possible to determine the palaeomagnetic pole responsible for the alignment of magnetic particles, and then a succession of pole positions for rocks of different age from one area. When the resulting polar wander curve of one continent differs from that of another there must have been independent drift of the continents. In this chapter we shall see how palaeomagnetic data have been used to chart the course of continental movements throughout the Phanerozoic. For recent reviews of the application of palaeomagnetism to problems of continental drift see Irving (1977, 1981); Morel and Irving (1981).

Palaeomagnetic Poles

It is not intended to give here either a review of the historical developments in the field of palaeomagnetism or of the basic principles of the method: these aspects are well treated elsewhere. But it will be worthwhile to mention briefly some of the main points about palaeomagnetic poles which are relevant to continental drift.

1. Secular variations of the geomagnetic field can be followed from records in observations for the last 400 years, in pottery furnaces for 1000 years, in deep-sea sedimentary cores for 10 000 years, in lake sediments for 15 000 years and in lavas for up to 20 Ma. These observations show that the average pole positions are centred on the earth's present rotation axis around the geographic rather than the geomagnetic pole. This pattern suggests that the field has been one of a geocentric axial dipole; in other words, like a dipolar magnet centred at the middle of the earth and aligned parallel to the earth's spin axis. This is important because the method of determining palaeomagnetic pole positions is based on the theory that the earth's magnetic field is aligned with the rotation axis. As a consequence, palaeoequators can be calculated to lie normal to palaeorotation poles and palaeogeographic maps with their palaeolatitudes can be constructed.

Two measurements are made. Firstly, one of palaeomagnetic declination giving the direction of the magnetic pole which, as argued above, defines the geographic pole; secondly, one of the magnetic inclination from which the palaeolatitude can be calculated because the tangent of the inclination equals twice the tangent of the latitude. A consequence of the method is that ancient latitudes relative to the present geographic pole can easily be determined from the inclination, but ancient longitudes cannot be determined because ancient declinations may be explained either by a rotation of the continent about a point within it or by translation of the continent along a small circle, or by some combination of the two. Although this is unfortunate, it is not too serious since it is only the palaeolatitudinal position that controls the formation of climatically sensitive sediments and glacial deposits (except for

157

alpine types) and for the most part faunal and floral distributions. Also, the geometrical fit of continental shapes and the matching of geological structures from one continent to another can be used to assist the reassembly of past continents, and oceanic magnetic anomaly patterns can help to reposition the continents as far back as the early Jurassic.

2. By measuring the angles of inclination and declination in a magnetized igneous or sedimentary rock, a palaeopole can be determined and if a sequence of poles are calculated from earlier and later rocks these can be plotted on a projection to define a polar-wander curve for the area concerned. At this point it may not be clear whether the locus of points making the curve was caused by actual wandering of the pole or by lateral movement of the sampling site. However, when the poles are plotted for two continents for rocks of known age from, shall we say, the late Mesozoic to the present, it is clear that two independent curves are defined that converge and meet at the present pole. Because the geomagnetic field is only dipolar this means that drift of the continents must have taken place, and this is the crux of the palaeo-magnetic argument validating the continental drift theory.

Polar-wander Paths

Palaeomagnetism has become a cogent force substantiating the continental drift hypothesis. In order to follow the movements of continents throughout the Phanerozoic it is necessary first to consider the palaeomagnetic evidence. In a review of this type it would, of course, be possible to treat the individual polar wander paths for each continent in turn; but we shall take a broader approach and consider, via their polar wander curves, the relative movement of continents belonging to the two landmasses of Laurentia and Gondwanaland.

Laurentia

S. K. Runcorn first realized in 1956 that the polar paths for North America and Europe were appreciably offset, that they eventually converged on the present pole and that this was significant palaeomagnetic evidence for drift. Fig. 11.1(a) gives an up-to-date portrayal of Runcorn's two curves, and Fig. 11.1(b) shows that with the two continents rotated 40° they became essentially congruous. This amount of rotation has the effect of closing the North Atlantic ocean and compares remarkably well, as McElhinny (1973) notes, with the 38° required by Bullard *et al.* (1965) to obtain the best geometrical fit between North America and Europe using the 500 fathom submarine contour.

The current interpretation of the polar paths in Fig. 11.1(b) is that their convergence from the Cambrian to the Silurian reflects the drift together of the two continents during the closure of the proto-Atlantic ocean. They remained in a co-polar situation from the Silurian to the Trias when the continents formed part of Pangaea, and then diverged in the Trias when the continents broke away from each other and drifted independently until the present. This sequence of events is consistent with information derived from magnetic anomaly patterns, palaeoclimatic indicators and geological correlations from one continent to the other.

For further debate on the most problematical Palaeozoic part of the Laurentian polar paths related to the history of the proto-Atlantic ocean, see Chapter 13.

Gondwanaland

Palaeomagnetic pole positions show that in the latest Precambrian there was an assembly of the continents which was progressively fragmented during the Phanerozoic. Pieces broken off in the Palaeozoic collided with one another forming the present northern continents and thus created the super-continent of Pangaea in the Permo-Triassic (Irving, 1981).

When Pangaea fragmented, the continents of Gondwanaland went their separate ways in every direction and their apparent polar wander paths are therefore very different through the Mesozoic and Cenozoic (Fig. 11.2); moreover, their paths are markedly dissimilar

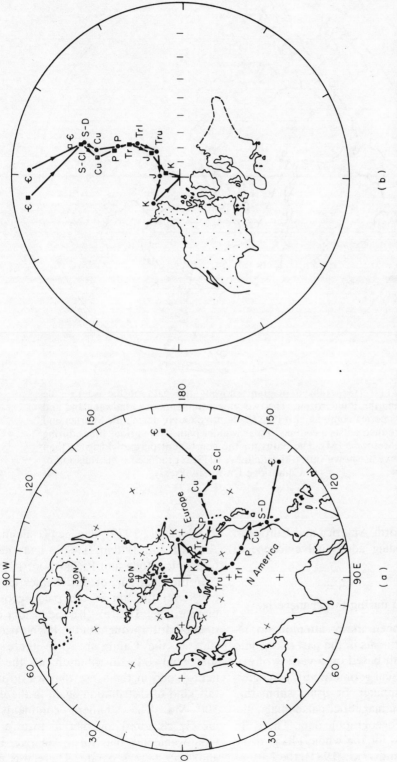

Fig. 11.1 (a) Comparison of the apparent polar wander paths for North America (circles) and Europe (squares) from the Cambrian to the present. (b) The two polar wander paths after rotation of 40° to the Bullard, Everett and Smith (1965) fit of the North Atlantic (after McElhinny, 1973; reproduced by permission of Cambridge University Press)

Fig. 11.2 Generalized diagram showing the path of the South Pole during the Phanerozoic. The two supercontinents of Gondwanaland and Euramerica combine into Pangaea approximately during the Silurian and there then exists a common polar wander path until various times during the Mesozoic (M). The paths for the constituent parts of Pangaea then diverge as shown (after McElhinney, 1973; reproduced by permission of Cambridge University Press)

from those of North America and Europe at this time, providing additional evidence of continental drift.

Continental Drift during the Phanerozoic

There have been many attempts to reposition the continents in the past taking into account their drift based on a variety of evidense such as visual geometric fit, geological correlations, computer fit using submarine contours, oceanic magnetic, palaeomagnetic, faunal and palaeogeographic data. There are recent base maps for the whole Phanerozoic (Irving, 1981; Smith et al., 1981), the Palaeozoic (Keppie, 1977; Scotese et al., 1979;

Ziegler et al., 1979), Pangaea (Hallam, 1980; Rickard and Belbin, 1980), and the early Cenozoic (Rona and Richardson, 1978).

In the final part of this chapter let us follow, with the aid of the palaeogeographic maps in Figs. 11.3(a)–(h), the history of continental drift throughout the Phanerozoic. We start in the Cambrian when there was a somewhat odd distribution of the continents—odd in the sense that we do not see this kind of distribution again in the ensuing 500 Ma. The southern continents were loosely clustered together to form a proto-Gondwanaland, but North America, Europe and Asia were separate. There was a small proto-Atlantic or Iapetus ocean between

Europe and North America. The south pole lay somewhere off northwest Africa. A striking feature is the relatively low latitude of the continents which gave rise to extensive carbonates in shelf regions (Ziegler *et al.*, 1979).

During the Ordovician, tillites were deposited in northwest Africa in the south polar region and evaporites in central Siberia close to the equator. The Silurian began with a sea-level rise caused by the melting of ice sheets with the result that shallow seas covered more of the continents in the Silurian than at any other time. An important development was contraction of the Iapetus ocean in the Silurian and its closure by the mid-Devonian, giving rise to the Caledonian–Appalachian fold belt. Continuing convergence caused the northwest corner of Gondwanaland to impinge on Euramerica, the final collision being responsible for the formation of the Hercynian–southern Appalachian belt in the Carboniferous, and Asia collided with Europe removing the intervening ocean and producing the Ural Mountains. But note the discrepancy in the age of formation of the Ural suture—between the Silurian and Devonian (Seyfert and Sirkin, 1979); between the early/mid and late Devonian (Irving, 1981); between the late Carboniferous and early Permian (Scotese *et al.*, 1979). By 250 Ma in the mid-Permian the Pangaea landmass was created together with the re-entrant Tethyan ocean.

Widespread deposition of evaporites and red beds took place in Laurasia during the Devonian, tillites in Gondwanaland in the Permian, and coal measures in Laurasia during the Carboniferous and in both continental masses in the Permian. These climatic indicators have a latitude-dependent distribution at these times, a fact largely consistent with the palaeomagnetically determined position of the continents. In particular, the tillites occur in high southern latitudes, evaporites and red beds are mostly in low altitudes between 35°N and 35°S, whilst coal measures occur predominantly in high latitudes, except for a minor coal belt that developed in humid tropical latitudes in the Permian.

Little happened to Pangaea for at least 50–70 Ma through the Permian and Triassic except for a small northward drift. The first continental fragments broke away from Pangaea in the mid–late Jurassic (North America from West Africa and Europe) and the last in the mid-Tertiary (South America from W Antarctica), i.e. it took about 130 Ma to fragment the supercontinent completely. The study of this break-up and subsequent dispersal of the continents takes into account the oceanic magnetic anomaly patterns, and therefore the relative positions of the continents for the last 200 Ma of earth history are known with a reasonable degree of confidence.

The first rifting took place in the North Atlantic and western Tethyan region about 200–180 Ma but the first creation of new ocean floor was not until 160–170 Ma (Hallam, 1980) whereas South America did not begin to break away from Africa until 130–120 Ma in the early Cretaceous by which time the North Atlantic ocean was half open (for details of the great variety of geological and geophysical evidence for the opening of the Atlantic, see Chapter 16).

Antarctica began to separate from India by 100–105 Ma, as evidenced by the age of the Rajamahal traps in India, and the Cenomanian age of the earliest marine sediments in Southeast India. Analysis of magnetic anomaly patterns indicates that there was a period of very rapid sea-floor spreading from 110–80 Ma in the Pacific and Atlantic oceans and between 75–55 Ma in the Indian Ocean. The latter motion assisted the separation of India from Antarctica (Norton and Sclater, 1979). Australia broke away from Antarctica about 45 Ma ago in the Eocene and finally South America separated from W Antarctica about 30 Ma ago in the Oligocene, and this allowed the creation of the circum-polar ocean current system (Hallam, 1980). By the mid-Tertiary Pangaea was totally fragmented and the consistent parts well on their way to their present positions. The youngest rifting event and the creation of new ocean floor occurred in the Red Sea and Gulf of Aden at about 3.5 my in the Pliocene (Le Pichon and Francheteau, 1978).

162

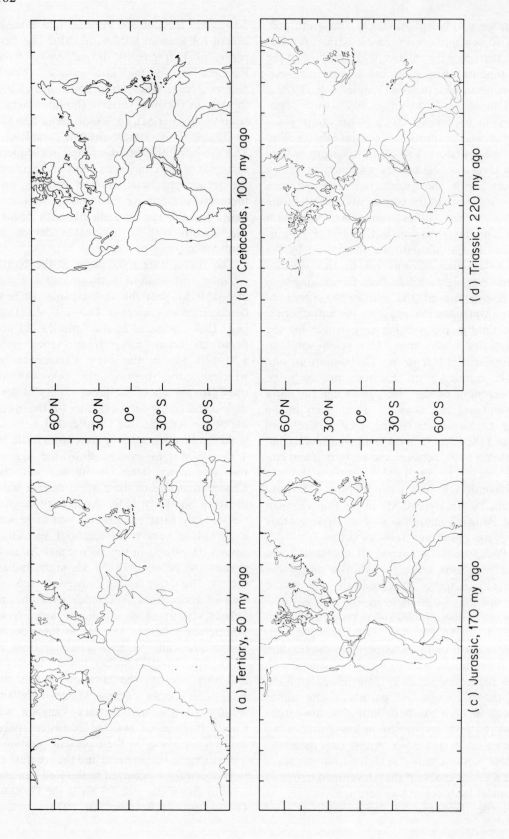

(a) Tertiary, 50 my ago

(b) Cretaceous, 100 my ago

(c) Jurassic, 170 my ago

(d) Triassic, 220 my ago

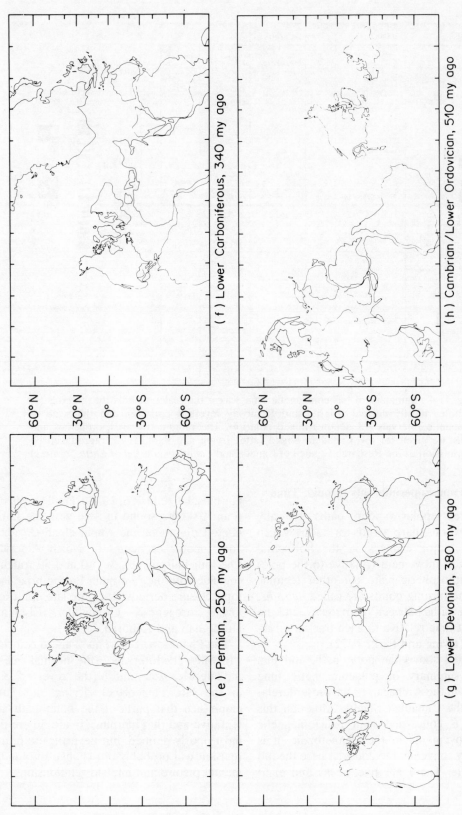

(e) Permian, 250 my ago

(f) Lower Carboniferous, 340 my ago

(g) Lower Devonian, 380 my ago

(h) Cambrian/Lower Ordovician, 510 my ago

Fig. 11.3 Continental drift from the Cambrian to the Tertiary illustrating the formation and break-up of Pangaea (maps based on Smith *et al.*, 1973; reproduced by permission of The Palaeontological Association)

164

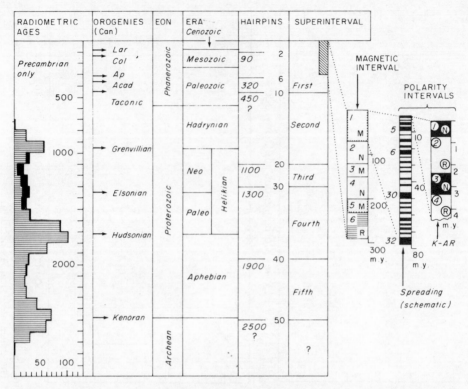

Fig. 11.4   Summary of palaeomagnetic time scales: the scales of Polarity Intervals (N denotes mostly normal polarity and R mostly reversed polarity), and the scale of Magnetic Intervals (M denotes mixed polarity). The radiometric ages, orogenies and eras are given for the Canadian Shield (after Irving and Park, 1972; reproduced by permission of the Research Council of Canada and *Canadian Journal of Earth Sciences*)

Hairpins and Superintervals through Time

Apparent polar-wander paths typically make several sharp bends or loops which describe major changes in the horizontal direction of movement (relative to the pole) of the lithospheric plate (or other tectonic unit). The turning points are called *hairpins*, the polar paths between them *tracks*, and the period of time relative to each track a *superinterval* (Irving and Park, 1972).

Fig. 10.2 shows the polar path, and Fig. 11.4 a summary of palaeomagnetic time scales from the Archaean to the present, relative to the Canadian Shield. Although this chapter is concerned with palaeomagnetic polar movements in the Phanerozoic, it is obviously convenient to consider here the full time scale. There are five tracks and eight

hairpins, labelled 2, 6, 10, 20, etc. Hairpins 2, 6 and 10 correspond in time with the Laramide–Columbian, the Appalachian–Acadian and the Taconic orogenies of North America. The four hairpins 20, 30, 40 and 50 roughly coincide with the four main Precambrian isotopic closure periods reflecting late uplift following orogenesis—Grenvillian, Elsonian, Hudsonian and Kenoran.

The Phanerozoic hairpins 2 and 6 coincide with Magnetic Intervals 2 (the 'normal' one in the late Mesozoic) and 6 (the 'reversed' one in the late Palaeozoic). Irving and Park emphasize that parts of the polar path are tentative and that hairpins 10 and 50 are the most poorly defined, but nevertheless future revision will probably not change the fundamental picture and implied relationships.

Chapter 12

Palaeoclimatology and the Fossil Record

Palaeomagnetic data tell us the relative positions of the continents in the past; the drift of these continents across the palaeolatitudes was responsible for, and so can be double-checked by, the deposition of climatically sensitive sediments such as carbonates, evaporites, desert sand dunes and red beds in warm and/or arid climates not far from the palaeoequator, and the formation of tillites in glacial climates in the palaeopolar regions (Briden and Irving, 1964; Donn and Shaw, 1977; Frakes, 1979; Habicht, 1979; Hay *et al.*, 1981).

However, one should be aware of the limitations of these palaeoclimatic indicators. Sedimentary palaeodistribution does not *prove* the theory of continental drift. But, making the basic assumption that continental drift did occur we can predict that certain sets of circumstances must have existed which controlled different types of palaeoenvironments. The probable existence of *regular* sediment–latitude patterns in the past, as we shall see in this chapter, does provide some confirmation of the continental drift theory.

Whilst the continents were adrift during the Phanerozoic, life was evolving first in the sea and later on the land and in the air. The development of life was clearly affected by continental drift. The fragmentation and reassembly of the continents and their latitudinal movements affected a large number of life-sensitive phenomena, such as circulation of the oceans, supply of nutrients, climatic changes, formation and destruction of

favourable habitats, transgressions and regressions, etc. In the second part of this chapter we shall review these variables and then follow the fossil record through the history of the Phanerozoic.

Carbonates and Reefs

Most modern carbonates and particularly reefs form in warm water ($>20°C$) symmetrically within 30° of the equator especially in tropical and subtropical latitudes, and most Palaeozoic–Mesozoic limestones, dolomites and organic (coral) reefs fall predominantly within 30° of the palaeoequator (Figs. 12.1, 12.4) (Briden and Irving, 1964). Carbonate deposition and reef growth are favoured by a high-energy environment in the shallow-water wave-agitated zone on the windward side of low banks or islands such as off the Bahama Banks, Persian Gulf and the Great Barrier Reef, NE Australia. An increase in water temperature decreases the solubility of calcium carbonate, thus carbonate deposition is favoured by shallow warm water. As the solubility constants of calcium carbonate are unlikely to have changed with time, it is probable that both thick and/or pure carbonate sediments and reefs throughout the Phanerozoic are a good indication of tropical–subtropical low latitudes, although in periods with no ice-caps, such as the Cretaceous, warm seas and carbonate deposition may have extended to higher latitudes.

Simple calcareous reefs, constructed by stromatolitic algae, flourished in the Protero-

166

zoic but more complex types evolved in the early Palaeozoic due to the contribution of organisms such as sponges, archaeocyathids, bryozoans, corals, crinoids, stromatoporoids and brachiopods. Because these fossil organisms required an enormous amount of $CaCO_3$, it is concluded that all the fossil reefs were built in warm, tropical and subtropical waters (Habicht, 1979). They formed sporadically in the Cambrian and Ordovician but they underwent a spectacular development in the Silurian, especially in North America and western Europe. This burst in carbonate reef construction depended not only on the presence of a multitude of suitable organisms but also on the immense shallow-water sea that extended over the cratons in mid–late Silurian time in the aftermath of the Taconic orogeny. An excellent example of a late Silurian barrier reef is in the Michigan Basin where it is associated with both red beds and evaporites.

Evaporites

The principal salts precipitated as evaporites are anhydrite ($CaSO_4$) and gypsum ($CaSO_4 2H_2O$), halite (NaCl), and sylvine (KCl)—this sequence indicates progressively more severe evaporative conditions. The only requirement for evaporite deposition is that evaporation be sufficient to cause a significant concentration of brine. However, this requirement is favoured by a restricted set of circumstances such as limited circulation of sea water in a basin bordering the sea or in an enclosed basin with a high temperature and low rainfall. The limited inflow of marine water is often caused by a carbonate barrier reef and thus there is a common carbonate reef–evaporite association.

These conditions are met today in arid climates but not near the equator. It is well known that there is a strong excess of evaporation over precipitation in the present-day oceans in two zones between 5°S and 35°S and between 15°N and 40°N, whilst there is a reverse relationship between 5°S and 15°N. Thus modern evaporites have a bimodal distribution with maxima in the sub-tropical

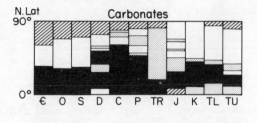

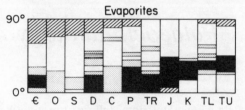

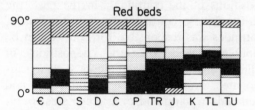

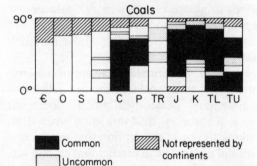

Fig. 12.1 Palaeoclimate spectra through the ages for northern continents (from Habicht, 1979, *Amer. Ass. Petrol. Geol. Studies in Geology*, No. 9, reproduced with permission)

high-pressure zones within 50° of latitude of the equator and a minimum in the low-pressure equatorial zone between 20°N and 20°S (Fig. 12.1). Most fossil evaporites, especially throughout the Palaeozoic, have a similar bimodal distribution with respect to their relevant palaeoequators (Fig. 12.1) (W. A. Gordon, 1975). The essential constancy of the evaporite regime provides a cogent argument that the atmospheric and oceanic circulations and the distributions of deserts and marine salinity have changed little throughout Phanerozoic time (Habicht, 1979). Also the better correlation of ancient evaporites with their respective palaeolatitudes than with the present-day latitudes supports the lateral movement of continents during the Phanerozoic.

The earliest known evaporites are of late Precambrian age but these are rare. They generally increase in number with time throughout the Phanerozoic, reaching a peak in the Permian–early Jurassic when the continents were assembled into the supercontinent of Pangaea. The earlier and later periods of lower evaporite accumulation correspond to the major episodes of continental dispersal (Gordon, 1975).

In many places evaporite sequences range in thickness up to 300–500 m and three of the thickest older onland deposits are in the Michigan Basin (1000 m, late Silurian), the Paradox Basin, Utah (1330 m, Pennsylvanian), and in SE New Mexico (1500 m, Permian). They reach 7 km in thickness in the Red Sea (late Tertiary). There was important evaporite deposition in late Triassic/early Jurassic, early Cretaceous and Tertiary rift valleys associated with incipient stages of sea-floor spreading of the North and South Atlantic (Van Houten, 1977; Evans, 1978). For a review of evaporite formation in marginal basins see Chapter 16.

Red Beds

Red beds are arkoses, sandstones, shales and conglomerates which contain haematite as a visible pigmenting agent. The main requirements for their formation are an oxidizing environment and an adequate supply of ferric oxide.

In most red beds the red pigment probably formed *in situ* by oxygenated pore waters during weathering and diagenesis from ferromagnesian detrital minerals. A hot climate, in which feldspars are relatively stable, is needed for limonite to be dehydrated to haematite (Seyfert and Sirkin, 1979); thus low latitudes are the most favourable. Present day red soils form at latitudes of less than 30°

from the equator, and all known red bed flora indicate a warm temperature or tropical climate (Van Houten, 1964). Most Phanerozoic red beds are situated within 30° of their palaeoequators and thus. are commonly associated with evaporite deposits (Fig. 12.1).

Red beds formed in particular in Devonian–Trias times when extensive land areas resulted from coalescence of the continents. The construction of Pangaea gave rise to the ideal conditions for the formation of evaporites and red beds. In fact, the association with evaporites provides perhaps the most compelling argument for the formation of red beds in arid to semi-arid environments; excellent examples are the Permian red beds, interstratified with halite, in a belt from Texas to North Dakota, and the Lower Trias red beds associated with gypsum deposits in Arizona and Utah. The red beds of Connecticut, W USA formed in a semi-arid climate in late Triassic/early Jurassic rift valleys during the incipient stage of opening of the present Atlantic.

The close association between, and distribution of, many Palaeozoic carbonate sequences, organic calcareous reefs, evaporites and red beds provides strong support for the approximate position of the relative palaeoequators suggested by palaeomagnetic data.

Coal

The accumulation and degradation of vegetation eventually leads under appropriate conditions to the formation of coal. Conditions most favourable for coal formation are warm and humid climate and shallow water; the presence of suitable vegetation is essential.

The earliest coal formed from algae in the early–mid-Proterozoic, in N Michigan and SW Greenland. The first land-plants appeared in the late Silurian, but it was not until the early Devonian that forests existed (although by then trees were at least 10 m high). The first land-plant-derived coal beds are of late Devonian age, in northern Russia and Bear Island near Spitzbergen, but the optimum conditions for coal formation appeared in the Carboniferous and Permian when extensive marine transgressions, caused by continental drift and closure of the oceans, gave rise to vast swamps on several continents. The Carboniferous coals formed in a tropical climate, whereas the Permian coals with their Gymnosperm woods formed in a cool climate. Coal beds continued to form on a smaller scale throughout the Mesozoic.

Permian to Miocene coal beds tend to be concentrated in belts lying polewards of the contemporary evaporites (Fig. 12.1). This distribution is just what one would expect from the fact that these belts are characterized more by precipitation than evaporation. Such symmetrical evaporite–coal belts are recognizable for the last 290 Ma. One might well ask then why more coal did not form in the high precipitation belt near the equator; the answer may be supplied by the observation that decomposition of vegetation in such tropical, wet belts is so rapid that insufficient is left over for significant coal formation.

Fossil tree rings tell an interesting story (J. M. Schopf, 1973). Carboniferous tree trunks from low latitudes lack entirely or have thin growth rings, suggesting an absence of distinctive seasonal changes in humidity/temperature and a rapid growth in a subtropical environment. However, in the late Permian many thick coal beds, associated with tillites in Gondwanaland, were formed within 5–30° of the south pole and they contain tree trunks with prominent growth rings, implying slow growth in an environment comparable with that in present day muskeg swamps in Siberia or in coniferous forests in temperate latitudes.

The distribution of coal beds seems to be a particularly useful palaeoclimatic indication of palaeolatitude, especially if the coals are taken in conjunction with the contemporary equatorward evaporites, red beds and carbonate sediments.

Phosphorites

Palaeomagnetic data have confirmed that most Phanerozoic sedimentary phosphate deposits formed within 45° of their palaeo-equators, like those of today; for phosphorite to form, the coastal part of a continent must drift into a low latitude position (Cook and McElhinny, 1979). The phosphate deposition occurs preferentially along the western sides of continents due to the upwelling of nutrient-rich cold polar currents as they move equatorwards in major oceanic gyrals along western continental margins (Fig. 12.2); alternatively, it may occur in arid zones along low latitude east–west seaways in response to current upwelling (Sheldon, 1981).

The first major period of phosphate accumulation was in the Cambrian (Fig. 12.3) contemporaneous with the fast development of shelly faunas (Ilyin, 1980a). The abundance of phosphorus dissolved in cold waters welling up from the deep oceans stimulated the production of organisms. There was a minor phase of deposition in the Ordovician, but a major hiatus from the Silurian to the Carboniferous which may have been a consequence of the lack of continental/coastal areas at low latitudes. Some Cambrian phosphorites immediately overlie evaporites. It can be envisaged that the evaporites formed in a rift valley, and that the phosphorites were deposited as soon as a narrow seaway formed and dynamic upwelling initiated (Cook and McElhinny, 1979).

The largest concentration of phosphorites in the world is in the Phosphoria Formation in the western USA, which was laid down at low latitudes in the Permian on the west side of the late Palaeozoic Pangaea which stretched longitudinally.

Major evaporites formed in the mid-Cretaceous (Aptian) in rifts associated with the initial breakup of the South Atlantic, but it was not until the late Cretaceous, and in particular in the middle Eocene some 55 Ma later, that important phosphorites were deposited on the west coast of Africa (Ilyin, 1980b). Computed spreading rates indicate that the ocean was 3000 km wide at this time, which was probably the minimum oceanic width necessary for the formation of major oceanic gyrals and associated upwelling. This in turn implies that phosphorites may develop at any time from 15 Ma to 200 Ma after the

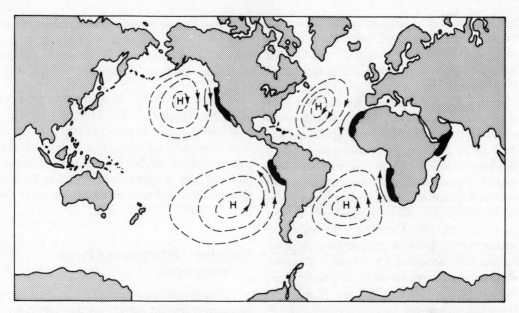

Fig. 12.2 The five major coastal upwelling regions and sites of important phosphorite deposition in the world (black), sea level atmospheric pressure systems and major currents (arrows) that influence them. The dashed circles represent mean idealized positions of isobars during the season of strongest upwelling (from Thiede and Suess, 1983, *Episodes*, **1983**, 15–18)

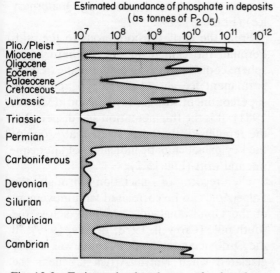

Fig. 12.3 Estimated abundance of phosphate (expressed as metric tons P_2O_5) in phosphate deposits throughout the Phanerozoic. The fixed time interval in the histogram is 25 Ma. It should be noted that a logarithmic scale is used for P_2O_5 abundance (from Cook and McElhinny, 1979, *Econ. Geol.*, **74**, p. 316, reproduced with permission)

initiation of sea-floor spreading along a N–S axis, depending on a slow (1 cm yr^{-1}) or fast (6 cm yr^{-1}) spreading rate, respectively (Cook and McElhinny, 1979).

In general, phosphorite genesis correlates with elevated sea-level resulting from transgression, allowing deposition on shallow shelves, and with warm climate which may be related to increased phosphorus flux to the oceans during times of increased chemical weathering on land, and to the development of widespread oxygen-depleted waters because rates of oceanic circulation and O_2-solubility are reduced (Arthur and Jenkyns, 1981). The increased nutrient supply gave rise to coeval development of widespread biogenic silica deposits.

Oil in the Tethyan Seaway

In a stimulating analysis (which forms the basis of the following account), Irving *et al.* (1974a) identified four factors that controlled the generation and preservation of oil during the Phanerozoic: palaeoclimate, especially temperature, supply of mineral nutrients, plate tectonic factors that controlled initial basin formation and, later, the preservation of the basins.

For the Phanerozoic as a whole, over 80% of all oil occurs in reservoir and source rocks whose latitude at the time of deposition was less than 30°, and over 60% was less than 10°. Cenozoic oil is uniformly distributed with respect to palaeolatitude, Mesozoic oil shows a marked maximum at the equator, and in the Palaeozoic most oil is within 30° of the palaeoequator (the Cenozoic and Mesozoic estimates include the disproportionately large Persian Gulf deposits). Of the world's known oil, 60% was generated in 5% of Phanerozoic time, between 110 and 80 Ma (Albian to Turonian, inclusive).

The basic reason for the oil generation was that dispersal of the continents in the 110–80 Ma interval gave rise to the following set of favourable circumstances:

1. A narrow Tethyan seaway developed (see Fig. 12.4) between Laurasia and Gondwanaland within 30° of the equator (Smith *et al.*, 1973) through which there was a strong westerly flow of wind and ocean currents (Luyendyk *et al.*, 1972).
2. There was very rapid sea-floor spreading in the Atlantic and Pacific Oceans. Fast-spreading ridges displaced ocean waters onto continental platforms (Valentine and Moores, 1970) giving rise to the greatest of all post-Devonian marine transgressions.
3. A dramatic increase took place in the number and variety of organisms, especially plankton, and the cosmopolitan distributions of the early Mesozoic were being replaced by a greater provinciality.
4. About 100 ± 20 Ma there was a peak in the intrusion of carbonatites, kimberlites, alkalic and ultrabasic rocks, this peak reflecting a major renewal of mantle convection which was, in turn, responsible for the formation of the Tethyan seaway and for an increase in the supply of mineral nutrients in the oceans facilitating the rapid development of life.

5. According to oxygen isotope ratios in calcareous fossils, the temperature of ocean waters reached about 21°C—the highest level since the early Mesozoic.

Many oil basins were obliterated by later plate movements if, for example, they lay near leading plate edges or at points of continent–continent collision. The Gulf of Mexico was protected by the northeastward movement of the Antillean arc and the Persian Gulf by the rapid northward movement of the Indian plate.

Glaciation: Ordovician and Permo-Carboniferous

There is well-documented evidence of early Palaeozoic glacial activity in the Saharan region (Harland, 1972) which was close to the Ordovician south pole as determined palaeomagnetically. Other likely occurrences of this age are in Brazil, South West Africa, Nova Scotia, Spain, France and Scotland (Macduff Group of the Dalradian); all these lie within 30° of the pole (and 30° of latitude is near the limit of most drifting Quaternary ice) (Frakes, 1979).

From the early Carboniferous to the mid-Permian the climate in the southern hemisphere cooled to such an extent that extensive permanent ice sheets developed across the supercontinent of Gondwanaland (Martin, 1981). It is the fragmentation and dispersal of the remains of these ice sheets that is one of the key palaeoclimatic indicators of later continental drift (Fig. 15.2).

The record of glaciation through the Palaeozoic can be correlated with movement of the Gondwanaland continents across the south pole (Crowell, 1978; Frakes, 1979). In the Ordovician the Sahara region was widely glaciated when North Africa lay near the pole. By Silurian to Lower Devonian time central South America arrived over the pole causing minor glacial activity there. According to Smith *et al*. (1973) the pole was still situated in this region in the Lower Carboniferous when alpine glacial deposits were formed. Eastern Antarctica had moved over

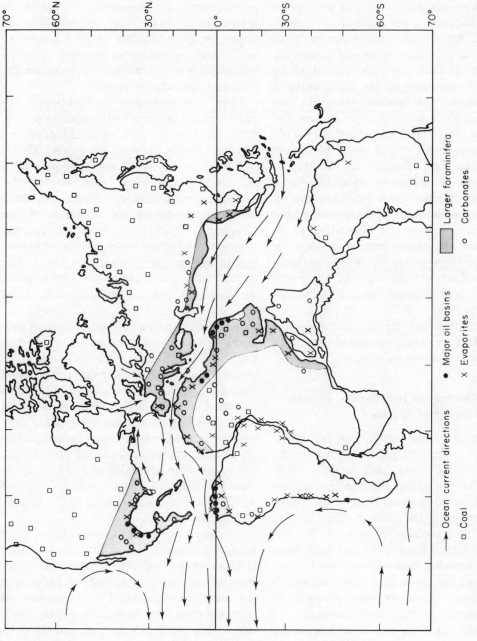

Fig. 12.4 The distribution of the continents in the Cretaceous, 100 ± 10 Ma BP (after Smith *et al.*, 1973; reproduced by permission of the Palaeontological Assoc.) showing the major Tethyan seaway. Oil basins after Irving *et al.*, 1974a (reproduced by permission of *Canadian Journal of Earth Sciences*), foraminifera after Dilley (1973), ocean currents from Luyendyk *et al.* (1972) and sediments after Seyfert and Sirkin (1979) and Habicht (1979)

→ Ocean current directions   ● Major oil basins   ✕ Evaporites   ☐ Coal

Larger foraminifera   ○ Carbonates

the pole in the late Carboniferous when the maximum glacial activity gave rise to several ice sheets scattered across South America, southern Africa and Antarctica. By the Permian the continents had moved westward so that Australia and Antarctica were glaciated, but ice centres had diminished or disappeared in Africa and South America. What brought the glaciation to an end before the Trias? The glaciation was not 'turned off' by the simple movement of the supercontinent away from the pole, because the south pole lay just off the edge of Antarctica from the Trias through to the Cretaceous (Smith *et al*., (1973), long after the end of glaciation. During the late Palaeozoic the Gondwana–Laurasia continent extended in a meridional direction from the south pole to northern high latitudes and this caused the oceanic and atmospheric circulation to be forced into meridional patterns around the world. The initial separation of Laurasia from Gondwanaland and the birth of the North Atlantic ocean enabled the air–ocean circulation to break into a more latitudinal pattern (that prevailed through the Mesozoic into the Tertiary) which prevented the growth and spread of further ice sheets.

Factors affecting the Distribution, Diversity and Extinction of Species

We saw in the last section how drift of the continents in the past influenced the formation of various climatically sensitive sediments. The 'new global tectonics' have beneficially affected palaeontology, like most other branches of the earth sciences, by providing a geographical framework for the understanding of fossil fauna and flora. Here we shall see what factors controlled the distribution of life forms and were responsible for the radiation and extinction of species at certain critical times in the Phanerozoic.

Continental drift was not the only influence on the distribution, diversity and extinction of past life forms; other factors include changes in continental area, sea level, seawater salinity, supply of nutrients to the oceans and, possibly, reversals of the earth's magnetic field. Let us look at these in turn as possible regulators of the fossil record.

If change of latitude is consequent upon continental drift, better or worse climatic conditions will create more or less favourable environments for organisms. Tolerance to temperature fluctuations is probably as important as availability of food resources, but climatic zonation is a major cause of latitudinal provincialization. Diversity usually increases towards the equator.

Plate tectonics enables continents to break-up and separate and eventually collide, thus changing the area of faunal provinces and the barriers to faunal migrations. This is well illustrated by the change from provincial to cosmopolitan biotic provinces in the Palaeozoic as Iapetus or the proto-Atlantic ocean closed during the formation of the Caledonian–Appalachian fold belt. But the destruction of ocean basins leading to extinction of marine species may be a boon to the development of new land plants and animals; such was the case at the inception of the Pangaea landmass in the late Carboniferous.

In pointing out that physical isolation produces its own adaptive radiation and that continental fragmentation therefore tends to increase biotic diversity, Kurtén (1969) noted that reptile radiation took about 200 Ma from the late Carboniferous to the Mesozoic when for a long period the continents were contiguous, whereas the mammals evolving often on isolated continents radiated in only 100 Ma in the late Cretaceous and Cenozoic—in fact the major advance took only about 10–20 Ma from the beginning of the Palaeocene to the Eocene.

The diversity of many marine fauna can be related to major fluctuations in sea level throughout the Phanerozoic time scale which were once related to periods of orogenesis but are currently related to the pattern of fragmentation and reassembly of the continents (Fig. 12.5) (Valentine and Moores, 1970, 1972; Ager, 1976). The plate tectonic mechanism is thought to control the world's marine transgressions and regressions in the following way. During periods of ocean-floor spreading the emergent active ridges have

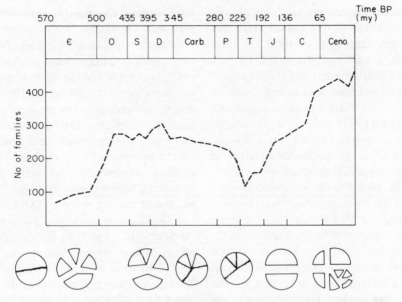

Fig. 12.5 A correlation of levels of faunal diversity with the patterns of continental assembly and fragmentation (indicated diagrammatically) throughout the Phanerozoic (after Valentine and Moores, 1970; reproduced by permission of Macmillan Press)

such a considerable volume that they form topographic highs with the result that oceanic waters are displaced to cause a transgression of the continents. Conversely, if spreading stops, the ridges subside (and their volume is more than that of the average trench at a subduction zone) so causing an increase in the volume of the ocean basin, emergence of the continents and, therefore, a marine regression. The volume of present ridges is about 2.5×10^8 km^3, enough to affect the ocean level by about 0.5 km around all the continents (Valentine and Moores, 1970). Thus, continental assembly favours regression and fragmentation transgression, possible examples being:

1. Regression—formation of late Proterozoic supercontinent; Caledonian–Appalachian and Hercynian suturing and formation of Permo-Trias Pangaea.
2. Transgression—fragmentation of late Proterozoic supercontinent; periods of break-up of Pangaea, especially in the early Jurassic and mid-Cretaceous.

Between 110 and 80 Ma (Aptian to Coniacian in the mid-Cretaceous) the South Atlantic ridge formed when Africa parted from South America. This was a period of particularly rapid sea-floor spreading in both the Atlantic and Pacific oceans resulting in a spectacular marine transgression that almost doubled the area of continental shelves (Reyment and Mörner, 1977). When the spreading rates returned to normal, near those of the present, about 60–50 Ma ago in the Palaeocene there was a corresponding major regression.

On transgression the continental shelf is covered with water and available habitats are considerably enlarged, increasing the total diversity of marine fauna. Alternatively a rapid fall in sea level eliminates most shore-beach and shallow-water habitats and so increases competition for space and nutrients, decreases faunal diversity and causes some extinctions.

Fluctuations in the salinity of ocean waters affect marine diversity; salinity reductions result from removal of salts from the water into evaporite deposits. The evaporites that formed in the Permo-Trias, in rifts and proto-oceans during the initial stages of continental separation (see Chapter 16), are commonly very large (up to 4000 km long, 600 km wide, 7 km thick, Kinsman, 1975a). The withdrawal of so much NaCl into evaporite sequences must have appreciably lowered the salinity of the world's oceans. Nakazawa and Runnegar (1973) therefore considered that changes in seawater salinity were responsible for the crises in molluscan bivalves in Japan at the Permo-Trias boundary.

The supply of nutrients to the oceans, a fundamental factor regulating faunal diversity, may be controlled in two ways. Firstly, formation of a supercontinent leads to an increase in the seasonality in shelf waters and so to an increased nutrient supply. Conversely, if a supercontinent is broken up, the climates of the smaller continental masses would be more equable with less distinct seasonality and vertical mixing, so leaving fewer nutrients in shelf waters (Valentine and Moores, 1970). Secondly, sea-floor spreading releases mineral nutrients into the oceans. About 112 Ma ago this release was increased by a factor of two in proportion to the increase in sea-floor spreading rate, one of the many possible reasons for the generation of the mid-Cretaceous oil deposits (Fig. 12.4) (Irving et al., 1974a).

More carbon was formed during the Cretaceous than that which is present in all the known world reserves of coal and petroleum. The storage of this excess carbon with sulphur led to a significant global increase in atmospheric oxygen and facilitated the formation of the unusual calcium-rich evaporitic salts in the South Atlantic in the Aptian. Such stagnant episodes effectively destroyed all benthic life underneath the hydrogen-sulphide-bearing anoxic bottom waters (Schlanger and Jenkyns, 1976; Ryan and Cita, 1977).

The most sweeping biotic extinction was at the Permo-Triassic boundary; other critical extinctions were at the end of the Cretaceous (Hallam, 1979), Cambrian, Devonian and Triassic. At least 17 hypotheses have been advanced to account for these dramatic episodic biological events (Herman, 1981). We have already discussed the importance of regressions resulting from plate tectonics. Although popularized by some (see 1st edition), the correlation between extinctions and polarity reversals of the earth's magnetic field is not supported by Plotnick (1980). The terminal Cretaceous event was linked by Gartner and Keany (1978) to the flooding of the world's oceans with cold brackish water on the opening of the Greenland-Norwegian passage. Currently popular are extra-terrestrial causes such as meteorite bombardment and wave action (McLaren, 1970), but the most remarkable discovery was by Alvarez et al. (1980) of the manifold enrichments in iridium and osmium in worldwide clay horizons at the Cretaceous–Tertiary boundary. They suggested that the earth captured an Apollo asteroid about 10 km in diameter. The impact crater discharged a large amount of pulverized rock dust into the atmosphere, which dimmed the sun, suppressed photosynthesis and upset the food chain, leading to the extinction of the dinosaurs and many other animal groups. Hsü (1980) and Emiliani et al. (1981) accept the cometary event, but suggest a splash down in the ocean rather than an impact on land. The first drew attention to the possible contamination of the environment by the constituents of the comet, and the latter to the fact that the water injected into the atmosphere could have led to an increase in the global surface temperature of more than $10°C$ and that modern reptiles cannot survive a temperature increase of more than $2°C$ because they live close to the upper limit of temperature tolerance. Hsü (1981) proposed a cometary impact on land in S Russia, accounting for the Kamensk and Gusev craters; calcareous plankton were poisoned by cyanide released by the comet, and by a rise in the calcite undersaturation depth in the oceans and terrestrial animals were killed by thermal stress. However you interpret it, the iridium factor is of momentous significance, although the validity of the analytical data was questioned by Surlyk (1980). Whatever the cause, an increase in temperature on land above biotic threshold limits is widely favoured, and estimates of the duration of such a terminal event vary from essentially instantaneous to 10^4 to 10^5 years (Herman, 1981).

The Phanerozoic Fossil Record

A correlation by Valentine and Moores (1970) of species' diversity (based on benthonic shelf families of nine major invertebrate phyla) throughout the Phanerozoic with the patterns of continental fragmentation and assembly is shown in Fig. 12.5. In general terms, the graph illustrates that faunal diversity was lowest when there was a single supercontinent in the early Cambrian and early Trias, that it was highest when this was broken into several smaller continental plates in the mid–late Palaeozoic and late Mesozoic–Cenozoic, and that there has been an overall increase in diversity over the last

600 Ma (Sepkoski *et al*., 1981). The times of low diversity (depressions in the graph) were periods of significant faunal extinction. Although there has been an overall trend towards specialization throughout the Phanerozoic, the major rediversification in the Mesozoic–Cenozoic is thought to be due to continental fragmentation and drift leading to higher provinciality and to climatic zonations caused by continental configurations giving rise to high latitudinal temperature gradients (Valentine and Moores, 1970, 1972) and to major marine transgressions and regressions (Ager, 1976).

We shall now take a more detailed look at the faunal radiations and extinctions during the Phanerozoic (summarized in Fig. 12.6). The time of appearance and extinction of major faunal and floral groups is indicated in Fig. 12.7.

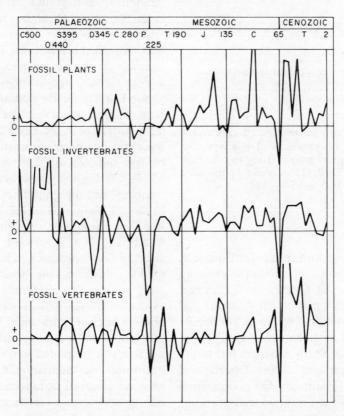

Fig. 12.6 Extinctions and appearances of new faunal and floral groups at stratigraphic boundaries during the Phanerozoic (modified after House, 1971, from I. G. Gass (Eds), *Understanding the Earth*, Artemis Press)

176

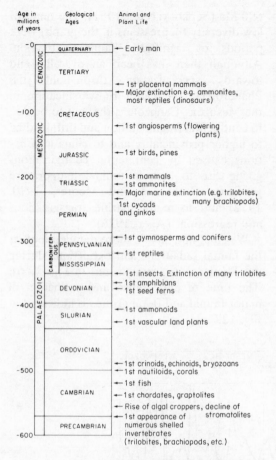

Fig. 12.7 The approximate time of appearance of major groups of animals and plants in the Phanerozoic (modified after Tarling and Tarling, 1971, *Continental Drift*; reproduced by permission of G. Bell and Sons Ltd.)

Early Palaeozoic

By the late Cambrian a triangular European plate was surrounded by ocean (to close up and give rise to the Caledonian–Hercynian and Uralian belts), and North America had moved away from the southern continents which remained in a coherent group. The equator lay across North America, just to the north of Europe and across Siberia and Antarctica and warm shallow continental shelves (caused by a Cambrian transgression) produced abundant limestones (especially in North America) and reefs with calcareous algae and archaeocyathids along the presumed Cambrian tropical belt; the groups

with calcareous hard parts were especially able to undergo rapid diversification. The overall climate seems to have improved considerably since the widespread glacial conditions of the late Precambrian. The combination of the climatic amelioration and the spread of shallow warm shelf seas probably greatly assisted the sudden diversification of life.

The Precambrian–Cambrian boundary records the time when organisms developed skeletons allowing their preservation as fossils, and when there was an explosive evolution in marine life. The following nine phyla of marine invertebrates appeared for the first time in the early Cambrian: Protozoa, Coelenterata (jelly fish, anemones), Archaeocyatha (sponges), Porifera, Bryozoa, Mollusca (gastropods, lamellibranchs, cephalopods), Brachiopoda, Arthropoda (trilobites) and Echinodermata (cystoids). Over 900 lower Cambrian fossil species are known. So rapid was the evolutionary explosion that all the remaining major groups of invertebrate fauna became evident by the middle Ordovician; in other words, it took about 120 Ma for the complete invertebrate range of phyla to evolve. The development of algae croppers in the early Cambrian led to the rapid demise of stromatolites which had been so common in the Proterozoic (Edhorn, 1977). During the Cambrian the rise of the trilobites and the graptolites was impressive (although two-thirds of the trilobite families were to disappear by the end of the Cambrian). The Ordovician, a time of extensive sea-floor spreading, was a major period of diversification—the brachiopods, graptolites, bryozoans, nautiloids and ostracods expanded considerably, corals and stromatoporoids became worldwide, bottom-dwelling graptolites evolved into planktonic types and the first fish continued to develop.

During the Silurian the marine transgression achieved its maximum extent giving rise to widespread reef limestones. A wide shallow-water sea covered much of Asia, Europe and North America and there were therefore no barriers to the migration of shelf faunas which by the Silurian had a cosmopolitan character. Also, this was the first time that attempts were made to colonize fresh water leading to the appearance of vascular land plants and true fish; these were the earliest indications of the great changes that were to come in the late Palaeozoic.

Late Palaeozoic

In the period 395–225 Ma the continents were drifting together to form Pangaea; this involved closure of the pre-Caledonian ocean (Iapetus) by the early Devonian (it began closing by the mid-Ordovician), of the pre-Hercynian ocean by the late Carboniferous and of the pre-Uralian ocean by the Permian. The effect on life forms was firstly to destroy the habitats of many marine invertebrates, causing the reduction in number of some and the extinction of others and, secondly, to provide a suitable terrestrial environment for the evolutionary advance of plants and vertebrates.

Transgressions near the lower/middle Devonian boundary led to the establishment of widespread equable carbonate regimes in tropical latitudes, enabling corals and stromatoporoids to diversify rapidly, joining algae as reef-forming complexes, and to become worldwide. In the early Upper Devonian, however, the reef complexes became progressively attenuated, resulting in the virtual extinction of the reef corals and stromatoporoids; and, once the protection of the platform edge carbonates was lost, organisms living on the shelf (especially suspension-feeding brachiopods) were so adversely affected that some of them became extinct.

Trilobites declined considerably in the Devonian, many groups dying out leaving only a few to continue into the Carboniferous and Permian. Graptolites also declined, surviving only until the end of the Carboniferous.

There was some degree of provinciality of faunas in the early Devonian but by the mid-Devonian continental masses had moved together to produce a semicontinuous landmass. One result of this convergence was the formation of the 'Old Red Sandstone Continent' extending across North America and NW Europe with the first widespread terrestrial sediments of the Phanerozoic. This tectonic evolution stimulated the rapid development in the Devonian of fish (freshwater types with lungs) and vascular plants (the first forests appeared in the middle Devonian), and the appearance of amphibians, winged insects and terrestrial vertebrates. Fig. 12.8 shows that the distribution of late Palaeozoic labrinthodont amphibia and of Triassic reptiles are all grouped systematically across the relevant palaeoequator; they have a random distribution with respect to present latitudes.

Major transgressions in the Carboniferous led to the widespread deposition in low latitudes of shallow-water limestones with diversified corals and crinoids, and Coal Measures with seed ferns and giant cockroaches (10 cm long) and dragonflies (0.75 m wing span). A high latitude equivalent of the Coal Measure flora developed across Gondwanaland with the well-known *Glossopteris* plants associated with glacial deposits.

Similarly, there were great advances on land during the Carboniferous. Amphibians flourished and gave rise to the reptiles, and forests of trees such as *Lepidodendron* and *Sigillaria* reached 30 m in height.

Regression began at the end of the Carboniferous and continued in the Permian. Although fusulinids (foraminifera) and echinoids were still diversifying, rugose and tabulate corals declined as reef builders, their place being taken by calcareous sponges and bryozoans. Permian brachiopod distribution enables the palaeoequator and geosutures to be defined. Amphibians declined as the reptiles flourished. The distribution of the terrestrial faunas of the late Carboniferous and early Permian supports the palaeomagnetic data, indicating that North America and Europe were contiguous at that time. Conifers proliferated and bugs, beetles and cicadas appeared in the insect world.

By the end of the Permian all the continents were sutured together and uplift of the land following the several periods of oro-

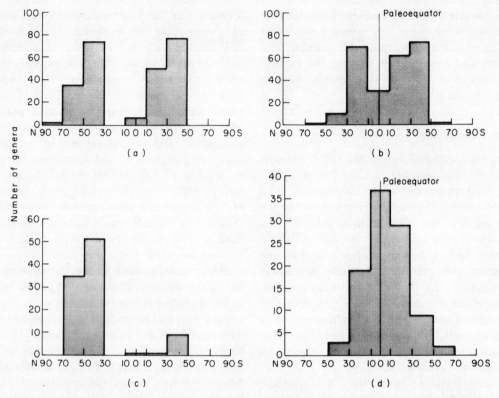

Fig. 12.8 Diversity of Triassic reptiles plotted against (a) Present-day latitudes and (b) Palaeo-latitudes; and diversity of late Palaeozoic amphibia against (c) Present latitudes and (d) Palaeo-latitudes (after Brown, 1968; reproduced by permission of *Australian Journal of Science*)

genesis was responsible for the withdrawal of seas from the Pangaea landmass. The result of these tectonic circumstances is that the Permo-Trias boundary at the end of the Palaeozoic represents the greatest period of extinction in the fossil record (Figs. 12.5, 12.6). Marine invertebrates were especially affected; trilobites, blastoids, fusulinid foraminifera, eurypterids, rugose corals, three-quarters of the bryozoan groups and several of the cephalopods, sponges, brachiopods and echinoderms, and many families of molluscs, all became extinct. Plant life was not seriously affected but vertebrates were. The agnatha vertebrates died out by the mid-Permian and between the Permian and the Triassic some 75% of amphibian and over 80% of reptile families disappeared.

Mesozoic

In the 30–40 Ma period between the mid-Permian and the mid-Trias the only new tectonic development was that the Pangaea landmass drifted northwards about 30°. In the late Triassic rift structures formed between North America and Gondwanaland, representing the incipient stages of break-up of Pangaea. The other continental masses separated during the Jurassic and Cretaceous. These tectonic conditions allowed widespread transgressions and the formation of extensive continental shelves along the trailing edges of dispersing plates, favourable habitats for the rediversification of many shallow-water invertebrates; the vertebrates continued to evolve and diversify on the separating continents.

In the Triassic, faunas were largely cosmopolitan and only a few major provinces can be differentiated. At this time the belemnoids, oysters, complex-sutured ammonoids, crustaceans and echinoderms became important. Amongst the reptiles dinosaurs evolved on the land, plesiosaurs and ichthyosaurs in the sea and gliding pterosaurs in the air. Mammals evolved from the therapsid reptiles in the late Triassic. Land floras flourished with important palm-like cycads, horsetails, conifers, ferns and ginkoes.

During the Jurassic those faunas that survived from the Palaeozoic or were established in the Triassic continued to diversify. Ammonoids and foraminifera flourished, as did corals and sponges in carbonate reefs on shallow shelves. The frog (an amphibian), coccoliths (planktonic plants), angiosperms (flowering plants) and the first bird (*Archaeopteryx*, with feathers developed from reptile scales) appeared. Terrestrial reptiles, such as dinosaurs, were prominent, the pterosaur reptiles learnt to fly, but mammals were slow to evolve. An important point is that by contrast with the cosmopolitan terrestrial faunas of the Triassic, by late Middle Jurassic times a Boreal Province among the ammonites and belemnites became distinguishable from a Tethyan Province in the northern hemisphere. Separation of the faunal provinces reached a peak during the late Jurassic and early Cretaceous, coinciding with a period of extensive regression, with the result that cephalopod correlations became impossible. In a similar way plant distribution shows a latitudinal–climatic zonation into a broad equatorial belt separating two east–west polar belts from the early Triassic to the early Cretaceous.

In the Cretaceous Tethys was closing and the North Atlantic ocean was opening (Fallaw, 1979). Some Gondwana continents began to separate but about 100 Ma ago the Gondwana landmass lay in the same latitudinal position as in the Jurassic, 70 Ma before (Fig. 11.3). There were no faunal–floral crises at the end of the Jurassic so that the life forms established in the Triassic and Jurassic continued to develop. Planktonic globigerinas, coccoliths and diatoms became so abundant as to form foramineral oozes, especially along the Tethyan seaway (Fig. 12.4) (Dilley, 1973), and calcareous sponges, jellyfish, corals, bryozoans, pelecypod molluscs, belemnoid cephalopods, echinoids and crinoids all flourished in Cretaceous seas. Salamander amphibians emerged and terrestrial, marine and flying reptiles attained a peak of development. Most spectacular was the radiation of the angiosperms which overtook the gymnosperms during this period. According to Colbert (1973) reptiles retained a cosmopolitan character throughout the Mesozoic because, in spite of the fact that continents were drifting, there was sufficient connection between blocks to allow the dinosaurs and their contemporaries to wander far and wide.

At the end of the Cretaceous, 65 Ma ago, about one-third of all living species became extinct, including the ammonites, belemnites, some plants, bryozoa, bivalve molluscs, echinoids and planktonic foraminifera, and most of the terrestrial, marine and flying reptiles (e.g. dinosaurs, ichthyosaurs and pterosaurs); of the reptiles only the crocodiles, snakes, turtles and lizards survived into the Tertiary. The Cretaceous–Tertiary boundary was a time of change in two other respects. Firstly, after the long transgressive period during the Cretaceous, there was a regression at the Maestrichtian–Danian boundary and, secondly, the normal geomagnetic polarity era of the Cretaceous moved into one of mixed polarity in the Tertiary. Possible causes to account for such relationships were reviewed in the last section.

Cenozoic

The continuing faunal and floral rediversifications during this period were a result partly of the increasing longitudinal provinciality created by fragmentation and separation of drifting continents (Fallaw, 1979; Fallaw and Dromgoole, 1980), and partly of the widening latitudinal spread of the continents causing cooling of shelf waters in high latitudes and of a consequent latitudinal climatic zonation into separate provinces with individual diversity patterns culminating in major glaciation from mid-Miocene onwards in Antarctica (Valentine and Moores, 1970).

The most important development in the history of life in the Tertiary was the extraordinarily rapid rise in about 10–20 Ma of the mammals which took over the environments vacated by the reptiles on the land and in the sea and air. By the Eocene, primates, elephants, rhinoceroses, pigs, rodents, horses, sea cows, porpoises, whales and bats, as well as most orders of modern birds and many families of plants, had all appeared.

The evolution of the vertebrates (fish, amphibians, reptiles and mammals) throughout the late Mesozoic and Cenozoic was strongly affected by the palaeogeography of the continents. Gondwanaland was more or less isolated from Laurasia in the late Mesozoic and this separation facilitated the development of two fairly distinct faunas.

Many foraminifera such as the globigerinids and nummulites evolved rapidly becoming abundant during the Tertiary. Gastropods, pelecypods and reef corals were the predominant invertebrates.

Chapter 13

Caledonian–Appalachian Fold Belt

The evolution of the continents during the Palaeozoic involved several phases of orogenesis. Within the North American/Eurasian continent there were five phases of orogeny during the Palaeozoic:

1. Caledonian: Upper Silurian–Lower Devonian (Britain, Scandinavia, E Greenland)
2. Acadian: end of Lower Devonian (Appalachians)
3. Hercynian: Upper Carboniferous (Europe)
4. Alleghanian: Upper Carboniferous or Lower Permian (Appalachians)
5. Uralian: Permian (Europe–Siberia)

The distribution of the three main tectonic events is illustrated in Fig. 13.1.

In this chapter we shall make a special study of the Caledonian–Appalachian belt, which extends from Britain to the eastern USA, as this is better documented than any other Palaeozoic fold belt. Important evidence in this belt in favour of a former proto-Atlantic ocean includes differences between European and American fauna, and ophiolite complexes which are remnants of that oceanic crust.

The final part of this chapter will review the mineral deposits in these Palaeozoic fold belts—some regional metal zonations provide corroboration for the plate tectonic models.

The Iapetus (proto-Atlantic) Ocean

Application of the plate tectonic theory to the formation of the Caledonide–Appalachian fold belt depends on the opening of an ocean in early Palaeozoic time and its closure during the late Palaeozoic (J. T. Wilson, 1966; Roberts and Gale, 1978). Let us look at the evidence for movement.

The Opening Stage

There is a lot of information in favour of the existence of an ocean along what is now the middle of the fold belt (Dewey, 1969; Bird and Dewey, 1970).

1. Faunal province data suggest that there was increasing separation of the American and European species during the Cambrian and Lower Ordovician.
2. As would be expected, there are no Cambrian, but only Ordovician and Silurian oceanic sediments preserved along the site of the former ocean; the earlier sediments were destroyed during subduction.
3. The facies types and evolution of the early sediments are remarkably similar to those of Mesozoic–Cenozoic Cordilleran-type geosynclinical belts, viz. miogeoclinal clastic wedge followed by carbonate platform and pelitic–clastic continental rise.
4. The late Precambrian–Lower Palaeozoic structural history on either side of the central 'oceanic' belt was markedly different.
5. There are slices of oceanic crust that were locally thrust on land as tectonic klippe (ophiolite complexes) which retain evidence of the nature of the early oceanic volcanics and sediments.

181

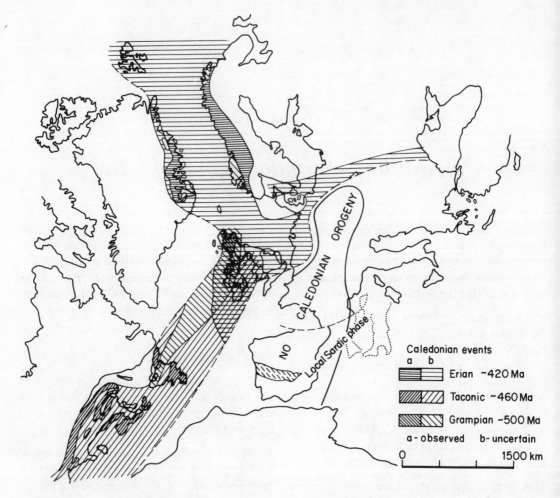

Fig. 13.1   The distribution of the three main tectonic events of Caledonian age in the North Atlantic region (after Zwart and Dornsiepen, 1980)

The Closing Stage

The following evidence is used to suggest that the Iapetus Ocean was contracting during the Silurian and was closed by the Middle Devonian:

1. The earlier distinctive American and European faunal provinces began to decline in the mid-Caradocian. Late Ordovician faunas show increasing evidence of mixing and, by Silurian times, the provinces are indistinguishable.

2. The formation of Benioff zones, and thus the contraction of the ocean, is indicated by the accumulation in the trench of scraped-up oceanic sediments and volcanics (the argille scagliose facies of the Dunnage mélange), the formation of high-pressure glaucophane-bearing assemblages and oceanic crust segments near the trench (e.g. Ballantrae ophiolite complex), and the intrusion and extrusion of calc-alkaline plutonic and volcanic rocks on the continental margins.

3. The gradual elimination of the ocean gave way in places to the onset of non-marine conditions in the Middle Silurian (Wenlock—the Midland Valley of Scotland). By the uppermost Silurian (Downtownian) fish were restricted to brackish environments and in the Devonian extensive Old Red Sandstone formed under desert (continental) conditions.

Now we shall look in some detail at the key evidence for the existence of that Iapetus Ocean: faunal provinces and ophiolite complexes.

Faunal Provinces

Faunal provinces are marine regions inhabited by a characteristic association of organisms and bounded by barriers preventing the spread and mixing of the characterizing species. It was once generally thought that the faunal differences arose from separation by intervening landmasses; however, the plate tectonic concept provides us with a firmer basis for reconstruction of the continents, confirmed independently by palaeomagnetic data, for example, and thus it can now be shown with some degree of sophistication that the barriers were tectonically controlled by continental movements (Valentine and Moores, 1970, 1972) (Fig. 12.5). Considerable provinciality arises when the continents are fragmented and widely scattered, providing topographical barriers such as landmasses and deep oceans, together with climatic barriers when the continents stretch over great latitudes. In particular it was the formation of an ocean that created a barrier to migration of shallow-water organisms. Lowest provinciality and highest cosmopolitan distribution is associated with a supercontinent or Pangaea.

We shall now consider the faunal province data which suggest that the American and European continents were widely separated by the early Ordovician and were close together in the Silurian (Holland and Hughes, 1979; Cocks and Fortey, 1982).

It is important to realize that it was not necessary for the ocean to be very wide for it to act as a barrier to migrating species. The ocean currents flowed parallel to the North Atlantic continental margins in the Ordovician and therefore quite a narrow ocean would have sufficed to prevent migration and faunal mixing.

The ocean was wide enough to separate the faunal provinces in Cambrian times. As the ocean closed, the pelagic animals crossed first, followed by animals with pelagic larval stages (trilobites and brachiopods). When the ocean had closed at one point, animals without a pelagic larval stage (benthic ostracods) were able to cross, and finally fish crossed when there were non-marine connections in fresh or brackish water. From estimates of pelagic larval migration (independently confirmed from subduction rates), McKerrow and Cocks (1976) calculated that the ocean had a minimum width of 2000–3000 km at 436 Ma in the late Ashgillian (latest Ordovician). However, Spjeldnaes (1978) concluded that by this time the ocean had ceased to act as a barrier to the dispersal of marine benthonic faunas; this is confirmed by D. J. Siveter (personal communication) from the distribution of ostracods. Also, Cocks and Fortey (1982) pointed out that the Iapetus ocean is well characterized by faunal differences from the Cambrian until the lower Ordovician, but from Caradoc times onwards these differences dwindle, until by the latest Silurian only the distribution of ostracods separates the two sides of the ocean on faunal grounds.

There is some disagreement about when Iapetus started to open—in the late Precambrian, the Torridonian clastics being related to the initial rifting at about 810 Ma (Dewey, 1969; Stewart, 1982), or in the earliest Cambrian (Anderton, 1980). By the early Cambrian two trilobite provinces were established.

The greatest degree of provinciality amongst Palaeozoic organisms reflecting the maximum separation of the continents occurred in the Lower Ordovician when the American (Pacific) and Acado–Baltic (Atlantic) provinces were prominent (Fig. 13.2).

There was, however, a drastic change in palaeofaunal boundaries by the latest Ordovician (Ashgillian) caused by a decline in provinciality. The faunal evidence is considered by many palaeontologists to be consistent with subduction throughout the mid-late Ordovician, resulting in contraction of the Iapetus ocean and the mixing of the previously independent fauna. The brachiopod data suggest that the beginning of faunal exchange, following on the initial continental collision, took place in the mid-Caradocian. Continents continued to get closer during the

184

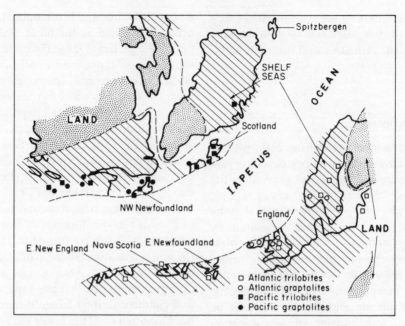

Fig. 13.2 Pacific- and Atlantic-type tribolites and graptolites in the shallow
shelf seas bordering the Iapetus ocean in Cambrian–early Ordovician time
(adapted from Cowie, 1974, with additional data from Cocks and Fortey,
1982)

Silurian with the result that there was a cosmopolitan or worldwide distribution of trilobites, graptolites, brachiopods, coelenterates, nautiloids, and crinoids; only ostracods show some sign of limitation to provinces. The Baltic Shield collided with the Canadian Shield in the late Silurian or early Devonian to produce the Caledonian Orogeny, but evidence from freshwater fish faunas suggests that not until the Middle Devonian was there a land connection between the Canadian–Baltic Shield and Gondwanaland. In early Devonian times the Old World faunal province (N America, Europe, Asia) was still separated from the Austral province (S Africa, S America, Antarctica). Likewise there was a high degree of provinciality amongst early Devonian brachiopods in three world provinces. However, by the Middle Devonian these continental blocks were joined, thereafter leaving a semicontinuous landmass composed of all the continents (proto-Pangaea). It can be no coincidence that this increase in land area in the Devonian took place at the same time as the explosive evolution of terrestrial verte-brates and plants which became similar the world over (see further in Chapter 12). During the Permo-Carboniferous the gradual elimination of shallow sea environments gave rise, by the end of the Permian, to extinction of many marine faunas (e.g. fusalinids, blastoids, rugose corals, eurypterids, trilobites) and to the marked decline of others (e.g. foraminifera, bryozoans, brachiopods, ammonoids, ostracodes). Finally, in the Upper Carboniferous (late Westphalian) central Europe collided with the Baltic Shield giving rise to the climactic deformation of the Hercynian Orogeny—the formation of Pangaea was completed by the end of the Palaeozoic.

Ophiolite Complexes

Ophiolite complexes are stratigraphic units ideally comprising the upward sequence: serpentinite and peridotite, gabbro, 'sheeted' basic dyke zone, pillow-bearing basic volcanic rocks, chert, pelagic limestone and argillite. Nowadays these are widely regarded as being slices of oceanic crust/mantle but there is a current tendency to label or misidentify some deformed mafic–ultramafic complexes as ophiolites. In the identification of Palaeozoic ophiolites much depends on the analogy with examples in the Alpine fold belt, such as Troodos in Cyprus and Vourinos in Greece, and in turn with present day oceanic crust. But amongst the many complexes in the Appalachian–Caledonian belt are some key examples, well preserved and documented, which leave little doubt that they are of oceanic crust/mantle derivation: the several bodies in Newfoundland (Fig. 13.3) and the Ballantrae Complex in Scotland. In fact the Bay of Islands Complex in Newfoundland is as good an example of transported oceanic crust/mantle as the Troodos Complex.

Newfoundland

In western Newfoundland there are several ophiolite complexes (Fig. 13.3): Bay of

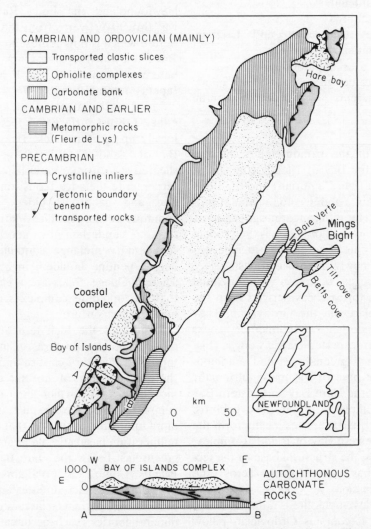

Fig. 13.3 The ophiolite complexes of W Newfoundland. The section shows the Bay of Islands Complex transported over a clastic wedge, itself thrust westwards over an autochthonous foreland. (Redrawn after Williams and Smyth, 1973; reproduced by permission of *American Journal of Science*)

Islands, Coastal, Baie Verte, Betts Cove–Tilt Cove and Hare Bay. Most of the successions consist of layered plutonic rocks underlain by metamorphosed and deformed basic rocks and overlain by little-deformed pillow lavas and sediments. The general succession, which may be up to 10 km thick, is as follows:

> *Top*
> Mafic pillow lavas with interlayered sediments (chert and greywacke)
> Meta-volcanic rocks cut by diabase dykes
> Massive gabbros, diorites, quartz diorites and trondhjemites
> Banded gabbros
> Peridotites (lherzolites and harzburgites), dunites, pyroxenites
> Garnetiferous amphibolites
> Meta-volcanic greenschists and minor meta-sediments (pelitic–psammitic schists)
> *Bottom*

Zircon ages of the trondhjemites indicate formation of the Bay of Islands complex at about 508 Ma (late Cambrian) and the Betts Cove complex about 463 Ma (mid-Ordovician). The difference in age supports the idea that each ophiolite belt is the product of a separate spreading episode in independent, narrow, newly rifted ocean basins.

The complexes occur in subhorizontal allochthonous thrust slices derived from the east and emplaced in the Ordovician. They are mostly underlain by sedimentary rocks belonging to transported clastic wedges (Fig. 13.3). Although greenschists and amphibolites form the base of the mafic–ultramafic successions in the Hare Bay and eastern Bay of Islands Complexes, peridotites, gabbros, volcanics and quartz diorites locally form the base of the western Bay of Islands Complex; in other words, the structural base of the tectonic slices has cut obliquely across the ophiolite succession.

The Betts Cove ophiolite is noteworthy for the fact that it contains Ordovician pillow lavas and dykes of peridotitic and pyroxenitic komatiite—such rocks are rare outside of the Archaean (Upadhyay, 1978). The Baie Verte ophiolite occupies a narrow lineament (Fig.

13.3) which is interpreted as the remains of a narrow (50 km) marginal basin (Kidd, 1977) and which is probably the root zone of other transported ophiolites (Williams, 1977). The Mings Bight ophiolite was generated as the floor of a small rear-arc or intra-arc basin (Kidd *et al.*, 1978). The Bay of Islands Complex originated on the non-transform side of a ridge termination, whilst the adjacent Coastal Complex acquired its tectonic complexities during its movement along a ridge–ridge transform and past a ridge termination (Karson and Dewey, 1978). In the SW corner of Newfoundland the Long Range ophiolite was thrust westwards from the main Iapetus Suture which is now represented by the Cape Ray Fault which separates high-grade gneissic basement from the W and E margins of the Iapetus ocean (Brown, 1976).

The basal contact of the successions is either a thrust surface or a tectonic mélange, locally up to 70 m thick. Below part of the Bay of Islands Complex there is *c*. 30 m of sheared serpentinite underlain by *c*. 17 m of serpentinite mélange (consisting of serpentinite and gabbro boulders in a finely-comminuted serpentinite matrix) which, in turn, is underlain by another 17 m of sedimentary mélange containing sandstone and serpentine boulders in a shaly matrix. These mélange zones are a clear indication that the ophiolite complexes are sitting on major thrust surfaces.

Sometimes the high temperature peridotites are situated on thrust planes directly on top of sedimentary clastics and the fact that there is no sign of normal contact metamorphism is a further indication that the complexes must have been transported. But more interesting is the fact that there is a high temperature basal aureole of amphibolite and greenschist below the Hare Bay and Bay of Islands Complexes (Williams and Smyth, 1973). The contact aureoles are very odd as they have regional metamorphic/tectonic mineral fabrics (such as lineated schistosity and rotated garnet porphyroblasts). The proposed plate tectonic model envisages a subhorizontal sheet of hot, recently accreted oceanic crust being thrust over supracrustal rocks at the continental margin which became metamorphosed and deformed by the over-riding thrust sheets. Subsequent to this the aureole rocks became welded against the basal peridotites, and the ophiolite slice moved along a lower structural base so that the aureole was included with it as a structur-ally underpinned slab. Tectonic transport then took place by cold gravity sliding giving rise to the characteristic mélange zone, and the aureole rocks moved with the ophiolite slice for at least 80–105 km.

Ballantrae

The Ballantrae ophiolite is described by Bluck (1978). It has the following critical fea-tures:

1. Serpentinites at the base of the body are underlain by foliated greenschists and amphibolites (Spray and Williams, 1980).
2. The basal ultramafics contain lherzolites, amphibole spinel-bearing eclogitic rocks.
3. Unfoliated gabbros associated with trondhjemites and diorites.
4. The gabbros are cut by a few basic dykes, but a sheeted dyke zone is absent.
5. A volcanic pile with spilitic pillow lavas, volcaniclastic rocks, black shales and cherts that was originally at least 5.0 km thick. REE patterns in the pillow lavas, the lack of a sheeted dyke swarm, and the very thick lava sequence suggest to Lewis and Bloxam (1977) formation of the 'ophiolite' in an island arc rather than a mid-oceanic ridge environment.
6. Overlying the ophiolite is an olistostrome containing fragments up to 7 m across of glaucophane schist, chert, volcanic pillows, gabbro, pyroxenite, amphibolite and serpen-tinite, etc.; the silty shale matrix to the frag-ments has a highly sheared texture. These conglomerates are interbedded with cherts and black shales containing mid-Arenig grap-tolites. Evidently the whole ophiolite was undergoing erosion at the time of conglomer-ate deposition. Zircons from the trondhje-mites have a U–Pb age of 483 ± 4 Ma, which is mid-Arenig, close to the age of the overly-ing conglomerates. These data indicate the formation and closure of a short-lived margi-nal basin (Bluck *et al.*, 1980).

Anglesey

In his 1969 paper Dewey suggested that a late Precambrian southward-dipping subduc-tion zone was sited in the area of Anglesey on the basis of the following evidence. In the Mona complex there are flysch-type sedi-ments, an olistostrome tectonic mélange, spilitic pillow lavas, chert, jasper, gabbro, ser-pentinized ultramafic rocks (dunite, harz-burgite, pyroxenite, lherzolite) (Maltman, 1977) and lawsonite–pumpellyite glauco-phane schists (Gibbons and Mann, 1983). These rocks do not occur in a single strati-graphic succession but are scattered through-out the island and there is no sheeted dyke complex. Nevertheless, they may represent fragments of oceanic crust-mantle tectonically emplaced in an accretionary terrain at a sub-duction margin (Thorpe, 1978). However such an origin is rejected by Gibbons (1983) who suggests the Mona complex is a 'suspect terrain'. Barber and Max (1979) put these rocks into a plate tectonic complex. The late intrusive Coedana granite has a Rb–Sr whole rock isochron age of 603 ± 34 Ma (Beckinsale and Thorpe, 1979).

Palaeomagnetism and the Position of the Continents

One of the main limitations of the palaeomagnetic technique is that it defines the palaeolatitude, but not the palaeo-longitude, of a particular continent. In other words, a particular continent may have a var-iety of positions relative to another at a cer-tain palaeolatitude.

In spite of these problems there is palaeomagnetic evidence that continental drift did occur in the Palaeozoic. Hailwood and Tarling (1973) and McElhinny and Bri-den (1971) found that rapid polar shifts rela-tive to various continental masses took place particularly in the Ordovician, but that the magnitudes and timing of these shifts varied for different continents; thus continental drift probably took place. The wide distribution of palaeomagnetic poles as plotted on the map of Pangaea is not consistent with the existence of that single landmass in the Ordovician. It seems to be difficult to date the actual time of closure of the Iapetus ocean. According to Phillips and Forsyth (1972) these continents were in contact from the Devonian, but Hailwood and Tarling (1973) concluded that final closure of the ocean occurred in the Carboniferous. Briden *et al*. (1973) calculated on palaeomagnetic evidence that there has been 1000 (±800) km of closure across the British Caledonides since early Ordovician times. A systematic discrepancy in palaeolatitudes led Piper (1979) to conclude that 3500 km of strike–slip displacement took place in lower/mid-Devonian times (*c*. 395 Ma) on the impingement of the Gondwanaland and Laurentian supercontinents during the Acadian orogeny.

The British Caledonides

The first major paper describing the evolution of the British Caledonides according to plate tectonic theory was by Dewey (1969). The most recent comprehensive review is by Harris *et al*. (1979).

Fig. 13.4 Structural subdivisions of the British Caledonides with position of the Ballantrae, Highland Boundary and Anglesey ophiolite complexes

Subdivisions

The orogen is divisible into three zones (Fig. 13.4).

1. *Northern zone*. This comprises:

(a) The unmetamorphosed late Precambrian Torridonian sandstones.

(b) Cambrian–Ordovician limestones and quartzites (unmetamorphosed) forming a thin, shallow-water shelf sequence now preserved only west of the Moine Thrust.

(c) The Moinian–Dalradian series of late Precambrian to Cambrian age occurring east of the Moine Thrust. This is a thick sequence of psammites, pelites, semipelites, marbles and basic volcanics which were deformed in recumbent folds and highly metamorphosed in the late Cambrian to middle Ordovician with a climax in the early Ordovician.

2. *Central zone*. This consists of a thick Ordovician volcanic pile and an Ordovician to Silurian sequence of shales, cherts and greywacke turbidites which underwent only low degrees of metamorphism and deformation in the late Silurian or Devonian (mostly in the Lake District and southern Uplands).

3. *Southern zone*. This consists of Precambrian rocks in the Mona Complex of Anglesey (including serpentines, cherts, pillow lavas and glaucophane schists) and the Church Stretton area (Uriconian volcanics and Longmyndian sediments), and a Cambrian to Silurian succession which was deformed and recrystallized to a low grade in late Silurian or Devonian times. It is equivalent to the Avalon terrain of the northern Appalachians (Williams, 1978a,b) The zone

can be divided into two:

1. In North Wales there is a deep-water 20 km thick monotonous sequence of shales and turbidites with prominent volcanics in the Ordovician and few unconformities.

2. On the Midland Shelf (Church Stretton area) there is a thin sequence of shallow-water shales and silty mudstones with three major unconformities, indicating transgressions and regressions in a platform environment.

Tectonic Evolution

Firstly, a synopsis (Fig. 13.5). The sedimentary framework was established in an ocean spreading stage during the late Precambrian–early Ordovician. The ocean development was symmetrical with thin shelf, or miogeosynclinal, deposits bordering on the continents in the extreme north and south (Durness Group of limestones and quartzites and Church Stretton shales and mudstones), followed inwards by thick accumulations in deeper water on each continental rise (Moine–Dalradian in the north and North Wales in the south). In the central zone the Ordovician ocean floor was covered with a thin veneer of pelagic deep-water cherts and shales, but in the Silurian during advanced stages of closure it was covered with a thick deposit of terrigenous turbidites and siltstones derived from the nearby landmasses.

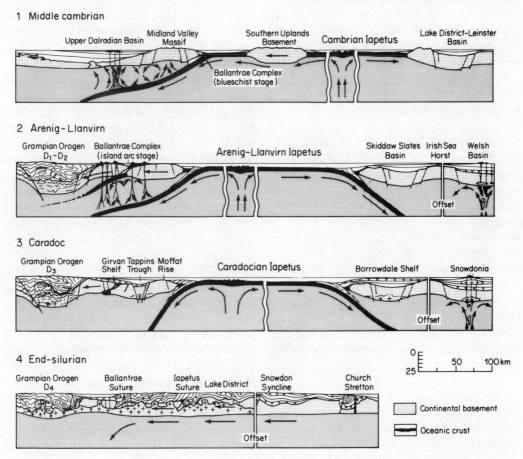

Fig. 13.5 A plate tectonic model to explain the evolution of the British Caledonides (after Watson and Dunning, 1979, in A. L. Harris *et al.* (Eds), *The Caledonides of the British Isles*, Scottish Academic Press, Edinburgh, Fig. 5, pp. 78–79)

Evidence of subduction lies in:

1. The late Precambrian glaucophane rocks and ophiolite suite in Anglesey on the south side, and in the early Ordovician Ballantrae Complex on the north.
2. Arc-type volcanics of Ordovician age in the Ballantrae, Lake District and North Wales areas.

Advanced movement on the Benioff zones was responsible for rise of geothermal gradients with production of regional metamorphism and for contraction and deformation of the sedimentary wedges along each continental rise. Early stages of deformation and metamorphism (sillimanite grade) in the Scottish Highlands reached a peak in the Lower Ordovician between 480 and 510 Ma. The two continents finally collided in early Devonian times when the last granites were intruded. After this, Caledonian orogenic activity ceased, the fold belt undergoing final uplift and erosion.

Now let us look at the evolution of the Caledonian belt step by step, from the late Precambrian to the Devonian.

Late Precambrian In the southern zone the Wentnor Series, consisting of red sandstones, siltstones and conglomerates, was deposited as a 5 km thick clastic wedge on the continental shelf; it may have been a post-orogenic molasse sequence following the deformation of the Mona Complex. The Charnian Group of water-lain tuffs and pyroclastics, and conglomerates, quartzites and shales were laid down in Leicestershire in a NW-trending aulacogen according to Dewey and Kidd (1974).

We can consider the late Precambrian supracrustal rocks of the northern zone in two parts.

The Torridonian This consists of a 9 km thick succession of predominantly red arkoses and cross-bedded sandstones which lie unconformably on the western foreland of the Caledonian mobile belt (A. D. Stewart, 1982). Two groups are distinguished: the Stoer Group, that locally infills valleys in the

Lewisian land surface, and the Torridon Group which overlies it unconformably; their magnetization (and depositional) ages are c. 1100 Ma and 1040 Ma (Smith et al., 1983) and their isotopic closure ages are 968 ± 25 Ma and 777 ± 25 Ma, respectively (Moorbath, 1969). According to Dewey (1969) the Torridonian accumulated as a sedimentary wedge on the northern continental terrace, but Dewey and Kidd (1974) suggest that, together with its north–south facies and thickness changes, it was deposited in an aulacogen in the early rifted continental margin. Significantly, the Torridonian rocks are coeval with the infilling of the higher parts of the Keweenawan rift in central USA.

The Moinian/Mid-Dalradian The Moinian comprises a rather monotonous succession about 8 km thick of metamorphosed arenaceous and argillaceous rocks with minor impure dolomites (Johnstone, 1975). Terms commonly used to describe the present rocks are psammites or psammitic granulites, semi-pelites and pelitic schists. Various sedimentary structures (current bedding, ripple marks, slump folds) show that Moinian sediments were deposited in shallow water to form thick, locally pebbly sandstones and thinner mudstones and shales. The rocks have been overprinted by the Caledonian 'orogeny'. Several phases of folding are recognizable, including early recumbent folds, and large-scale interference patterns are well developed. Extensive chronostratigraphic correlations are made difficult by the lack of well-defined stratigraphic markers in a monotonous succession, by the lack of response of granitic gneisses to changes in metamorphic grade and by the complication of the folding.

The Morar Division of the Moinian is older than 1002 ± 94 Ma (Brook et al., 1977) and the Ardgour granitic gneiss has a Rb–Sr whole-rock isochron age of 1030 ± 50 Ma (Brook et al., 1977). These ages suggest equivalence of some Moinian and Torridonian (Soper and Barber, 1979), and show that the Grenville (Canada)–Gothian (SW Scandinavia) orogenic belt extends through NW Scotland (van Breemen et al., 1978). Later

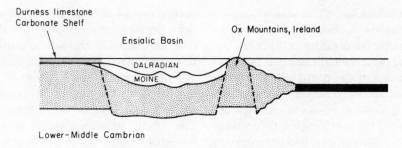

Fig. 13.6   The Moine–Dalradian sequence deposited in an ensialic basin (redrawn after Phillips *et al.*, 1976; reproduced by permission of The Geological Society of London)

isotopic events recorded in the Moines include the formation of pegmatites in the range 718 ± 19 Ma to 573 ± 13 Ma (Piasecki and von Breemen, 1979) and the main Caledonian metamorphism between 470 Ma and 410 Ma (Brewer *et al.*, 1979).

The Lower and early Middle Dalradian sediments are of Precambrian age (Harris *et al.*, 1978). The Precambrian–Cambrian boundary occurs somewhere within the Middle Dalradian. The major glacial tillite horizon that occurs intermittently over a distance of at least 600 km through Scotland to Ireland forms the base of the Middle Dalradian (Spencer, 1971). The Precambrian sediments consist largely of mature quartzites, pelites and limestones deposited in shallow water.

The Moine–Dalradian sequence has a total thickness of about 35 km. Dewey (1969, 1974) interpreted this sequence as a miogeoclinal continental rise deposit that derived its terrigenous material from the continent to the northwest, but Phillips *et al.* (1976) and Harris *et al.* (1978) conclude that it was laid down in an ensialic basin (Fig. 13.6) because clastic debris has a southerly derivation (e.g. from the Ox Mountains in Ireland). The latter conclusion is supported by the discovery that Carboniferous agglomerates in the Scottish Midland Valley have brought up fragments of granulite facies gneisses from a pre-Palaeozoic basement (Upton *et al.*, 1976).

Cambrian   In this period, when the Iapetus ocean was expanding, there was sediment deposition on the two continental margins (northern and southern zones), but none in the central oceanic zone.

The Durness Group of limestones and quartzites and the Church Stretton quartzites were laid down as thin shallow-water shelf facies on the two continental margins. At the same time great thicknesses of terrigenous sediments were deposited in deeper water as flysch wedges on each continental rise. In the north the Upper Dalradian (partly Cambrian age on the basis of the agnostid trilobite, *Pagetia*, in the Leny Limestone) consists mostly of greywacke-like rocks and in the south (North Wales) a 5 km thick pile of shales and turbidites was deposited. But also in North Wales there are important Cambrian welded ignimbrites, agglomerates and tuffs (Arvonian Volcanic Series) and intrusive potash granites (Coedana, Sarn and Twt Hill) with ages of 580–610 Ma.

Along the Highland Boundary Fault there is an ophiolitic complex consisting of serpentinites, gabbros, spilitic lavas, ferruginous and manganiferous cherts, phyllites, clastics and amphibolites (Henderson and Robertson, 1982). This complex originated in a narrow Cambrian intra-Dalradian oceanic basin separate from the Iapetus ocean to the south and it closed by northward subduction.

Finally, one should note the formation of an uplifted block, the Irish Sea horst, on which no Cambrian sediments were laid down (Fig. 13.4).

Ordovician   The most complicated phase in the evolution of the British Caledonides was in the Ordovician; there was sedimentation, subduction, obduction, volcanism, deformation, metamorphism and uplift.

The faunal evidence indicates that the maximum separation of the continents was in the early Ordovician. In Arenig times the obduction of the Ballantrae ophiolite complex and the extrusion of the overlying arc-type volcanics show that a Benioff zone had developed on the north side of the ocean. Movement on this consuming plate was intense, as the main deformation and metamorphism of the Scottish Highlands took place during a relatively short episode from 510–480 Ma—late granites have a Rb–Sr whole-rock isochron of 460 Ma (Pankhurst, 1974). Collapse of the continental rise and deformation of the Moine–Dalradian Series was diachronous in Ireland and Scotland from the early to late Arenig. This deformation was polyphase and marked by the early development of recumbent folds several kilometres across (the Iltay and Ballachulish nappes) prior to the onset of the metamorphism (Thomas, 1979). This early Ordovician tectono-metamorphic event in the British Isles is termed the Grampian Orogeny (Lambert and McKerrow, 1976).

On the south side of the ocean there was extensive volcanic activity in two areas:

1. North Wales. Most of the Llanvirn–Llandeilo succession in the Dolgelley–Arenig area and the Caradoc succession in the Snowdon area consists of andesites, ignimbrites and tuffs, with prominent gold and copper mineralization suggesting that the former Anglesey subduction zone was still operating.
2. The Lake District. Here great volumes of andesite and thin tholeiitic basalt were extruded in late Llanvirn to mid-Caradoc times, suggesting the formation of a new subduction zone. If one compares these andesites and tholeiites derived from a shallow source with the Ordovician alkaline basalts in the Welsh border areas derived from a deep-seated origin, the subduction zone must have dipped to the south (Fig. 13.5) (Moseley, 1977).

In the mid-Ordovician the northern zone was undergoing uplift and erosion. The occurrence of detrital glaucophane in Llandeilo conglomerates shows that the Ballantrae complex was exposed to erosion by this time, and detrital andalusite in Caradocian sediments indicates that at least 3 km of uplift had exposed the higher level nappes with Buchan assemblages in northeast Scotland. The result of this erosion was increased deposition of terrigenous sediment in a flysch wedge on the north side of the trench—there are at least 6 km of mid–upper Ordovician greywackes, conglomerates, sandstones and limestones with trilobites and brachiopods in the Girvan area within an imbricated accretionary prisim (Leggett et al., 1979). Meanwhile, on the oceanic crust on the south side of the trench, only a thin 50 m sequence of pelagic cherts and graptolite-bearing shales accumulated. With increasing erosion the flysch wedges infilled the trench and encroached progressively southwards onto the oceanic plate during the Caradocian and early Llandoverian times when 8 km of turbidites were deposited.

The Final Stages   In the Silurian there was further sedimentation, but no volcanic activity, indicating movement on the subduction zones had ceased. By mid-Silurian times the continents were close together and thick turbidite–flysch sediments covered the remaining intervening ocean; up to 3 km of turbidites were laid down in fairly deep water in the Girvan area (Cocks and Toghill, 1973). These thinned southwards with the result that only 1 km of limestones and shales (Upper Llandovery, Wenlock, Ludlow) with corals, brachiopods and trilobites was deposited on the continental shelf in the Midlands of England. Uplift, erosion and thermal instability continued through the late Silurian when the continents finally collided giving rise to deformation and metamorphism. The timing of collision seems to have been progressively later towards the southwest. Phillips et al. (1976) accordingly advocate an oblique collision model, the unstable triple point migrating southwestwards across the British Isles from the late Ordovician to the late Silurian. Collision finally reached eastern Canada in the middle Devonian (Mitchell and McKerrow, 1975). The presence of Lower Old Red Sandstone andesites and of granite plutons (380–400 Ma) in Scotland suggests that some subduction continued in the early Devonian (Thirlwall, 1981). Steady state conditions were not reached until middle Devonian times; the Iapetus ocean had opened and closed to give rise to a new fold belt that had taken 500 Ma to form.

Finally, a considerable advance in our knowledge of the subsurface structure of the British Caledonides comes from the N–S deep seismic profile experiment LISPB (Bamford *et al.*, 1978). Three major discontinuities indicated by velocity changes are interpreted to separate the mantle, the lower crust, pre-Caledonian basement, and the upper crust of Caledonian metamorphic rocks. Also the offshore MOIST reflection profile has defined the Moho and Flannan thrust which appears to transect the Moho (Smythe *et al.*, 1982). A thick-skinned model for the structure of the Moine thrust zone was proposed by Soper and Barber (1982).

The Appalachians

The Appalachian fold belt extends along the eastern side of North America from New-foundland to Alabama (Hatcher, 1978; Williams, 1979). It has a history ranging from the late Cambrian to the Carboniferous and the main tectonic features can be matched with those of the British Caledonides (Dewey, 1969; Williams, 1978a). It is a classic fold belt as the original concept of the formation of a geosyncline was worked out here by Hall and Dana in the last century and it has become a key model for the evolution of a pre-Mesozoic fold belt by plate tectonic mechanisms.

Subdivisions

In 1978b H. Williams published a unique tectonic lithofacies map of the whole Appalachian orogen and Williams and Hatcher (1982) made an interpretation of it. The orogen is divisible into the following zones (Fig. 13.7).

1. The miogeoclinal belt constitutes the North American continental shelf and includes allochthonous thrust slices that have been transported westwards. Some allochthons are sedimentary-volcanic composites and contain ophiolites (e.g. Bay of Islands Complex). The allochthons have come from the ancient continental slope and rise. The

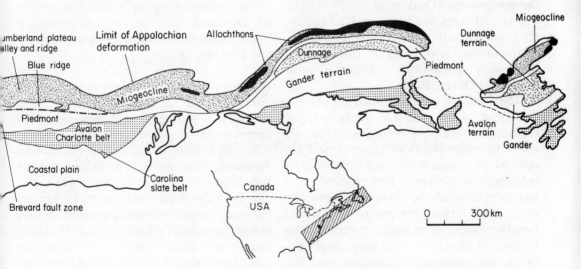

Fig. 13.7 The zonation of the Appalachian orogen showing the position of major geological and tectonic boundaries (modified after Williams and Hatcher, 1982)

miogeocline includes a major Cambrian carbonate bank deposited on the continental margin and a sedimentary succession up to the Carboniferous which has been subjected to westward thrusting. In general, there was a progressive clastic transgression from east to west across the subsiding continental margin. West of the limit of Appalachian deformation are flat Devonian–Carboniferous rocks at the surface in the south and a more deeply eroded Precambrian Shield in the north.

2. The Piedmont terrain consists mainly of late Precambrian–early Palaeozoic metaclastics lying on Grenville basement seen as gneiss domes in the Blue Ridge belt, and near New York, Philadelphia and Baltimore. The Piedmont rocks are metamorphosed to upper greenschist and amphibolite facies with intense deformation directed towards the miogeocline and were derived from off the eastern edge of the miogeocline, possibly from a separate microcontinent (Cook *et al.*, 1979). Ophiolitic mafic–ultramafic bodies are common along the active belt, having been derived from the oceanic crust-mantle off the continental margin (Misra and Keller, 1978). The orogenesis was Taconic (mid-late Ordovician). Seismic reflection studies show that the entire Piedmont in the southern Appalachians belongs to a single subhorizontal crystalline slice that was emplaced above the miogeocline (Cook *et al.*, 1979, 1981). The crustal transition from the Piedmont to the Charlotte belt of the Avalon terrain may represent the autochthonous late Precambrian–early Palaeozoic continental edge (Cook and Oliver, 1981).

3. The Dunnage terrain is made up of early Palaeozoic mafic volcanic rocks, marine sedimentary rocks, and mélanges of volcanic and sedimentary blocks which are up to 1 km in width in a shale matrix and which overlie ophiolitic sequences. Deformation and metamorphism of the terrain is uniformly weak. The terrain represents vestiges of Iapetus, including volcanic material from Ordovician island arcs that were caught between the colliding miogeocline and the Avalon terrain (Kean and Strong, 1975). Palinspastic restoration of the widest segment of the Dunnage terrain in Newfoundland implies a minimum oceanic width of 1000 km before the Taconic deformation (Williams, 1980).

4. The Gander terrain consists mainly of pre-middle Ordovician arenaceous rocks in Canada and of volcanic rocks and shales in New England. Some granitic gneisses may form a basement to these cover rocks. Intense metamorphism and deformation producing widespread gneisses and many granitic intrusions are mainly Acadian. The terrain includes early–mid-Ordovician back-arc basin meta-sediments that separated the island arc sequences of the Dunnage from the Avalon terrain (Colman-Sadd, 1980). The terrain was developed far from the North American miogeocline; it records the formation and destruction of a continental margin that lay to the east of Iapetus.

5. The Avalon terrain comprises mainly late Precambrian sedimentary and volcanic rocks that are relatively unmetamorphosed and undeformed compared to those of nearby terrains. Locally these rocks pass upwards conformably into Cambrian shales with Atlantic realm tribolite faunas. In the south this terrain contains the Charlotte and Carolina Slate Belts. The Avalon terrain was largely unaffected by Taconic (Ordovician) orogeny. It is a 'suspect' terrain accreted to the remainder of the orogen during the Acadian orogeny in the Devonian. There was an easterly migration of accretion across the orogen (Williams and Hatcher, 1982).

A major recent breakthrough has come from the COCORP seismic-reflection profiling across the southern Appalachians which demonstrates that the crystalline Precambrian and Palaeozoic rocks of the Blue Ridge, Piedmont, Charlotte Belt, and Carolina Slate Belt constitute an allochthonous sheet generally 6–15 km thick which overlies relatively flat-lying autochthonous lower Palaeozoic sedimentary rocks (Cook *et al.*, 1979, 1981; Harris and Bayer, 1979). The crystalline rocks have been thrust at least 260 km over the miogeocline to the west, on a master décollement which may be rooted in the

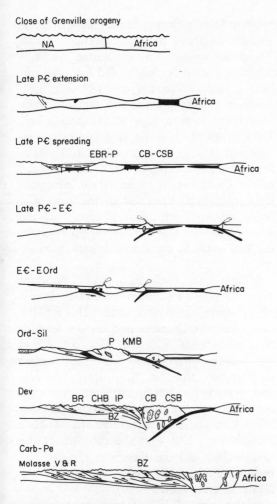

Close of Grenville orogeny

NA | Africa

Late PЄ extension

Africa

Late PЄ spreading

EBR-P   CB-CSB

Africa

Late PЄ - EЄ

EЄ - E Ord

Africa

Ord - Sil

P KMB

Dev

BR CHB IP   CB CSB

BZ

Africa

Carb - Pe

Molasse V & R   BZ

Africa

Fig. 13.8 Sequential sections showing a possible mode of development of the southern Appalachian orogen. EBR: eastern Blue Ridge, P: Piedmont, CB: Charlotte Belt, CSB: Carolina Slate Belt, BR: Blue Ridge, V & R: Valley and Ridge, BZ: Brevard zone (after Hatcher and Odom, 1980)

lower crust beneath the Charlotte Belt (Iverson and Smithson, 1982). The location of the Palaeozoic Iapetus suture is likely to be off the present shore (Zen, 1981). A section across the southern Appalachians of Fig. 13.7 is provided by a tectonic model for the development of the belt in Fig. 13.8.

Tectonic Evolution

Late Precambrian The evolution of the orogen started in the late Precambrian;

according to the plate tectonic theory the earliest sediments mark the initial opening of the Iapetus ocean.

On the northwest side of the belt there is an 8 km thick sequence of Eocambrian clastics (quartzites, arkoses, siltstones, sandstones) in the Great Smokey Mountains of Tennessee and North Carolina, and on the southeast side of the belt there is a similar 6 km sequence in the Avalon Peninsula of Newfoundland (Rodgers, 1970). These terrigenous sediments were derived by erosion of nearby mountainous Precambrian landmasses and they were deposited on miogeoclinal shelves as prograding continental terrace wedges, the detritus being derived consistently from the cratons. The sediments are locally associated with tholeiitic volcanic rocks and mafic dykes.

In the southern Appalachians a rifted block of continental crust (the Piedmont microcontinent) was separated from North America by a narrow marginal sea (Fig. 13.8). The Carolina Slate Belt island arc began to form by westward subduction (Whitney *et al.*, 1978).

Cambrian–Early Ordovician During this period, when the ocean went through its main 'opening' stage, there were five sedimentary-tectonic environments:

1. A shallow-water marine carbonate-rich sequence developed on a continental shelf. During the Cambrian, basal sands followed by carbonates encroached landwards in a gradual marine transgression until, during the late Cambrian and early Ordovician, a shallow-water carbonate bank up to 3 km thick extended on the western margin of the fold belt, from Texas in the south to Newfoundland in the north; its continuation can also be traced in NW Scotland and E Greenland (Swett and Smit, 1972). Rodgers (1970) compared this carbonate bank with the modern carbonate platform in the Bahamas off eastern USA (Dietz, Holden and Sproll, 1970). The carbonate sequence contains brachiopods (thus referred to as the shelly facies), and contains worm burrows, oolites, stromatolitic algal structures, desiccation

cracks and flake conglomerates, indicative of an intertidal to subtidal environment (Swett and Smit, 1972).

2. A deep-water pelitic to clastic sequence developed during the Cambrian and early Ordovician on the continental rise to the east of the carbonates on the continental shelf (see Fig. 13.8. This is the Piedmont zone, now mostly metamorphosed and deformed, consisting of greywackes, pyrite- and graptolite-bearing black shales (thus referred to as the graptolite facies), and andesitic volcanics. The sediments coarsen and thicken eastwards reaching 15 km in thickness.

3. Offshore island arc volcanic activity in the Dunnage terrain took place in the early–middle Ordovician. The presence of pyrite in shales, of turbidite structures in greywackes, and the association with andesites suggests deposition in relatively stagnant deep troughs bordering volcanic islands—an environment comparable with that in the vicinity of Quaternary island arcs in the East Indies.

4. During the subduction, sediments and volcanics were scraped up off the oceanic floor and dumped in the active trench—seen today as the Dunnage mélange in Newfoundland which resembles the argille scagliose of the Appenines and the Franciscan mélange of California (Kay, 1976). Alternatively the mélange may be an olistostrome deposited in a back-arc basin (Hibbard and Williams, 1979).

5. In the southern Appalachians the offshore marginal sea began to close by east-dipping subduction along the west side of the Piedmont microcontinent. The environment was not very different from the present NW Pacific where the Kamchatka and Korean peninsulas and the Japanese and Kurile island arcs form a complex of partly rifted continental blocks separated by marginal seas with oceanward island arcs (Hatcher, 1978).

The boundary between the carbonate platform and the pelitic–clastic trough defines an important tectonic zone. It is marked by carbonate blocks in the continental rise deposits at several stratigraphic levels within the Appalachians, ranging from Middle Cambrian to Middle Ordovician–a classic example is the Cow Head Breccia in Newfoundland. These are regarded as submarine landslide breccias, the blocks having slid from the carbonate bank into the deeper water to the east; Rodgers (1970) pointed out that the steep to vertical eastern edge of the present-day Bahama Banks may be an analogous environment. He also suggested that the platform–trough boundary marks the true edge of the Cambro-Ordovician North American sialic continent, the crust to the east of it being oceanic, just as it is today off the Bahama Banks. This continental edge may be marked today by the Brevard fault zone.

Middle–Late Ordovician An important orogenic phase (Taconic) took place in this period along the Appalachians. This was the time when the Dunnage and Gander terrains were juxtaposed with the miogeocline and with each other—this was the earliest accretionary phase in the orogen (Williams and Hatcher, 1982). In the southern Appalachians the Taconic orogeny is identified with the final closure of the offshore marginal sea and with the collision of the miogeocline and the Piedmont terrain. The metamorphic peak occurred at 480–450 Ma (Hatcher and Odom, 1980).

In the early Middle Ordovician the previous tectonic and sedimentary pattern underwent drastic changes. Fragmentation and partial foundering of the carbonate shelf was caused by block faulting as black, mantling shale sequences were deposited on the subsiding carbonate rocks of the shelf. Continued subsidence of the former shelf led to gravity sliding of sediments in allochthonous slices into the subsiding miogeocline which now had the character of an exogeosyncline. Emplacement of the allochthons took place in the Middle Ordovician, during the main phase of the Taconic Orogeny in the northern and central Appalachians.

Continued compression in the Taconide Zone, caused by interaction between the North American margin and the Piedmont microcontinent resulted in further northwest thrusting and nappe transport (but now of crystalline rocks including Precambrian basement gneisses), and also in folding and regional metamorphism.

Late-orogenic clastic sediments (shales, sandstones and conglomerates) were deposited in eastern Pennsylvania during final stages of uplift.

Early–Mid-Silurian In the early Silurian post-orogenic alluvial clastics were laid down in the exogeosyncline. Sandstones coarsen and conglomerates increase eastwards showing that they were derived from the uplifted terrain to the east: they gradually transgressed further to the east as the relief of the mountainous area was reduced.

By Middle Silurian time there was an immense shallow-water sea over much of the North American craton in which carbonates with algal reefs and evaporites accumulated, particularly in the region of the Michigan Basin.

Early Devonian The Acadian orogeny occurred about 380 Ma ago throughout most of the orogen. It was most intense east of the Taconic deformation zone, because it resulted from the westward accretion of the Avalon terrain. In the south this was caused by the collapse of the back-arc basin to the west of the Charlotte–Carolina Slate Belt island arc (Fig. 13.8).

Middle–Late Devonian Uplift and erosion gave rise to the enormous volume (3 km thick) of clastics laid down in the exogeosyncline of this period. The molasse sediments even spread westwards to form an extensive subaerial alluvial plain, the Catskill delta, where continental arkosic red sandstones, conglomerates and shales were deposited in an arid climate about 20° from the equator.

Carboniferous–Permian Continued uplift and erosion of the mountain range produced a great thickness of coarse clastics in the exogeosyncline during the Carboniferous and early Permian. Locally extensive coal measures developed in shallow-water swamps during the Upper Carboniferous (Pennsylva-

nian). In the Permo-Carboniferous (300–250 Ma) the Alleghenian orogeny caused continent-directed folding and thrusting of the entire miogeocline along the western margin of the south and central Appalachians giving rise, for example, to the Blue Ridge thrust with a minimum displacement of 150 km, and many thrusts in the Valley and Ridge Province and the Cumberland Plateau. It also caused regional metamorphism and plutonism in the Avalon terrain above the eastern margin of the orogen. The Alleghanian orogeny resulted from the collision between Africa and North America. With its termination the formation of the Appalachian orogen was complete.

Mineralization

The question that concerns us here is: Do the mineral deposits in the Appalachian Caledonide fold belt have a systematic regional distribution and, if so, is it similar to that in Mesozoic–Tertiary fold belts? Such a correspondence in metallogenic provinces would provide us with additional evidence that the fold belt was situated at lithospheric plate junctions.

Newfoundland

Strong (1974) and Swinden and Strong (1976) give a review of the Newfoundland mineral deposits, in relation to possible plate tectonic environments. The general features of the deposits are similar to those of 'modern' accreting and consuming plate margins. The evolutionary development of the ores is summarized in Fig. 13.9.

Southern Appalachians

Across the Appalachian fold belt there are many distinctive ore types which correlate well with the main tectonic belts (Fig. 13.10). Some mineral deposits formed in the late Precambrian and early Palaeozoic pre-accretionary stage of development. They include Mississippi Valley-type Pb–Zn deposits in carbonates of the miogeocline

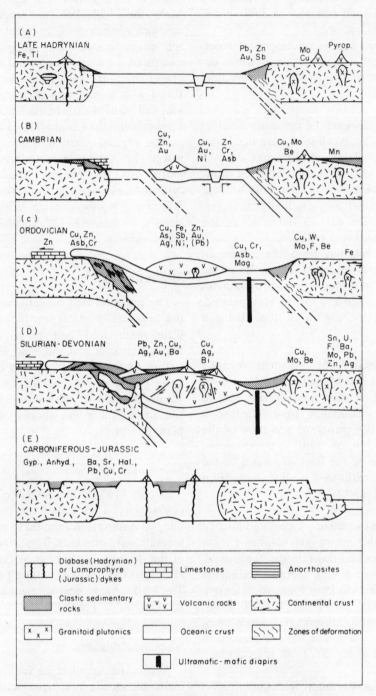

Fig. 13.9 Plate tectonic model to explain the zonation of mineral deposits of the Newfoundland Appalachians (after Swinden and Strong, 1976; reproduced by permission of The Geological Association of Canada)

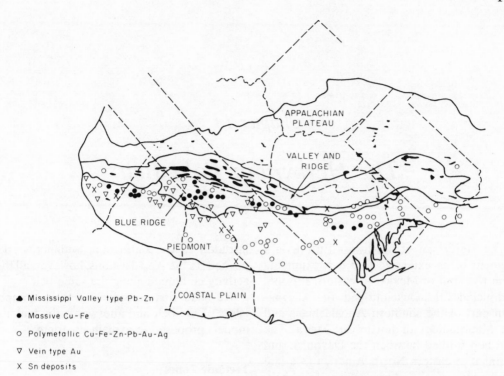

Fig. 13.10   Palaeozoic mineral zonation of the southern Appalachians (after Strong, 1974; reproduced by permission of the Association of the Institute of Mining Engineers)

(Scott, 1976), Cyprus-type Cu-pyrite deposits (Tuach and Kennedy, 1978) and asbestos occurrences (Williams *et al.*, 1977) occur in ophiolitic rocks of the Dunnage terrain, and polymetallic deposits in the Gander terrain. Also the Blue Ridge province contains largely Cu–Fe deposits, followed to the southeast by a narrow belt with vein-type Au along the border of the Piedmont which contains, in its main part, polymetallic Cu–Fe–Zn–Pb–Au–Ag as well as some Sn deposits. In the Piedmont there are no economic gold occurrences today although this was the main source of gold in the USA in the first half of the nineteenth century. There are several porphyry-type Cu and Mo deposits in the Avalon terrain of Maine and New Brunswick (Hollister *et al.*, 1974).

British Caledonides

In the southern British Caleonides metal occurrences can be related to the three sub-duction zones passing through Ballantrae, the Lake District and Anglesey. Whilst Pb and Zn are common throughout most of the region, Cu occurs with them as a prominent ore only in the two belts immediately overlying the two southerly-dipping subduction zones. In these belts Au occurs in southeast Ireland and in the Harlech Dome of Wales, minor Ag in Parys Mountain, Anglesey, and Fe oxide (magnetite) in the Avoca area of southeast Ireland. Further to the south there is a belt with Pb, Zn and minor Cu and Ba, followed to the south by Mo, F, Ba, Pb and traces of W (scheelite) in the Mountsorrel Granodiorite in Leicestershire. In Scotland there is weak galena mineralization in the Durness limestone-Cambrian carbonate shelf (Scott, 1976), and Cu–Mo mineralization in deeply eroded calc-alkaline plutons (Ellis, 1977; Evans, 1977). Stratabound sulphide deposits in the Caledonides and Appalachians are reviewed in Vokes and Zachrisson (1980).

Chapter 14

The Hercynian Fold Belt

The Hercynian of northwest Europe is a segment of an extensive fold belt extending from the Gulf of Mexico to eastern Europe, and includes the Ouachita belt, the Alleghanian part of the southern Appalachians and the Mauritanides of northwest Africa. The fold belt formed between the Devonian and Permian in eastern North America as a late expression of the Appalachian fold belt, whereas in Europe it developed as a separate entity, physically apart from the Caledonian fold belt to the north (Fig. 14.1). The Uralides of eastern Europe emerged at approximately the same time.

For a variety of reasons knowledge of the belt is scanty and tectonic models to explain its origin have been slow to advance. In the first place, the general poor degree of exposure makes correlation difficult, the European belt is extensively covered by Mesozoic–Cenozoic sedimentary rocks, the Alleghanian was superimposed on and is not easily separated from the Appalachian events, the Hercynian of southern Europe is heavily overprinted by the Alpine belt and, not least of all, publications on the subject are in many languages.

Remember that whilst the initial stages in the evolution of the Hercynian were taking place in Europe during the early Devonian, so also were the final stages in the evolution of the Caledonides in Scotland. This means that about 380–400 Ma ago marine sediments were being deposited only a few hundred kilometres south of Scotland where andesites were being extruded, granites

intruded and continental sediments deposited. In the Appalachians, however, all this activity continued along the same belt.

We shall first consider the main components of the belt and afterwards the tectonic models proposed to explain its origin.

Tectonic Zones

Fig. 14.2 shows the distribution of the three main zones of contrasting stratigraphy and tectonic history (Johnson, 1978; Rast, 1983) and Fig. 14.3 some of the important features of the belt.

Immediately north of the fold belt is the Northern Shelf, which formed as a foreland in the Devonian during continental sandstone deposition, a shelf in early Carboniferous times and the site of discontinuous external (paralic) coal basins in the late Carboniferous (Johnson, 1982). No orogenic activity occurred until the end of the Carboniferous (Asturic) when the shelf was folded by compression from the south. This deformation is correlated with plate collision at the final closure of the Mid-European ocean to the south.

The Rheno-Hercynian Zone (Behr et al., 1980) contains Devonian and Carboniferous sediments that generally show a northward transition to a more shallow facies. Tectonic activity includes northward-verging late Carboniferous (Asturic) thrusts and overfolds (Shackleton et al., 1982; Coward and McClay, 1983) and huge horizontal nappe displacements (Meissner et al., 1981). This

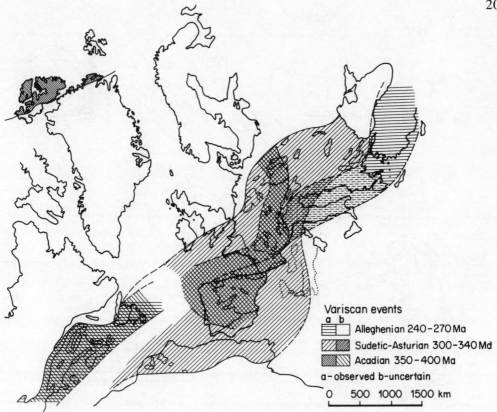

Fig. 14.1 Distribution of tectonic events along the Hercynian fold belt (from Zwart and Dornsiepen, 1980)

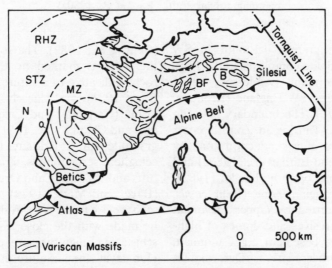

Fig. 14.2 Zonation of the Hercynian belt of Europe (from Rast, 1983, Variscan orogeny, in P. L. Hancock (Ed.), The Variscan Fold Belt in the British Isles; reproduced by permission of Adam Hilger Ltd). A: Armorican Massif, B: Bohemian Massif, C: Massif Central, V: Vosges, BF: Black Forest, RHZ: Rhenohercynian zone, STZ: Saxothuringian zone, MZ: Moldanubian zone. – – – – zonal boundary, O/C Oporto–Cordoba lineament

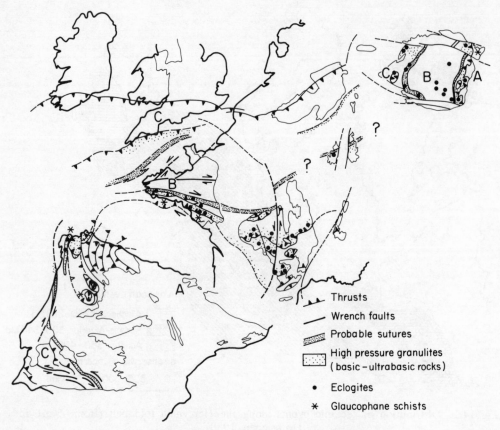

Fig. 14.3 Some key features of the Hercynian belt of NW Europe. A, B, C: three proposed continental plates (after Bard *et al.*, 1980a)

zone contains a granite batholith in SW England and ophiolite complexes in the Harz Mountains in Germany (Anderson, 1975) and in the Lizard in SW England (Bromley, 1976; Kirby, 1979). The boundary between this and the Saxo-Thuringian Zone is interpreted as the site of a subduction zone by Johnson (1978) and Bard *et al.*, (1980a) and a back-arc marginal basin by Leeder (1982).

The Saxo-Thuringian Zone (Behr *et al.*, 1980) is characterized by Upper Devonian volcanics and intrusives and domes of high-grade Precambrian basement, e.g. Cadomian, in Brittany (Mobus, 1977). Deformation occurred twice in the late Devonian (Bretonic) and mid-Carboniferous (Sudetic) and there is no evidence of earlier Caledonian activity. An important pile of Carboniferous nappes in Bavaria was rooted in the boundary with the Moldanubian Zone to the south

(Behr *et al.*, 1982). The southern border of this zone is marked by the occurrence of rocks indicative of a suture zone (Matte and Burg, 1981), viz. blueschists, on the Ile de Groix (Carpenter and Civetta, 1976; Quinquis and Choukroune, 1981), basic-ultrabasic granulites in S Brittany (Hanmer, 1977), eclogites, serpentinites, granulites, gabbros, tuffs and cherts at Cabo Ortegal in NW Spain (Bayer and Matte, 1979) with glaucophane schists close to the west. A direct analogy may be made with the close association of blueschists and high-pressure granulites on the Himalayan Indus Suture in Pakistan (Tahirkheli *et al.*, 1979); note, however, that Pin and Vielzeuf (1983) ascribe an age of 450 Ma to this high-pressure granulite-grade metamorphism. The trace of the suture is marked by a geophysical anomaly indicating the presence of a major mafic body which may be a relic of oceanic crust (Lefort *et al.*, 1982).

The Moldanubian Zone contains several windows of Precambrian basement, such as the Bohemian Massif, Black Forest, Vosges and Massif Central, below folded lower Palaeozoic. There is some evidence of Caledonian orogenic activity in this zone in the Devonian interval 400–365 Ma (Autran, 1978). The Moldanubian is characterized by early Carboniferous (Sudetic–Visean to Namurian) deformation, widespread syntectonic and post-tectonic granites (Losert, 1977; Suk, 1977) with an important stage at 331 ± 4 Ma (Rb–Sr whole-rock isochron) in Bohemia and high-temperature metamorphism (granulites have a U–Pb zircon age of 345 ± 5 Ma) superimposed on a widespread Cadomian (late Precambrian-Cambrian) basement (Van Breemen *et al.*, 1982). By the late Carboniferous (late Westphalian and Stephanian) at a high level in the orogenic belt there was potassic ignimbrite volcanism and deposition of limnic continental deposits and coal in intermontone basins. The Asturic late Carboniferous deformation phase is absent and Permain sedimentation continued unbroken from the Carboniferous.

Sedimentation and Volcanism

Uplift on the site of the Caledonian orogenic belt in the Devonian gave rise to an erosional source area in Spitzbergen, East Greenland, western Norway and northeastern Scotland which has been termed the 'Old Red landmass' because it was an area of continental red sandstone deposition in inter- and extra-montane basins. To the south of this landmass there was an Atlantic-type continental margin bordering the Mid-European sea to the south (Johnson, 1978).

The Devonian period began with a transgression. The early Devonian rocks in the Rheno-Hercynian zone belong to a continental facies, clastic detritus being derived from the landmass to the northwest; greywackes, sandstones, orthoquartzites and shales predominate. A shallow-water marine facies evolved in the mid-Devonian with conspicuous reef limestones. There was widespread extrusion of albite-rich soda-keratophyres and spilitic lavas and tuffs in the Rhenisch Schiefergebirge in the mid-Devonian and of spilitic lavas in north Cornwall and south Devon from the mid-Devonian to the early Carboniferous.

The Rheno-Hercynian zone passes northwards in southwest England onto the shelf by an increase in the proportion of continental (sandstones, arkoses and conglomerates) at the expense of marine (shales and limestones) sediments. The Old Red Sandstones of Wales are typical of the shelf facies.

Sedimentation in the Moldanubian zone was controlled by the distribution of basement uplifts. Generally sediments are thinner than further north and clastics are less important than limestones and marls.

Following a regression at the end of the Devonian the Carboniferous period began with a transgression that may have been caused by growth of a new mid-oceanic ridge. The sea advanced across the Old Red Sandstone plain giving rise to a shallow-water carbonate shelf; by late Carboniferous times only a few islands, such as St George's Land in Wales, stood above sea level. There is a southward facies change from the shelf limestones via a marginal reef facies to the deeper water, partly detrital, sediments (sandstones, cherts, limestones and black shales) of the Culm trench of southwest England with which are associated pillow lavas and the Lizard ophiolite complex.

The Culm facies was also developed in southwest Ireland, Belgium and the Ruhr in the Rheno-Hercynian zone. In central and western Europe there was a major phase of tholeiitic lava extrusion in the mid-Lower Carboniferous. The Sudetic deformation in the mid-Carboniferous was associated with granitic intrusion and with acid-to-intermediate explosive volcanism in the Moldanubian zone. By the early Permian the formation of the Hercynian was complete, except for localized post-orogenic intrusion of lamprophyres, extrusion of acid and basic volcanics and local graben formation.

Granites, Migmatites and Metamorphism

The Hercynian belt of NW Europe has an abundance of granitic rocks, especially in the Moldanubian and Saxo-Thuringian zones, but few have been well dated. The Moldanubian zone is characterized by the presence of much pre-Hercynian basement, an abundance of granites and migmatites, and a low-pressure/high-temperature facies series of regional metamorphism. The large number of granites in this zone range from mesozonal to high-level types, the deeper seated granites being associated with migmatites and probably formed by remobilization of gneissic basement rocks.

Vitrac *et al*. (1981) showed that there is a systematic variation of lead isotope compositions across the granites of the Hercynian belt, especially in the Moldanubian zone, and concluded that all the granites were produced by remobilization of older crust. It is an interesting challenge to explain the cause of so many granites in the Hercynian belt. Wong and Degens (1983) suggested that the predominance of clastic deposits, in contrast to carbonates as in the Alpine belt, released a high proportion of H_2O rather than CO_2 and this meteoric water initiated partial melting and the production of voluminous granitic intrusions.

The Cornubian granite batholith in SW England is *c*. 250 km long, 40 km wide, and extends to a depth of *c*. 10–12 km (Exley *et al*., 1983). The most likely age of emplacement was in the period 303–295 Ma. The granites have high K/Na ratios (giving alkali feldspars), low Fe_2O_3/FeO ratios (inhibiting magnetite), a deficiency in hornblende and sphene, and intermediate-to-high initial Sr isotope ratios—*c*. 0.7086 (Floyd *et al*., 1983). These features suggest that these granites formed by partial melting of crustal material. A problem arises in relating the origin of the granites to the current thin-skinned interpretation of the structures of SW England because the present crust is only 27 km thick and not more than 5 km of late Palaeozoic sediments can have been eroded (Shackleton *et al*., 1982). These authors suggest the gran-

ites were generated further south and were injected northwards as a sheet-like body.

Mineralization

During the formation of the Hercynian fold belt there evolved several important types of mineral deposit (Fig. 14.4) including lead–zinc in shelf limestones, copper–lead–zinc in rifts and tin–tungsten–uranium in association with post-orogenic granitic plutons. Although both types are known from earlier times, these Hercynian examples represent a peak of development reached late in earth history.

Pb–Zn in Limestones

Galena–sphalerite deposits formed in Carboniferous carbonate sediments in shelf areas or in interior basins landward of passive continental margins (Mitchell and Garson, 1981) such as the Pennine field of England, and in Eire (Russell, 1976). They are commonly referred to as the 'Mississippi Valley-type ores'. The deposits typically occur as lenses conformable with bedding and as infillings of discordant fractures often associated with major fault sets in limestones or dolomitized limestones. The common association with biohermal reefs substantiates a generally accepted shallow-water shoreline or shelf environment, a direct result of the widespread Carboniferous transgression.

Cu, Pb, Zn in Rifts

Several massive sulphide deposits of Devonian and possibly early Carboniferous age occur in the Hercynian belt–in Silesia (Cu, Pb, Zn) in felsic pyroclastics, at Rammelsberg (Cu, Zn, Pb) in slates with volcanic components, at Meggen (Zn) in slates with a tuff horizon, and at Rio Tinto in SW Iberia (pyrite with minor Cu, Zn, Pb) in felsic or mafic volcanics and slates. Sawkins and Burke (1980) suggest all these deposits occur in Hercynian rifts that formed as a result of either late Caledonian or Acadian collisions.

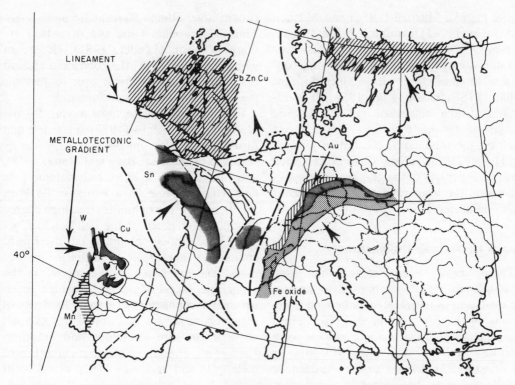

Fig. 14.4 Hercynian mineralization in western Europe (after Gabelman, 1976; reproduced by permission of The Geological Association of Canada)

Sn, W, U etc. in Granites

Tin-tungsten–uranium mineralization associated with Carboniferous post-orogenic granite plutons is well known in Europe (Fig. 14.4), i.e. Cornwall, Brittany, Massif Central, Germany and the Iberian peninsula.

There are several important Carboniferous (Westphalian, c. 300 Ma uranium-enriched, two-mica granites in the Massif Central. Le Roy (1978) suggested that the Saint Sylvestre granite is similar to the two-mica granites of Manaslu-type in Nepal (which have no uranium deposits but very high background values of uranium—P. Le Fort, in Mitchell and Garson, 1981) which were generated by partial melting of the lower crust as a result of Miocene thrust-nappe stacking (see Chapter 21). Such Himalayan-style thrusts in the Massif Central are described by Bard et al. (1980b). The primary uraninite mineralization in the granites may have been released by weathering in the dry, vegetation-free Permian climate and redeposited as pitch-blende, vein-type deposits (Barbier, 1974).

In Portugal economic tungsten–tin (wolframite–cassiterite) deposits occur in a 290 Ma two-mica granite at Panasqueira. Kelly and Rye (1979) suggest that the granite is similar to the anatectic S-type granites of Chappell and White (1974).

The mineralization in the late Carboniferous post-tectonic granites of SW England is well known. There are major tin–copper deposits and minor associated tungsten and fluorite mineralization with complex ore-mineral parageneses in hypothermal, mesothermal, epithermal and greisen-flanked veins (Exley et al., 1983). The plutons are andalusite/cordierite-bearing adamellites with minor two-mica adamellites and granites, commonly enriched in tourmaline and uranium. Simpson et al. (1979) suggest that these are Andean-type granites with an initial Sr isotope ratio of 0.706, but Dewey and

Burke (1973), Mitchell (1974) and Mitchell and Garson (1981) favour a collisional origin. It is important to note that the granites occur in a thin-skinned thrust-nappe stack on a continental foreland close to the obducted Lizard ophiolite (Shackleton *et al.*, 1982). It is difficult to avoid a comparison with the Miocene tourmaline two-mica granites in the thrust stack on the foreland of the Indian plate in the Higher Himalaya (see Chapter 21). Such ideas on the origin of the granites have implications on the source material and origin of the mineralizing fluids.

Models for Tectonic Evolution

There have been some interesting developments in ideas proposed to explain the evolution of the Hercynian belt since my writing of the first edition in 1976. Before that time most plate tectonic models were produced by British geologists with little or no first-hand knowledge of the Andean or Himalayan belts, but since then the main ideas have come from French geologists and are based on direct experience of Himalayan geology (Cogné, 1977; Bard *et al.*, 1980a,b; Matte and Burg, 1981). Until 1976 many workers favoured a model with a major shallow-dipping northward subduction zone situated south of France, but more recently the location of a northward-dipping subduction zone active during the mid–late Devonian has been recognized in S Brittany, extending eastwards north of the Massif Central and south-westwards near the Cabo Ortegal complex in NW Spain (Lefort, 1979; Bard *et al.*, 1980a,b; Matte and Burg, 1977) (Fig. 14.3). Today this is most commonly regarded as the main Hercynian suture. On the north side in S Brittany there is an Andean-type volcanic arc and a low pressure/high temperature metamorphic belt (Le Fort, 1979), glaucophane schists are accompanied by high pressure granulites on the suture (as in the Indus Suture, Pakistan) and in the Massif Central there is a thrust stack in gneisses with inverted metamorphic isograds (Bard *et al.*, 1980b).

According to this model the boundary between the Rheno-Hercynian and Saxo-Thuringian zones is the site of a back-arc marginal basin (Leeder, 1982), closure of which gave rise to the obducted Lizard ophiolite which was thrust over its tectonic mélange onto the thrusted foreland.

In criticizing plate tectonic models for the Hercynian belt Krebs (1977) suggested that thrust nappes are uncommon (this hardly seems to be the case: Burg and Matte, 1978), and noted the lack of paired metamorphic belts (there is one on S Brittany: Le Fort, 1979), and of true ophiolites (the high degree of uplift over much of the belt would obviously have eroded ophiolites, but the Lizard complex is acceptable, and there are not many true ophiolites on the sutures in the Himalayas (see Chapter 21), of high pressure metamorphism (there are 4 localities of glaucophane schists and abundant examples of high pressure granulites and eclogites: Bard *et al.*, 1980a), of mélanges (the Lizard mélange!) and andesites (again much uplift would have removed high-level andesites, and there are no andesites in the Alps!). Krebs' criticisms clearly do not stand up to the facts.

In recent years the importance has been realized of transcurrent movement on faults and sutures in many collisional orogenic belts. Arthaud and Matte (1977) first pointed to the possibility of major strike–slip faulting in S Europe and N Africa. Lefort and Van der Voo (1981) compiled data on mid–late Carboniferous strike–slip faults from which they suggested the idea that the West African craton acted as a rigid indenter against the North American craton (following the Tapponnier and Molnar, 1976, Himalayan slip line indentation theory), and as a consequence the Hercynian European Craton was forced sideways (northeastwards) on major wrench fault systems with 2000 km sinistral offset along the Great Glen fault in Scotland (Van der Voo and Scotese, 1981), and equally important dextral displacements in N Africa and S Europe (Fig. 14.5). In contrast, Badham (1982) proposed a strike-slip model for the Hercynides of Europe which involved the same dextral interaction between Europe

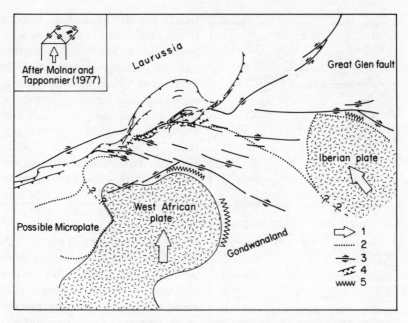

Fig. 14.5 Structural and kinematic interpretation of the Middle to Upper Carboniferous motions during and after the Gondwana–Laurussia collision. Explanation: 1: possible motion of indenting plates during Carboniferous time, 2: suture, 3: strike-slip faults, 4: grabens, 5: fold belts active during the Late Carboniferous (compiled after Lefort and van der Voo, 1981 and Lefort *et al*., 1982. © by the University of Chicago)

and Africa but which was founded on the extensional zone outlined by Sawkins and Burke (1980). Bachtadse *et al*. (1983) used palaeomagnetic data on mid-Devonian to early Carboniferous rocks from central Europe to interpret the Hercynian belt in terms of a large-scale shear zone. Oblique collision caused the dextral megashear in the model of Wong and Degens (1983).

There is an interesting problem as to the origin of the Ibero-Armorican arc. Matte and Ribeiro (1975), Burg *et al*. (1981) and Matte and Burg (1981) proposed that the arcuate form was acquired after collision by impingement of a southern continental promontory into the northern continent (Fig. 14.6), as proposed by Tapponnier and Molnar (1977) for the syntaxis in the western Himalayas. Accordingly, the arc is bordered symmetrically by dextral shear zones in S Brittany and a sinistral shear zone in W Spain and Portugal (the Oporto-Cordoba lineament). Note, however, that Brun and Burg (1982) present evidence for sinistral wrench

movement along the northern shears which they consider invalidates the impingement model.

Rast (1983) points out that the Rheno-Hercynian zone is equivalent to the Valley and Ridge Province of the southern Appalachians with over-thrusting being directed towards the craton, that the Saxo-Thuringian and Moldanubian zones with their cores of earlier basement correspond to the Piedmont, and that the majority of Alleghenian granites are concentrated in the Piedmont, as in the Moldanubian zone. Clearly, the North Americans have led the way with the interpretation of the Appalachian belt in terms of the plate tectonic model, and this bodes well for future progressive insights into the Hercynian belt.

Yet the fixists are still here. Krebs (1977) and Zwart and Dornsiepen (1978) prefer to interpret the Variscan belt in terms of vertical tectonics.

There is a final aspect to consider which bears on the problem of the mode of evolu-

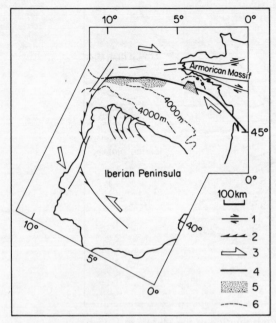

Fig. 14.6 Relationship between the South Armorican arcuate structure and the Ibero-Armorican shear zones (after Matte and Ribeiro, 1982 and Burg *et al.*, 1981). 1: strike-slip faults, 2: strike-slip thrusts, 3: major sense of shear, 4: suture zone, 5: mafic bodies, 6: South Armorican arcuate structure (from Le Fort *et al.*, 1982; reproduced by permission of Elsevier, Amsterdam)

tion of the Hercynian belt: palaeogeography. Irvine (1977, 1981) and Scotese *et al.* (1979) published a series of palaeomagnetically derived palaeogeographic maps for the late Palaeozoic which indicated the presence of oceanic plates between various segments of western European plates; Raü and Tongiorgi (1981) discussed the relevant problems and their implications for the production of the Hercynian fold belt. They favoured a collision between Africa and Europe rather than between Africa and N America and concluded that the preferred model implied a late opening of Tethys between southern Europe and Africa in the Permo-Triassic rather than in the Devonian-Carboniferous, as proposed by Johnson (1978). Johnson (1981) gives a fascinating and comprehensive evaluation of the palaeogeographical implications of the collisions between the three continental plates (Laurasia, Gondwanaland and Angaraland) that led to the assembly of the supercontinent of Pangaea; we look into these questions in the next chapter.

Chapter 15

Pangaea: Permian–Triassic

All the continents had drifted together to form a continuous landmass by the end of the Carboniferous—the final stage of fusion is marked by the union of Europe and Asia and the formation of the Urals orogenic belt (Ivanov *et al*., 1975). This Pangaea remained intact for about 100 Ma, until the early Jurassic. The formation of such a supercontinent gave rise to a new set of tectonic, climatic and biogeographic conditions, very different from those in the Cambrian–Devonian when continents were drifting and oceans opening and closing. Thus it is not surprising that different types of sediments accumulated, fauna and flora fluorished, and a vast new glacial epoch evolved in this period. We shall look at some of the geological evidence that supports the concept of such a Pangaea, including some of the pre-separation geology, because if the continents formed a single plate, their geology prior to fragmentation should be broadly similar.

There are three historic landmarks in the evolution of ideas on the formation of Pangaea. Observing the distribution of the Glossopteris flora and glacial sediments in the southern hemisphere, Suess proposed in 1885 the existence of a supercontinent called Gondwanaland incorporating India, Africa and Madagascar and he later included South America and Australia. Realizing the problem of migration of the plant species, many geologists then thought that the intervening oceanic areas between the present continents had sunk or that land bridges had existed between them; but in 1924 Wegener proposed the revolutionary idea that a former supercontinent had broken up and the individual continents had drifted to their present position. In 1937 du Toit produced ten lines of geological correlation supporting juxtaposition of South America and Africa and he proposed the existence of two supercontinents, Laurasia and Gondwanaland, separated by a seaway, the Tethys, which prevented mixing of the northern and southern flora. Du Toit also produced a reconstruction of Gondwanaland based on general morphological fit and geological correlations. In the last two decades reassembly of the continents has been helped by computerized techniques using, for example, the 500 or 1000 fathom isobaths (e.g. Smith and Hallam 1970), but there is still some disagreement about the construction of Gondwanaland (Fig. 15.1).

Now let us consider some of the geophysical and geological evidence that is consistent with the geometrical arrangement of a late Palaeozoic–early Mesozoic supercontinent.

Palaeomagnetic Data

As considered briefly in Chapter 11 by determining palaeomagnetically the pole positions and the palaeolatitudes for sedimentary and igneous rocks in different continents and by comparing their polar-wander curves, it is possible to get a fix on the relative positions of the continents for any given period. Such palaeomagnetic data indicate that the continents of Gondwanaland and Euramerica came together to form a first supercontinent in the Silurian, and then in Permian times

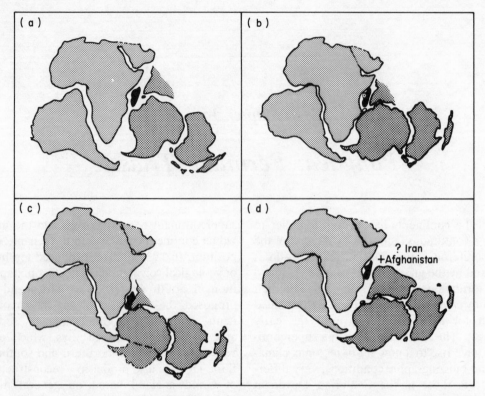

Fig. 15.1(a)–(d)   Four possible reconstructions of Gondwanaland. In particular, note the different positions of Malagasy and of the southern tip of South America with respect to Antarctica (from Powell *et al*., 1980; reproduced by permission of Elsevier, Amsterdam)

Asia combined with Europe to form the Laurasia plate; and so Pangaea was formed. All the continents were clustered closely together from the late Carboniferous to the early Jurassic.

From palaeomagnetic data Morel and Irving (1981) constructed a Pangaea (Fig. 15.2) which is valid for the latest Carboniferous and early Permian (c. 290–260 Ma). However, this reconstruction differs from that of Wegener, as quantified by later workers, which is most reliably relevant for a short interval from the latest Triassic to early–mid-Jurassic (200–170 Ma). The transition from the first to the second Pangaea may have been achieved by a 3500 km dextral megashear between Laurasia and Gondwana during the late Permian and Triassic (c. 250–200 Ma) without the formation of a new ocean (Irving, 1977). Morel and Irving (1981) suggest that Pangaea was not an immobile configuration as envisaged by Wegener and most subsequent workers, but that it evolved more or less continuously. Rickard and Belbin (1980) produced a new continental assembly for Pangaea which is marked for its relatively small Tethyan ocean, in contrast to other assemblies, and extensive sinistral shear between Gondwana and Laurasia. A relatively small Tethys is also favoured by Smith and Woodcock (1982), based on the palaeomagnetic data of Smith *et al*. (1981); this obviates the need for an expanding earth as postulated by Crawford (1979) to account for the lack of geological evidence for a large Tethyan ocean. Note that in the Smith *et al*. reconstruction for the late Permian (240 Ma) South America is situated south of western Europe; substantial dextral transform motion is required to position South America south of North America.

The palaeomagnetic maps of Smith *et al*.

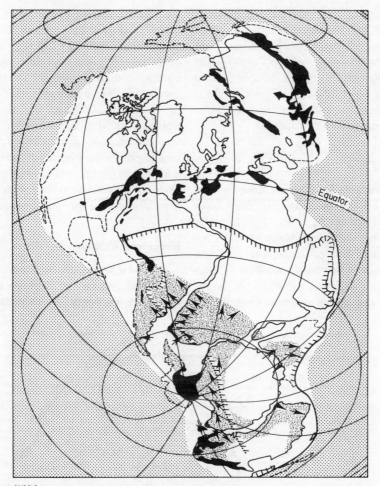

Area affected by Permo-Carboniferous glaciation

Area inhabited by Glossopteris flora

Edge of Precambrian Shield

Direction of ice movement

Fig. 15.2 Pangaea 280 Ma ago. Palaeolatitudes are based on palaeo-magnetic evidence (Morel and Irving, 1981). The late Palaeozoic fold belts (e.g. Appalachians, Hercynian, Urals) are shown in black. Areas affected by glaciation after Crowell and Frakes (1970) and inhabited Glossopteris flora after Plumstead (1973). Edge of Precambrian Shield from Powell *et al*. (1980)

(1973), Scotese *et al*. (1979) and Irving, (1981) show that the continents underwent very little change in position for over a quarter of Phanerozoic time (Fig. 11.3). Gondwanaland was intact for 450 Ma from the early Cambrian to the mid-Cretaceous, when it fragmented. It drifted and rotated as a supercontinent until the Trias (220 Ma), after which its position relative to the south pole

did not change until the mid-Cretaceous (100 Ma) see Fig. 11.3.

Gondwanaland Glaciation

This subject has been dealt with in Chapter 12 on palaeoclimatic indicators. Fig. 15.2 shows the area of Gondwanaland affected by the ice sheets. The close grouping of the glaciated areas of the individual continents provides one of the clearest indications of subsequent continental drift. For an account of the palaeoclimatology of the Permo-Trias period, see Robinson (1973). Frakes (1979) gives a detailed picture of the major ice sheets and flow directions, and Hambrey and Harland (1981) present many details.

Glossopteris Flora

The Glossopteris land plants flourished in the areas south of the Tethys in the Lower Gondwana Formations. The first species evolved during the later stages of the Carboniferous or Permian glaciation in some areas, and soon afterwards in others (Plumstead, 1973); their distribution is larger than that of the glacial deposits (Fig. 15.2); clearly the stimuli provided by the cold climate favoured the growth of this unique flora. Plumstead (1973) suggested that the cause of this plant evolution may have been cosmic radiation, which is known to be greater in polar regions and which did not affect the northern coal plants of Laurasia; she also pointed out that the special climatic conditions favourable for development of the flora could not have operated independently in each of the (at present) widely separated continents, but could have done in a supercontinent dominated by enormous ice sheets.

Relief and Climate

It is well known that there is a positive correlation between the average elevation and area of a continent. To estimate the area of

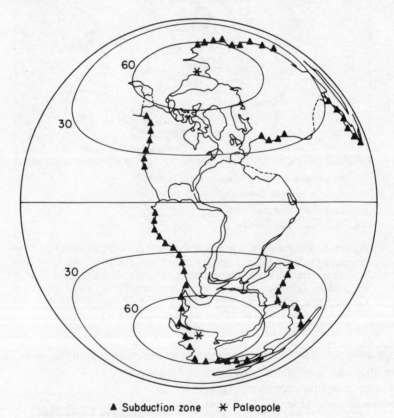

▲ Subduction zone   ✳ Paleopole

Fig. 15.3 Reconstruction of Pangaea with respect to a Jurassic palaeomagnetic reference frame indicating areas of the margins for which there is geological evidence for subduction. Lambert Equal Area projection with 30° paleolatitude lines indicated (reproduced with permission from Hay *et al.*, 1981, *Geologische Rundschau*, **70**, 302–315)

Pangaea one has first to consider the state of the continental margins. Fig. 15.3 shows the active subducting margins around the periphery of the supercontinent. Such margins imply continental drift, but the palaeomagnetic data referred to earlier show that Pangaea did not move far during the Permo-Trias—one would therefore not expect to find very high Andean-type mountain belts against these subduction zones. There is currently much dispute about the volume of sediment entering subduction zones (Karig and Kay, 1981) and this question would affect calculations of the continental elevation. Another factor to consider is that along most passive margins the continental crust has been attenuated by stretching by a factor of 3 during the rifting, and thus the area of Pangaea was reduced by 5×10^6 km^2 during break-up by the attenuation process.

The average elevation of Pangaea is estimated by replacing all the sediment presently residing on the passive margin slopes and rises formed since break-up and the pelagic sediments accumulated in the Mesozoic–Cenozoic. Hay *et al.* (1981) calculate that these sediments would form a layer 1.76 km thick over the whole supercontinent which, after accounting for pore space volume, would give rise to an increased average elevation of 320 m higher than the present average elevation of 740 m. However, sea level has risen by 265 m during the Mesozoic–Cenozoic and therefore we arrive at an average elevation of Pangaea of at least 1320 m above the sea level of the early Mesozoic—almost double the present elevation.

For such calculations we can envisage that the supercontinent had extensive high plains,

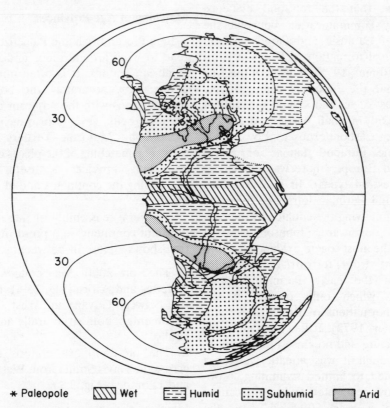

* Paleopole   ⟍⟍ Wet   ⊟ Humid   ⋮⋮ Subhumid   ▨ Arid

Fig. 15.4   A hypothetical climatic map of Pangaea plotted with respect to a Jurassic palaeomagnetic reference frame. Lambert Equal Area Projection with 30° palaeomagnetic lines indicated (reproduced with permission from Hay *et al.*, 1981, *Geologische Rundschau*, **70**, 302–315)

plateau areas and an internal drainage system rather like central Asia today. Fig. 15.4 presents a hypothetical climatic map of Pangaea. It is interesting to compare these climatic zones with the area covered by the Permo-Carboniferous glaciation (Fig. 15.2) and the palaeogeography of the continents described by Ziegler *et al.* (1979) and Bambach *et al.* (1980).

Tetrapods

Terrestrial vertebrates obviously require a land connection in order to wander from one continent to another, and since the idea of land bridges is no longer tenable, contiguity of the continents gives us a viable model. According to Cox (1974) the Triassic is the only period during which tetrapods show clearly that land connections existed between every one of today's continents. There seems little dispute that the tetrapod evidence favours easy communication between the continents of Laurasia and Gondwanaland during the final formation of Pangaea (Colbert, 1973; Romer, 1975); a result of this free access is that the fauna is comparatively homogeneous or cosmopolitan throughout the world. Only a small barrier existed between Upper Carboniferous to Lower Permian aquatic tetrapod faunas of North America and Europe (there was no barrier for the terrestrial types). In Laurasia Carboniferous and Permian tetrapods migrated southwards following the equator as the continent drifted northwards (Johnson, 1980).

Probably the most cogent evidence for continental drift from terrestrial vertebrates comes from the early Permian reptile, Mesosaurus, which only occurs in the Gondwana Formations of Brazil and South Africa (Romer, 1973). It has a peculiar distinctive structure and no close relatives elsewhere. Although it was aquatic, it was a reptile adapted to limited swimming, probably in shallow freshwater, and so one cannot conceive that it swam 5000 km across the Atlantic; the most likely answer is that the two landmasses were joined.

It is difficult to quantify the palaeontological data in favour of Pangaea and subsequent continental drift, but a notable attempt by Brown (1968) produced a significant result. It is well known that the maximum degree of diversity of present day fauna occurs in the tropics and that the diversity decreases polewards. Brown plotted the number of genera of late Palaeozoic amphibia and Triassic reptiles against the present day latitudes of their localities (Fig. 12.8) and found no correlation. But when the number of genera were plotted against their palaeolatitude (Fig. 12.8), they have a symmetrical distribution. Also the diversity of invertebrates decreased to a major low in Permo-Trias times coinciding with the period of maximum assembly of the continental fragments (Fig. 12.5). Anderson and Schwyzer (1977, part 4 and earlier parts) presented a detailed synthesis of the biostratigraphy of the Permian and Triassic.

Matching of Age Provinces

Late Precambrian and Palaeozoic fold belts tend to wrap around older undisturbed Precambrian cratons. It is often not difficult to define fairly accurately the boundaries of these cratons with the adjacent reworked or accreted belts, several of which were fragmented by Mesozoic–Tertiary continental drift. The matching of the pieces across facing continents provides a useful means of regrouping the continents and of demonstrating drift.

A general reassembly of these cratons, on different continents in a pre-drift reconstruction shows a close fit between:

1. The pre-2000 Ma cratons in South America and Africa (Fig. 15.5); in particular that between Guyana and West Africa is well documented, both structurally and radiometrically.

2. The pre-2500 Ma cratons that extend across the Davis Straits from West Greenland and Labrador (Fig. 2.1).

Many of the internal major features of the Precambrian fold belts can be followed across facing continents. For example, there is the line of mid-Proterozoic anorthosites across

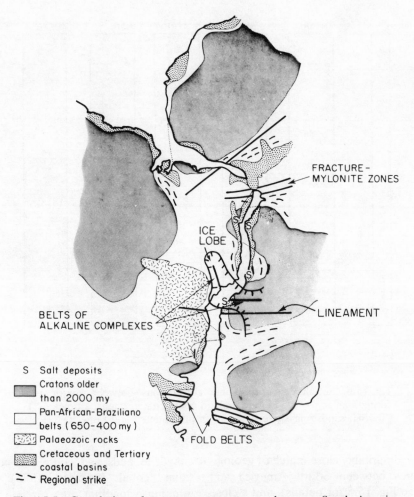

FRACTURE-MYLONITE ZONES

ICE LOBE

LINEAMENT

BELTS OF ALKALINE COMPLEXES

FOLD BELTS

S Salt deposits

Cratons older than 2000 my

Pan-African-Braziliano belts (650-400 my)

Palaeozoic rocks

Cretaceous and Tertiary coastal basins

≃ ∼ Regional strike

Fig. 15.5 Correlation of cratons, structures, etc. between South America and Africa (compiled from Asmus and Ponte, 1973; Nairn and Stehli, 1973; Neill, 1973; Frakes, 1979; Torquato and Cordani, 1981)

both Laurasia and Gondwanaland which, in the former, are accompanied by rapakivi granites, alkaline complexes and acid volcanics (Chapter 7), and the line of early Proterozoic banded iron formations crossing every continent.

Linear Phanerozoic fold belts provide an excellent means of checking the reassembly of the continents suggested by other criteria. There are two important regions: Gondwanaland and the North Atlantic part of Laurasia.

A. L. Du Toit in 1937 recognized that the Palaeozoic fold belts of the southern hemisphere form an integrated pattern when the supercontinent is reconstructed. A map of Gondwanaland (Fig. 15.2) shows that the late Palaeozoic fold belts extend from South Africa across Antarctica to Australia (Engel and Kelm, 1972). When NW Europe is joined to North America on a pre-Trias fit, three fold belts fall into alignment: Grenville–Gothide, Caledonian–Taconic and Acadian–Hercynian.

If the outline of the Phanerozoic fold belts can be followed across the continents, so too should their internal geological features. For example, in the Palaeozoic belts there is the carbonate platform extending from East Greenland, through Scotland, to eastern USA and there are many other stratigraphic correlations between the British Caledonides and the Appalachians (see Chapter 13).

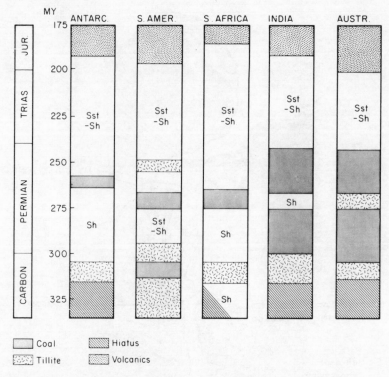

Fig. 15.6 Simplified stratigraphic correlations between major rock formations across the Gondwanaland continents in Pangaea times. SST–SH: sandstone–shale (adapted from Doumani and Long, 1962)

There is a remarkably close match of geological features between South America and Africa (see Fig. 15.5 for details), and a close correlation between volcanics, coal deposits and tillites across all the continents of Gondwanaland between the Upper Carboniferous and the Lower Jurassic (Fig. 15.6).

Metallogenic Provinces

When the continents were aggregated to form Pangaea conditions were ideal for the deposition of continental red sandstones, siltstones and argillites which were hosts for uranium, vanadium and copper mineralization. Copper deposits formed mainly during the Devonian–Trias and to a lesser extent in the Jurassic (they are often associated with anhydrite beds and contain important subsidiary cobalt–Jacobsen, 1975), whilst the uranium–vanadium ores reached their peak of development in the Triassic and Jurassic (Rackley, 1976). The Colorado Plateau region of western USA is the principal area in the world for such ores. The clastic sedimentary copper deposits mostly formed in shallow-water lacustrine or lagoonal continental depressions or basins.

The uranium concentrations formed under arid conditions in fluviatile sediments as fossil stream channel deposits.

In the Permian, and more particularly in the Triassic, the first rifts formed in the supercontinent, some of which led to its break-up. Important mineralization is related to some of the failed rifts and we shall consider these in the next chapter.

Chapter 16

The Break-up of Pangaea: Mesozoic–Cenozoic

There is good evidence for believing that Pangaea broke up into the present landmasses largely during the Mesozoic. The first indications of rifting actually came in the Permian (Oslo Graben), and of continental separation in the Trias; but the main separation stage was during the Jurassic and Cretaceous, with some late rifting continuing into the Tertiary and Quaternary (Red Sea and East African Rift Valley). The total time span of this fragmentation process was at least 225 Ma. In this chapter we shall look at the whole gamut of geological evidence for this mega-fragmentation, which was the incipient stage in the great drift of the continents which dominated the most recent geological history of our planet.

Triple Junctions, Aulacogens, Domes and Rifts

A new approach to the study of continental fragmentation came from the discovery of aulacogens by N. S. Shatski in 1961 and this idea evolved into the dome–triple junction–aulacogen–rift concept of Burke and Dewey (1973a). These sorts of structure apparently began to appear about 2500 Ma ago when stable continental platforms had evolved. Prominent examples are found in Proterozoic and Palaeozoic terrains, but the development of this kind of structure was most notable during the Mesozoic and Tertiary when Pangaea fragmented. All the various stages in this break-up are in evidence

today on or around the continents, but it was not a simple progression with time (some of the best preserved early stages in the rifting process are of Tertiary age). It will be convenient for us to study the full time range of these structures, as together they provide us with the complete break-up story.

Fig. 16.1 shows the development sequence from early domes, triple junction rift networks (rrr junction) based on deep-seated axial dykes, followed by various modifications. Spreading may take place on any of the three arms. If it develops on all three, then three plates are formed, but if only on two, then the formation of two plates leaves an abandoned rift or aulacogen, which is defined as a linear trough extending from a continental margin or geosyncline into a foreland platform or continent (Burke, 1977).

For examples of this kind of structure look at Fig. 16.2 which shows, firstly, the Cretaceous triple junction joining South America and Africa—two arms separated to form the South Atlantic, the third was abandoned and remains as the Benue Trough (Olade, 1978; Petters, 1978; Fitton, 1980)—and, secondly, the Tertiary triple junction centred around the Afar Depression (Mohr, 1978)—two arms have just begun to separate (the Red Sea and the Gulf of Aden), whilst the third (the Ethiopian Rift) failed to move.

But let us start at the beginning of the development sequence and look at the crustal doming, warping and flexure.

As long ago as 1939 H. Cloos realized the

217

218

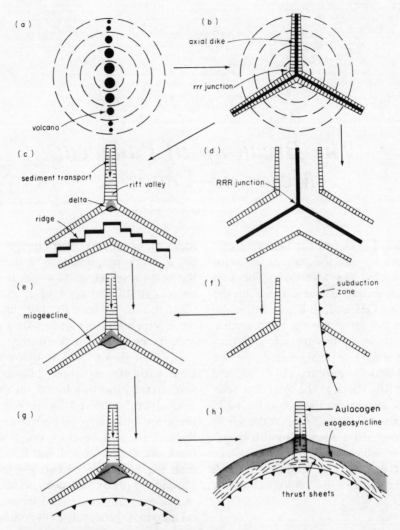

Fig. 16.1 Schematic inception and evolution of plume-generated triple junctions (after Burke and Dewey, 1973a). (a) Uplift develops over plume with crestal alkaline volcanoes (e.g. Ahaggar). (b) Three rift valleys develop meeting at an rrr junction (e.g. Nakuru). (c) Two rift arms develop into a single accreting plate margin (ridge) and continental separation ensues, leaving the third rift arm (failed arm) as a graben down which a major river may flow and at the mouth of which a major delta may develop (e.g. Limpopo). (d) Three rift arms develop into accreting plate margins meeting at an RRR junction (e.g. early Cretaceous history of Atlantic Ocean/Benue Trough relationship). (e) Atlantic-type continental margin evolves with growth of delta at mouth of a failed arm and miogeoclines (e.g. Mississippi). (f) One arm of RRR system of D begins to close by marginal subduction; if ocean is sufficiently wide, a chain of calc-alkaline volcanoes will develop along its margin; any sediments in the closing arm will be deformed (e.g. Lower Benue Trough). (g) Atlantic-type continental margin with miogeoclines and failed rift arms approaches a subduction zone. (h) Continental margin collides with subduction zone, collisional orogeny ensues, sediment transport in the failed arm reverses polarity, and failed arm is preserved as an aulacogen striking at a high angle into an orogenic belt (e.g. Athapuscow)

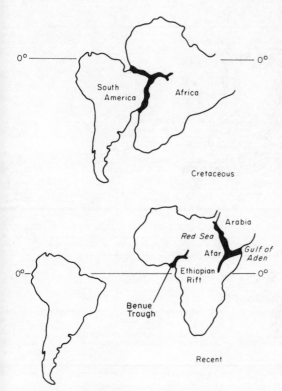

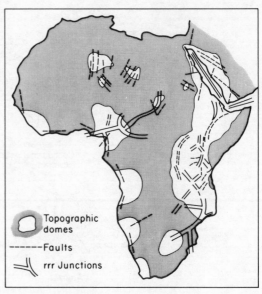

Fig. 16.2 Maps showing origin of Benue Trough as rift arm abandoned during continental rift separation of Africa and South America in Cretaceous time and, similarly, Ethiopian Rift valley during separation of Africa and Arabia in last 25 Ma (after Hoffman *et al*., 1974; reproduced by permission of The Society of Economic Palaeontologists and Mineralogists)

Fig. 16.3 Areas of late Mesozoic and early Tertiary domal uplift in Africa with rifts and triple junctions [reproduced by permission of the Geological Department, University of Newcastle upon Tyne], developed from Le Bas, 1971, and Gass, 1972a) with related triple junctions (after Burke and Whiteman, 1973; reproduced by permission of the Academic Press; London, New York and San Francisco)

importance of doming to rifting when he experimentally demonstrated their relationship with respect to the Rhine and Red Sea rifts. But in recent years most information on this subject has come from Africa (Gass, 1975; Thiessen *et al*., 1979; Kazmin, 1980; Le Bas, 1980a; Girdler, 1983). The domal uplifts (Fig. 16.3) are usually about 1 km high, about 500 km across and are characterized by negative gravity anomalies reaching about 100 mgal and by alkaline vulcanicity (Le Bas, 1980b).

There is a linear crustal arch related to the rift between South America and Africa. The Palaeozoic–Mesozoic sedimentary basins of Brazil (the Parana basin and the Sao Fransisco–Parnaíba–Salitra group) and of the Congo and Kalahari are separated from the

Atlantic by a coastal basement strip 300–600 m wide, which marks a linear Cretaceous arch, the precursor of the protorift zone (Neill, 1973).

The next stage in development was the formation of three rifts which ideally were symmetrically orientated at 120° to each other. Burke and Dewey (1973a) recognized 45 triple junctions throughout the world and Burke (1976) identified over 100 graben around the Atlantic ocean which are linked in a complex network of triple-rift systems (Fig. 16.4).

The oldest rift systems concerned with the break-up of Pangaea are the Midland Valley of Scotland, with its Carboniferous alkalic basalts and Permian tholeiites (Francis, 1978), the Oslo graben with its Permian alkaline rock suite (Oftedahl, 1978a,b; Russell and Smythe, 1983) and the Triassic graben of western and central Europe (Ziegler,

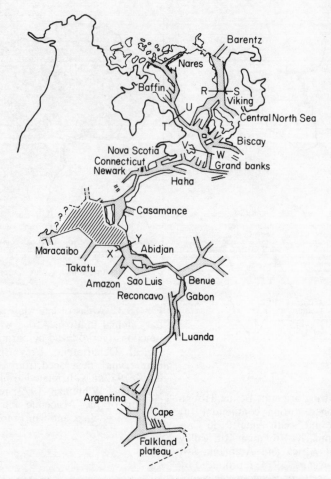

Fig. 16.4 Map illustrating the distribution of major graben around the Atlantic Ocean that formed in association with the early stages of continental rupture. Graben between lines V–W and X–Y formed between 210 and 170 Ma ago, those south of line X–Y between 145 and 125 Ma ago, those north of T–U formed about 80 Ma ago and those north of R–S 80–60 Ma ago (after Burke, 1976; reproduced by permission of Elsevier, Amsterdam)

1982a,b). These rifts did not reach a spreading stage and they represent the first attempt at mega-fracturing of Pangaea soon after its coalescence.

As Gondwanaland began to break up into separate continents during the Jurassic and Cretaceous (200–130 Ma), there was extensive intracontinental rifting in eastern Africa (Fig. 16.3) associated with widespread igneous activity. The rift valleys are located across several domes: a Jurassic one in Tanzania and Tertiary domes in Somalia, Ethiopia, Kenya and the Red Sea. The development of the domal uplifts was related to mantle plume activity during stable periods in plate motion: before 80 Ma, at 60–35 Ma and at 20–14 Ma (Kazmin, 1980). Part of the East African rift valley has opened up 10.0 km since the Miocene, that is, 12–30 Ma ago (Khan, 1975). This is an interesting example of a rift system still undergoing mild tectonic activity. For many papers on the geodynamic evolution of the Afro–Arabian rift system, see *Accad. Naz. dei. Lincei*, Vol. 47, 1980.

Structural Control of Rifting

The questions we need to ask ourselves here are: 'Do new rifts tend to follow old structures?' and 'Why have continents rifted where they have?' In other words, to what extent do former mobile belts/sutures/lineaments exert a structural control on the location of younger rift systems and continental margins?

There must be doubt that a fracture pattern the length (5550 km) and breadth (1000 km) of the Rift System of eastern Africa could be due simply to pre-existing lineaments in the basement; it must have formed in response to a definite stress pattern of its own, and there are several places where the rift valleys cut obliquely across Precambrian mobile belts to demonstrate this.

Nevertheless, there are so many places where young rifts or continental margins follow the grain of old fold belts that local structural control may be suspected. Three examples will suffice to illustrate this relationship:

1. The Rift System of eastern Africa commonly parallels the trend lines of Precambrian mobile belts (e.g. the Mozambique) and in places follows the border of older cratons (e.g. Tanzania Shield, Zambian Block), a conformability distance of at least 4000 km (McConnell, 1977).
2. The join between South America and Africa between the Niger delta and Walvis Bay for the most part follows the grain of the Pan-African/Braziliano mobile belts for at least 2400 km (Nairn and Stehli, 1973). Leonardos and Fyfe (1974) suggest continental rifts require a refractory lower crust of granulites formed in an earlier orogenic episode.
3. The North Atlantic ocean opened in the Mesozoic along roughly the suture of the proto-Atlantic that closed in the Palaeozoic (J. T. Wilson, 1966). More recent geophysical data of Ballard and Uchupi (1975) suggest the presence in the Gulf of Maine, off the eastern coast of North America, of a 250 km long linear Carboniferous–Permian basin that developed as part of a major right-lateral shear zone caused by the sliding past each other of the two continental masses in the final stages of closure of the Iapetus ocean. The structural weakness created by this basin probably controlled the subsequent major crustal rifting in the Trias in this area.

Bahat (1979) disputed the Burke and Dewey (1973a) model and maintained that the East African Tertiary rift junctions were not the loci of initial fracture (rrr junctions), but were derived as lateral fracture bifurcations that followed old structural lineaments and that were due initially to tension exerted on the lineated crust by diapirism (a type of vertical magmatic indentor). Another relationship to earlier structures is seen in Tanzania where there is a major WSW–ENE chain of Cenozoic volcanics that cross-cuts the axis of the East African rift system. This volcanic chain may represent an early *en echelon* leaky transform fault which controlled the magmatism and which prevented crustal extension on the main Kenya rift (Fairhead, 1980).

According to the concept of J. T. Wilson (1965) transform faults, which control offsets of mid-oceanic ridges and are matched by offsets of the rifted continental margins, were initiated by lines of weakness in the original single continental mass. But such fractures within the continents are not well documented. Good examples are demonstrated by Garson and Krs (1976) in the Red Sea where a great many transform faults extend offshore into ENE-trending transverse fractures in Precambrian rocks. The direction of sea-floor spreading was clearly guided by the original continental fracture patterns. Similar transverse structures controlled the initial splitting of the continent in the Afar rift of Ethiopia (Barberi *et al*., 1974) and two major faults in Nigeria coincide with and probably controlled initiation of the Romanche and Chain fracture zones in the South Atlantic (Wright, 1976). The White Mountain plutonic-volcanic series of New England, E USA was emplaced mostly in the period 180–120 Ma along a linear continental extension of a transform fault (Foland and Faul, 1977).

The Timing of Break-up

It takes a long time to fragment a supercontinent. The break-up of Pangaea began in the early Jurassic, 180 Ma ago, and the last continent was not separated until the early Tertiary (45 Ma). In other words some continents broke away long after others: the separation of North America was very advanced, but Gondwanaland fragmentation was rather late (Irving, 1977).

The sort of criteria used to date the time of separation of the landmasses are geologically meagre in quantity, and subject to some uncertainty. Nevertheless they do demonstrate distinctive environmental changes that can be dated by isotopic or fossil methods. The most common types are basic dykes, alkaline complexes often associated with faults, the first appearance of marine sediments along continental margins after a long period of continental sedimentation or lack of it, and affinities and differences between marine fauna on facing coasts suggesting contiguity or separation of the continents (Smith and Hallam, 1970).

The first landmasses to break apart were North America and Africa/South America. Rifting began in the Triassic, but separation did not begin until the early Jurassic—its important to distinguish these two stages (Hallam, 1980). For a discussion of the early history of the North Atlantic, see Noltimier, 1974). Masche (1977) reviewed the criteria for dating the union of the North and South Atlantic oceans. Evidence for the early history of the North Atlantic is as follows:

1. The onset of rifting is expressed by the formation of fault-bounded late Triassic basins with continental sediments and volcanics of the Newark Group on the eastern coast of the USA; these sediments indicate that insufficient separation had taken place by this time to allow ingress of marine waters. Rifting episodes are indicated by the intrusion of the Palisade sill at 190 Ma (Dallmeyer, 1975), diabase dykes 175 Ma (Sutter and Smith, 1979), coast-parallel dykes in Liberia (Dalrymple et al., 1975) 180 ± 10 Ma, the Freetown Igneous Complex (Umeji, 1983)

193 ± 3 Ma on the coast of Sierre Leone (Beckinsale et al., 1977), and the Messejana dolerite dykes in Spain at 184 ± 5 Ma (Schermerhorn et al., 1978). From palaeomagnetic data Smith and Noltimier (1979) concluded that active spreading in the central Atlantic began in the interval 175–170 Ma; however, recent deep-sea drilling results of Sheridan and Gradstein (1981) suggest that the first oceanic opening did not start until 165–160 Ma, in the mid-Jurassic.

2. Palaeontological evidence of the age of sediments overlying layer 2 basalts allows the determination of the sea-floor spreading rate, which, if extrapolated westwards, gives an age for the American continental margin of 190–200 Ma (Phillips and Forsyth, 1972). From sedimentary distributions and the spread of clastic deposits Jansa and Wade (1975) concluded that sea-floor spreading between the African and North American plates was initiated about 170 Ma.

3. Off the continental shelves of North America and North Africa there are 'quiet zones' in which the linear magnetic anomalies are weak or absent. The 400 km wide western zone is twice the width of the eastern. They are regarded as congruent isochrons. The western zone probably formed in the early Jurassic period of constant normal polarity (190–165 Ma), the edge of the continental shelf having an age of about 200 Ma (Phillips and Forsyth, 1972). In order to account for the asymmetry of the quiet zones Noltimier (1974) suggested that the axis of the active spreading ridge shifted to the east by several hundred kilometres 160 Ma ago after which spreading continued symmetrically until the present.

There is more detail known about the timing of the break between South America and West Africa than between any other two landmasses; the sediments and fauna in marginal basins on the facing coasts are comparable and ideal for erecting a chronology (Van Andel et al., 1977). The following sequence of events took place.

The earliest rift phase is expressed by linear dyke swarms along the Namibian coast with an isochron age of 134 Ma and effusion of extensive rift-plateau basalts in Brazil and Namibia occurred simultaneously at 121 Ma (Siedner and Mitchell, 1976). The presence of identical non-marine ostracods of late Jurassic to early Cretaceous age in Brazil and West Africa shows that the two continents must have been contiguous at this time and terrestrial crocodiles were able to make the crossing in the early Cretaceous (Buffetaut and Taquet, 1979). Geophysical work on the South Atlantic oceanic crust indicates that the spreading ridge axis formed by 127 Ma in earliest Cretaceous time (Larson and Ladd, 1973). During the Aptian (106–112 Ma) seawater flooded into part of the graben and extensive evaporite deposits accumulated which are preserved on both the Brazilian and West African coasts (Burke, 1975; Evans, 1978) (Fig. 15.5). Palaeomagnetic poles on lavas from SW Africa and Brazil suggest that the South Atlantic 'had not opened appreciably by 112 Ma' (Gidskehaug et al., 1975). The detailed ammonite biostratigraphy of Reyment and Mörner (1977) documents the oscillatory inundations (transgressions and regressions) of sea water into the narrow gulf during the Albian (106–112 Ma), Cenomanian (94–100 Ma) and Turonian (88–94 Ma) (Table 16.1). The first passage of marine organisms between the two arms of the Atlantic took place in the late Middle Albian (c. 103 Ma) during a phase of high eustatic sea level, as indicated by ammonite faunas (Kennedy and Cooper, 1975; Förster, 1978), whilst interchange of surface waters took place in the Gulf of Guinea in the early Cenomanian, about 97 Ma (Mascle, 1977). The development of a deep-water system of currents in the Guinean zone took place in the interval 90–80 Ma (Reyment, 1980).

There is a controversy regarding the timing of break-up of Gondwanaland (Hallam, 1980). The discrepancy between the post-late Carboniferous polar-wander paths for the South American-African block and Australia suggests that the fragmentation of Gondwana started in Permo-Carboniferous or early Permian times (Valencio, 1974). Successive stages in the separation of the different parts of Gondwanaland through the Mesozoic and Cenozoic are shown by Irving (1981). Fig. 16.9 indicates the timing of separation of all the main components of Pangaea.

Sedimentation

There were four phases of sedimentation during progressive break-up (Hay, 1981):

1. The elevated rift valley graben: sedimentation in swamps, freshwater lakes and streams. Initiation of sedimentation in the rift starts 30 Ma before separation and lasts about 20 Ma. Extensive freshwater lake deposits may have high organic carbon contents, so that these sediments contain as much as one-eighth of the earth's sedimentary carbon. At 10 Ma before separation the rift floor reaches sea level so the rift may be filled with marine water or dessicated basins below sea level, like the present-day Afar and Dead Sea.

Table 16.1 The early opening of the South Atlantic

Successions	Age	Environment
Marine limestones and sandstones	Albian (100–106 Ma) and younger	Freely circulating sea-water in transgressions and regressions—comparable to the Red Sea
Evaporites (up to 2 km thick)	Aptian (106–112 Ma) (mid-Cretaceous)	Inflow of sea-water restricted by local barriers—comparable to the Afar Depression and the Red Sea
Non-marine clastic sandstones and shales	Pre-Aptian (Upper Jurassic–Lower Cretaceous)	Continental lagoons, alluvial plains, flood plain deposition on levelled basement surface

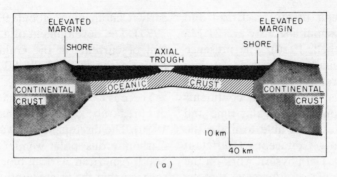

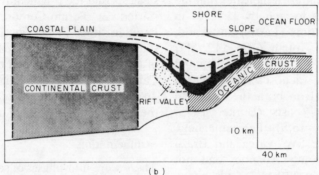

Fig. 16.5 (a) Proto-ocean evaporites in the southern Red Sea. The upper 3–4 km of evaporites are proved by drilling, the lower section from seismic data. (b) Proto-ocean evaporites deeply buried at the base of a trailing continental margin sediment prism. Similar sections seen in Mississippi Delta and northern Gulf of Mexico (after Kinsman, 1975; reproduced by permission of Macmillan Press)

2. The narrow sea between high walls: hypersalinity and evaporites. The walls of the graben may reach 2–3 km above sea level and the graben width may increase to 100–200 km. As dry air flows down into the graben it is adiabatically heated and develops a high evaporative capacity; this phase lasts about 20 Ma.

3. The narrow ocean. This phase occurs in the interval from 10–30 Ma after separation, when the ocean widens from 200 to 600 km or more. Sediments are largely terrigeneous detritus and reef or back carbonates.

4. First flooding of the shelf: reversal of drainage. This phase begins about 30 Ma after separation of the continental fragments and it lasts for 20 Ma. The uplifted margin subsides below sea level and so becomes a site of sediment accumulation on the developing shelf, slope and rise. The sediment is mostly detrital and very large in volume. Fifty Ma after separation the continental margin reaches maturity, when the ocean is wide enough to dominate the climate of the marginal landmass. We see evidence of these phases of sediments along the present-day passive continental margins.

Marginal Basins

In the initial stages of break-up fractures and faults formed and down-faulted basins and structural depressions were infilled with continental sediments. With increasing separation of the continental 'plates' more and more marine water invaded the basins, until narrow seaways were produced which eventually gave rise to the present oceans. Today this course of events is recorded in marginal basins along the trailing edge of the continents in which the progression from non-marine to marine sequences is characteristic within sedimentary piles that may reach up to 16–17 km in thickness. The Atlantic-type continental shelves throughout the world mostly originated in the Permian and showed a history of fault-controlled subsidence in the Trias, Jurassic and early Cretaceous (Kent, 1976).

Along the Atlantic margin of the USA there are several fault-controlled linear basins infilled with up to 8 km of Late Trias non-marine sediments and some volcanics. The sediments include sandstones, shales, and arkosic red beds with mudcracks and rainpit impressions suggestive of subaerial deposition in flood plain depressions, whilst the remains of dinosaur footprints, and fossil freshwater fish confirm a non-marine environment. Basaltic sills and intrusives (such as the famous Palisade Sill of New York) in these sequences result from magma injection into early rifts along the continental margin. Seismic reflection profiles have delineated three comparable major Triassic crustal rifts off the eastern coast of North America in the Gulf of Maine (Ballard and Uchupi, 1975). Between Florida and Newfoundland there is a total of eight offshore basin structures filled with 7–14 km of Triassic–Tertiary sediments (Sheridan, 1976). Their boundary faults formed during the initial rifting of the Atlantic margin and later differential block tilting resulted from stresses produced by continued spreading of the Atlantic Ocean. Some basins contain Triassic to early Jurassic red beds and evaporites (Jansa and Wade, 1975).

The continental break-up and the development of Mesozoic marginal basins around southern Africa are reviewed by Scrutton and Dingle (1976) and Scrutton (1976). The first rifting took place 180 Ma BP along the eastern side of southern Africa (separation from Antarctica and formation of proto SW Indian Ocean), and between 130 and 125 Ma the western side separated from South America. On the western rifted margin sedimentary basin formation was by upbuilding of a sediment terrace during the rift phase (180–130 Ma) and the subsequent 30 Ma,

with outbuilding of the terrace especially in the Cenozoic.

The features outlined in this and the last section show that marginal sedimentary basins, formed along the plate accretion boundaries and preserved along the present continental coasts, provide key evidence for the sequence of events and timing of the early stages of continental break-up.

Evaporites

Evaporite deposits often form, given certain limitations, in the early rifting stage of the fragmentation of continents and of the formation of proto-oceans.

The following evaporites occur in rifts, proto-oceans or on the margins of separated continents:

1. The North Sea troughs contain Permian evaporites (Whiteman et al., 1975) and the Cheshire graben has Triassic (Lower Keuper) evaporites. The trilete trough patterns with their triple junctions are seen as failed rift arms that have undergone no active spreading, sited on an inner continental margin and related to the initial rifting of the North Atlantic (Ziegler, 1982a,b).
2. The Red Sea is a 20 Ma old proto-ocean with evaporites up to 100 km wide and 7 km thick overlying new oceanic crust (Fig. 16.5a) (Kinsman, 1975). There are also continental-based 1 km thick Pleistocene–Pliocene evaporites in the marginal graben of the Danakil Depression of Ethiopia.
3. Late Permian to late Trias evaporites (up to 1.5 km thick in Greece and 1 km in Italy) border the western proto-Tethys (see Chapter 20).
4. There are evaporites in the nearshore or onland mid-Carboniferous to Triassic graben of the Canadian Maritimes, the Appalachian Piedmont and in NW Africa. Upper Triassic to mid-Jurassic evaporites also occur off the North Atlantic continental margins and they originally extended from the Gulf of Mexico via the Old Bahama Channel to the coast of Senegal (Burke, 1975).
5. Aptian evaporites up to 2 km thick occur

on the continental margins (Fig. 15.5) of Brazil and West Africa (Evans, 1978) and they extend for 250 km offshore (Roberts, 1975).

The above examples show that evaporites occur in rift systems, marginal basins, down the continental slope and on oceanic crust, in all cases having formed during the early stages of continental rifting. Kinsman (1975) calculated that evaporite deposition occurs during the initial 10–20 Ma of spreading and in proto-oceans the masses may reach up to 4000 km long, 600 km wide and 7 km thick. The breakup of Pangaea may have resulted in extraction of up to $10 \times 10^6 \text{ km}^3$ of evaporites from the ocean (Hay et al., 1981). Along trailing continental margins such evaporites are likely to be buried at the base of miogeoclinal sedimentary piles (Fig. 16.5b).

Besides structural control the development of evaporites depends on (1) the climatic palaeolatitudinal position and (2) the correct balance between restriction and availability of saline ocean waters (Kinsman, 1975; Burke, 1975).

1. As pointed out in Chapter 12, evaporites can only form where evaporation exceeds rainfall and the present-day Hadley cells controlling atmospheric circulation limit the location of such conditions to two zones 5–35°S and 15–40°N; modern evaporites thus have a bimodal distribution, being absent in a zone close to the equator where equatorial rains are high. The distribution of evaporites and other climatically sensitivie sediments by Seyfert and Sirkin (1979) shows that since the Mesozoic the high pressure subtropical belts of net evaporation have roughly maintained their present width and position relative to their palaeoequator. It is therefore likely that since the Mesozoic, most evaporites formed within 35° (Kinsman, 1975) or 50° (Gordon, 1975) of the equator. Thus evaporites could only form on the margins of fragmented continents that were located within these latitudinal limits.

2. There has to be a delicate balance between the influx of sea water and the total evaporation and this is best achieved by the presence of physical barriers. Burke (1975) interpreted the salt deposits as the products of evaporation of oceanic waters repeatedly spilled over structural sills into low latitude sub-sea level graben formed in the early stages of continental rupture.

The introduction of fresh water would be inimical to salt deposition. Early rifts are located on domal uplifts and so rivers at this stage generally flow down the slope of the domes. Later spreading is associated with subsidence of the rift margins and a reversal of the drainage pattern, so preventing further salt formation.

Igneous Activity

Continental rifting was commonly associated with three types of igneous activity: intrusion of basic dykes and alkaline complexes, and extrusion of lavas. It is usually possible to obtain an isotopic age on these rocks and they therefore provide another means of dating the break-up of continents.

We shall consider two aspects of these igneous rocks: their tectonic and age relationships.

Basic Dykes

During the early stages of opening of the North and Central Atlantic many parallel sets of dolerite dykes were intruded in late Triassic to early Jurassic times in the coastal regions of eastern North America, West Africa and northeastern South America (May, 1971) (Fig. 16.6). On a reassembly of the continents for the Triassic the dykes form radial alignments interpreted by May as being parallel to lines of tension in a principal stress trajectory field imposed on the continental margins by movements in the upper mantle at the onset of Atlantic sea floor spreading.

In the Skaergaard area of East Greenland there is classic evidence of the mechanism of continental break-up 55 Ma ago in the North Atlantic, namely a coastal flexure and associated dyke swarm (Myers, 1980). The sequence of events was as follows: the extrusion of

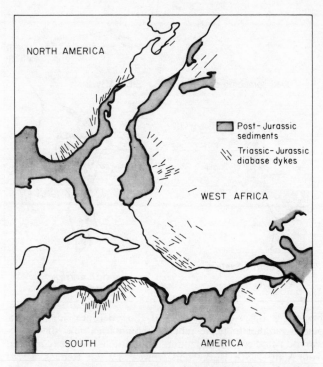

Fig. 16.6 Basic dyke swarms intruded during the incipient stages of opening of the Central Atlantic ocean in
early Mesozoic times (redrawn after May, 1971; reproduced by permission of The Geological Society of
America)

lavas and the intrusion of coast-parallel vertical dykes, flexure of the coast over a steep
shear zone in the lower crust or basement,
intrusion of layered complexes during and
after the flexuring. Aeromagnetic data show
that offshore the swarm extends for over
780 km parallel to the coast and is one of the
biggest post-Precambrian dyke swarms in the
world (Larsen, 1978).

The East Greenland flexure is similar to
the Lebombo Monocline in SE Africa. Burke
and Dewey (1973a) suggest that coastal flexures of this sort mark rifted arms of junctions
that have developed to the ocean opening
stage, although they are not often preserved
along present coasts. Similar flexures are
known on the Bombay coast, the east side of
the Afar Depression and the north side of the
Gulf of Aden. Other coast-parallel basaltic
dykes which are a useful indication of early
activity along rifted continental margins are
the 162 Ma dykes in SW Greenland, the

dykes parallel to the axis of the coastal flexure of Brazil and the 134 Ma old dyke swarm
along the Namibian coast (Siedner and
Mitchell, 1976).

There are thousands of tholeiitic dykes in
the Karroo swarms of southern Africa (Vail,
1970). They have an interesting tectonic setting as they lie parallel to the arms of the
lower Zambesi–lower Limpopo triple junctions of Burke and Dewey (1973a) and, in
particular, parallel to the Lebombo Monocline which is situated along the eastern
north–south arm (Fig. 16.7). This 700 km
long monocline is a major crustal flexure near
the margin of the continent with an enormous
difference in structural level on either
side—estimates vary from 9–13 km. The
dykes were intruded into the zone of maximum extension or warping located along the
rifted continental margin and into a well-
defined failed spreading axis or aulacogen
(Reeves, 1978).

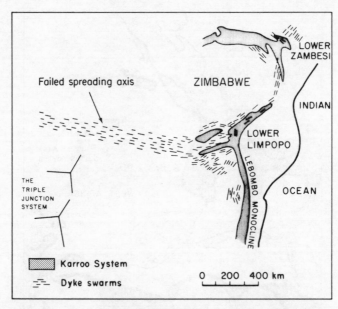

Fig. 16.7 Basic dyke swarms of Karroo (Jurassic) age emplaced parallel to triple junction boundaries near the downwarped rifted continental margin of southeast Africa (compiled from Vail, 1970; Burke and Dewey, 1973a; Reeves, 1978)

Tholeiitic Flood Basalts

There were enormous volcanic outpourings of predominantly quartz-tholeiitic lava in Gondwanaland during the Jurassic and Cretaceous (Fig. 16.8). The Jurassic lavas extend in a zone across the centre of Antarctica and no doubt reflect an abortive attempt at fragmentation, but the Cretaceous lavas are located along or near separated continental margins. As would be expected, most of the volcanicity was associated with faulting, some faults being slightly older and others slightly younger than the lavas.

What is remarkable is the scale of lava expulsion at this time. The Parana basalts of Brazil cover an area of 1 200 000 km^2 which is larger than that covered by the Deccan Traps and the Columbia basalts put together. In places the Karroo lava pile reached 9 km in thickness and the lava field, probably covering 2 000 000 km^2, is the largest on earth.

There was continental distension and rifting in the western Tethys prior to the accreting ridge activity associated with the early Jurassic plate motions. Late Triassic flood basalts were extensively erupted in a region extending from Morocco to Greece, the Caucasus and Iran (Dewey *et al.*, 1973).

Alkaline Volcanic Rocks and Intrusive Complexes

The close association between continental rifting, doming and alkaline magmatism is well known (Bailey, 1974; Le Bas, 1980b). The East African rift system with its early Jurassic–Recent igneous activity is the most pertinent to consider.

Of the volcanic rocks alkali basalts are the most common, especially in Ethiopia, although phonolites and trachytes are locally extensive, as in Kenya. Plutonic ring complexes, especially south of the Rungwe junction, contain syenites and nepheline syenites, which are deep-seated equivalents of the trachytic volcanoes, and the critical association of carbonatites and nephelinites. The keynote of the rift magmatism is the abnormal enrichment of volatiles (and alkalis) expressed by the highly explosive and frag-

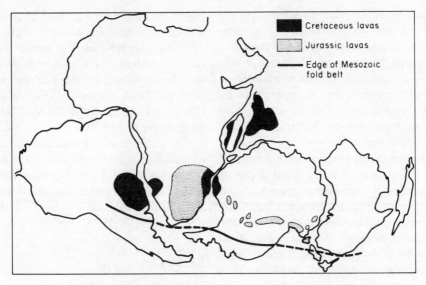

Fig. 16.8   The distribution of Mesozoic lavas plotted on a map of Gondwana-land. The Jurassic lavas broadly parallel the margin of and lie in the foreland of the Mesozoic fold belt, whilst the Cretaceous lavas are localized near separated continental margins (compiled from Cox, 1970; Smith and Hallam, 1970)

mental nature of the volcanism (Bailey, 1974). Rift volcanic rocks are also characterized by light REE enrichments and Sr^{87}/Sr^{86} ratios in the range 0.703 to more than 0.710 (Condie, 1982a).

The first appearance of three magmatic episodes are of paramount importance and correlate with critical tectonic phases:

3 Pliocene	Flood trachytes in rift floor	Main rifting
2 Upper and Middle	Flood phonolites and fissure eruptions	Main doming
1 Miocene	Regional nephelinites and carbonatites; alkali basalts	Early uplift

This sequence represents a progression from incipient melting of the upper mantle associated with early uplift, through extensive partial melting of the uppermost mantle and lower crust giving rise to phonolites, to pervasive partial melting of the lower crust due to lowering of the central rift segment into the heated zone causing trachyte outpourings (Bailey, 1974).

The volume and composition of the volcanics depend on the length of the stable period in motion of the African plate over the mantle plumes (Kazmin, 1980). Subalkaline basalts appeared only during long stable periods, while during short stable periods only alkaline magmas were generated. The formation of rift fractures initiates magmatism independent of the mantle plume activity, the volume and composition of the volcanics being controlled by the opening rates. At rates of less than 1 mm yr^{-1} there is little or no volcanism. At rates of 1–10 mm yr^{-1} alkaline to mildly alkaline basalts appear accompanied by large volumes of peralkaline silicics, and rates of 1–2 cm yr^{-1} correlate with large volumes of transitional basalts with subordinate peralkaline silicics.

The first eruptive rocks in the cycle, which we see in Ethiopia today, were extensive alkali flood basalts generated from melts at depths greater than 35 km. More saturated basalts were then erupted in the rifts and generated in the order of 10 km depth and, finally, with incipient separation of the continental blocks, dykes were injected and new oceanic crust formed as volcanic islands in the Red Sea with the composition of oversaturated low-K tholeiitic basalt generated at very shallow depths of about 3–5 km. The sequence of igneous rocks related to this early rift development was controlled by the progressive rise of geoisotherms in the crust.

Geophysical data support the concept that a diapir of relatively low density material is situated below parts of the rift system formed from some higher density material either by heating and expansion or by change in bulk chemical composition. Seismic data suggest that the East African rift system is a zone of shallow earthquakes (average focal depth of 20 km) similar to the Rhinegraben, Lake Baikal rift and the mid-Atlantic ridge. Contrary to early conclusions that the rifts are associated with negative Bouguer anomalies, recent gravity surveys indicate a more complicated picture. In Ethiopia (and the Red Sea) the anomaly is positive, in the Gregory rift there is a narrow positive within a broader negative and to the south of Kenya there is only a negative which decreases in amplitude southwards, eventually disappearing in North Tanzania (Khan, 1975). In other words, in the Red Sea, where crustal separation has clearly taken place and dense material has intruded upwards, the anomalies are highest. A progressive southward decrease in anomaly intensity is correlated with an expected decrease in the quantity of upwelled material. A combination of seismic refraction and gravimetric data suggest the presence below the Gregory rift of a body of material of low density, 3.15 g cm^{-3} (P wave velocity c. 7.5 km s^{-1}) from 20 to 60 km depth and 200 km wide. This is consistent with seismic results indicating that the crust below the rift is thinned to 20 km (from a normal 36 km for surrounding East Africa) and has an anomalously low sub-Moho velocity. There is a striking similarity between the model proposed for the structure below the Gregory rift based on the above data and that suggested for the mid-Atlantic ridge (Khan, 1975).

It is well known that peralkaline (soda-rich) subvolcanic ring complexes are associated with continental rift valleys. The rocks are silica-undersaturated and include feldspathic types (nepheline syenites, phonolites), non-feldspathic types (ijolites, nephelinites) rich in nepheline and pyroxene, carbonatites and fenites formed by *in situ* alkali metasomatism of the country rocks (Le Bas, 1980b). The relationship between the rifts, igneous activity and high elevation of Africa may be explained by the fact that the continent has been stationary relative to mantle plumes for the last 25 or 40 Ma (Gass, 1975; Thiessen *et al.*, 1979).

Fig. 15.5 shows the location of linear belts of alkaline complexes and carbonatites in Brazil and Angola; the lineaments may be failed arms of triple junctions formed in association with the opening of the South Atlantic (Neill, 1973).

Kimberlites

The majority of kimberlites are of Mesozoic–Cenozoic age and are confined to the interior and margins of stable continental cratons (Dawson, 1980); they rarely occur within the late Tertiary–Recent rift valleys of eastern Africa. The intrusions took place during uplift or dilation of the cratonic areas along deep-seated fractures; some are located on domes, others in graben or transform lineaments. The 1200 km long belt of alkaline complexes in Angola contains 94 Middle Cretaceous kimberlite pipes, especially in the Lucapa graben near the Zaire border. Crough *et al.* (1980) inverted the volcanic traces formed by three hot spots and so determined the post-Triassic positions of Africa, S America and N America, and discovered that the majority of kimberlites must have formed within 5° of a mantle hotspot.

Age relationship between continental rifting and igneous activity

Scrutton (1973) suggested that the continental break-up usually occurred about 25 Ma after the beginnings of igneous action. From his review of the subject (Fig. 16.9) he demonstrated that the main period of igneous activity lasted on average about 30–40 Ma and that the break-up occurred at or just after, but not before, a peak in the activity. He proposed the following typical sequence of events from the rifting of a continent to the

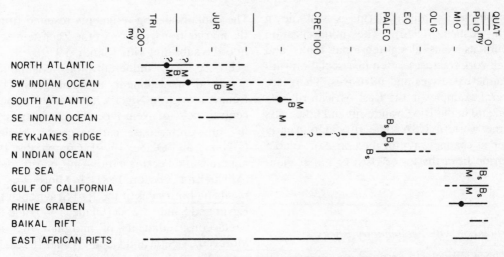

Fig. 16.9 A summary of data relevant to the timing of break-up of Pangaea. —●— duration of igneous activity with peak if known; – – – possible continuation of activity; B_s: time of break-up based on sea-floor spreading data; B: approximate time of break-up based on projected sea-floor spreading data, first appearance of marine sediments and other criteria; M: earliest marine sediments known (after Scrutten, 1973; reproduced by permission of the *Geological Magazine*)

formation of a new ocean:

1. Arching of a continental area with precursory igneous activity.
2. Rifting and onset of more continuous igneous activity; first marine incursions.
3. Igneous activity reaches a peak.
4. Break-up occurs at, or soon after, the igneous peak.
5. Fully marine conditions established.
6. Igneous activity on the continental margins decreases as an active mid-ocean ridge develops.

Igneous rocks related to the continental break-up range in age from mid-Trias (200 Ma) to the present (Fig. 16.9). They had a diachronous development as different oceans opened at different times. We are fortunate in having every stage of igneous activity preserved today, from rifts that failed to spread, to proto-oceans like the Red Sea, to broad oceans with continental margin volcanics, dykes and plutonic complexes.

Mineralization

Some mineral deposits formed in connection with intracontinental hot spots and plume tracks prior to any rift formation or separation of plates (e.g. Sn and U in anorogenic granites). In our study of the break-up of Pangaea we shall not consider these, but only those deposits found in intracontinental rifts and aulacogens and along present continental margins where plate separation has taken place (Olade, 1980; Mitchell and Garson, 1981; Sawkins, 1982).

Association with Magmatic Rocks

Carbonatites in the Neogene East African rift systems contain deposits of apatite ± vermiculite, niobium (pyrochlore) and rare earths ± strontianite. Examples in the Malawi rift and Lucapa graben in Angola are Jurassic–Cretaceous.

The Permian Oslo graben has economic reserves of apatite in jacupirangite associated with alkaline complexes, porphyry-type molybdenum mineralization related to biotite granites, vein deposits of molybdenite-bearing quartz in the biotite granites, and contact metasomatic deposits of Zn, Pb, Cu, Fe and also W occurrences in thermal aureoles.

The close association between fluorspar

deposits and continental rifts is well-known (van Alstine, 1976). The mineralization occurs as veins along tensional faults and stockwork fractures, often in association with alkaline extrusives and intrusives. There are several examples in the East African rift system, and in the Rio Grande rift and other rifts in the western USA. The rift faults clearly acted as channelways for the uprise of volatile fluorine from the lower crust or upper mantle.

Association with Sedimentary Rocks

There is much stratabound copper mineralization in the sediments of rifts. Very similar Cretaceous deposits with minor lead–zinc occur in Aptian terrigenous clastics in Morocco, Gabon and Angola (Caia, 1976). These sediments were deposited during initial subsidence of the intracontinental rift of the Atlantic. Most deposits are overlain by evaporites formed during the first marine incursions into the opening, subsiding rifts. Because mineralization post-dates sedimentation and the overlying evaporites were deposited in early stages of sea-floor spreading, deposition of the copper itself took place in the initial stages of the development of the passive continental margin, rather than in the intracontinental rift (Mitchell and Garson, 1981).

Copper deposits occur in the Kupferschiefer within the Upper Permian Zechstein basin that extends from N England to the USSR. The Cu–Pb–Zn sulphide ores (Rentzsch, 1974) are in a bituminous dolomitic marine shale that lies on the non-marine Rotliegende red sandstone and sabkha deposits that were laid down in the initial intracontinental rifts of the North Sea region. The copper-bearing sediments resulted from the marine transgression of the Zechstein Sea into the subsiding rifts (Smith, 1979).

Pb–Zn–(Cu) mineralization commonly occurs in the sediments of aulacogens. The Benue Trough in Nigeria has a 560 km long central belt of vein-type Pb–Zn deposits in mid-Cretaceous carbonaceous shales (Ukpong and Olabe, 1979). Comparable Pb mineralization occurs in the Amazon rift zone (Mitchell and Garson, 1981), in Miocene sediments on both sides of the Red Sea graben in Egypt and Saudi Arabia (Olade, 1980) and in Cretaceous sediments of marginal rifts in Morocco, Algeria and Gabon (Caia, 1976). It seems likely that the metals of most of these occurrences were derived from leaching of host or basement rocks by remobilized brines from nearby evaporites set in motion by geothermal heat associated with rifting (Olade, 1980). A modern analogue is in the Red Sea geothermal brine which is recirculated seawater which acquired its high salinity due to interaction with Miocene evaporites and which is supersaturated with galena, sphalerite, chalcopyrite and iron monosulphide derived from hot ocean-floor lavas in the axial rift zone (Shanks, 1977).

There are a few occurrences of 'red bed' uranium mineralization in sandstones and conglomerates in the Benue Trough and in Karroo sediments of the Zambesi rift, and of stratabound manganese deposits in red bed clastics and shallow marine carbonates in Cretaceous graben-infills in Morocco and on the Red Sea coast of Egypt and Sudan (Olade, 1980).

In summary, intracontinental rifts and passive continental margins have a wide variety of mineralization types which tell us a great deal about processes that operated during rifting and plate separation.

Chapter 17

Plate Tectonics and Sea-Floor Spreading

After the break-up of Pangaea individual or groups of continents began to drift apart. Throughout the Mesozoic and Cenozoic they were dispersed in different directions until they took up their present positions. They did this because the intervening ocean floor began, and continued, to grow. Since the early 1960s intensive exploration of the ocean floor in the form of magnetic and seismic surveys, dredging, coring and deep-sea drilling (and isotopic dating of the samples) has enabled the growth history of the oceans to be worked out in surprising detail and, as a result, a time scale for the drift of the continental plates.

In this chapter we shall look at several aspects of this new oceanic crust (in particular its spreading history and structure) because its evolution strongly affected that of the continents.

The Plate Mosaic

The theory of global plate tectonics evolved from that of sea-floor spreading, the main advance coming from the field of seismology. Using data from 29 000 earthquakes between 1961 and 1967 Barazangi and Dorman (1969) outlined the seismic belts of the world. Fig. 17.1 shows the dip direction of the subduction zones towards the areas of intermediate and deep focus earthquakes; shallow focus earthquakes are not shown. The seismic belts are narrow, continuous and never transect each other, the interiors of blocks are relatively stable, and the seismic zones follow the mid-oceanic ridges, island

arcs and Mesozoic–Tertiary mountain belts, the divergent margins having less activity than convergent ones.

Formulation of the plate tectonic theory embracing continental drift and sea-floor spreading can be ascribed to Wilson (1965) who described the complex network of ridges, subduction zones and transform faults within an evolving plate mosaic Wilson (1966) introduced the concept of the opening and closing of oceans and this has since been referred to as the Wilson Cycle.

Morgan (1968) and Le Pichon (1968) outlined six major and about fourteen minor lithospheric plates bordered by tectonic belts (rises, trenches and transform faults), and Sykes *et al*. (1970) made a detailed study of the relations between seismology, the plates and their movements. By this time the seismic data were combined with the ocean-floor magnetic data and the hypothesis of global plate tectonic was on a sure footing.

Although some are solely oceanic, most of the plates consist of continental and oceanic parts. The reason for this combination within one plate is that new oceanic has been attached to old continental crust by sea-floor spreading. Some continent–ocean boundaries occur within plates and are seismically stable. The margins of the present Atlantic and Indian Oceans are seismically inactive and thus contrast with the active margins of the Pacific Ocean (Fig. 17.1).

Morgan (1968) adapted J. T. Wilson's (1965) transform fault concept to a spherical earth. The tensional mid-ocean ridges are generally oriented radially to a pole and the

234

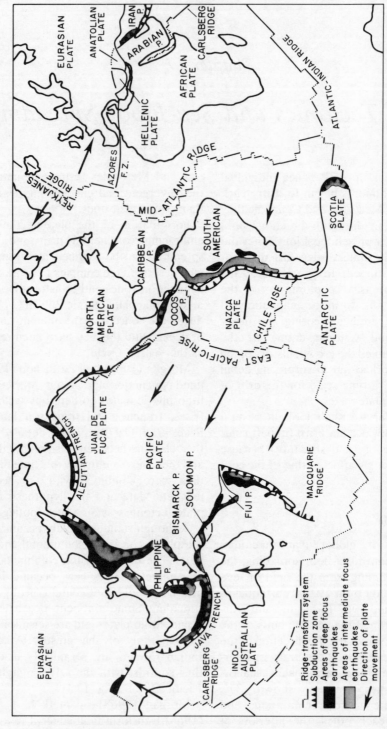

Fig. 17.1   The major lithospheric plates (each named) of the world and their accreting, transform and converging boundaries (partly after Dewey, 1972). Areas of deep and intermediate focus earthquakes (based on Barazangi and Dorman, 1969) and movement directions of plates (after Le Pichon, 1968)

simple shear transform fractures follow orthogonal small circles; thus the pole of spreading lies roughly perpendicular to the transform faults. The rate of spreading, therefore, increases progressively away from the poles where it is zero to a maximum at the equator of rotation. Following these principles Le Pichon (1968) established the poles, rates and direction of the six major lithospheric plate movements.

The increasing rate of spreading with distance from the pole of rotation is reflected in the progressively greater distance between particular magnetic anomalies and the ridge axis. Similarly, the rate of plate convergence at subduction zones increases away from the pole of rotation (Dewey, 1972). A corollary of this is that the intensity of development of fold belts, including the generation of calc-alkaline magmas, should be at a maximum near the equator of rotation.

Broad Structure of the Plates

Fig. 17.2 shows that the outermost part of the earth is made up of a number of layers. The plates consist of the rigid lithosphere, which includes both the continental or oceanic crust, and the underlying denser layers of mantle down to the top of the asthenosphere. The continental crust averages 40 km in thickness and increases to more than 70 km under the Andes and 80 km under the Himalayas, while the oceanic crust averages 7 km. A lithospheric plate is 100–150 km thick under the continents and 70–80 km under the oceans; it is characterized by appreciable strength or rigidity with resistance to shearing stress and thus undergoes little significant internal deformation during lateral drift. The lithospheric plates ride or drift over the asthenosphere which extends to about 700 km depth, the topmost part having no significant strength and is thus plastic and capable of internal flow with lower S and P wave velocities than the overlying or underlying mantle.

Types of Plate Boundary

The plates are bordered by three types of seismically and tectonically active boundaries: divergent or spreading zones, transform faults, and convergent or subduction zones.

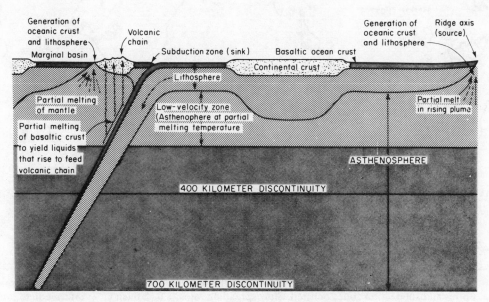

Fig. 17.2   Cross-section of the upper mantle and crust showing a lithospheric plate riding on the asthenosphere. The continent is embedded in the plate and moves with it (after Dewey, 1972; reproduced by permission of *Scientific American*)

Divergent Boundaries

These are mid-ocean ridges (Toksöz *et al.*, 1980) radial to poles of rotation; they are narrow seismic belts, totalling 80 000 km in length, characterized by very shallow (<70 km) earthquakes (Fig. 17.1). The seismic activity is typically far smaller than at major convergent zones. According to Sykes, *et al.* (1970), the largest earthquake known in the ocean-ridge system has about one eighth the energy of the largest shock in an arc. The seismic data are consistent with a thin lithosphere at the ridges subject to tensional stresses and lack of a shear component. The ridges are also zones of high heat flow (up to ten times the average) and active tholeiitic magmatism responsible for generation of new oceanic lithosphere.

Topographically the ridges stand up above the surrounding ocean floor as 500–1000 km wide and 2–3 km high mountain chains.

Spreading rates (Figs 17.3 and 17.4) control the gross morphology of ridges (Macdonald, 1982). At slow *total* opening rates of 1–5 cm yr^{-1}, a narrow 1.5–3.0 km deep rift valley with a discontinuous chain of central steep volcanoes and with rugged faulted topography marks the axis (e.g. Atlantic and Indian oceans). At intermediate rates of 5–9 cm yr^{-1}, the rift valley is only 50–200 m deep; this shallow rift is superposed on a broad axial high and the flanking topography is smooth (e.g. East Pacific Rise at 21°N and the Galapagos spreading centre). At fast spreading rates (9–18 cm yr^{-1}) there is no rift valley and central volcanoes have gentle slopes (e.g. East Pacific Rise south of 15°N). Fig. 17.3 shows that the opening rate of the East Pacific Rise changes along its length, and the same is true, in finer detail, for the Mid-Atlantic Ridge (Le Douaran *et al.*, 1982). Atwater and Macdonald (1977) (see also Stein, 1978) suggested that oblique spreading occurs at ridges which spread slowly, while the spreading is perpendicular at faster spreading ridges. Spreading rate also controls the nature of magma supply and fractionation processes at the ridge (Flower, 1981). Slow spreading ridges are characterized by polybaric magma

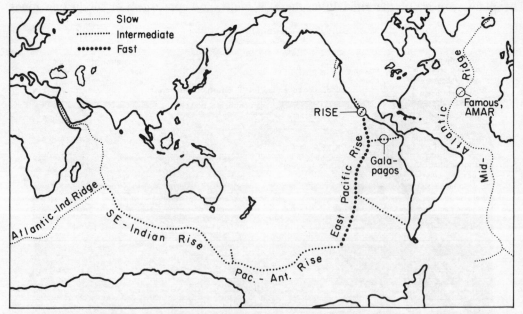

Fig. 17.3 Major sea-floor spreading centres shown schematically; transform faults, back-arc spreading centres and subduction zones omitted. Slow spreading rates, 1.0–5.0 cm yr^{-1}; intermediate rates, 5.0–9.0 cm yr^{-1}; fast rates, 9.0–18.0 cm yr^{-1}. Mid-ocean ridge diving-expeditions indicated (reproduced with permission from Macdonald, 1982, *Ann. Rev. Earth Planet Sci.*, **10**, 155–190. © 1982 by Annual Reviews Inc.)

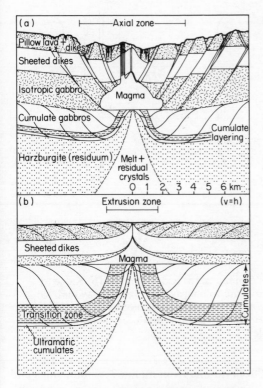

Fig. 17.4 Cross-sections of oceanic spreading ridges (after Dewey and Kidd, 1977). (a) Rifted, slow-spreading ridge. (b) Non-rifted, fast-spreading rise crest (from Burke and Kidd, 1980, in D. W. Strangway (Ed.), *The Continental Crust and its Mineral Deposits*, *Geol. Ass. Can. Sp. Pap.*, **20**, Fig. 2, p. 508)

fractionation in complex reservoir systems reflected in the widespread accumulation of calcic plagioclase in low viscosity melts. Fast spreading axes show low-pressure basaltic fractionation trends to iron-rich compositions with little plagioclase accumulation in stable shallow-level magma chambers.

Although much has recently been learnt about surface topography of ridges from deeply towed instrument packages, we still lack direct information on the deeper levels of ridges, and thus require the constraints provided by ophiolite complexes in order to develop a steady-state model for plate accretion at oceanic ridges. Dewey and Kidd (1977) suggest that such a model involves four main components: (1) a convectively cooling lid accelerating from the ridge axis across a zone of decreasing dyke-injection rate and thickening by the addition of extrusive basalts above and gabbro underplating beneath; (2) a wedge-shaped magma chamber with a flat floor; (3) a differentially subsiding wedges of cumulates; (4) a narrow axial, partially melting, lherzolite-derived diapir from which basalts are liberated over a narrow axial melt into the magma chamber and from which residual harzburgites are plated near the axis below the base of the subsiding cumulates. The uppermost portion of oceanic crust (extrusive basalts underlain and fed by basaltic dykes) represents a chilled roof zone or lid to the spreading centre magma chamber. Rosencrantz (1982) presents a model for the formation and evolution of the uppermost crust by lid flexure and rotation. Mid-oceanic ridges are also the site of extensive hydrothermal activity (Fyfe and Lonsdale, 1981; Mottl, 1983).

Transform Faults

Although transcurrent faults were long known from the continents, it was not until 1952 that extensive fracture zones were recognized across the ocean-floor—in the Mendocino area of the Pacific. Vacquier in 1959 demonstrated that the zones displaced magnetic anomaly patterns and J. T. Wilson (1965) called them transform faults.

Transform faults are shear boundaries parallel to the movement direction vector between two plates and they lie on small circles normal to poles of rotation. Along them there is no generation or destruction of lithosphere. The only part of the fault that is tectonically active is that lying between two ridges, and this is characterized by shallow focus earthquakes and strike–slip movement along a sub-vertical plane. Many transforms were only located by the offset of the oceanic ridge seismic belts.

The motions of transforms, sinistral or dextral, are opposite to those across transcurrent faults on the continents. There are twelve main types (J. T. Wilson, 1973): six dextral and six sinistral, named according to their terminations, e.g. ridge–ridge,

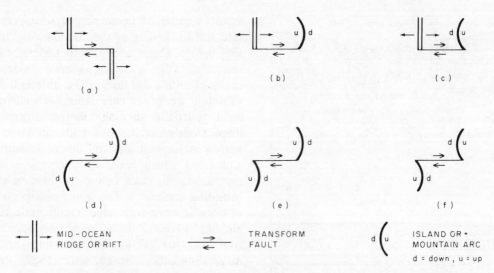

Fig. 17.5 The six possible types of dextral transform faults. (a) Ridge to ridge. (b) Ridge to concave arc. (c) Ridge to convex arc, (d) Concave arc to concave arc. (e) Concave arc to convex arc. (f) Convex arc to convex arc. Note that the direction of motion in (a) is the reverse of that required to offset the ridge (after Wilson, 1973; reproduced by permission of J. Wiley)

ridge–concave arc, ridge–convex arc, etc. The six possible dextral transforms are illustrated in Fig. 17.5.

Fig. 17.6 illustrates the steady-state (average) geometry of a finite-width transform zone that offsets two spreading centres.

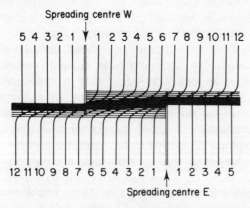

Fig. 17.6 Geometry of a finite-width transform fault that offsets two spreading centres. Lined zones indicate sea floor with transform deformation history. Numbered isochrons illustrate the asymmetrical evolution of the sea floor in a transform zone (after Schouten et al., reprinted with permission from *Nature*, **288** 470–473. Copyright © 1980 Macmillan Journals Ltd)

Observations from manned submersibles and deep-towed instruments show that the width of crust affected by deformation in the fault zone may be as much as a few tens of kilometres (Schouten *et al.*, 1980). The geometry in Fig. 17.6 predicts that the average age of the sea floor in the transform zone will be older than that of the surrounding sea-floor. The inactive trace of the transform zone is only half the width of the active transform. Sea-floor spreading is commonly asymmetrical and discontinuities in one-limb rates result when asymmetrical and symmetrical spreading occurs in adjacent zones, and this produces ridge-crest offsets and transform faults (Hayes, 1976). Ridge transforms may also be initiated as thermal contraction joints in the cooling oceanic lithosphere (Collette, 1974).

Plate tectonic theory predicts that the oceanic lithosphere cools, contracts and subsides as it moves away from an accretion boundary, in other words, the subsidence rate is inversely proportional to the square root of the age of ocean-floor (Menard, 1969; Tréhu, 1975; DeLong *et al.*, 1977). This rule means that along an oceanic fracture zone the higher, younger side will subside more rapidly than the lower older side (Bonatti, 1978; DeLong et al., 1979). Thus fracture zone scarp height changes because of dip-slip motion along the fracture zone and this relation can be constrained from evidence from ophiolite complexes (Karson and Dewey, 1978). Hence different petrological levels may be juxtaposed across a fracture zone, and as a consequence sampling is likely to recover rocks of many different types, such as basalts, gabbros, amphibolites, peridotites and serpentinites (Bonatti and Honnorez, 1976). However, it has also to be borne in mind that in contrast to Wilson's (1965) original definition, some crust is accreted in some transforms. For example petrological evidence of Bonatti (1973) suggests that serpentinized peridotites in the Vema fracture zone intruded as a mantle-derived diapir; such diapirism is also considered by Fox et al. (1976) and Francheteau et al. (1976).

Around the present-day Pacific Ocean there are several on-land transforms, the Philippine Fault, the Alpine Fault in New Zealand and the Atacama Fault in Chile, but the most well known is the San Andreas Fault in California.

Convergent Boundaries

These occur where two plates have come together and one has overridden the other. Four types of collision are possible between:

1. Oceanic/oceanic plates along island arcs (see Chapter 18).
2. Oceanic/continental plates along island arcs (see Chapter 18).
3. Oceanic/continental plates along continental margins (see Chapter 19).
4. Continental/continental plates (see Chapters 20 and 21).

On collision one plate becomes bent and is thrust down a subduction zone into the mantle where it is melted. Where continental and oceanic plates converge, the latter is subducted due to its higher density. The convergent boundaries are the regions of maximum shortening and highest shear motion on the earth's surface, and thus it is not surprising that of all seismic belts they are the widest, have the maximum activity and the largest earthquakes (Sykes et al., 1970). They are characterized by oceanic trenches, shallow (<70 km), intermediate (70–300 km) and deep (>300 km) earthquakes. The deep foci locate the position of the subducting surface that dips at an average of 45° and extends to depths of up to 700 km.

Dewey and Burke (1974), having analysed continental margins involved in plate separation and collision, showed that ideally they should be irregularly shaped, i.e. not straight, because the initial fracturing and break-up developed along interlinked triple junctions tends to give rise to a series of projections and embayments (Fig. 17.7). Upon progressive plate convergence the impinging projections become points of high strain, ophiolite obduction and basement reactivation, and within embayments (which tend to be zones of lower strain) remnants of the former ocean may be preserved. The difference in timing between early and late collision zones causes diachronous development of structures such as basement nappes and flysch fans within the orogenic belt.

Molnar and Atwater (1978) pointed out that cold, thick, old lithosphere sinks at a subduction zone faster than hot, thin, young lithosphere. These relations may cause a fast roll-back or oceanward retreat of the subduction hinge or trench line, in the former case giving rise to an extensional arc with possible back-arc or intra-arc spreading, as seen in the arcs of the West Pacific, and a slow roll-back in the latter case resulting in back-arc compression, as in the arcs of the East Pacific. Extensional arcs, such as Marianas (Fig. 17.8), are intra-oceanic, have plentiful, mainly basaltic, magmatism, few earthquakes in the overriding plate, back-arc basin plate accretion, a thin basaltic crust, low relief, small or absent sedimentary fans, an ophiolitic fore-arc basement, deep trenches and steep Benioff zones. In contrast, compressional arcs such as the Peruvian Andes lie on continental margins, have little, mainly silicic magmatism, many large earthquakes in the overriding plate, back-arc thrusting, a thick continental

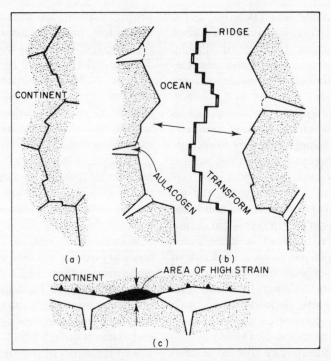

Fig. 17.7   Idealized plate boundaries during the opening and closing of oceans. The triple point configuration in (a) leads to an irregular plate margin with projections and embayments in (b). In (c) a projection impinges against another colliding continent causing high stress and possible ophiolite obduction and basement reactivation. In the embayments former oceanic crust may be preserved (redrawn after Dewey and Burke, 1974; reproduced by permission of The Geological Society of America)

crust, high relief, thick sedimentary dispersal fans, continental fore-arc basements with inner trench wall tectonic erosion, shallow trenches and flat Benioff zones (Dewey, 1980). Neutral arcs with intermediate characteristics are seen in Sumatra, Alaska, Japan, central America and Barbados. They typically have a tapering subduction-accretion prism above a very flat Benioff zone (Fig. 17.8), the prism being widest where the sediment supply to the trench is greatest.

The island arcs and Andean-type arcs are described separately in Chapters 18 and 19, respectively.

Oceanic Magnetics and Sea Floor Spreading

When lavas containing magnetic particles, such as oxides of titanium and iron, crystallize below their Curie point (just below 500°C) in the presence of an external magnetic field, the particles align themselves in the direction of the prevailing magnetic field of the earth. The magnetic anomalies, found by trailing a magnetometer from a research ship across the oceans, reflect variations in the intensity of the earth's magnetic field. By subtracting the regional geomagnetic field of the earth the local magnetic deviations are obtained, locked into the upper 0.5 to 2 km of the basaltic layer of the oceanic crust. The anomalous field due to a normally magnetized strip reinforces the earth's field and gives rise to a positive anomaly and conversely for reversedly magnetized strips. The sort of magnetic anomaly pattern discovered in the ocean floor is shown in Fig. 17.9 where it can be seen that the axis of the Reykjanes

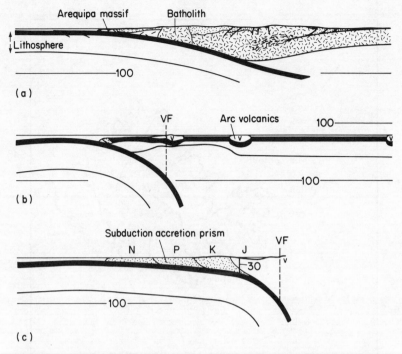

Fig. 17.8   True scale sections across the Pacific margin (scale in kilometres). (a) Central Peru, (b) Marianas, (c) Alaska. Age of subduction-
accretion prism: J: Jurassic, K: Cretaceous, P: Paleogene, N: Neogene
(from Dewey, 1980, in D. W. Strangway (Ed.), *The Continental Crust
and its Mineral Deposits*, *Geol. Ass. Can. Sp. Pap.*, **20**, 553–573)

Ridge pattern points directly towards the active graben that extends across Iceland. There are similar patterns throughout most of the world's oceans (Fig. 17.10). The remarkable feature of such a pattern is its almost perfect symmetry about a central positive anomaly (labelled no. 1) parallel to the mid-oceanic ridge. The positive anomalies are numbered outwards on each side of the ridge anomaly.

The significance of these anomalies was not realized until Vine and Matthews (1963) proposed that they reflected strips of oceanic lava which had been magnetized in normal (as today) or reversed directions. Positive anomalies, like no. 1, have a normal polarity and negative ones a reversed polarity. Two important results of the Vine–Matthews hypothesis were:

1. It confirmed the sea-floor spreading hypothesis. If the symmetrical pattern was caused by the emplacement of magma at the ridge axis, and older reversally magnetized lava was carried further away from the ridge, then the ocean floor must be accreting.

2. It added the time factor to the process. The anomaly patterns enabled a magnetic stratigraphy to be erected for the oceans, comparable to the fossil stratigraphy of the continents.

The application of K/Ar dating first to magnetized lavas from the continents and then to the oceanic lavas gives rise to two further results:

3. The erection of a time scale for the stratigraphy of normal and reversed magnetic strips. The present normal polarity goes back about 700 000 years—this is the longest duration of any polarity interval in the last 4.5 Ma and the shortest is 20 000 years. Beyond 4.5 Ma the radiometric dating method is imprecise; nevertheless, by assuming that the

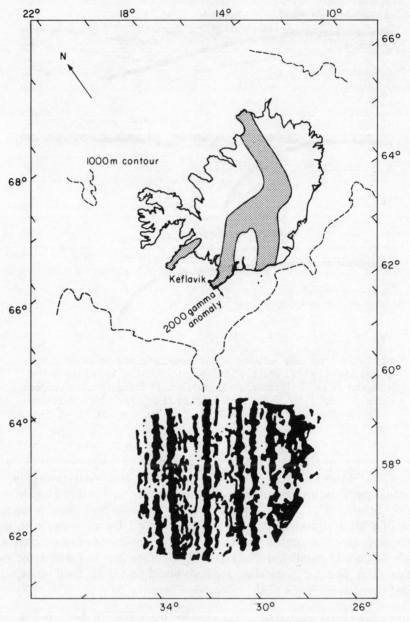

Fig. 17.9   Positive magnetic anomalies (black) across the Reykjanes Ridge
(see Fig. 17.1) pointing towards the rift structure on Iceland (after Heirtzler
et al., 1966; reproduced by permission of Pergamon Press)

spreading rate for the last 4.5 Ma was constant
for earlier time, the age of earlier anomalies
can be worked out—that is, back to at least
180 Ma in the early Jurassic (Fig. 17.11).
Extrapolation of the time scale enables isochrons to be drawn on the ocean floor up to
2000 km from the ridge axis. Such isochrons
have been drawn right across the North
Atlantic. Because new material at ridge crests
in different oceans was similarly affected by
polarity reversals, it is possible to correlate
anomalies of specific number and age from
one ocean to the other.

4. The rate of sea-floor spreading can be cal-

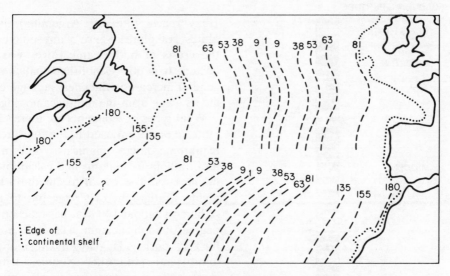

Fig. 17.10 Magnetic anomalies (numbered) in the North Atlantic (redrawn after Pitman and Talwani, 1972; reproduced by permission of The Geological Society of America)

culated by comparing the age of an anomaly with its distance from its mid-oceanic ridge. Wide anomalies reflect a longer duration of polarity than narrow ones and the spreading rate decreases towards the spreading poles.

The net result of the information outlined so far not only provides compelling evidence for sea-floor spreading and therefore the drift of continents, but also enables us to follow the history of the oceans since the early Jurassic. As Fig. 17.10 shows, the evolutionary pattern of the North Atlantic is complete from side to side.

Stratigraphy of the Oceanic Crust—Layers 1, 2, 3

There are two ways to find out how the oceanic crust varies with depth: firstly, by sampling on the surface and by drilling, and, secondly, by seismic surveys. With this information we can find out about the age, thickness, composition and physical properties of the layers that make up this new oceanic floor.

Sampling has difficulties. Much dredging has been done but it is not always sure which layers the samples come from, and drilling is expensive, particularly for penetration to the deeper layers. Sampling has revealed that the top layer is sediment and that this is underlain by basalt. But current knowledge of the overall structure and composition of the oceanic crust has gained much from studies of ophiolitic complexes (Rosencrantz, 1982).

The seismic methods obviously have greater potential in this field and we are fortunate here that the rather simple structure of the oceanic crust (as shown by magnetics, for example) is suitable for a method, the main assumption of which is that the crust must be formed of uniform layers with different seismic velocities. Earlier models consisted essentially of three layers, but today with more sophisticated techniques and results a more complicated structure is envisaged with each main layer being divided into several sublayers, as shown in Fig. 17.12 (Harrison and Bonatti, 1981).

Layer 1 consists of unconsolidated sediments with a low seismic P wave velocity of about 2.0 km S^{-1}. One of the remarkable features about the sediments is their variation in thickness across the oceans. It was no surprise to find great thicknesses of terrigenous sediment near the continental margins

244

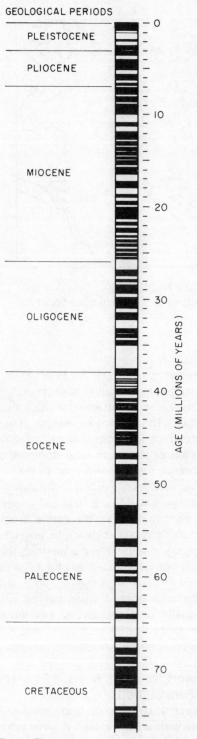

GEOLOGICAL PERIODS

PLEISTOCENE

PLIOCENE

MIOCENE

OLIGOCENE

EOCENE

PALEOCENE

CRETACEOUS

AGE (MILLIONS OF YEARS)

0

10

20

30

40

50

60

70

Fig. 17.11 Chronology of geomagnetic field reversals extrapolated from sea-floor spreading data. Periods of normal and reversed polarity are in black and white respectively (after Heirtzler, 1968; reproduced by permission of *Scientific American*)

(they reach 7 km off Argentina), but the almost complete absence of any sediment over the crests of mid-oceanic ridges was unexpected. It is now established that there is a general increase in thickness and age of sediments away from the mid-oceanic ridges.

What types of sediments are there? Important are oozes enriched in the four groups of planktonic micro-organisms, viz. foraminifera, radiolaria, diatoms and calcareous nannoplankton. The use of foraminifera to subdivide stratigraphy of Cretaceous to Recent sediments (about 140 Ma) has reached a high degree of sophistication. There are also zeolite clays enriched in palygorskite, montmorillonite and phillipsite. Nodules with ferromanganese coatings are locally important on the ocean floor.

Layer 2 is 1–3 km thick and its seismic velocity ranges from 3.5 to 6.1 km s^{-1}, depending on the nature of the basalt and the presence or absence of dykes (Fig. 17.12). It is commonly believed that the lower part of layer 2 has a high proportion of sheeted dykes (Christensen and Salisbury, 1975). It outcrops extensively on the sediment-free oceanic ridges from where it decreases in thickness towards the ocean basalts. Most mid-ocean ridge basalts (MORB) are olivine (Fo_{88-90}) tholeiites with calcic plagioclase (An_{75-85} in cores to andesine in rims). They are calcic (10–12%), aluminous, potash- and soda-poor, and are notably impoverished in incompatible elements, particularly Rb, K, U, Th and Ba (Cann, 1981). Chondrite-normalized REE patterns are depleted in LREE (Sun *et al.*, 1979). Values of $^{87}Sr/^{86}Sr$ are very low (0.7025–0.7030) while $^{143}Nd/^{144}Nd$ ratios are high (Zindler *et al.*, 1982), indicating that the depletion in the more incompatible elements took place more than 1000 Ma. The geochemical features of MORB are believed to have resulted from the extraction of the continental crust, from the segment of the mantle now forming the source for MORB, throughout geological time (O'Nions *et al.*, 1979).

The basalts are often weakly metamorphosed, belonging now to the zeolite and greenschist/amphibolite facies. Most of the

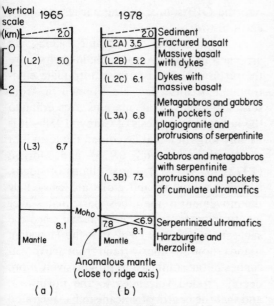

Fig. 17.12 Development of models of the oceanic crust at about 1965(a) and 1978(b). Vertical scale is approximate. The current thinking illustrated by the right-hand model is that the identified layers do not have sharp boundaries but grade into each other. The Moho is a sharper boundary, as reflections of sound waves have been seen which originate from it. Seismic P wave velocities are shown in km s^{-1} (reproduced with permission from Harrison and Bonatti, 1981, in C. Emiliani (Ed.), *The Oceanic Lithosphere*, Wiley, New York, pp. 21–48)

original igneous textures and chemical compositions are preserved, except for introduction of H_2O and Na_2O (spilitic types) and decrease in CaO. The zeolite facies rocks occur in normal oceanic basins, but the greenschist- to amphibolite-grade rocks only occur at mid-oceanic ridges where there is the highest heat flow and a temperature of more than 230°C (Elthon 1981). This metamorphism is caused by hydrothermal circulation of seawater (Fyfe and Lonsdale, 1981). Low temperature alteration by seawater–crust interaction is also common, increasing progressively away from the ridges; this ranges from bottom water temperature to the start of the zeolite facies (Honnorez, 1981).

Layer 3 is about 6 km thick and has seismic velocities of 6.8–7.3 km S^{-1} (Fig. 17.12). Dredging samples show that layer 3 makes up

the plutonic foundation of the oceanic crest and consists of a complex intercalation (not known in detail) of gabbro, metagabbro and minor plagiogranite, serpentinite, anorthosite and diorite (Fox and Stroup, 1981). But note a considerable bias, because almost all samples have been recovered from slowly accreting boundaries. Seawater seems to have penetrated to the depth of emplacement of the plutonic rocks. In places there is a linear relationship between metamorphic grade and intensity of deformation indicating a dependance on ductile shearing and later brittle faulting (dynamic metamorphism) and introduction of sea water along faults and dykes (hydrothermal metamorphism).

The Life Cycle of Ocean Basins

If we look at all the ocean basins in the world, we see that they are in different stages of growth or decline. They can therefore be arranged in an order representing the life cycle of a typical ocean basin.

The East African Rift System may be considered to be in an embryonic stage of development; it is an arm of a triple junction which has failed to open or lags behind the other two arms, the Red Sea and the Gulf of Aden, which represent the incipient stage of growth of an ocean. The Atlantic Ocean began to open in the early Jurassic and is still growing in an advanced stage of maturity. The Indian Ocean floor has begun to be subducted below Java on its eastern side, whilst the Pacific is being subducted on both sides. Although it is still the largest ocean, the Pacific is smaller than when it surrounded Pangaea in the late Carboniferous to early Jurassic and J. T. Wilson (1973) regards it as now being in the first stage of decline. In the Mediterranean region today there are few remnants of the former Tethys Ocean that existed between Laurasia and Gondwanaland (see Chapter 20); the Indus fault line in the Himalayas represents a cryptic suture between the colliding Indian and Asian landmasses, the intervening Tethyan oceanic plate having been destroyed, except for a few distinctive ophiolites (see Chapter 21).

Oceanic Islands

There are probably at least 7000 submarine volcanoes higher than 1 km in the Pacific and Atlantic, but only a few stand above sea level as islands for ready examination. Nevertheless, these islands provide one of the most important clues to magma generation in post-early Mesozoic times. Only a few islands are not volcanic, e.g. Seychelles (granitic), St Paul Rocks (mantle ultramafics) and Macquarie (uplifted oceanic crust). Oceanic islands are shield volcanoes 20–70 km in diameter at sea level. The shield-building phase was short lived, commonly less than 1 Ma duration (McDougall and Duncan, 1980).

The oceanic islands developed after the oceanic crust on which they sit. The lower parts of the island volcanics, mostly submarine, are presumed to be tholeiitic and the upper parts, commonly seen as subaerial islands, consist of alkali basalts enriched in Na and K and their subordinate differentiates—andesine and oligoclase andesite (hawaiite and mugearite), trachyte and phonolite. In other words, the alkaline rocks represent the later stages of evolution. Large volcanoes, such as Hawaii, may ideally consist of tholeiite capped by alkali lavas. Alkali-rich basalt comprises 85–99% by volume of the accessible part of most islands and large submarine volcanoes.

These ocean island basalts (OIB) have high concentrations of incompatible elements, a feature which is particularly striking for the more alkaline lavas. Nb and Ta are more enriched than other elements. Strontium isotope ratios are higher than MORB, while Nd isotope values are lower. Lead isotope values are radiogenic (Sun, 1980) and indicate that the source region of IOB has been separate from that of MORB for a period of the order of 1500 Ma. Hypotheses to account for the difference between the source regions of OIB and MORB include the involvement of deep mantle material which has not suffered the extraction of continental crust (Dupré and Allègre, 1980) and the subduction of sediment and seawater-altered basalt into the OIB source during earlier stages of earth history.

A large number of oceanic islands are arranged in linear volcanic chains, which are the result of volcanism in an intraplate environment because they are situated far from seismically active plate margins. According to the hot spot hypothesis of Carey (1958—it is usually forgotten that he coined the term 'hot spot'), Wilson (1963, 1973) and Morgan (1972, 1973, 1981), linear chains of ridges, islands, seamounts and atolls are caused by the movement of the lithospheric plates over mantle plumes. It has been known since the suggestion of James Dana that the Hawaiian islands show an orderly trend from active volcanoes in the southeast to progressively more deeply eroded volcanoes to the northwest, and that the centre of volcanism has migrated along the chain to the southeast with time. Other subparallel chains in the Pacific Ocean with a similar history include the Macquesas, Society and Austral Islands. The average migration rate for the volcanism of the Hawaiian chain is 9.66 ± 0.27 cm yr^{-1} over the last 27 Ma and for that of the other three chains is 11 cm yr^{-1} (McDougall and Duncan, 1980). The Easter volcanic chain is in the SE Pacific on the eastern side of the East Pacific Rise (Fig. 17.13), but extends across the extinct Galapagos ridge which was active up to 10 Ma ago. Mantle plume activity that gave rise to this chain occurred along a line parallel to the direction of motion of the Nazca plate and Bonatti et al. (1977) explain this in terms of location on an upwelling limb of a Richter (1973) mantle convective roll which developed below fast moving plates with their axes parallel to plate motion (Fig. 17.13).

There is some dispute about whether the hot spots are mutually fixed. For example, Molnar and Atwater (1973) suggested that hot spots can move in relation to each other at speeds of up to 2 cm yr^{-1}. However, in a detailed analysis Morgan (1981) shows that all hot spots, except those in the Pacific, are fixed relative to one another (the Pacific plates are ignored because their motions are decoupled from the motions of other plates by the subduction zones surrounding the

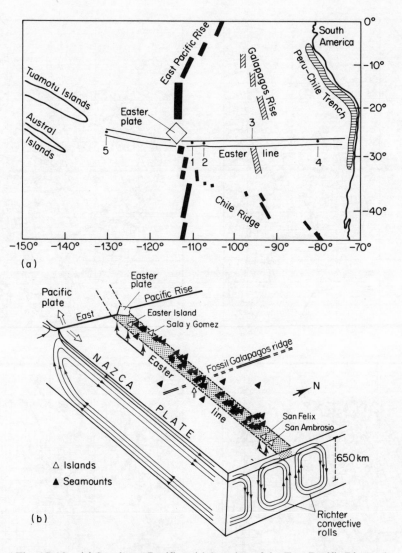

Fig. 17.13 (a) Southeast Pacific, with location of the East Pacific Rise and of the fossil Galapagos spreading centre. The location of the Easter line is indicated; arrows point to hot areas along the Easter line. 1: Easter Island, 2: Sala y Gomez, 3: the intersection of the fossil ridge and Easter line, 4: San Ambrosio and San Félix, 5: Pitcairn Island. (b) Schematic model of the Easter hot line in the Nazca plate showing the distribution of islands (open triangles) and of known seamounts (solid triangles) in the area. The vertical solid arrows indicate sites of young (<2 Ma) volcanism; the vertical open arrow indicates the site of old (~10 Ma) volcanism at the intersection of the Easter line and the fossil ridge. The Easter hot line is shown above upwelling limbs of Richter mantle convective rolls; upwelling spouts may actually be responsible for the hot line (from Bonatti *et al.*, 1977, *J. Geophys. Res.*, **82**, 2457–2478, copyright by the American Geophysical Union)

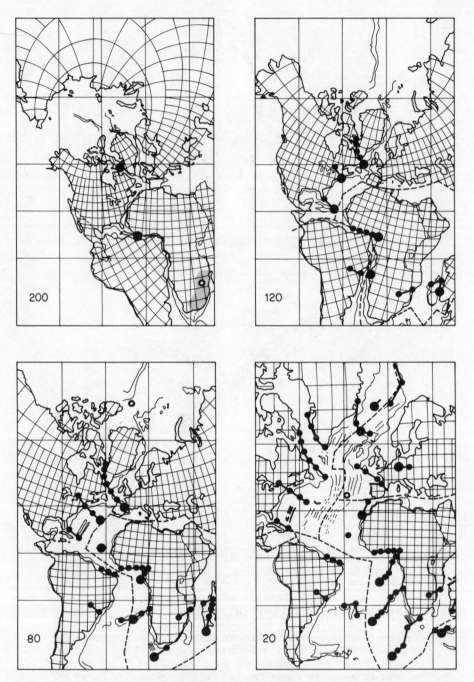

Fig. 17.14   Reconstruction of the continents around the Atlantic for 200, 120, 80 and 20 Ma ago in a fixed hotspot coordinate system. The hotspot tracks are indicated. See discussion in the text (reproduced with permission from Morgan, 1981, in C. Emiliani (Ed), *The Oceanic Lithosphere*, Wiley, New York, Fig. 1 (a, b, c) pp. 460–462)

Pacific plates on three sides). According to Morgan the hot spots are stationary to within 5°, allowing a relative motion at an average rate of only 5 mm yr^{-1} over a 100 Ma interval. Significantly many hot spot tracks are along present continental margins, suggesting that the hot spot heat weakened the lithosphere prior to the continental rifting and plate separation. Several hot spot tracks appear to be related to extensive flood basalt fields, i.e. the track passed between the areas before eruption. Fig. 17.14 illustrates the hot spot tracks in the expanding Atlantic ocean.

Mineralization in Oceanic Environments

Here we shall review briefly the mineralization in the extant oceanic crust; in Chapters 20 and 22 we shall consider the mineralization in ophiolite complexes which are widely regarded as on-land fragments of oceanic crust. It is useful to treat these separately so that one can distinguish the observed-oceanic from the interpreted-oceanic (ophiolitic) types. Useful reviews are by Bonatti (1981) and Mitchell and Garson (1981).

On the ocean floor hydrogeneous deposits (Table 17.1) form by precipitation of metals, mainly from seawater, whilst hydrothermal deposits form by precipitation of metals from hydrothermal solutions injected from below. Ferromanganese nodules and encrustations occur both overlying the lavas at ridges (hydrothermal) and overlying the sediments off the ridges (hydrogeneous). The nodules have high contents of Cu, Ni and Co.

Metal-enriched sediments (Rona, 1978) have now been recovered from the Salton Sea and Red Sea (sulphide-rich muds and metalliferous brines) and ridges in the Pacific, Atlantic and Indian oceans, mostly Fe and Mn oxides and hydroxides, except for sulphide ores on the East Pacific Rise (Francheteau et al., 1979; Oudin et al., 1981) and on the Juan de Fuca ridge (Normark et al., 1983). These sulphide ores are enriched in Zn and Fe. The metalliferous brines and muds in the Red Sea have remarkable concentrations of Fe, Zn, Mn and Cu (Hackett and Bischoff, 1973; Bignell, 1975).

Two types of mineralization are known from transform faults. Major baryte deposits

Table 17.1 Mineral deposits characteristic of oceanic settings (from Mitchell and Garson, 1981, Table VIII, in *Mineral Deposits and Tectonic Settings*; reproduced by permission of Academic Press, London)

Tectonic setting	Association	Genesis	Type of deposit/ metals	Examples
Mid-ocean ridges and basins	Pelagic red clays and basalts	Hydrogeneous (authigenic) sedimentary	Oxide and hydroxide nodules and encrustations Mn Ni Co Cu	Atlantic, Pacific, Indian oceans (Recent)
	Ocean ridge basalts	Hydrothermal exhalative sedimentary	Mn Fe oxides and hydroxide nodules and encrustations	Mid-Atlantic Ridge; E Pacific Rise (Recent)
	Ocean ridge basalts	Sea water hydrothermal exhalative sedimentary	Cu Fe Zn sulphides	E Pacific Rise; Red Sea deeps (Recent); Troodos, Cyprus (Cretaceous)
Oceanic transform faults	Fan sediments, high Ba basalts, Mn	Sea water hydrothermal exhalative sedimentary	Ba	San Clemente Fault Zone (Recent)
	Ocean crust igneous rocks	Sea water hydrothermal sedimentary	Fe Mn oxides and hydroxides	Romanche Fracture zone (Recent)

occur on the San Clemente transform off San Diego in California (Lonsdale, 1979), and FeMn oxide- and hydroxide-rich sediments on the Romanche and Vema fracture zones in the Atlantic (Bonatti *et al.*, 1976); but note also that the Salton Sea metalliferous brines are located in the transform section of the San Andreas fault system of California and the metalliferous sediments of the Red Sea are concentrated at intersections of transform faults with the axial zone (Mitchell and Garson, 1981).

Chapter 18

Island Arcs

An island arc is a tectonic belt of high seismic activity characterized by a high heat flow arc with active volcanoes bordered by a submarine trench (Talwani and Pitman, 1977; Toksöz *et al*., 1980; Leggett, 1982). It forms where a plate of oceanic lithosphere collides with, and is subducted beneath, another oceanic or a continental plate along a Benioff or subduction zone which extends to about 700 km depth and is the focus of shallow to deep earthquakes. Partial melting of the downgoing slab at 150–200 km depth gives birth to magmas that rise and are extruded in volcanoes located 150–200 km from the trench axis and 80–150 km vertically above the Benioff zone. Other important features of island arcs are sedimentary rocks including, in particular, turbidites, tectonic mélanges on the inner wall of trenches, paired metamorphic belts with high-pressure glaucophane-bearing types, plutonic intrusions (granodioritic) in the deeper roots of the magmatic arcs, Kuroko-type mineral deposits, and a systematic chemical variation of volcanics, plutonics and mineral deposits across the arcs.

In this chapter we shall review some of the main features of island arcs which are important belts of continental construction.

Location and Age

The active arcs of today are situated predominantly on the west side of the Pacific from the Aleutians in the north to the Kermadec–Tonga arc in the south (Fig. 18.1); important also are the Scotia and Antillean (Caribbean) arcs in the Atlantic. Most arcs are intra-oceanic (Marianas, Scotia), some lie close to continental margins (Japan), and the Aleutian arc passes laterally into a Cordilleran-type fold belt. Thus, there may be transitions from suboceanic to subcontinental lithospheric junctions.

Many of the arcs are active today—there are over 200 Quaternary volcanoes in Japan and the world's earthquakes are largely confined to these areas. The most recent volcanic arcs have been active during the last 10 Ma since the early Pliocene. The following were developed primarily in the Tertiary: Macquarie, Tonga–Kermadec, Fiji, New Hebrides, Solomons, Marianas, Aleutians and the Lesser Antilles. The formation of the Puerto Rico and Cuba arcs was completed in Cretaceous to Palaeogene times.

There is evidence that some arcs were constructed on older subducting plate boundaries. The history of the East Indies arcs ranges from the Mesozoic to Recent, whilst in Japan the presence of glaucophane-bearing Ordovician, Carboniferous, Permian-Jurassic and Jurassic–Cretaceous successions is indicative of a long history of lithospheric subduction beneath this arc system and tectonic activity continues today.

Subduction Zones

Seismic foci show that subduction zone dips mostly range from 30° to 70° (Fig. 18.2), the seismic activity being concentrated on the upper surface of the downgoing slab of lithosphere, thus defining the 'seismic plane'

252

Fig. 18.1 Distribution of trenches and, in particular, marginal basins in the western Pacific showing the threefold classification based on increasing age, depth and crustal heat flow (based on Karig, 1974; reproduced by permission of Annual Reviews Inc.)

or Benioff Zone which may be up to 20–30 km wide (Isacks and Barazangi, 1977). Fig. 18.2 shows a vertical section through the northern Isu–Bonin arc.

Individual subduction zones tend to dip more steeply with depth. Luyendyk (1970) showed that the dip of the slab is inversely related to the velocity of convergence at the

trench. Because the downgoing slab is heavier than the asthenosphere, it tends to sink passively to an equilibrium at a near-vertical position. With time, therefore, the dip at depth increases.

Earthquakes vary from shallow to deep. Focal mechanisms for shallow earthquakes below the seaward slope of the trench indicate

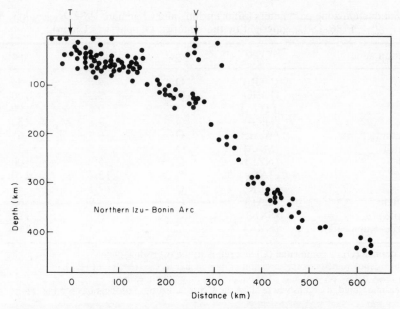

Fig. 18.2   Vertical cross-section of hypocentre distributions beneath the
northern Izu–Bonin arc. T: trench, V: recently active volcanic chain.
Distance measured from trench axis (after Condie, 1982a; reproduced by
permission of Pergamon Press)

the presence of extensional stresses normal to
the trench axis in the sediments and crust in
the upper 75–100 km of the lithospheric slab.
A tensional zone exists where the layer bends
sharply to plunge beneath the arc trench. The
most active shallow seismic zone lies mainly
in the crust on the landward side of the
trench, where the slip vector of earthquake
foci lies on the plane of shallow dip of the arc
and normal to the trench, indicating underthrusting of one arc slab beneath the other.
The focal mechanisms of intermediate-to-
deep earthquakes indicate that the maximum
compression axis is usually parallel to the dip
of the seismic zone and that the downgoing
slab at depth is under compression and meeting resistance (Sykes *et al.*, 1970).

There is a close relationship between the
amount of seismic activity, depth, seismic
velocities, depth of trench, and the rate of arc
convergence. Earthquake activity in the
upper 100–200 km decreases progressively
with depth. This decrease correlates with a
decrease in seismic velocities, and viscosity
and variations in partial melting temperatures
and ductile/shear deformation ratios. There is

a positive correlation between the depth of
seismic activity and the slip rate (cm yr^{-1}) of
plate convergence: the deepest earthquakes
occur in arcs with the greatest rate of underthrusting, (Sykes, Oliver and Isacks, 1970).
When the aseismic slip component of plate
motion is high, there are less earthquakes and
more volcanic eruptions (Acharya, 1981).

Kanamori (1977) suggested that the degree
of inter-plate (oceanic and continental) coupling and decoupling are the major factors that
affect the mode of subduction and evolution
of island arcs. Strong inter-plate coupling
results in great earthquakes and disrupted
Benioff zones; partial and incomplete decoupling results in aseismic slip, continuous
Benioff zones, gravitational sinking of the
subducting lithosphere and retreating subduction zones. Kanamori also showed that
the seismic slip rate is of about the same order
of magnitude as the rate of plate motion.
McKenzie (1977) demonstrated that more
than 130 km of slab must be subducted
before a subduction system becomes self-
sustaining. The depth of penetration of the
oceanic slab into the mantle is dependent on

Table 18.1   Subduction-zone parameters (after Furlong and Chapman, 1982, *J. Geophys. Res.*, **87**, B3, 1786–1802, copyright by the America Geophysical Union)

Subduction zone	Code	Type[a]	Velocity[b]	Age[c]	Dip[d]
New Hebrides	(NHB)	O	9.8	50	68
Kermadec	(KER)	O	7.7	100	66
Tonga	(TGA)	O	9.5	100	58
Izu-Bonin	(IZB)	O	9.8	130	68
Western Aleutians	(WAL)	O	6.7	65	62
Middle America	(MAM)	O	8.5	30	70
New Zealand	(NZL)	C	5.8	100	60
Japan	(JAP)	C	12.7	110	40
Alaska	(ALK)	C	7.1	45	60
Central Aleutians	(CAL)	C	8.9	65	70
Southern Kurile	(SKR)	C	10.7	115	44
Northern Kurile–Kamchatka	(NKK)	C	6.3	115	45

[a]Type is either oceanic (O) or continental (C) and refers to the overriding plate.
[b]Velocity is convergence velocity in the direction perpendicular to the trench (cm yr^{-1}).
[c]Age is million years before present and is determined by magnetic lineations.
[d]Dip is in degrees measured at depths where slabs show constant drop. For locations see Fig. 18.3.

its age (Vlaar and Wortel, 1976). The maximum absolute depth of the trench increases with the age of oceanic lithosphere (Grellet and Dubois, 1982). Table 18.1 lists some basic parameters of subduction zones in oceanic and continental arcs located on Fig. 18.3.

Fig. 18.4 shows the relationship between the structure of a typical arc system and gravity anomaly profiles. There is a gravity minimum reaching about -200 mgal on the inner side of the trench, and a positive anomaly of the same amplitude on the island side reaching a peak in the arc–trench gap. The positive anomaly is widely ascribed to the higher density of the cool downthrust lithospheric slab, whilst the negative is attributed to the accumulation and subsequent under-

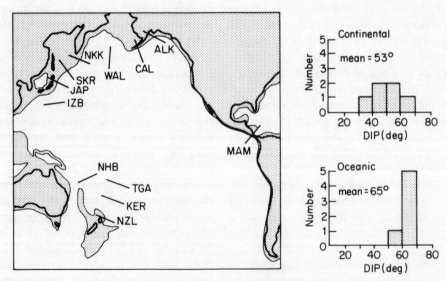

Fig. 18.3   Location map of subduction zones used in Table 18.1 (from Furlong and Chapman, *J. Geophysical Research*, **87**, B3, 1786–1807, 1982. Copyright by the American Geophysical Union)

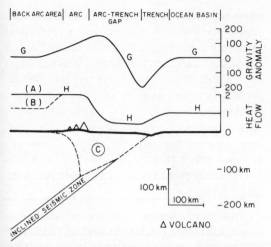

Fig. 18.4 Sketch showing characteristic geo-physical features of arc systems in profile. Heavy line, with volcanoes, is topographic–bathymetric surface. Line G is gravity anomaly profile. Line H is heat-flow profile; in retroarc area line A is the observed pattern for marginal seas, line B is the inferred pattern for thrust belts of foreland basin behind fringing arcs. Triangular space C, bounded within dashed lines, is region of diffuse shallow seismicity beneath main arc and arc trench gap (from Dickinson, 1973b; reproduced by permission of the University of Western Australia Press)

thrusting (especially below the inner slope) of sediments in the trench. Fig. 18.4 also shows the typical heat-flow values across arc systems.

Types of Arcs and their Structure

There are several types of island arc. Some are clearly intra-oceanic (Tonga, Solomons, New Hebrides, Macquarie Ridge, Scotia, Marianas and Lesser Antilles); some are separated from sialic continent by narrow semi-ocean basins (Japan, Kuril, Banda, Andaman islands and Sulawesi); some pass laterally into a Cordilleran-type fold belt (Aleutian); and some are built against continental crust (Sumatra–Java). Thus there are transitional types between intra-oceanic arcs and Andean-type continental margins in which the volcanic belt is an integral part of the continental landmass. Fig. 18.5 illustrates diagrammatically a cross-section through a typical intra-oceanic arc.

There is a systematic variation in structural zones as one passes across the strike of an island arc. The trench is typically 50–100 km wide and may be up to 11 km deep (Mariana), and contains pelagic and turbidite sediments, an accretionary prism of thrusted oceanic and terrigenous sediments and layer 1–3 ophiolites, an inactive fore-arc 50–400 km wide with diverse sediments, and a magmatic arc 50–100 km wide composed of high-level volcanics and deep-level plutonic intrusions. The back-arc or inter-arc basin

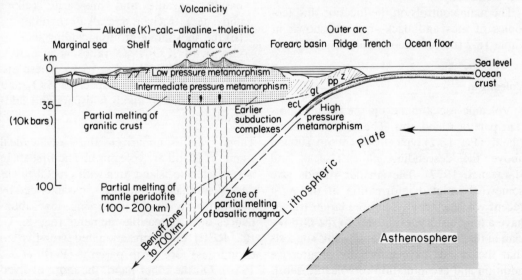

Fig. 18.5 Diagrammatic crustal section across a mature island arc, based on the Japan model of Barber, 1982, *Proc. Geol. Ass.*, **93**, 131–145, reproduced with permission

(often called a marginal sea) is sited on thin oceanic crust which may have grown by back-arc spreading.

There are two types of intra-oceanic arc:

Those with thin sub-arc crust, e.g. 15–25 km total thickness in the Kuril Islands and 15 km in the Marianas and Tonga–Kermadec arcs; these have no pre-arc sialic basement.

Those with thick sub-arc crust. The Japanese arcs have a crustal thickness of 30–35 km and are underlain by a layer with seismic velocities typical of continental crust. This may contain a large component of intrusive plutonic arc rocks and/or a segment of detached continental basement (Barker, 1982).

The width of the arc–trench gap is dependent on four factors (Dickinson, 1973a; Cross and Pilger, 1982). Subduction of young lithosphere produces two opposing tendencies: (a) low-angle subduction and increased arc–trench distance owing to its low density and (b) decreased arc–trench distance owing to its higher temperature. Accretion of sediment in trenches depresses the upper portion of the subducting oceanic plate and causes the trench axis to migrate seaward. Prolonged subduction thickens the upper plate, depresses the isotherms in the subducted plate, and may create a broader arc, thus increasing the arc–trench gap.

The main controls on the location and tectonics of arcs and back-arcs are shown in Table 18.2 (these include Andean-type arcs).

Volcanism

Volcanic rocks make up the most important part of island arcs. The line of volcanic islands (Fig. 18.2) typically lies about 100 km above the descending plate (Isacks and Barazangi, 1977). The smaller islands give somewhat isolated information of the most recent volcanic activity, but the larger islands have a long history, even back to the Ordovician in the Honshu arc of Japan. This suggests that the islands have grown by progressive addition and accumulation of volcanic debris from many successive arc systems. The height above sea level of intra-oceanic volcanoes is constant, whereas the height from their base is variable; the converse is true for volcanoes situated on continental crust in the E Pacific (Ben-Avraham and Nur, 1980). Volcanic arcs are characterized by a high proportion of pyroclastic rocks (vol % of fragmental material ranges from 83 to 93) interbedded with thick deposits of greywackes and mudstones and by overprinting by high T/low P metamorphism (Garcia, 1978).

There are three series of volcanic rocks in island arcs:

1. Tholeiitic Series, including tholeiitic basalt, icelandite (andesite), and some dacite. SiO_2 contents range from 48–63% by wt. The rocks contain groundmass augite, pigeonite and sometimes orthopyroxene but no hornblende or biotite, and show little or no increase in SiO_2 but considerable increase in total FeO with crystallization.
2. Calc-alkali Series with high alumina basalt and abundant intermediate rocks (andesite and dacite) with some rhyolite. SiO_2 contents are in the range 52–70% by wt. They contain orthopyroxene but no pigeonite in the groundmass and may have hornblende and biotite but show little iron-enrichment during fractional crystallization.
3. Alkali Series, divisible into:
   (a) the sodic group with alkali olivine basalt, hawaiite and mugearite (alkali andesites), trachyte and alkali rhyolite;
   (b) the shoshonite group including shoshonite (with K_2O/Na_2O ratios near unity), latite and leucite-bearing rocks. These are undersaturated with respect to SiO_2 and typically contain alkali feldspars and feldspathoids.

Table 18.3 summarizes the geochemical trends of island arc volcanism. The first stage is restricted to island arcs with crustal thicknesses less than 20 km and is characterized by the tholeiitic series possessing low abundances of incompatible elements (e.g. K, U, Ba, REE) and unfractionated (relative to chondrites) rare earth patterns (Perfit et al., 1980). On the other hand, the more evolved stages of volcanism in arcs with crustal thicknesses more than 20 km are characterized by

Table 18.2 Factors affecting the geometry of subduction zones from Cross and Pilger (1982), *Bull. Geol. Soc. Amer.*, **93**, 545–62, reproduced with permission

Factor	Possible effects	Contemporary examples	Associated phenomena
A. Convergence rate	Increased rate decreases angle of subduction, depresses isotherms, and increases width of arc-trench gap	Trans-Mexican volcanic belt	Increased rate increases down-dip length of inclined seismic zone
B. Absolute motion of upper plate	Increased motion towards the trench decreases angle of subduction. Arc-trench separation either increases or the arc is extinguished and a new arc develops 600 to 1000 km inland from the trench. Slow or retrograde motion permits steeper subduction and seaward migration of the trench	Trans-Mexican volcanic belt versus Central American arc	Rapid overriding and low-angle subduction creates compressional stress regime in upper plate; crustal shortening (Cordilleran or Lara-mide style) results. Retrograde motion creates extensional stress regime in upper plate; back-arc and/or intra-arc extension results.
C. Subduction of aseismic ridges, intraplate island-seamount chains or oceanic plateaus	Reduced average density and consequent relative buoyancy of lithosphere reduces subduction angle. Very low-angle subduction is common. Volcanic arc is extinguished, but a new one may form 600 to 1000 km inland from the trench	Aseismic ridges: Nazca and Cocos Ridges Intraplate seamount chains: Juan Fernandez Ridge, Louisville Ridge, and Kodiak-Bowie seamount chain	Buoyant lithosphere resists subduction. Increased area of interface increases coupling between upper and lower plates. Compression-al stress regime usually is produced in upper plate, and basement-rooted thrusting (Lara-mide style) may result. Isostatic subsidence above subducted ridge creates pericratonal basins. If absolute upper-plate motion is retro-grade, back-arc spreading rate is retarded
D. Age of descend-ing plate	Young lithosphere is relatively buoyant and subducts at reduced angle. In various combin-ations with other factors, subduction of young lithosphere will cause volcanism to migrate trenchward, landward, or cease entirely	Trans-Mexican volcanic belt Andean arc Sandwich arc Sandwich arc	Subduction of young lithosphere generally results in back-arc and intra-arc compression. Subduction of old lithosphere generally results in back-arc and intra-arc extension. Down-dip length of inclined seismic zone decreases with decreasing age
E. Accretion of sediment in trenches	Flattens the inclined seismic zone at shallow levels only. Arc-trench separation is increased by seaward migration of the trench	Circum-Pacific and northern Indian Ocean arcs	Weight of accretionary prism depresses oceanic plate prior to subduction
F. Duration of subduction and age of arc(?)	Additive effects of accretion (E) and depres-sion due to prolonged subduction of old lithosphere increase the arc-trench separation	Circum-Pacific arcs	Thickens upper plate?

Table 18.3   Some characteristics of island arc tholeiites and calc-alkaline suites. (From table 1 of Coulon and Thorpe (1981), *Tectonophysics*, **77**, 79–93. Reproduced by permission of Elsevier, Amsterdam)

	Island arc tholeiites	Calc-alkaline series
SiO_2 range	45–70%	53–70%
SiO_2 mode	53%	59%
FeO_T/MgO_{55}	3.5–1.5	2–1
FeO_T/MgO_{60}	4.5–2.5	3–1.5
K_2O_{55}	0.4–1.5	0.8–2
K_2O_{60}	0.6–1.5	1–3
Mineralogy	groundmass pigeonite(?)	orthopyroxene groundmass(?)
Fe–Ti oxide phenocrysts	common	rare
Hydrated minerals	generally absent	more common
Fe(+V, +Ti) enrichment	yes	no
Incompatible elements	10–30 × chondritic	30–100 × chondritic
K/Rb	500–1300	220–560
Na_2O/K_2O	4–6	2–3
Rare earths	unfractionated	strongly fractionated
La/Yb	0.7–3.5	3.5–20

the calc-alkali series with higher abundances of incompatible elements for a given SiO_2 content and with fractionated rare earth patterns (Coulon and Thorpe, 1981).

In mature island arcs such as Japan (Le Bas, 1982) there is a progression in two belts from the tholeiitic to calc-alkali and alkali series from the oceanic to the continental side (Fig. 18.6). In these and some less mature island arcs (i.e. Indonesia and the Kuril Islands) the most characteristic variation is an increase in the alkali content towards the continent in island arc tholeiites and calc-alkaline suites; that is, for rocks with a similar SiO_2 content an increase in K_2O, $Na_2O + K_2O$ and K/Na ratio. This is controlled by the presence of sanidine in high pressure eclogites which causes lavas with higher K_2O to equilibrate at greater depths (Marsh and Carmichael, 1974).

The variation in composition of the eruptive products of the primary magmas correlates closely with the pattern of isobaths of mantle earthquake foci. This compositional zonality may be explained by the difference in the depth of magma generation. Thus the primary magma composition is most siliceous and least alkaline at the volcanic front where the depth of magma generation is the shallowest.

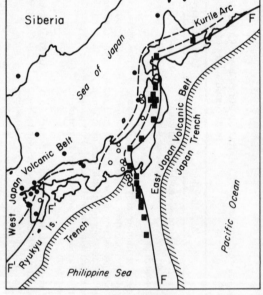

Fig. 18.6   Map of Japan showing the subduction trenches, the two volcanic fronts F-F and F′-F′ of the East and West Japan Volcanic Belts, respectively, and the three petrographic zones: (1) the tholeiitic or P series (squares), (2) the calc-alkali or H series (open circles) and (3) the alkali or A series (dots) (from Le Bas, 1982, *Proc. Geol. Ass.*, **93**, 179–194, reproduced with permission)

There is also a close correlation between the three chemical series and the degree of tectonic activity of different types of arcs. In the youngest (least mature) arcs, with the deepest earthquakes and highest rate of plate convergence, the tholeiitic series is typical and the calc-alkali and alkali series only occur in the mature stages of development. In more mature arcs the calc-alkali series is typical, the alkali series present but unimportant and the tholeiitic volcanics uncommon. In arcs with the slowest rate of plate convergence and shallowest earthquakes, alkali volcanics are common and the other two series atypical. The systematic abundance variation in elements, elemental ratios and isotopes across a typical arc are as follows:

many of the youngest arcs consist of chains of volcanic islands (Kuril, Marianas, Tonga–Kermadec). Uplift and erosion has given rise to larger islands which contain a long history of older arc complexes (Japan).

The petrological evolution of island arcs is reviewed by Ringwood (1974, 1977). There are two stages of development, as summarized in Fig. 18.7:

1. Amphibolite from the subducted oceanic crust is dehydrated at 70–100 km depth. The water produced causes partial melting in the upper mantle wedge above the Benioff zone,

---

continent   ocean

$\longrightarrow$

Fe (max), heavy REE, Y, K/Rb, Na/K, Sr^{87}/Sr^{86}, FeO/MgO, SiO_2

K, Rb, Sr, Ba, Cs, P, Pb, U, Th, light REE, Th/U, Rb/Sr, La/Yb

$\longleftarrow$

In island arcs separated from continental crust by marginal basins there seems to be a correlation between the composition of erupted lava and the type of crust beneath the enclosed seas. When the crust is closer to the oceanic type (like the Bering Sea behind the Aleutian arc) the lavas are gradational towards the tholeiitic type. On the other hand, where the back-arc crust has more continental affinities (as in the Indonesian region), the lavas are of more continental type, gradational towards the calc-alkaline series common along continental margin fold belts like the Cordillera of western America. Thus, there is a continuous spectrum between the early tholeiitic series of island arcs and the calc-alkaline and alkaline series of Andean-type continental margins (Green, 1980; Coulon and Thorpe, 1981).

Older, more eroded, arcs have some granitic rocks exposed; they account for 15% of the surface area of Japan. The calcium content of these granodiorites decreases from the oceanic to the continental side of the arcs, just like the Quaternary volcanics.

In intra-oceanic arcs the earliest volcanics were probably submarine but they soon built up large strato-volcanoes with the result that

so giving rise to magmas that differentiate to produce the early tholeiitic series.

2. With further subduction of oceanic crust to 100–150 km depth serpentinite is dehydrated. The quartz eclogite oceanic crust is partially melted giving rise to rhyo-dacite magmas which react with overlying mantle pyrolite to form pyroxenite, diapirs of which rise upwards and partially melt to produce magmas which fractionate by eclogite crystallization at 80–150 km depth and amphibole crystallization at 30–100 km thus creating the calc-alkaline series. Note that both processes operated at high water vapour pressures.

The combination of low relative Ti-group elemental concentrations (e.g. La/Ta > 30, Ti/V < 20) plus radiogenic Nd ($^{143}Nd/^{144}Nd < 0.5131$) and Pb ($^{207}Pb/^{204}Pb$ above the mantle array or $^{206}Pb/^{204}Pb > 18.6$, or both) is both unique to volcanic arcs (in contrast to mid-ocean ridge basalts) and resistant to the metamorphism (Gill, 1983).

Sedimentation

Sediments are an important component of island arcs (Underwood and Bachman,

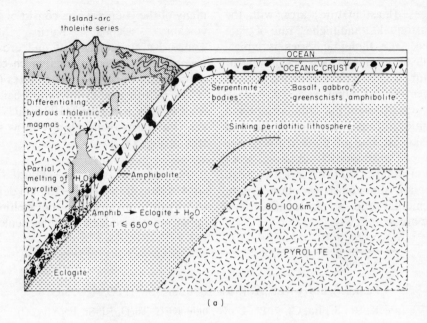

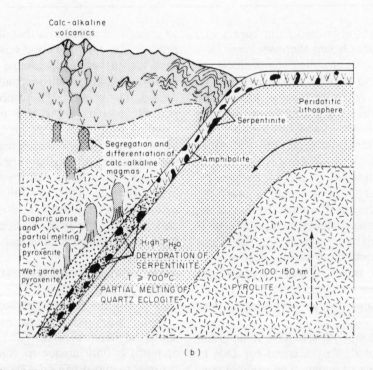

Fig. 18.7   Early and late stages in the petrological evolution of an island arc. (a) Involves dehydration of subducted amphibolite, introduction of water into the overlying wedge and generation of the island arc tholeiite igneous series. (b) Involves partial melting of subducted oceanic crust and reaction of liquids with mantle above Benioff zone leading to diapiric uprise and formation of calc-alkaline magmas (after Ringwood, 1974; reproduced by permission of the Geological Society of London)

1982). Within the plate tectonic framework they are deposited in four areas: the trench–subduction complex, fore-arc, arc and back-arc.

The Trench–Subduction Complex

According to plate tectonic theory the trench should contain pelagic oceanic sediments (and thrusted slices of ophiolite) off-scraped into the mouth of the subduction zone. However, most trenches, especially in the intra-oceanic arcs of the western and northern Pacific, contain less than 300 m of pelagic sediment and hardly any terrigenous sediment. Scholl *et al*. (1977) concluded that any deep-sea sediments have been subducted and that offscraping of terrigenous deposits

into a trench is only possible where a thick (>500–1000 m) sequence is swept against an island arc (Fig. 18.8b).

The outer-arc ridge (or trench–slope break) marks the inner edge of the subduction complex which formed by tectonic accretion and which comprises a thrust stack of off-scraped sediments from the trench; it emerges as Barbados Island in the Lesser Antilles (Westbrook, 1982). Lundberg (1983) emphasizes that intra-oceanic arcs like the Marianas and Costa Rica lack thick clastic deposits and mélange complexes because of low terrigenous sediment input and thus also lack high-grade exotic components such as blueschists and eclogites (Fig. 18.8a).

The tectonic accretion zone exhibits a decrease in depositional and metamorphic

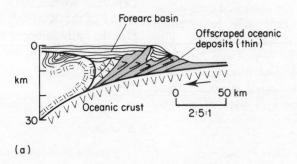

(a)

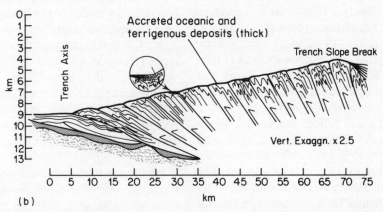

(b)

Fig. 18.8 Conceptual models of accretion and deformation at trench margins. (a) Stacking of thrust slices of a relatively thin sequence of oceanic deposits (after Scholl *et al*., 1980, *Geology*, **8**, Fig. 1A, p. 565). (b) Accretion and tectonic rotation of a thick thrust stack of oceanic and terrigenous material between the trench and the trench–slope break (or outer-arc ridge) (after Karig and Sharman, 1975, *Bull. geol. Soc. Amer.*, **86**, Fig. 7A, p. 386)

age proceeding towards the trench. The tectonic model in Fig. 18.8 shows how higher grade, more deeply subducted, older, landward mélange units tend to thrust oceanward over the more weakly recrystallized, nearer-surface, younger, seaward mélanges. In other words, underplating results in the elevation and exposure of older, once deeply subducted slices of the trench mélange. This style of successive underthrusting is well demonstrated in many arcs, such as the Aleutian, Lesser Antilles and Kodiak Island in Alaska (Moore and Wheeler, 1978). Such tectonic accretion beneath the inner trench slope causes progressive loading of the descending plate and a flattening of the subduction zone (Karig et al., 1976). Also, the progressive underthrusting tends with time to rotate the accreted prisms from a low to a high angle. Where there is high sediment input the upturned ends of the thrusted turbidite slices may be overlain unconformably by a thin veneer of sediment—Fig. 18.8b (the slope-basin turbidites of Dickinson, 1983). The wide spectrum in island arcs is thus reflected in considerable variation in the type and style of tectonic accretion.

The Fore-Arc

Between the volcanic arc and the trench–slope break is the fore-arc basin (Dickinson and Seely, 1979). Some modern basins are submarine (central Aleutians, Marianas, Tonga–Kermadec) and some partly submarine and partly subaerial (Japan and Costa Rica). The fore-arc basin (or outer-arc trough) contains flat-lying undeformed sediments that may reach 5 km in thickness, and that reflect progressive shoaling as the basin fills up: turbidites, shallowing marine sediments and fluviatile-deltaic-shoreline complexes (Dickinson, 1982). The sediments are mostly derived from the adjacent arc volcanoes (volcaniclastic) or by erosion of uplifted basement metamorphic-plutonic rocks; a classic example is in the Sunda arc off Sumatra (Moore et al., 1982).

The Arc

In intra-oceanic arcs the volcanoes are commonly submarine and the sediments mostly of volcanic derivation. According to Dickinson (1974b) three sedimentary facies types may be expected in intra-arc basins:

1. A central facies near the eruptive centres of pyroclastics with andesites and pillow lavas, bounded by a proximal facies such as biogenic reefs on the flanks of partly submerged volcanoes;
2. A dispersal facies as aprons and blankets of clastic sediment;
3. A basinal facies of volcaniclastic turbidites, marine tuffs and submarine ash flows.

Examples of intra-arc sediments include late Tertiary marine beds on the Tonga Ridge, late Tertiary marine sediments capped with lacustrine deposits in Honshu, Japan, late Cenozoic volcaniclastic marine beds (10 km thick) in the Kuril arc, and similar rocks in Fiji, the early Miocene Matanui Group in the New Hebridean arc and the late Cretaceous–early Tertiary beds of the Greater Antillean island arc.

Back-Arc

There is no simple pattern of sedimentation in marginal basins adjacent to continents except for large inputs of terrigenous material deposited by rivers, but in basins remote from such terrigenous debris there is a regular pattern of sedimentation dominated by volcanic input (Karig and Moore, 1975). Against the volcanic chain accumulate volcaniclastic aprons several kilometres thick, beyond the distal ends of which is deposited either pelagic brown clay rich in montmorillonite, glass and phenocrysts, or calcareous biogenous ooze.

Within the back-arc Dickinson (1974a) distinguishes *inter-arc basins*, floored by oceanic crust which receive turbidite wedges of volcaniclastic beds shed from the main arc, from *retro-arc basins*, floored by continental basement that contain fluvial, deltaic and marine beds up to 5 km thick mostly derived from the uplifted area in the fold–thrust belt behind the arc.

Paired Metamorphic Belts

Miyashiro (1972) suggested that metamorphic belts in Japan are arranged in parallel pairs, with an inner low-pressure belt with andalusite and cordierite associated with granitic and volcanic rocks in the arc, and an outer high-pressure belt with glaucophane associated with serpentinites on the ocean-ward side along the inner wall of the trench. In the outer-arc (fore-arc) ridge with increasing depth there may be zeolite, prehnite–pumpellyite, glaucophane-lawsonite and eclogite facies metamorphism, and in the deeper levels of the arc there may be kyanite-bearing intermediate pressure metamorphism (Fig. 18.5). The paired metamorphic belt concept has been widely applied to many arcs and orogenic belts throughout the world—there are at least 14 paired belts around the Pacific. However, Barber (1982) has suggested that the high- and low-pressure belts in Japan are not contemporaneous and also that the main tectonic belts there are allochthonous slices that have been brought into juxtaposition by major transcurrent sinistral faults as a response to the oblique subduction of the northwards-moving Pacific and Kula plates. In short, they are 'suspect terrains' (Coney et al., 1980) or 'allochthonous terrains' (Ben-Avraham et al., 1981). Whilst there are positive factors in support of the paired metamorphic belt model, such as differences in heat flow from trench to arc, the re-evaluation of the type example in Japan must throw doubt on the viability of other paired belts around the Pacific and in earlier orogenic belts. All other paired belts now await re-assessment.

Marginal Basins

Because marginal basins are such an important component of island arcs, we shall consider them here in some detail. Marginal basins lie behind island arc systems and are best seen along the western side of the Pacific, from the Aleutian Basin to the Tasman Sea (Fig. 18.1) (Karig, 1974). These basins are underlain by oceanic crust and are situated between the main magmatic arc and the corresponding continental margin (e.g. Japan Basin) or another island arc–trench system (e.g. Philippine Basin and Fiji Plateau). The term marginal basin is also used to describe basins formed along complex ridge-transform fault boundaries, the Andaman Sea (Eguchi et al., 1980) and possibly the Cayman Trough (Lawver and Hawkins, 1978).

The first attempt to interpret marginal basins in terms of modern plate tectonic theory was by Karig (1970) who explained the origin of the Lau–Havre trough west of the Tonga–Kermadec trench by extensional rifting. In 1971 Packham and Falvey, and Matsuda and Uyeda proposed from indirect evidence that marginal basins formed by a type of sea-floor spreading which caused the opening of the area behind the arcs. The extensional origin for marginal basins has gained wide acceptance in recent years, in contrast to the early idea of Umbgrove, Beloussov and others which postulated an origin through the subsidence of continental crust.

Karig (1974) divided marginal basins into actively spreading and older (mostly early to late Tertiary) inactive types (Fig. 18.1). Active basins have a very high heat flow and occur in five arc systems (Tonga, Kermadec, Mariana, Bonin and New Hebrides). The remaining basins are inactive and are subdivided into two groups: an older one with normal heat flow (e.g. South Fiji, West Philippine and Aleutian), and a younger with high heat flow that has yet to cool back to normal (e.g. South China, Kamchatka, Okhotsk, Japan and Parece Vela).

The presence of symmetrical magnetic lineations in many marginal basins (Weissel, 1981), such as the Shikoku, S Fiji and W Philippine (Watts et al., 1977), the Lau basin (Lawver et al., 1976) and the Scotia Sea (Barker, 1972), demonstrates that they have formed by sea-floor spreading and the generation of new oceanic crust. The basins are characterized by diffuse magnetic anomalies making correlations very difficult. High thermal anomalies, thin lithosphere, locally-thick sedimentary cover, abundant short ridges offset by many long transform faults, many point-source magma leaks (seamounts), slow spreading (Lau basin is <2 cm yr^{-1} half rate) and prolonged hydrothermal alteration all serve to diminish magnetic intensities (Lawver and Hawkins, 1978); most of these features contrast with those found in 'normal' oceans. Also, the basin lithosphere does not cool, thicken and subside according to (time)$^{\frac{1}{2}}$, as is observed at 'normal' spreading oceanic ridges.

Sedimentation rates in active marginal basins are very high. Accumulation rates of volcaniclastic aprons extending from the volcanic chain are in excess of 100 m Ma^{-1} (Deep Sea Drilling Project Reports quoted in Karig, 1974). Montmorillonitic clays, fine-grained volcanic detritus, foraminiferal and calcareous nannofossil oozes are the predominant sediments. Inactive marginal basins have a basement morphology similar to the active basins, but older basins have a thick veneer of sediment that ranges from 100 m of pelagic mud in those protected from continental sources (e.g. Philippine Sea) to several kilometres of terrigenous turbidites in those flanked by a continent (e.g. Bering Sea).

The depth to basement in marginal basins increases with age from about 2.5 km in active to more than 5.5 km in older basins. The greater depth of many marginal basins compared with oceanic crust of the same age is a likely consequence of hydrous remelting about a subduction zone of the relatively cool mantle beneath a terrain of old ocean crust and superimposed island arc (Dick, 1982). Anderson (1975) shows that the active Mariana Basin has a central zone of high heat flow (4.50 HFU to 8.50 HFU at a fracture zone intersection) corresponding to an axial topographic high and a flank zone of low heat flow (0.00 to 0.90 HFU). These results suggest that the axial high with its high thermal anomaly is a narrow zone of magmatic intrusions which may be the centre of sea-floor spreading activity.

Back-arc basalts are more vesicular than their normal ocean ridge equivalents and their corresponding glasses have higher water contents, they tend to be less depleted in incompatible elements than mid-oceanic ridge basalts, and they have geochemical characteristics transitional towards arc magmas, particularly during early stages of back-arc extension. This is manifest as higher ratios of large ion lithophile elements (K, Rb, Ba, Sr, Th, U and light RRE.) to high-field strength elements (Nb, Ta, Ti, Zr). These features suggest that a component from the subducted slab may be involved in their petrogenesis (Tarney et al., 1981). Hawkins (1977) suggests that the range in basalt chemistry may be related to differences in thermal gradients under the basins.

There is a strong correlation between the opening of marginal basins and the relative motions of the main converging plates (Jurdy, 1979). The motion of the Pacific plate relative to Eurasia was first westward (80–59 Ma), then northward (59–36 Ma) and then westward again (36 Ma to present). The Pacific marginal basins active during the time period of northward motion of the Pacific plate opened in a N–S direction, and most of those which were active during the recent time period of westward motion of the Pacific opened in an E–W direction. Toksöz and Bird (1977) calculated that there is a time interval of 20–40 Ma between initiation of subduction and the opening of a marginal basin and, with Hsui and Toksöz (1981), they suggested that induced convective current in the mantle wedge immediately above the downgoing slab is the most likely mechanism responsible for the back-arc spreading. The opening of the marginal basin can also be correlated with the roll-back of the subduction hinge in extensional arcs with steep subduction, as in the western Pacific (Molnar and Atwater, 1978; Uyeda and Kanamori, 1979; Dewey, 1980). Zonenshain and Savostin (1981) calculated that for the past 10 Ma the Eurasian and Indian plates have been moving away from the western Pacific island arc, both rotating clockwise, and this fact provides for the opening of the back-arc basins, whereas the opening of the South China and Andaman seas is ascribed by Tapponnier et al. (1982) to propagating extrusion tectonics in Indonesian consequent of the indentation of

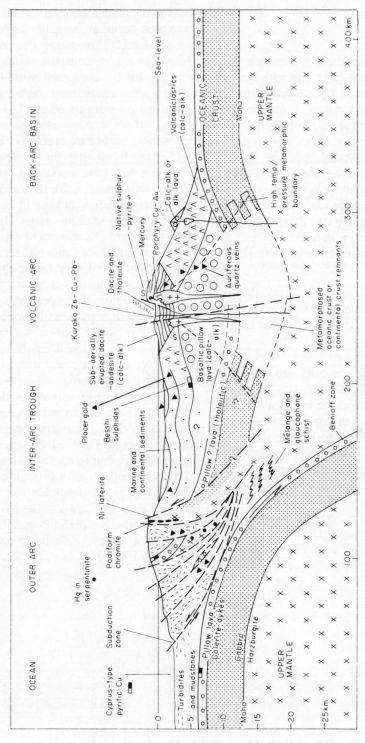

Fig. 18.9 Schematic cross-section through an island with a well-developed outer arc showing principal types of mineralization (after Mitchell and Garson, 1976; reproduced by permission of *Minerals, Science and Engineering*)

the Indian into the Asian plate (see further in Chapter 21).

Mineralization

Across island arcs there is a succession of mineral deposits that correlates with the variation of rock types and geological structure; much of the mineralization is characteristic of, or indeed unique to, consuming plate boundaries of the island arc type. There is considerable complexity because of the superimposition of arc systems, and thus to understand the metallogeny one may have to separate in a mature arc the successive developments from the late Palaeozoic to the Quaternary. The mineralization can be related to three stages of arc growth (Evans, 1980):

1. *Early tholeiitic stage*. Although they are expected, no massive sulphide deposits have yet been found in the early tholeiitic lavas of island arcs.

2. *Main calc-alkaline stage*. Besshi-type stratiform sulphide deposits are associated with andesitic or basaltic, largely pyroclastic volcanics and deep-water sediments such as shales, mudstones and greywackes in the arc–trench gap (Fig. 18.9). There are three varieties: compact pyrite, chalcopyrite, sphalerite; banded sulphides and silicates; copper-rich ore with chalcopyrite and minor pyrite. They all tend to have high Pb, Zn, Ag and Ba contents. Examples occur in the high-temperature Sanbagawa, Hida and Abukuma metamorphic belts of Japan (Ishiara, 1978) and in many Palaeozoic arcs (Sawkins, 1976b).

Porphyry copper deposits occur at this stage in association with calc-alkaline intrusions in the main arc (Sillitoe, 1980). Tertiary examples occur at Bougainville in the Solomon Islands, Taiwan and Puerto Rico, but these ores are scarce in the island arcs of the western Pacific compared with continental margin arcs where their host calc-alkaline intrusions are more abundant (Uyeda, 1981).

Also, there are pyrometasomatic skarn deposits at carbonate-magmatic contacts in Indonesia, Japan and the Philippines. The intrusions may have associated vein deposits of tin (Indonesia), gold (Solomon Islands) and mercury (Japan and the Philippines).

3. *Waning calc-alkaline stage*. In the late stages of arc development Kuroko-type stratiform sulphide mineralization (Sato, 1977) is associated with dacitic-rhyolitic lava domes erupted in the main arc under submarine conditions (Sillitoe, 1982) or, more rarely, andesitic pyroclastic rocks together with shallow-water marine sediments. The main ores are enriched in Zn–Pb–Cu–Ag and are typically associated with barites. They are probably volcanic exhalative deposits formed by mixing of seawater with rising hydrothermal solutions from hot springs (Urabe and Sato, 1978). Their type locality is in the Miocene Green Tuff belt of Japan in which the ores were emplaced in shallow water up to 500 m deep within a time interval of 0.2 Ma. They formed about 150 km above the Benioff zone in intra-arc basins related to cauldron subsidence (Sillitoe, 1980, 1982). There are many Palaeozoic occurrences (Sawkins, 1976b).

Chapter 19

Continental Margin Orogenic Belts—the Western Americas

Continental margin orogenic belts such as the Cordilleran–Andean belts of western America evolve along tectonically–seismically active margins where an oceanic lithospheric plate is consumed beneath a continental plate; they are convergent plate junctions adjacent to subduction zones. Their main features include a trench with turbidite-type sediments, high heat flow and regularly arranged metamorphic zones, calc-alkaline volcanics (especially of the andesite–rhyolite type), monzonitic–granodioritic plutons and batholiths, molasse-type successor basins, and extensive sulphide mineralization including the well-known porphyry copper deposits. The evolution of these belts may also involve the accretion of allochthonous oceanic and continental fragments.

In this chapter we shall review the main features of these belts and consider some of the complexities of their evolution.

Tectonic Evolution

Analysis of the general structure of the orogenic belts in western North and South America is complicated by several main factors.

1. The belt has a history ranging from the late Precambrian to the Quaternary, and so one must distinguish between tectonic belts developed at different times, e.g. in the Palaeozoic and Mesozoic.

2. Any one 'orogenic phase' has a history ranging from the formation of early basins (oceanic or continental) with sedimentation and volcanism to later episodes of magmatism, metamorphism, deformation and uplift.

3. It has only recently been recognized that the Cordilleran belt of North America is a tectonic collage of more than 50 fragments (Fig. 19.1); these are the suspect terrains of Coney et al. (1980) or the allochthonous terrains of Nur and Ben-Avraham (1982a,b). These terrains are of several types: (a) some contain ophiolitic sections of oceanic crust mostly of late Palaeozoic–early Mesozoic age; (b) some consist of volcanic arc material subsequently swept against the Cordilleran margin, mostly of late Palaeozoic to Jurassic in age; some have no known basement, and some rode passenger on older basement terrains; (c) some terrains are fragments of unknown continental margins; (d) several contain much basalt and gabbro of late Triassic age, suggesting that they are fragments from rifting events, intraplate volcanism or oceanic plateaus. It is noticeable that there are rather many ophiolites in the Cordilleran belt; an explanation for this may be that obduction of oceanic lithosphere is facilitated by the injection of non-subductable packages of light buoyant bodies, such as continental slivers, island arcs, old hot spot traces, seamounts, and oceanic plateaus into the subduction zones. As these allochthonous terrains are accreted, slices of intervening oceanic lithosphere are easily obducted with them (Ben-Avraham et al., 1982).

In contrast, the Andean belt of South America still remains relatively simple in structure and evolution, having formed by eastward subduction of oceanic lithosphere below a fairly straight continental margin which is marked by a prominent calc-alkaline batholith belt. In spite of the fact that Nur and Ben-Avraham (1982a,b) 'suspect' that the Andean belt might contain some allochthonous terrains, in this chapter the term Andean belt will refer to the simplest type of continental margin orogenic belt with few or no accreted fragments, whereas the term Cordilleran belt will refer to the complicated collage of mini-plates that make up the orogenic belt that extends from California to Alaska.

4. An extra complication is introduced by the fact that the suspect terrains were swept in eastwards as a result of eastwards subduction, became attached in turn to the western margin of the North American plate, and thus they themselves each behaved as part of a leading Andean-type margin during later eastwards subduction and so were likely themselves to be intruded by calc-alkaline batholiths and overlain by associated volcanics. In this respect some of the suspect terrains potentially have a three-stage history: formation as an oceanic or continental fragment, deformation and metamorphism associated with collisional accretion with the main continental margin and intrusion by batholiths and extrusion of lavas.

The Cordilleran belt of the western USA is further complicated by the presence of the extensional Basin and Range Province, and the San Andreas transform fault system.

The Cordilleran belt of North America

In North America the Belt–Purcell Supergroup (1400–900 Ma old) was overlain by the Windermere rocks 800–600 Ma ago (see Fig. 8.1 and Chapter 8). These strata consist predominantly of shallow-water clastic sediments which pass westwards (in the Purcell) into deep-water turbidites and eastwards thin progressively onto the craton. They are interpreted as a miogeoclinal wedge formed on a continental shelf–slope–rise prior to the main decoupling of the oceanic–continental plates in the early Palaeozoic (J. H. Stewart, 1972). This environment may have been interrupted by a short-lived subduction giving rise to the Racklan–East Kootenay orogenies.

During the earliest Palaeozoic the Cordilleran belt continued to be the site of a continental terrace miogeocline (Figs. 19.1 and 19.2). In North America from the Cambrian and until the mid-Triassic a carbonate–orthoquartzite bank extended from Alaska to California; it formed a broad continental shelf marginal to the landmass analogous to the Bahama Banks off Florida today.

This miogeoclinal terrace is particularly important because it can be argued that all Palaeozoic and younger geological terrains outside (west of) it are inherently suspect as being allochthonous fragments which must have accreted to the continental margin during Mesozoic–Cenozoic time (Coney et al., 1980). Over 70% of the Cordilleran orogenic belt is composed of more than 50 suspect terrains.

Fig. 19.2 shows two belts west of the miogeoclinal terrace. The eastern has two components: (1) chert, shale, basalt and ultramafic rocks which represent oceanic ophiolite deposits deformed in two periods—Cambrian through middle Devonian and Carboniferous through Triassic; (2) mixed volcanic and plutonic rocks, being volcanic arc deposits of Upper Palaeozoic age, including Permian and Triassic (Churkin and Eberlein, 1977). The western belt consists of mixed volcanic and plutonic rocks and volcaniclastic beds, the volcanic- and remnant-arc deposits of two ages—late Precambrian through Devonian and Carboniferous through Triassic with no known older basement. Within these belts 9 allochthonous terrains can be recognized which represent at least 6 lithospheric plates, some from widely differing provenance, as indicated, for example, by the faunas, and by discordant palaeomagnetic data. Stable miogeoclinal sedimentation persisted through mid-Palaeozoic time until the first fragments of the outer belts were thrust eastward upon it in Mississippian and again in Permo–Triassic time (Speed, 1979). The outer-belt accretion then proceeded throughout Mesozoic time, leaving the miogeocline sandwiched in its inner-belt position (Saleeby, 1983). Most of the accreted fragments are variously interpreted (Churkin and Eberlein, 1977) as (a) segments of Asia or features of unknown provenance that collided with North America (Monger et al., 1972), (b) allochthonous terrains dismembered by large-scale strike-slip faulting (Jones et al., 1972) or (c) para-autochthonous microcontinental fragments and volcanic arcs that moved outboard and inboard (away from and towards North America) to accommodate a succession of marginal ocean basins opening and closing

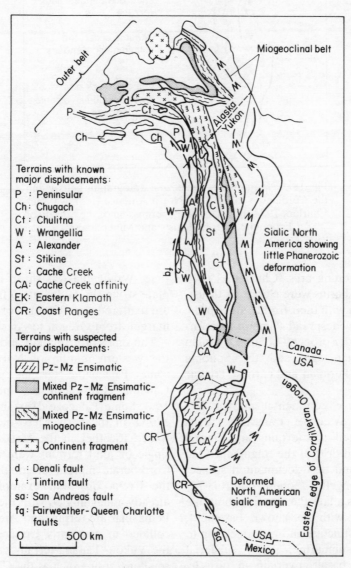

Fig. 19.1 Map showing generalized distribution of the main allochthonous terrains and the miogeocline of the North American Cordillera. Pz: Palaeozoic, Mz: Mesozoic (from Saleeby, 1983, *Ann. Rev. Earth Planet Sci.*, **11**, Fig. 1, p. 47; modified after Coney et al., 1980)

270

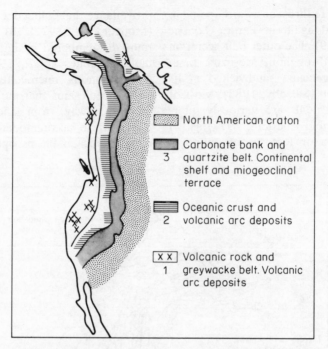

Fig. 19.2   Major lower Palaeozoic stratigraphic belts within
the Cordilleran fold belt of North America (redrawn after
Churkin, 1974; reproduced by permission of the Society of
Economic Paleontologists and Mineralogists)

behind the migrating arcs (Churkin, 1974).
These marginal basins were closed by thrusting and collision with their frontal arcs during
two periods of widespread Palaeozoic deformation: the Antler orogeny of late Devonian
and Mississippian age, and the Sonoma
orogeny of late Permian and earliest Triassic
age.

The Palaeozoic continental margin was
transected by extensive transverse faults
which fragmented the terrains and caused
major northward drift of the fragments. One
of the largest and best-documented suspect
terrains is Wrangellia (Jones *et al.*, 1977)
which comprises a late Palaeozoic submarine
arc assemblage with no known basement,
overlain by a distinctive and very thick (up to
6 km), partly submarine, partly subaerial
Upper Triassic basalt, overlain in turn by
sedimentary rocks which extend into the early
Jurassic. Palaeomagnetic data record a low
Triassic palaeolatitude. After accretion,
mostly in late Cretaceous to early Tertiary

time, Wrangellia was sliced by major intra-
plate strike-slip faults into 5 fragments which
are scattered over 2000 km of the Cordilleran
margin, from Oregon to Alaska (Fig. 19.1).

The times of formation and of accretion to
the western margin of North America of
some Palaeozoic terrains, mostly in the
Canadian Cordillera, are indicated in Fig.
19.3. Some Palaeozoic terrains continued to
evolve in the early Mesozoic, several were
joined together in the Mesozoic, and these
single or joint terrains were accreted to the
Cordilleran margin from the Jurassic to
the Cretaceous. For example, Wrangellia,
already amalgamated with the lower Jurassic
Peninsular arc terrain and Alexander terrain,
collided in post-early Cretaceous time with
the Yukon–Tanana terrain, entrapping some
smaller terrains such as the Chulitna within a
flysch-filled suture zone and, finally, docked
against the Cordilleran margin in mid-
Cretaceous time. The entire process of terrain
accretion and intra-plate slicing lasted for at

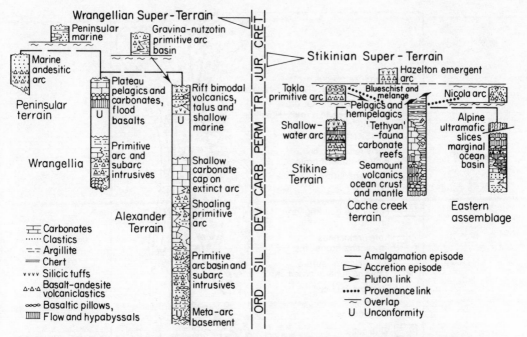

Fig. 19.3 Diagram showing the petrotectonic, amalgamation and accretion histories of the Wrangellian and Stikinian super-terrains of British Columbia and southern Alaska (after Saleeby, 1983, *Ann. Rev. Earth Planet Sci.*, **11**, Fig. 2, p. 52)

least 120 Ma, from the mid-Jurassic to the early Tertiary (Coney *et al.*, 1980). Fig. 19.4 is a palaeotectonic map for the Palaeogene of part of the North American Cordillera.

Let us now consider the development of the Cordilleran belt from a different standpoint in Mesozoic–Cenozoic time.

A major change in the world's plate movements took place in the early Jurassic marked by the beginning of fragmentation of Gondwanaland. During this Mesozoic cycle from 180–80 Ma ago the world's spreading systems developed, there was an ubiquitous marine transgression of continents and major plutonism took place in the Cordilleran orogenic belts. The end of the cycle in the late Cretaceous was defined by widespread erosion and the cessation of major plutonic activity within the Cordilleran belts, and the beginning of a major continental regression.

In the Cordillera of the USA there were four Mesozoic 'orogenic episodes':

Laramide (late Cretaceous)
Columbian (latest Jurassic to early Cretaceous)
Sevier (late Jurassic to late Cretaceous)
Nevadan (late Jurassic)

The tectono–plutonic activity occurred in a western eugeosynclinal belt and was similarly and synchronously developed along much of the eastern Pacific from Canada to the central Andes. There was a general tendency for plutonism to migrate eastwards towards the continental margin where it continued in the late Tertiary. This Mesozoic plutonic–volcanic arc truncated the Palaeozoic–Proterozoic basement in the southwest USA.

The two most characteristic products of this activity were the andesite-dominated volcanics and the granitic plutons and batholiths which can reasonably be ascribed to processes above eastward-dipping subduction zones (Fig. 19.4). During early stages of uplift of the Cordillera in the Mesozoic and early Tertiary clastic rocks were deposited in successor basins and easterly foredeeps and exogeosynclines.

In the late Mesozoic and early Tertiary deformation migrated eastward into the foreland, mainly by thrusting and overriding of

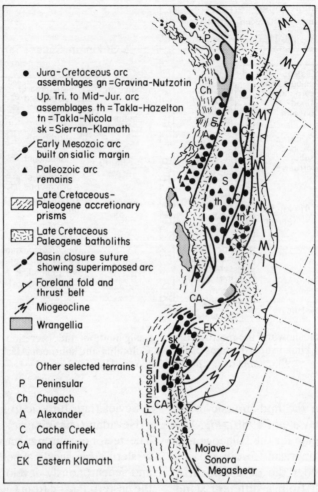

Fig. 19.4 Palaeotectonic map for the Palaeogene of the North American Cordillera (reproduced with permission from Saleeby, 1983, *Ann. Rev. Earth Planet Sci.*, **11**, Fig. 4B, p. 67. © 1983 by Annual Reviews Inc.)

the foredeeps. In the southern Rocky Mountains there was a 140 km shortening of sedimentary beds between the early Cretaceous (136–100 Ma ago) and the early Eocene (54–49 Ma ago)—a gross mean rate of 2 mm per year of crustal shortening (Wheeler *et al.*, 1974). The stacking of the thrust sheets caused 8 km of tectonic thickening of cover rocks above a passive but depressed basement.

About 80 Ma ago there was another dramatic change in tectonic pattern in the Cordilleran belt. This period marks the end of major plutonism in the Sierra Nevada and of Franciscan deposition. Starting in the early

Tertiary there was extensive erosion of the mountain belt and the planation surfaces were preceded and accompanied in the early to middle Tertiary by major volcanism. This significant volcanism–planation relationship is recognizable along much of the Cordillera from Canada to Chile and Bolivia. Following the eastward migration in the Mesozoic, tectonic activity during the Laramide orogeny continued in the early Tertiary to move towards the cratonic interior.

There was a further important period of tectonic reorganization about 40 Ma ago (middle to late Eocene) which marks the end of Laramide compressive deformation in the western United States. Apart from the localized Cascadan orogeny in British Columbia, plutonism, volcanism and deformation ceased 40 Ma ago in much of North America (NW Canada to the Caribbean), after which there was only ignimbrite eruption in the Rocky Mountains and Great Basin.

To account for the two tectonic transitions Coney (1972) suggested that the differences in Cordilleran tectogenesis may be correlated with the variable motions of the North American plate which were due to the spreading history of the North Atlantic oceanic crust. The Nevadan and Sevier–Columbian deformations in the period 180–80 Ma were related to the northwesterly rotation of North America away from Africa, the Laramide deformation was a reflection of the southwesterly rotation of North America from Europe, and in the post-40 Ma period North American plate motion was reduced as indicated by Pacific submarine data. According to this model the Cordilleran–Andean orogeny developed along the leading edges of westward-moving continental plates actively driven by spreading in the Atlantic to the east.

Alternatively, the position, presence or absence of Cenozoic magmatic arc activity in the western United States may be related to the geometry of the subducted plate (Cross and Pilger, 1978). Chemical data suggest that the dip of the Benioff zone in SW North America was almost constant from 135 to 80 Ma BP and the convergence rate was 7 to 8 cm yr^{-1}. Decreasing alkalinity of magma types with decreasing age for Laramide age rocks shows that about 80 Ma ago the dip began to flatten and reached about 10° at 55 Ma BP and the convergence rate was about 14 cm yr^{-1}. At such a shallow angle no asthenospheric mantle is present between the two converging plates, so magmatism ceases—this explains the Eocene igneous gap. Increasing alkalinity of magmas with decreasing age indicates steeping of the subducting slab from 20° to 60° from 40 to 15 Ma BP when the convergence rate decreased substantially (Keith, 1978, 1982).

It is appropriate here to mention the 25 or so Cordilleran 'metamorphic core complexes' which occur in a sinuous belt extending from Canada to Mexico (Fig. 19.5) (Coney, 1979; Davis and Coney, 1979; Crittenden et al., 1980; Armstrong, 1982). These are complicated structures, each with their own individual characteristics. At the simplest they can be described as diapiric upwellings of gneissic lower crust over which brittle extensional features have been superimposed. They developed in the back-arc area between the magmatic arc and the eastern thrust front, and were produced during post-Laramide to pre-Basin-and-Range time (i.e. 55–15 Ma ago). They are up to 400 km^2 in area and 2 km in domal relief. They typically consist of a core of augen gneiss, a décollement of lineated mylonitic breccias, and an unmetamorphosed cover of subhorizontal sediments with extensional faults. It is generally considered that boudinage-like compression, tectonic thickening and stretching of the lower crust induced by plate convergence and underthrusting above the subduction zone (Brown, 1978) was responsible for the diapiric uprise of crystalline material which became plated to the base of a layered cover along a décollement zone which is commonly near the base of the Phanerozoic pile. Above the décollement extensional basins were filled with lower–middle Tertiary sediments (Fig. 19.6). Gravitational layer instability may have been responsible for the upwelling of individual gneiss domes within the core complexes (Brown, 1978). Davis (1983) produced an interesting extensional ductile shear-zone model for the core complexes which, he points out, formed at depths of about 10 km.

Finally, we come to the San Andreas fault system that segmented the continental margin of California. About 29 Ma ago a transcurrent slip margin was developed in western California when the Pacific plate contacted North America after subduction of the intervening Farallon plate (Fig. 19.7). Initially, the slip was located along the continental margin but, as the North American plate overrode the East Pacific Rise and its associated transform faults, slip between the Pacific and North American plates moved eastwards to

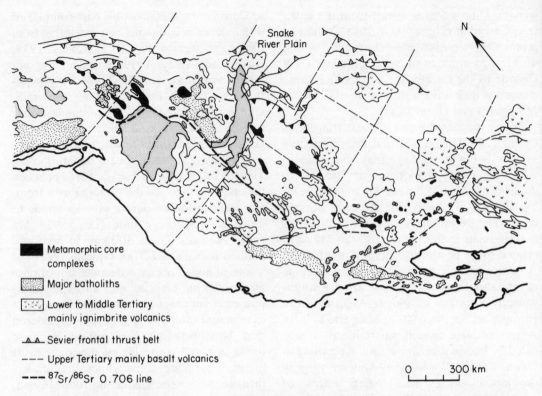

Fig. 19.5 Distribution and general regional tectonic setting of Cordilleran metamorphic core complexes (from Davis and Coney, 1979, *Geology*, **7**, Fig. 1, p. 121)

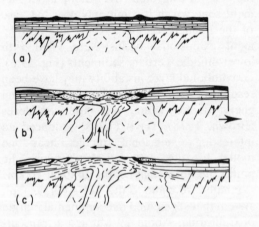

Fig. 19.6 Model for Tertiary evolution of metamorphic core complex in southern Arizona (a, thin cratonic shelf). Basement extends and is intruded by Early to Middle Tertiary pluton. The extension produces a 2/3 thinning of the Phanerozoic cover. This creates a basin which fills with Tertiary sediments and volcanics (b). The structure then domes about 3 km from isostatic adjustment and thermal input (c) (reproduced with permission from Coney, 1979, *Soc. Econ. Pal. Mineral.*, 15–28)

intracontinental dextral faults of the San Andreas system. About 300 km of right lateral movement has occurred along the San Andreas fault in the last 10–15 Ma. A series of Neogene elongate sedimentary basins (Fig. 19.7) developed, commonly in extensional pull-apart gaps associated with the right lateral slip between adjacent blocks (Blake *et al.*, 1978). Continued shear on the faults caused folding and faulting of the Neogene beds and this deformation produced numerous anticlinal and fault traps for the accumulation of oil and gas.

The Andean Belt of South America

Compared with the Cordilleran belt of North America the Andes has a relatively simple structure with little or no evidence of accreted mini-plates. Its characteristic features are (Clarke *et al.*, 1976; Zeil, 1979):

1. A basement of Palaeozoic, or rarely

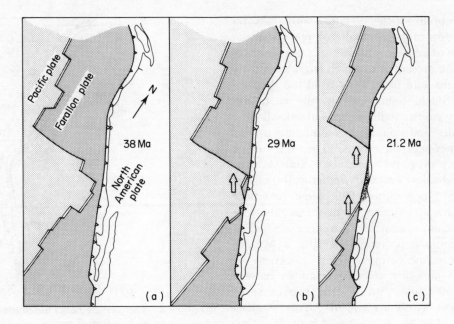

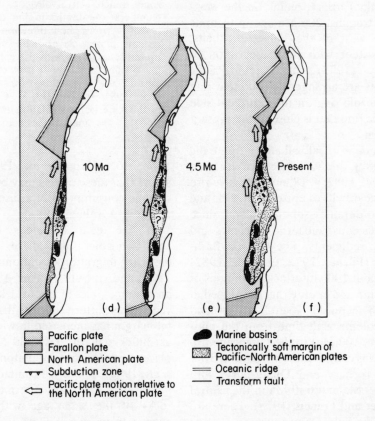

Fig. 19.7 Schematic model of interaction of Pacific, Farallon and North American plates for six Tertiary time intervals, showing time of initial development, location, and general shape of Neogene basins (from Blake *et al*., 1978, *Am. Ass. Petrol. Geol. Bull*., **62**, Fig. 2, pp. 346–347; reproduced with permission)

Precambrian rocks. There were several Palaeozoic orogenies that took place at a succession of active continental margins (not far from the present one) which migrated mostly westward with time, away from the Precambrian Shield boundaries to the east, and which gave rise to linear belts of calc-alkaline plutonics and volcanics, sedimentary basins and metamorphic rocks. The still active Andean 'orogeny' began about 200 Ma ago.
2. Mesozoic–Cenozoic supracrustal rocks lie on the block-faulted Palaeozoic basement and show no evidence of significant shortening or compression. In Peru there is a pair of Cretaceous depositional troughs, a western eugeosyncline with a 7 km thickness of marine andesitic volcanics, and an eastern miogeosyncline with 6 km (thickness) of sediment (Cobbing, 1976). The Andean granitoid batholith developed in the late Cretaceous to mid-Tertiary, mainly in the western volcanic troughs, 200 Ma ago in tectonic belts to the west of the Palaeozoic belts. This is consistent with the observation of Puscharovsky (1973) that circum-Pacific orogenic belts are arranged in a double ring with a Palaeozoic ring on the continent-side of a Mesozoic ring that is close to the present Pacific margin.
3. The Andean calc-alkaline plutonic (Pitcher, 1982) and volcanic activity (see *Earth Sci. Rev.*, 1982, v. 18, no. 3/4) gave rise to parallel linear belts of epizonal plutons and batholiths, bordered by submarine and, since the mid-Cretaceous, subaerial volcanics and volcaniclastic sediments. The Coastal Batholith of Peru (Pitcher, 1978; Cobbing, 1982) consists of about 1000 interlocking plutons of gabbro, tonalite and granite that developed in rhythms with increasing overall acidity and decreasing volume with time from the mid-Cretaceous (*c.* 100 Ma) to the Oligocene (30 Ma). There was considerable volcanic activity in the Neo-Tertiary and Quaternary, with peaks at 12–9 Ma and 6–0 Ma in the central Andes (Baker and Francis, 1978). There are two gaps in the volcanic products in N Peru and central Chile (Fig. 19.8) which correspond to regions where the Benioff Zone is anomalously shallow, whereas in the volcanically active parts the seismic planes dip at

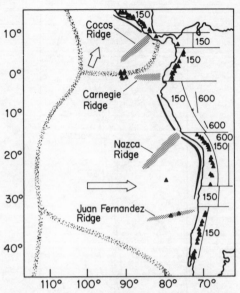

Fig. 19.8 Tectonic elements along the western South and Central America consumption zone: trench, active volcanoes and seismicity. Numbers are depths in km of the seismic planes. Arrows show direction of motion of oceanic plates. Several aseismic ridges are presently colliding with the continents, causing volcanic and seismic gaps on land (reproduced with permission from Ben-Avraham *et al.*, 1981, *Science*, **213**, 51. Copyright 1981 by the American Association for the Advancement of Science)

about 30° (Jordan *et al.*, 1983). Fig. 19.8 shows that these effects may be related to the oblique consumption of aseismic ridges (Nur and Ben-Avraham, 1981).

In the central Andes during Mesozoic–Cenozoic time there has been an eastward migration of plutonic (Farrar *et al.*, 1970) and volcanic (Baker, 1977) centres at an average rate of about 1 km per million years. This pattern suggests that the continental margin has increased in width by as much as 200 km in the last 200 Ma by the emplacement of successive plutonic belts (Drake *et al.*, 1982). There is a common geochemical polarity of the main intermediate volcanic rocks of the same age with an eastward change from calc-alkaline to shoshonitic or alkaline (Déruelle, 1978; Thorpe and Francis, 1979). There has been a marked westward migration of volcanic centres in the last 1–2 Ma (Drake *et al.*, 1982). Such a change

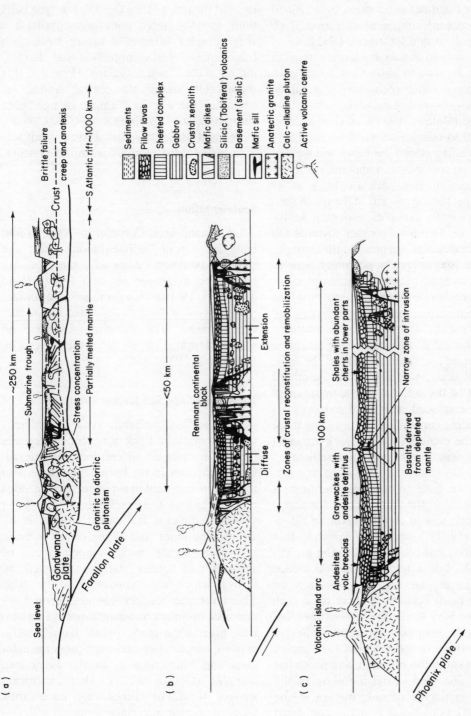

Fig. 19.9 Tectonic setting in southernmost South America during the late-middle to early-late Jurassic (a) and during the early Cretaceous in northern (b) and southern (c) portions of the back-arc basin. The sequence a–b–c represents the different stages in the evolution of the back-arc basin, showing the transition of the back-arc crust from continental through intermediate to oceanic in character (from De Wit and Stern, 1981; reproduced by permission of Elsevier, Amsterdam)

(a)

~250 km

Sea level

Gondwana plate

Submarine trough

Brittle failure

creep and anatexis

—Crust

S Atlantic rift ~1000 km

Stress concentration

Partially melted mantle

Granitic to dioritic plutonism

Farallon plate

Sediments
Pillow lavas
Sheeted complex
Gabbro
Crustal xenolith
Mafic dikes
Silicic (Tobiferal) volcanics
Basement (sialic)
Mafic sill
Anatectic granite
Calc–alkaline pluton
Active volcanic centre

(b)

<50 km

Remnant continental block

Diffuse

Extension

Zones of crustal reconstitution and remobilization

(c)

100 km

Volcanic island arc

Andesites volc. breccias

Graywackes with andesite detritus

Shales with abundant cherts in lower parts

Basalts derived from depleted mantle

Narrow zone of intrusion

Phoenix plate

in polarity may be a result of a westward jump in the subduction zone caused by the blocking of the older subduction by an island arc or an oceanic fragment of similar high relief (Toussaint and Restrepo, 1982).

In the southern Andes back-arc extension in the late Jurassic to early Cretaceous gave rise to the ophiolitic 'rocas verdes' complex which is a gabbro-sheeted dyke-pillow lava assemblage (Dalziel, 1981). A change from extensional to compressional deformation in the Cenomanian closed the basin and welded the arc to the continental mainland. An interesting aspect of this back-arc basin is its wedge shape, the narrow rift in the north seen in the Sarmiento Complex widening southwards via the Tortuga Complex towards the extant central Scotia Sea near South Georgia. Thus there is a unique opportunity here to follow the successive stages in the evolution of a back-arc basin (De Wit and Stern, 1981). In the northern narrow extremity of the original basin, mafic melts intruded into the continental crust over a diffuse zone causing extensive remobilization and reconstitution of the sialic continental crust, whereas in the southern wider part of the original basin, mafic magmas were emplaced at a localized oceanic-type spreading centre. Fig. 19.9 shows three stages in the evolution of the back-arc crust from continental, through intermediate to oceanic.

Finally, we have to ask the question: is there evidence in the Andean belt for accretion of allochthonous terrains? Nur and Ben-Avraham (1982) 'suspect' on principle that such accretion has contributed to the growth of the belt, but agree that firm evidence is so far largely lacking. It is certainly an interesting point because as there is no such evidence we have to ask the further question, why is the Andean belt apparently different from most other orogenic belts? One aspect that is relevant to this problem, and which has been little discussed in the literature, is the difference in degree of deformation in the Cordilleran belt of North America compared with the Andean belt. Clearly, when an allochthonous terrain is driven against a continental margin it suffers folding and thrusting; when the next allochthonous terrain is driven into its back, then it again suffers folding and thrusting. Thus Cordilleran-type belts built up of accreted mini-fragments have a very complex internal structure since each fragment suffered compression and shortening on at least two occasions. However, it is remarkable that in the central Andes, for example, 'there is no evidence of significant shortening or compression' (Clark *et al.*, 1976) and the little deformation present was a result of vertical-uplift tectonics (Pitcher, 1978).

Sedimentation

Like island arcs, Cordilleran–Andean fold belts have four sedimentation areas: the trench-subduction complex, the fore-arc basin, the magmatic arc and the foreland basin (Fig. 19.10). However they also contain molasse-type sediments deposited in late successor basins. The following discourse on sedimentation is largely from the reviews by Dickinson (1974a and b, 1976).

The Trench and Subduction Complex

Trenches in Cordilleran–Andean fold belts typically contain thick terrigenous deposits derived by erosion of continental material. On the east side of the Pacific they contain up to 2 km of sediment in two layers, terrigenous mud, turbidites and conglomerates overlying pelagic sediments and basalt (Moore *et al.*, 1979). The inner wall of the trench is made up of imbricate wedge-shaped slices of sheared and folded material bounded by thrusts which are subparallel to bedding. The imbricate wedges contain terrigenous deposits, oceanic crust-mantle rocks (radiolarite, greenschist-grade pillow basalt, metagabbro, serpentine) and high-pressure eclogites and blueschists as in the Franciscan complex (Cowan, 1978). These imbricate wedges represent successively underthrust slices of oceanic material intermixed with terrigenous material accreted tectonically to the base of the accretionary prism or tectonic mélange beneath the trench slope break (or

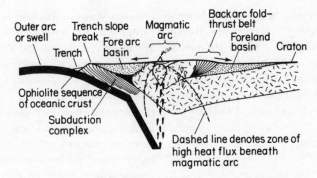

Fig. 19.10 Schematic diagram of continental-margin arc-trench system showing key tectonic elements discussed in text. Lateral extent depicted is roughly 1000 km, but the tectonic features shown may span twice that distance in some cases. Vertical exaggeration is about ×3; hence, dips of thrusts and imbricate slices in subduction complex and back-arc fold-thrust belt are drawn much steeper than actual. Stipples indicate undeformed sediments and open arrows show main direction of sediment delivery into fore-arc and foreland basins, although patterns of sediment dispersal within basins may be longitudinal. The near-vertical fault zone between the fore-arc basin and the magmatic arc is intended to show the upper-slope discontinuity in the sense of Karig and Sharman (1975). Solid droplets denote rise of main arc magmas, at least in part from below the crust and associated with subduction of the oceanic lithosphere that descends into the mantle. Open droplets denote potential rise of crustal melts associated with partial subduction of continental lithosphere in the infrastructure of the orogenic belt between the main magmatic arc and the back-arc fold-thrust belt (after Dickinson, 1976, *Can. J. Earth Sci.*, **13**, Fig. 3, p. 1271, reproduced by permission of the National Research Council of Canada)

outer-arc ridge), which defines the inner limit of the active subduction zone (Dickinson, 1976). As new material is successively thrust under older material, thrust wedges in the accretionary prism are gradually rotated to a steep attitude near the trench–slope break (Fig. 19.10), and the subduction complex grows laterally as the trench axis and trench–slope break migrate.

The Fore-Arc

Fore-arc basins located between the trench–slope break and the volcanic arc front are the depositional site of fluviatite, deltaic and pro-deltaic sediments belonging to progradational coastal plain complexes built upon subsiding shelves against the open sea. Most sediments are derived from the volcano-plutonic arc, they do not contain intercalated lavas, and they are only mildly deformed.

Along the eastern Pacific Mesozoic–Tertiary arc–trench sequences have been uplifted and are locally preserved in the present mountain belt. Their original tectono-stratigraphic relationships can still be seen as they are ideally situated between the contemporaneous mélanges representing deformed oceanic and trench material and the coeval metavolcanic–plutonic rocks representing the magmatic arc belts. Probably the best known example is the Late Mesozoic Great Valley sequence of California, comprising largely sandstones and shales, situated between the coastal Franciscan assemblage with its blueschists, eclogites and mélanges, and the Sierra Nevada Batholith. The Great Valley arc–trench depositional site lay across the margin of the continent, because on its west side it consists of a deep trough facies overlying an ophiolite complex, and on its east it includes 12–15 km of turbidites with longitudinal (coast-parallel) palaeocurrent structures and further to the east this passes into a thinner shelf facies (Dickinson, 1974b).

The Arc

Sediments within arc basins are largely of volcanic origin and ideally may consist of sub-aerial tuffs and volcaniclastic intermontane red beds in extensional graben-like troughs, together with unstable shelf facies and conglomerates. However, in continental margin fold belts like the Cordillera and the Andes that have undergone substantial uplift they are unlikely to be extensively preserved; here the underlying, deeply eroded, granitic plutons are best displayed.

Examples of preserved intra-arc sediments include the Upper Triassic Nicola Group in central British Columbia, early Jurassic sediments in western North America and continental red beds intercalated with Tertiary volcanics in the central Andes. Along the back of the arc is a thrust belt with vergence towards the craton (Fig. 19.10) exemplified by the belt in the foothills of the Cenozoic Andes, and the Cordilleran thrust belt in the Rocky Mountains. New thrusts propagate below the older ones on the east side of the belt, and in this way the zone of most active thrusting migrates with time away from the arc (Jordan et al., 1983).

Foreland Basins

The back-arc area in continental margin belts varies from the shallow seas with marine beds on the Sunda Shelf behind Sumatra and Java to the continental facies in the Amazon Basin behind the Andes. Ideally, the latter may consist of piedmont or deltaic clastic wedges that thicken towards the magmatic belt. Clastic sequences may reach 5–10 km in thickness. Sedimentary sources are mostly in the non-volcanic highlands in the thrust belt rather than the magmatic arc, thus sandstones are more quartzose and less feldspathic than in fore-arc basins (Dickinson, 1982).

According to Churkin (1974) extensional marginal ocean basins developed behind the frontal volcanic–greywacke arc in the Ordovician, Silurian and Lower Devonian from Alaska to California (Fig. 19.2). To the east was a carbonate–quartzite belt interpreted as a continental shelf–upper continental slope deposit. The marginal basins contain graptolitic shales, cherts and minor quartzites regarded as deep-sea pelagic sediments deposited on oceanic crust. The marginal basins were closed by deformation in the Late Devonian and Mississippian.

Later back-arc sedimentation took place in the Cordillera from Alaska to the Gulf of Mexico within a foreland basin in which marine Cretaceous beds grade westwards to a deltaic and coastal-plain facies and, further west, to piedmont aprons of clastics that lie against the thrust belt.

Successor Basins

Successor basins are deeply subsiding troughs with predominantly clastic greywacke–arkosic sediments initiated by the uplift of a mountain belt and which overlie the partly eroded, deformed and intruded rocks of the orogenic belt.

During the Mesozoic and early Tertiary successor basins were developed in the Intermontane, Rocky Mountain and Insular belts of the Canadian Cordillera. In the Intermontane Belt there are three successor basins (Eisbacher, 1974).

The development of the successor basins records the progressive continentalization of the crust in the mobile belt. Gradual uplift and erosion unroofed the adjacent crystalline complexes which contributed the clastic components of the younger basins.

Igneous Activity

Continental margin orogenic belts have a

long and complicated history of igneous activity. For example, in the belts along the eastern side of the Pacific magmatism has a life span equal to the tectonic history extending from the late Precambrian to the Quaternary; it was a complex development dependent on the multiple stages of tectonic evolution of the belts.

Broadly, the igneous products are of two types, volcanic and plutonic. Because these belts typically undergo extensive post-orogenic uplift, many (but not all) of the volcanic rocks formed in the main arc zone are eroded and the underlying granitic rocks in the arc core are exposed. Thus in western North and South America both magmatic types are available for inspection today. The intrusive and extrusive arc rocks are predominantly mantle-derived and represent new additions to the continental crust which is correspondingly growing at a rate close to $0.5 \text{ km}^3 \text{ yr}^{-1}$ (Brown, 1982).

We shall now consider in turn the volcanic and granitic rock within the eastern Pacific belts.

Volcanic Rocks

Rocks of volcanic affinity are abundant, but where many of them formed depends on individual interpretation of the tectonic environment. The following types are given in an evolutionary sequence.

The earliest volcanics are late Precambrian tholeiitc basalts, near the bottom of the Belt Supergroup, which were considered by Souther (1972) to be related to the thinning and rifting of the crust during initial continental separation. Some late Proterozoic and early Palaeozoic basalts were extruded as flows into the continental wedge near the carbonate–shale boundary at the slope–rise transition; examples occur along the east side of the Selwyn Basin in southern Canada. Volcanic rocks, clastics and chert accumulated on the oceanic crust in an offshore deep-water environment from the late Precambrian until the late Devonian; they form part of the central zone 2 of Fig. 19.2.

In late Palaeozoic times there was only a minor development of volcanic rocks. In Canada there are two groups:

1. Tholeiitic basalts with ultrabasics and gabbros, possibly extruded on the ocean floor.
2. Basalts, andesites and acid volcanics with abundant pyroclastics that formed in a volcanic arc environment. According to Churkin (1974) this arc extended from Alaska to California from the Ordovician to the Lower Devonian; volcanic products include abundant basaltic–andesitic breccias and tuffs, aquagene tuffs, pillow basalts and abundant broken pillow breccias and structureless submarine and subaerial lavas.

There are two groups of Mesozoic volcanics: meta-basalts in the tectonic mélanges and andesites with more silicic types and pyroclastics in the main arc.

1. *Ophiolites.* Rocks formed in the oceanic crust typically include meta-ultramafics and serpentinites, meta-gabbros and basalts, sheeted dykes (diabases), cherts possibly with radiolaria, and pelagic limestones. Because these are transported into a subduction complex at a trench they may be accompanied by eclogites, tectonic mélanges, greywacke flysch turbidites and blueschists. This suite is thus a mixture of rocks emplaced partly at a mid-oceanic ridge and partly in a trench.

Important examples in the Cordillera of these oceanic rocks either with or without their trench accompaniments include:

(a) The late Jurassic to earliest Cretaceous Franciscan mélange of California and S Oregon (Cowan, 1978). According to Roure and Blanchet (1983) the Franciscan was deposited in a marginal basin between an island arc to the west and the north American continent.

(b) The Del Puerto ophiolite underneath the Great Valley sequence in California (Evarts and Schiffman, 1983). This may have formed in or near an island arc. The Smartville ophiolite in California may also have formed in an intra-arc basin (Menzies *et al.*, 1980).

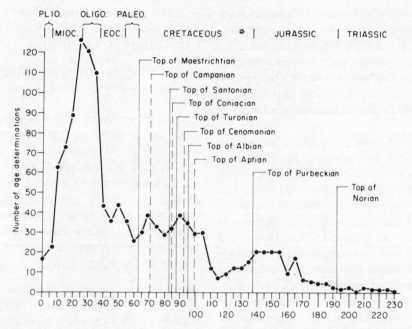

Fig. 19.11   Post-Palaeozoic dates on salic plutons and volcanics in the North American Cordillera between Alaska and Baja California (after Gilluly, 1973, reproduced by permission of Geological Society of America)

(c) The sequence underlying the Intermontane Belt of Canada—Carboniferous to Triassic (Price *et al.*, 1981).

(d) The accretionary terrains of central Alaska (Jones *et al.*, 1982).

2. *Arc volcanics*. Mesozoic volcanics along the eastern Pacific margin are dominantly fragmental augite andesites, ignimbrites, pyroclastics and subordinate dacites and rhyolites. Andesitic volcanicity began in the Permian in the western United States, the Trias in northern Peru and the Lias in southern Peru. It continued extensively during the Jurassic in the main arc zone that was intruded by granitic batholiths in the Cretaceous.

Whereas island arc volcanism largely produces andesites, dacites and high-alumina basalts, volcanism in the continental margin orogenic belts related to subduction activity produces these plus abundant silicic types. The shallow granitic batholiths developed along the main arc axis were roofed by intermediate-to-silicic calc-alkaline volcanic fields. Fig. 19.11 gives a compilation of the ages of salic plutons and volcanics in the North American Cordillera in Mesozoic and Cenozoic time.

In the central Andes magma generation in the Jurassic was at a depth of about 50 km and this depth increased to 200–300 km through the Mesozoic and Cenozoic. This may be correlated with the increase in the rate of plate convergence from 1 cm yr^{-1} in the Jurassic to at least 11 cm yr^{-1} in the Mio–Pliocene (Clark *et al.*, 1976).

Noble *et al.* (1974) suggested that episodes of Cenozoic volcanism in the Andes of Peru could be correlated with tectonic movements of the oceanic plates (Fig. 19.12). The late Eocene pulse of igneous activity may be related to the abrupt change in the rate and direction of rotation of the Pacific plate deduced by Clague and Jarrard (1973). The Oligocene was a period of volcanic and tectonic quiescence. The reinception of tectonism and volcanism at the start of the Miocene may correlate with the increase 20–25 Ma ago in the rate of rotation of the Pacific plate from less than 0.5° per Ma to 1.3° per Ma

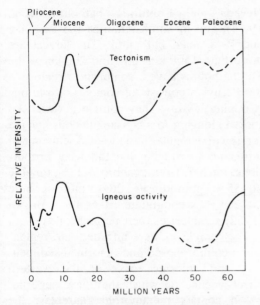

Fig. 19.12   Schematic diagram showing inferred relative intensities of igneous and tectonic activity in the Andes of Peru during the Cenozoic (after Noble *et al.*, 1974; reproduced by permission of North-Holland)

(Clague and Jarrard, 1973). This change may also have been responsible for on-land events of this age in the western United States and the Aleutian and Lesser Antilles arcs. The present-day rapid spreading of the East Pacific Rise began about 10 Ma ago and this timing coincides with the inception of extensive middle and late Miocene and Pliocene igneous activity and middle Miocene deformation.

In the Canadian Cordillera there was an abrupt change from explosive eruption of calc-alkaline lavas associated with sub-volcanic plutonic rocks in the Eocene to quiet effusion of alkali–olivine plateau basalts in late Miocene and early Pliocene times. This may mark a change from calc-alkaline magma generation above an eastward-dipping subduction zone to derivation of alkali–olivine basalt directly from the mantle at depths of 300–400 km, implying that oceanic plate consumption ceased prior to Miocene time and was substituted by simple shear motion along the Denali–Fairweather–Queen Charlotte transform fault that today occupies most

of the Canadian part of the North American coast.

During Pleistocene and Recent times central British Columbia has been the site of nearly 150 alkali–olivine basaltic cinder cones and stratovolcanoes arranged in linear belts associated with north–south graben— this may be the first expression of an opening rift system along this non-subducting plate margin.

This review of the volcanic rocks preserved in the Cordillera demonstrates that they vary from early tholeiitic basalts formed in a continental rift–miogeoclinal environment, to oceanic tholeiitic basalts largely subducted but locally preserved as in the Franciscan mélange, to predominantly andesitic calc-alkaline lavas developed above eastward-dipping subduction zones, to alkali–olivine basalts in extensional rift environments.

At this point we shall consider the chemical variation of the volcanic rocks across the continental margin orogenic belts. It is well known that across island arcs (see Chapter 18) the composition of lavas changes from tholeiitic to calc-alkaline and then alkaline and/or shoshonitic with increasing distance from the trench. Similar chemical variations take place in volcanic rocks across continental margin orogenic belts and these may be expressed in terms of K_2O/SiO_2 and K_2O/Na_2O ratios. Increasing Zr/Nb and La/Yb ratios with increasing distance from the plate boundary are the most consistent geochemical indicators of subduction polarity that are unaffected by metamorphism (Gill, 1982).

Along the Andes there are three linear zones of active andesite volcanism (Fig. 19.13): a northern zone in Columbia and Ecuador characterized by basaltic andesites, a central zone in S Peru and W Chile with andesite–dacite lavas and ignimbrites and a southern zone largely in S Chile dominated by high-alumina basalts, basaltic andesites and andesites. The northern and central zones are 140 km above an E-dipping Benioff zone, but the southern zone is 90 km above a Benioff zone. Continental crust is c. 70 km thick below the central zone, but is 30–40 km thick in the other two zones (Thorpe and

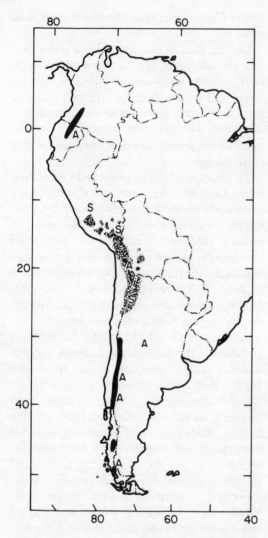

Fig. 19.13 The three active volcanic zones of the Andes. S: shoshonite, A: alkali–basalt (from Thorpe and Francis, 1979, *Tectonophysics*, **57**, Fig. 1A, p. 54; reproduced by permission of Elsevier, Amsterdam)

Francis, 1979). There are comparable correlations with trace element and isotope data.

Granitic Batholiths

Granitic rocks form extensive batholiths and plutons along the Cordillera from Alaska to Chile. Prominent in North America are the Coast Range, Boulder–Idaho, Sierra Nevada and California batholiths (Fig. 19.14), and in South America the Peruvian and Patagonian batholiths. The largest are more than 1500 km long and 200 km wide. The batholiths have a composite structure— the Peruvian consists of some 1000 plutons (Cobbing, 1982). They contain a variety of rocks belonging to a consanguineous magma series with a regular basic-to-acid sequence of intrusions in roughly the following proportions (Pitcher, 1978): gabbro 7–11%; tonalite 46–55%; granodiorite 20%; monzogranite 20–26%; syenogranite 0.6–4%. Brown (1982) summarizes their petrochemistry.

The complexes were intruded into regionally metamorphosed and deformed sedimentary and volcanic rocks of the western eugeosyncline on which they have superimposed contact metamorphic aureoles; the Peruvian plutons were emplaced at depths of 3–8 km beneath a cover of their own volcanic ejecta (Pitcher, 1978). They are regarded as the deep-seated equivalents of the arc volcanics, exposed today because of the major late uplift of the continental margin orogenic belts. The emplacement of the Peruvian batholith spanned 70 Ma, the Sierra Nevada batholith 60–80 Ma and the W Mexico batholith 55 Ma (Pitcher, 1975).

Since Lindgren made the suggestion in 1915, there has been much controversy about possible systemic variations in age and composition of the plutonic rocks from west to east across the Cordillera. Let us look at these two aspects in turn.

The granitic complexes vary in age from Triassic to Miocene with the main intrusive phase in the mid–late Cretaceous. The age problem is complicated because too many age determinations have been made by the K/Ar method. There were five periods of emplacement of the granitic rocks in the western USA, each lasting 10–15 Ma. There is clearly no simple progressive eastward decrease in age of the granitic rocks. The earliest are found on both east and west margins, but during the late, main and post-orogenic phases, from the mid-Cretaceous to the early Tertiary, there was an eastward migration of plutonic centres. However, according to Kistler

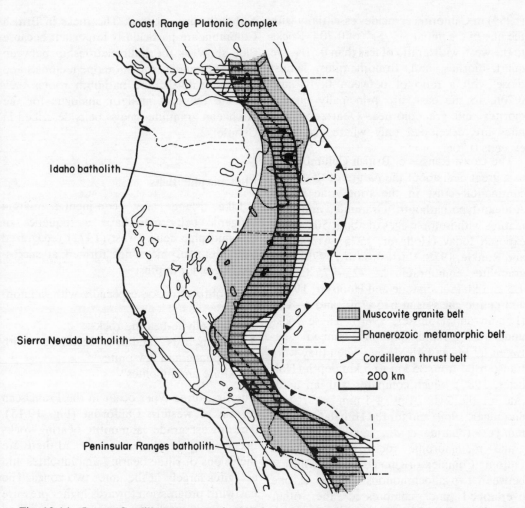

Fig. 19.14 Inner Cordilleran muscovite-bearing plutonic belt in relation to foreland of Cordilleran thrust belt and belt of Mesozoic and Cenozoic regional metamorphism (after Miller and Bradfish, 1980; reproduced with permission)

(1974) the youngest Oligocene–Miocene rocks in the Cordillera of the western USA occur in both the east and the west.

Moore (1959) demonstrated that in western USA a line separates quartz-diorites to the west from largely granodiorites and quartz monzonites to the east. Systematic increases in K_2O, SiO_2, K_2O/Na_2O and K_2O/SiO_2 away from the continental margin are reported across the batholiths of the Sierra Nevada (but with no age correlation), southern California, northern Chile and Peru. There are two contemporaneous magma series in the Boulder Batholith with the sodic

to the west, based on K_2O/SiO_2 and Na_2O/SiO_2 variations, but there is no systematic variation in $(K_2O + Na_2O)/SiO_2$ across the batholith in Baja California. There is an important inner belt of muscovite granites in the Cordillera of North America (Fig. 19.14). These are strongly peraluminous S-type granites with high Sr^{87}/Sr^{86} implying a great contribution by older crustal material.

An interesting aspect concerns the initial strontium isotope ratios which vary with geographic position but not with age from 0.703 to c. 0.708 (Sierra Nevada; Bateman, 1980). The quartz-diorite line of Moore

(1959) in California coincides essentially with the line of the initial Sr^{87}/Sr^{86} of 0.704. Rocks to the west, with a ratio of less than 0.704, are quartz-diorites and trondjhemites, whilst those with a ratio of between 0.704 and 0.706, to the east, are principally quartz-diorites and granodiorites. Quartz monzonites are developed only where the ratio exceeds 0.706.

The Coast Ranges of British Columbia tell us a great deal about the nature of the lower continental crust in the root zone of an Andean-type batholith. Granulites and migmatites with isotopic ages of about 50 Ma are exposed today (Hollister, 1975; Armstrong and Runkle, 1979; Harrison et al., 1979). The granulites equilibrated at 725–775°C and 4.2–5.5 kb (Selverstone and Hollister, 1980), and pelitic gneisses at 675–750° and 6–8 kb (Lappin and Hollister, 1980). These rocks underwent a rapid uplift of 2 mm yr^{-1} between 62 Ma and 48 Ma ago when they were transported from 35 km to 5 km depth (Hollister, 1982), which compares with an uplift rate for the Swiss Alps of 1 mm yr^{-1} (Clark and Jäger, 1969) and for the Himalayas of 0.8 mm yr^{-1} (Sharma et al., 1980). The high-grade metamorphic rocks of the Coast Plutonic Complex originated during collision between two allochthonous terrains as one overlapped and compressed the other (Monger et al., 1982). These rocks are excellently described by W. W. Hutchinson (1982) who shows that at this deep level of the crust there is a gradation from autochthonous to allochthonous, even within individual plutons—several plutons can be traced from their migmatitic roots into their recumbent and discordant heads. There is also clear evidence of the evolution of substantial volumes of plutonic rocks from pre-existing gneisses. The plutons consist mainly of granodiorite and quartz-diorite with a calc-alkaline chemistry. This crustal generation took place in a thickened crust and well-exposed, large-scale, thrust–nappe structures suggest that a horizontal style of tectonics contributed to the thickening of the Andean-type continental margin. There are few areas in the world where deep-level sections of Andean-type batholiths are exposed. The rocks in British Columbia are particularly important because they show us the interrelationship between young granulites, thrust–nappe tectonics and Andean calc-alkaline batholith roots: they provide the best modern analogue for the Archaean granulite–gneiss belts described in Chapter 2.

Metamorphic Belts

Like island arcs, continental margin orogenic belts may have a sequence of metamorphic belts. Ernst (1971) recognized four metamorphic zones formed at successively greater depths:

1. Zeolitized rocks, especially with laumontite.
2. Pumpellyite-bearing rocks.
3. Greenschists and/or blueschists with glaucophane and lawsonite.
4. Albite amphibolites.

The first three types occur in the Franciscan terrain of western California (Fig. 19.15). The lowest grade laumontite-bearing rocks occur along the present coast, and there are inclusions of rutile-bearing amphibolites and eclogites largely in the inner two zones. The eastward progression towards higher pressure parageneses marks the direction of presumed lithospheric descent.

Similar blueschist rocks occur in many parts of the Cordillera, but their metamorphic zonations are not so well defined as in the Franciscan.

Returning to the Franciscan section, which is the best documented, the predominant rocks are greywacke, microgreywacke, shale and serpentinized peridotite with minor mafic pillow lava and chert. But little stratigraphy is preserved as the rocks have been highly deformed into a tectonic mélange. The western zeolite facies zone was recrystallized at depths not greater than 5–10 km, but the mineral assemblages of the pumpellyite-, lawsonite-, and jadeite-bearing rocks must have been buried to depths of at least 35 km, equivalent to 9+ kb pressure. Oxygen

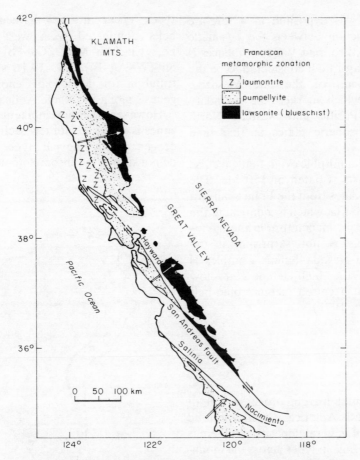

Fig. 19.15 Progressive metamorphic zonation in high-pressure/low-temperature assemblages in chiefly metaclastic rocks of the Franciscan Formation. Arrows indicate direction of presumed underflow (after Ernst, 1971; reproduced by permission of *Contrib. Mineral. Petrol.*)

isotopic and phase equilibrium data suggest pressures in excess of 8 kb. These figures lead to the conclusion that the blueschist metamorphics were presumably dragged down into the throat of the subduction zone and buried beneath the overriding lithospheric plate, but subsequently rebounded isostatically to shallow depths. The leading edge of the overlying plate seems to experience upward drag near the suture (Ernst, 1971).

Interpretations of the site of deposition of the Franciscan rocks vary slightly. They are generally thought to have been deposited in and adjacent to an oceanic trench but clearly they are a mixed series: the ophiolites formed at a mid-oceanic ridge, the radiolarites were deposited on the oceanic plain, and the flysch turbidites were poured into a trench. They may have been deposited, not against a major ocean, but in a marginal basin between an island arc to the west and the continental margin of N America (Roure and Blanchet, 1983).

The Franciscan rocks pass eastwards into the contemporaneous Great Valley sequence, and the interrelationships between them have been the subject of much controversy. The Great Valley sequence comprises conglomerates, lithic sandstones, siltstones and shales;

288

most likely it was deposited in a fore-arc basin. The junction between the ensimatic Franciscan and the Great Valley sequences is marked by a major fault considered to be the crustal expression of the late Mesozoic Benioff zone which marked the boundary between the Pacific and western North American lithospheric plates at that time (Ernst, 1971).

The main metamorphic event that gave rise to the blueschists is dated at 150 Ma. The range of K–Ar dates from the Franciscan as a whole is 150–70 Ma, which is identical to the period of extensive magmatism in the batholithic belt. This is not surprising if the Klamath–Sierra Nevada–Salinia Complex is interpreted as the high temperature–low pressure half of a paired metamorphic belt sequence (Ernst, 1971).

Mineralization

The continental margin orogenic belt of North and South America is well endowed with a variety of mineral deposits and there is a systematic change in types across the mountain belts, which reflects the increasing depth of the underlying subduction zone away from the continental margin. The metals were largely emplaced as components of the calc-alkaline magmas that gave rise to the granitic batholiths and andesitic–rhyolitic lavas in the main arc zone (Noble, 1976).

The generalized sequence of metal provinces across the mountain belt in terms of the plate tectonic–subduction model was summarized by Sillitoe (1976) and Frutos (1982). Ore types temporally, spatially and genetically related to intrusive and extrusive magmatism in the Central Andes include, eastwards from the coast, contact metasomatic Fe deposits, Cu–Au and Ag veins, Manto-type Cu deposits, stratiform Mn deposits, Cu-bearing breccia pipes, porphyry Cu–Mo deposits, Cu–Pb–Zn–Ag deposits of vein and contact metasomatic types, a volcanogenic Fe deposit, Sn–W and Sn–Ag veins, and porphyry Sn deposits. The magmatogenic ore

types occur in well-defined longitudinal belts dominated from west to east by: Fe, Cu–(Au–Mo), Cu–Pb–Zn–Ag, and Sn–(W–Ag–Bi). Fig. 19.16 shows the distribution of Mesozoic–Cenozoic metallogenesis along the whole Andean belt.

However, the development of the metal zones is variable with few sections across the orogen containing all types. Table 19.1 demonstrates the incomplete nature of many

Fig. 19.16 Mesozoic–Cenozoic Andean metallogenic-palaeogeographic scheme: 1: Andean foreland, 2: Andean geosynclinal area, 3: iron-apatite belt, 4: polymetallic deposits, 5: porphyry copper belt, 6: tin belt, 7: Peru–Chile trench, 8: plate movement direction, 9: Patagonian polymetallic belt, 10: Precambrian ranges (Pampean Arc and Patagonian Arc) and shield areas in the Andean foreland, 11: Altiplano polymetallic belt (Cu, Ag, Pb, Zn, and also sedimentary Cu), 12: Fe–P sedimentary deposits, mainly Cretaceous, 13: Au-placer deposits, 14: Mo-(U)-belt (mainly hydrothermal veins) (reproduced with permission from Frutos, 1982, in G. S. Amstutz *et al*. (Eds), *Ore Genesis: the State of the Art*, Springer Verlag, Berlin, Fig. 1, p. 497)

Table 19.1 Metallogenic variations across some Andean-type margins

Metals		Region
West	East	
Hg, Cu, Au, Ag, W, Pb, Mo		Western USA
Fe, Cu, Mo, Zn, Pb		British Columbia
Fe, Cu, Au, Pb, Zn, Ag, Cu, Au, Pb, Cu, Sn		Peru
Cu, Mo, Au, Pb, Zn		Ecuador and Columbia
Cu, Ag, Pb		Mexico
Fe, Au, Cu, Pb, Zn, Ag		Mexico
Cu, Cu–Pb–Zn–Ag, Sn–W–Ag–Au–Sb–Bi–Pb–Zn		Central Andes

sections, but nevertheless the ones present do not depart significantly from the generalized sequence.

Swinden and Strong (1976) emphasize the fact that the metallogeny in the Cordilleran and Andean belts is complicated because of the long succession of events from the late Proterozoic to the present. The coastal Fe occurs as an oxide but most of the remaining ores are sulphides. The formation of the metals in one traverse may be essentially contemporaneous. In Peru there are porphyry copper deposits of late Mesozoic and Pliocene–Quaternary and in Chile of Pliocene age; in the Cordillera and Andes as a whole they are largely restricted to the Cretaceous and early–middle Tertiary (Stanton, 1972), but in places there has been a repeated development of a particular metal. For example, in the Cu province of Chile copper deposits range in age from Jurassic to Pliocene and in the Sn province of Bolivia there was tin deposition in late Triassic, Miocene, Pliocene and perhaps Pleistocene times. This recurring Sn mineralization has been ascribed to the periodic tapping of a persistent Sn anomaly in the upper mantle rather than to subduction zone activity.

From the Triassic to the early Tertiary, mineralization was narrowly confined and dominated by Cu ores, but from 15 Ma ago diverse mineralization developed over a broader region (Fig. 19.17). The period from 15 Ma onwards, particularly from 10–4 Ma, accounts for a large part of the metals currently exploited in the central Andes (Sillitoe, 1976).

At an early stage of evolution of the Andean belt in the Jurassic–early Cretaceous, volcanic island arcs of an incipient, rather basic, partly tholeiitic, calc-alkaline nature were associated with metal deposits with a small atomic number (Fe, Ni, Mn, V, Cr, Co). In contrast, in the final stages of evolution of the belt in the middle–late Cenozoic when the continental crust was thick and calc-alkaline magmatism had reached the granitic extreme, high atomic number metal deposits (Sn, Bi, Sb, W) were developed together with high Mo/Cu ratio porphyry-type deposits (Frutos, 1982).

Porphyry copper deposits occur in some island arcs but are more common in continental margins (e.g. the Andes, the Rockies and Iran) due to the greater abundance there of suitable granitic host rocks. They are situated near the apex or around the margins of small porphyritic, predominantly quartz monzonitic–granodioritic plutons intruded into or below andesitic or dacitic volcanics (Jacobsen, 1975; Tiley, 1982). They were emplaced at only a few kilometres depth, almost contemporaneously with their intrusive or extrusive host rocks—only two million years difference in Utah. Association with calderas and strike-slip faults suggests they were emplaced late in the development of the magmatic arc. Two types of porphyry copper deposit are defined by their respective molybdenum/gold contents. Those richest in Mo pierced mainly a continental crust that was relatively thick and included at least Palaeozoic metamorphics (and often Precambrian basement); they are well developed in North and South America as well as in the Alpine belt from Yugoslavia to Pakistan,

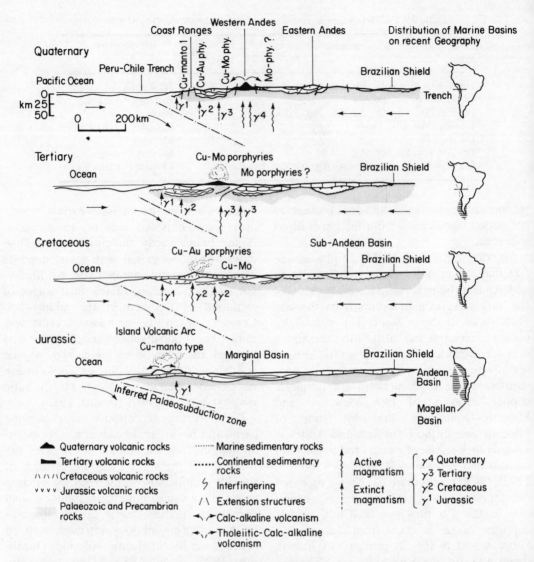

Fig. 19.17 Relationship between metallogeny and the magmatic evolution of the Andean belt (reproduced with permission from Frutos, 1982, in G. S. Amstutz *et al*. (Eds), *Ore Genesis: the State of the Art*, Springer Verlag, Berlin, Fig. 4, p. 501)

whereas those with appreciable Au intruded thick marine volcanics and are prominent in the island arcs of the SW Pacific.

There appears to be a relationship between the type of metallogenic belt and the dip angle of the Benioff zone. An increase in inclination may be related to a decrease in rate of lithospheric descent (Luyendyk, 1970). According to Mitchell and Garson (1981) Sn–W–Bi–Fe deposits and associated granites are emplaced above shallow-dipping Benioff zones, and porphyry copper–gold deposits in andesites and tonalites form above steeply dipping zones in island arcs and in silicic volcanic rocks above tin-bearing granites. To account for the metallogenic development of some continental margin belts it may be necessary to consider a major change in dip of the Benioff zones. For example, in Peru the late Mesozoic porphyry copper–molybdenum associated with andesitic volcanoes and tonalitic plutons formed above a steeply dipping subduction zone, but the Quaternary porphyry coppers in silicic volcanics and granitic plutons formed when the subduction zone had decreased in dip.

Major porphyry copper deposits in the Canadian Cordillera are spaced on average about 87 km apart, and in Peru and Chile they are 71 km apart (Sawkins, 1980). Interestingly, the average spacing of volcanoes along continental margin volcano–plutonic arcs is slightly higher than 70 km.

Chapter 20

The Alpine Fold Belt

In this chapter we shall consider the complex network of structures that make up the Alpine fold belt that extends from Gibraltar to the Middle East; further east it passes along strike into the Himalayan belt. The Alpine belt is far from simple as it consists of a multitude of small parts and, because of its accessibility, it is known in some detail. It began to form in the early Jurassic with the break-up of Pangaea and the movement of major plates, it reached its climactic deformation stage in the Tertiary, and tectonic activity still continues today, as is evidenced by the frequent earthquakes and volcanoes. In plate tectonic terms it is a continent–continent collision belt formed by interaction of the African and Arabian plates with the Eurasian plate. In the past there were several classical ideas to explain its formation, but we shall consider it in terms of the current plate tectonic model. This concerns not only the formation of the on-land structures, but also the disappearance of the Tethys Ocean.

The evolution of the Alpine belt is complicated since not only does it consist today of a complex mosaic of microplates, but also it evolved over a period of about 200 Ma by the continuous motion of a large number of plates and microplates (Fig. 20.1). There was thus a constantly evolving interconnecting network of mid-oceanic ridges, continental margins, island arcs, back-arc basins and transform faults. Unlike the Andean continental margin orogenic belt, which has a semi-continuous trench system and a main arc axis that can be defined for stretches of thousands of kilometres, the Alpine belt has an extremely complex and variable structure because most of the individual mini-belts of zones have different tectono-metamorphic and stratigraphic histories, each phase being largely diachronous. Thus no one evolutionary sequence is applicable to all parts of the belt. For example, according to the view of Dewey *et al.* (1973), there were six periods of basalt formation, seven of ophiolite formation, three of ophiolite obduction, eleven of deformation and seven of high T/P metamorphism taking place diachronously in different areas.

Another reason for the complexity of the Alpine belt is that many continental margins have been created and subsequently destroyed. Most geological complexities lie near such margins. Thus each kilometre of margin brings into existence a 150 km or so wide strip of potential complexity. Methods need to be devised to reconstruct the shapes of these areas prior to compressional deformation (Smith and Woodcock, 1982).

Nevertheless, it is possible to outline a generalized sequence of events applicable to an idealized complete tectonic cycle with respect to the history of a single plate from its birth at an accreting ridge to its disappearance at a consuming plate margin, with the consequent appearance of new material at the convergent juncture. Evidence of individual segments of this history can be found in different parts of the belt. The following is a synopsis of the principal events that contributed to the development of this orogenic belt.

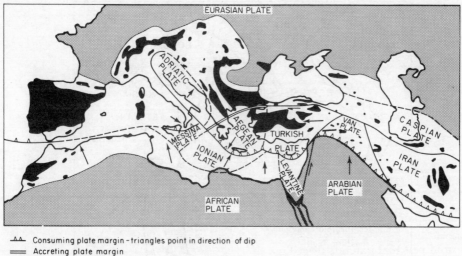

Fig. 20.1 The present lithospheric plates and types of plate boundary in the Alpine System of Europe together with the distribution of pre-Alpine continental basement. Arrows refer to plate motions relative to the Eurasian plate (compiled and redrawn after Dewey *et al.*, 1973; reproduced by permission of The Geological Society of America)

Synopsis

1. Contraction of the Permo-Triassic Palaeo-Tethys ocean took place at the same time as the growth of the Neo-Tethys oceanic plate in Triassic and later times.

2. Formation of late Triassic carbonate platforms often sited on evaporite sequences along the rifted continental margins. Red beds border the continents.

3. Collapse of the carbonate platforms began in the early Jurassic and was followed by deposition of deeper water pelagic facies with radiolarites and shales.

4. At plate-consuming margins deposition of Cretaceous flysch by turbidity currents in trenches associated with nearby ophiolite-bearing mélanges and blueschist metamorphism. Acid volcanism in the arc environment, closure of marginal basins, thrusting and obduction.

5. Upon continent–continent collision (Eocene–Oligocene) overriding of nappes and thrust-sheets with diachronous 'synorogenic' flysch deposition. High temperature regional metamorphism. Marginal

basins continued to open in the Neogene in the western Mediterranean.

6. Late orogenic uplift and consequent erosion led to deposition of mostly late Tertiary, non-marine, clastic wedges (molasse) in foredeeps or exogeosynclines along the cratonic margins. Evaporites laid down in several basins in late Miocene (Messinian). Continued tectonism in the Pliocene in some areas such as the Jura.

For recent reviews of Alpine palaeogeography and tectonic evolution, see Laubscher and Bernoulli, 1977; 1982; Roeder, 1978; Smith and Woodcock, 1982; Trümpy, 1982.

Rock Units

The rocks in the Alpine Belt will be dealt with in order of appearance, from the early to the late evaporites, after which the tectonic evolution of the mountain belt will be reviewed.

Late Triassic Evaporites

Evaporites started forming in the late Per-

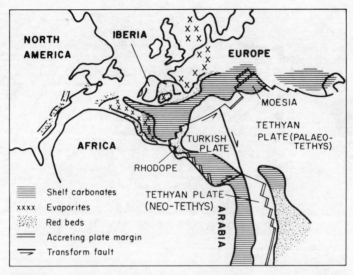

Fig. 20.2 Late Triassic palaeographic reconstruction and plate margins in the Alpine System (modified after Dewey *et al*., 1973; reproduced by permission of The Geological Society of America)

mian and reached a climax in the late Trias in the region bordering the Palaeo-Tethys (Jenkyns, 1980) (Fig. 20.2); they are now preserved in the North Sea (Ziegler, 1975) and in a belt extending to the present Mediterranean area. Salt layers are more than 1500 m thick in western Greece and 1000 m in Italy.

The evaporites are one of four Upper Triassic sedimentary facies in the Mediterranean region. On the bordering continental areas extensive red beds were deposited in terrestrial semi-arid conditions. Evaporites may have formed in inland salt lakes and on continental sabkhas. Shallow-water reef and shelf carbonates formed in tidal flats and lagoons, and in deeper water there were cherty micritic radiolarian limestones. Desiccation of the salt basins took place when they became isolated from the Tethyan Ocean by the reef and shelf carbonates; the formation of the basins may have been controlled by early rifts and graben structures. Alternatively, the evaporites may have been deposited on oceanic crust in a proto-rift Red Sea environment (Dewey *et al*., 1973; Kinsman, 1975), similar to that of the Aptian evaporites in the marginal basins along the coasts of eastern South American and West Africa

which formed at an early stage of the separation between these continents.

The evaporites are also important tectonically because they later formed the décollement (detachment) zone that enabled the overlying Mesozoic sediments to be thrusted over their basement in many parts of the Alpine System such as the Jura, the Atlas Mountains, the northern Apennines and the Subalpine chains of the western Alps.

Carbonate Platforms

A miogeoclinal, shallow water, carbonate platform is a characteristic feature of a rifted continental margin in a near-equatorial region. The growth of carbonate platforms in the Alpine belt began in the middle–late Trias and continued mostly to the early Lias. Subsidence appears to have kept apace with rapidly accumulating carbonates with the result that enormous thicknesses of carbonate deposits developed. Subsidence and sedimentation rates were about 100 mm 10^{-3} years. In the peri-Adriatic zone of the Dinarides exceptional carbonate formation continued from the early Triassic to the late Cretaceous and Tertiary giving rise to a sequence about 7 km thick.

The platform rocks consist of supratidal to shallow subtidal limestones and dolomites, containing stromatolitic, pelletal and oolitic varieties, commonly with gastropods and calcareous algae. Evidence that these sediments locally emerged to near or above sea level is given by intercalated coal seams and bauxites. The platform rocks are also associated with carbonate reef deposits containing many sponges, corals, hydrozoans and calcareous algae.

The carbonate platforms throughout the Alpine–Mediterranean region began to break up in the early Jurassic. Neptunian dykes and sills, formed by fissure infilling, penetrate the platform sediments and are related to larger scale extensional tectonics expressed by block faulting that caused differential subsidence of the platforms and formation of a submarine seamount–basin topography (Bernoulli and Jenkyns, 1974) (Fig. 20.3a). This block faulting was associated with widespread volcanic activity locally associated with lead-zinc mineralization. On the seamounts carbonate sedimentation continued at depths of less than 200 m with formation of condensed sequences (starved of sediment), crinoidal calcarenites, iron pisolites, ferromanganese nodules, the well-known nodular 'ammonitico rosso' facies and, locally, algal stromatolites that grew in the photic zone. Very different deeper water sedimentation took place in the basins, giving rise to marls and radiolarites and a noticeable lack of pure carbonates. As the carbonate compensation depth increases with the depth of water

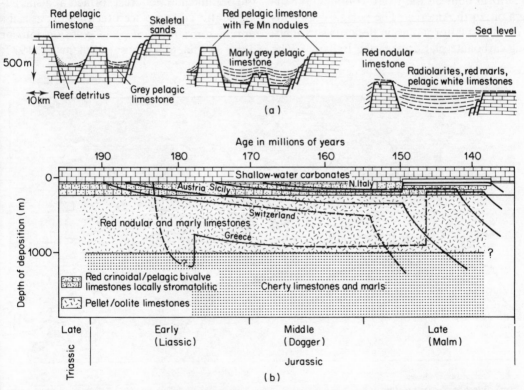

Fig. 20.3 (a) Diagrammatic palaeographic evolution of the collapse of the carbonate bank of the southern rifted continental margin of Tethys during Jurassic time (redrawn after Bernoulli and Jenkyns, 1974; reproduced by permission of the Society of Economic Paleontologists and Mineralogists). (b) Subsidence tracks for parts of the Tethyan continental margin. Dotted lines indicate period of non-deposition (after Jenkyns, 1980, *Proc. Geol. Ass.*, **91**, Fig. 5, p. 115; modified after Bosellini and Winterer, 1975)

(Fig. 20.3b), the presence of the Alpine radiolarites and sympathetic absence of carbonates in the basins suggests that the deposition depth there was of the order of 4.5 km. However, there is considerable disagreement about the depth of deposition on the seamounts in the early-middle Jurassic: less than 200 m (Bernoulli and Jenkyns, 1974) to 1000–2400 m (Bosellini and Winterer, 1975). Fig. 20.3a shows a possible sequence of events in the collapse, fragmentation and subsidence of a carbonate platform situated on a rifted continental margin. As blocks subsided, deeper water sediments took the place of carbonates, and submarine topography became smoother, so that by the late Jurassic a deep-water facies was widespread.

It is interesting to compare these Alpine structures with modern equivalents. The carbonate platforms are similar to the Bahama Platform built on the rifted continental margin of North America. The pelagic sediments resemble platform-margin deposits surrounding carbonate platforms in the Bahamas. The rifted environment may also be compared with the Red Sea where early clastics pass into a carbonate–evaporite facies, where block faulting and Recent volcanics are common and where lead-zinc enrichments occur in the geothermal brines. There are several modern seamount terraces with depths, fauna and sediments similar to those on the early Jurassic Alpine seamounts (Jenkyns, 1980).

Ophiolites—Remnants of Oceanic Crust

The Alpine belt is the homeland of ophiolites, so-called by A. Steinmann in 1926 for the serpentinite–spilite–chert trinity. Today the succession commonly found in ophiolite complexes (magnesian ultramafics, gabbros, dykes, pillow lavas and pelagic sediments such as radiolarian chert) is widely considered to represent a section through layers 3, 2, 1 of the oceanic mantle–crust formed at a spreading centre and emplaced tectonically as slabs during some form of plate collision into or onto continental crust (Coleman, 1977;

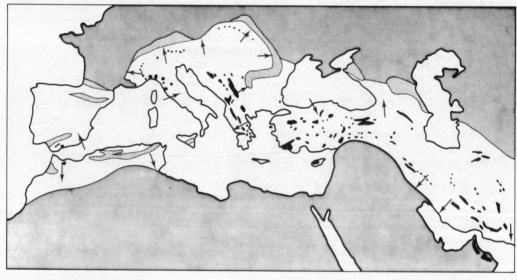

- ●   Ophiolite complexes
- ←   Direction of overthrusting
- ▨   External molasse troughs
- ▨   External forelands

Fig. 20.4 Distribution of ophiolite complexes and molasse basins along the foreland margin and direction of late overthrusting in the Alpine System (compiled from Dewey *et al.*, 1973; reproduced by permission of The Geological Society of America)

Panayiotou, 1980). Examples in the Alpine belt are common, especially in the eastern part (Fig. 20.4) and include the Troodos complex, Cyprus, the Oman, the Othris Mountains, Greece, the Vourinos complex, Greece, Masirah island, Oman and the northern Apennines.

There are significant differences between the ophiolites in the eastern and western Mediterranean. In the east there are harzburgites, chromite deposits and deep sea cherts, whereas in the west there are lherzolites, no chromite ores and terrigenous rather than pelagic sediments (Karamata, 1980). Geochemical evidence supports the idea that the western ophiolites were derived from a series of small Red Sea-type basins caused by fragmentation of continental crust, and the eastern ophiolites originated in back-arc basins behind a series of arcs (Pearce, 1980).

Evidence of early Jurassic rifting is widespread, especially in the western Mediterranean. Sea-floor spreading, giving rise to Neo-Tethys, began in the west in the mid-Jurassic (similar to the age of the onset of spreading in the Atlantic—Sheridan and Gradstein, 1981), but in the eastern Mediterranean oceanic basins are only of mid-upper Cretaceous age (Rocci *et al.*, 1980). The reason for this east–west difference in tectonic style lies in the wedge-shaped opening of Neo-Tethys with its early-opening apex and narrow rifts in the west (Sengör *et al.*, 1980). From the mid- to upper Cretaceous spreading ceased and crustal shortening began.

We shall now review the well-documented Troodos Massif which contains the main characteristics of a typical ophiolite complex.

Fig. 20.5 shows a map and cross-section of the complex, which has the form of an elongate dome, and Fig. 20.6 lists its major rock units, correlated with the seismic layers of oceanic lithosphere.

The lowermost unit of the plutonic complex consists largely of foliated harzburgite. The ultramafic cumulate zone comprises dunite, clinopyroxene dunite, poikilitic wehrlite and picrite, whilst the overlying zone above the seismic 'moho' contains meta-gabbro, olivine and pyroxene gabbro. The granophyre, often called quartz-diorite, plagiogranite or trondhjemite forms residual segregations in the gabbro and screens between basic dykes; similar rocks have been dredged from the ocean floors (Aldiss, 1981). The sheeted intrusive complex is made up of 90–100% dykes, mostly basaltic in composition, separated by thin screens of pillow lava (typically multiple dykes with chilled margins). They have moderate to high SiO_2(50–60%) and low alkali contents (less than 0.2% K_2O). The Lower Pillow Lavas characteristically contain 10–50% dykes with chilled margins; they are mainly oversaturated (45–65% SiO_2) basalts, often intensely silicified, with a K_2O peak around 0.25%. Erosional conglomerates associated with sedimentary sulphide deposits (ochres) occur in depressions on the Lower Pillow Lava surface. The Upper Pillow Lavas are generally free of dykes, undersaturated (40–50% SiO_2), are often olivine-bearing basalts with low to high K_2O values (up to *c*. 2%) and contain more basic (picrite) and even ultrabasic types. The pelagic sediments consists of Fe–Mn rich mudstones—umbers (Robertson, 1976), radiolarites, volcanogenic (bentonitic) clay and sandstone.

The lowermost harzburgite is regarded as depleted upper mantle; Moores and Vine (1971) include the dunite as part of the depleted mantle. The pyroxenite-gabbro may be a product of partial fusion segregated from parent mantle material and crystallized in cumulate intrusive bodies, the granophyres being their residual liquids (Moores and Vine, 1971). In contrast to stratiform intrusions the plutonic lithological sequence is not a time sequence: all products crystallized at the same time, but in different parts of the reservoir and under different P/T conditions and volatile fugacity.

The Sheeted Dyke Complex and the Lower Pillow Lavas (with similar chemistry) are genetically related and are the only direct products of the spreading axis process. There are four generations of dykes—diabase, dolerite, picritic lamprophyre and basalt, belonging to two families of dykes. The first was

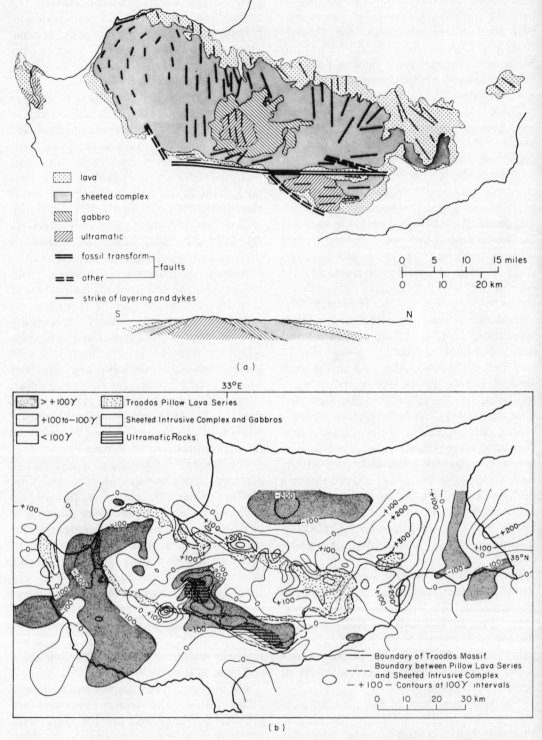

Fig. 20.5 The Troodos Complex, Cyprus. (a) Map and cross-section (after Moores and Vine, 1971; reproduced by permission of The Royal Society). (b) Magnetic anomaly map (after Vine *et al.*, 1973; reproduced by permission of Macmillan Press) in which an appropriate regional field has been removed from the total field aero survey

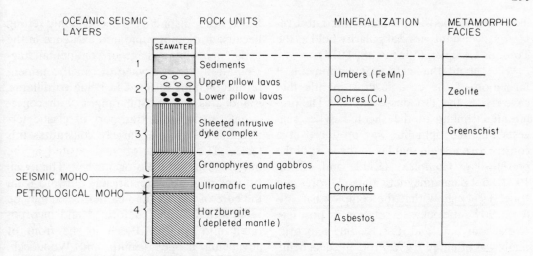

Fig. 20.6 The rock units, mineralization and metamorphic facies of the Troodos Complex, Cyprus (redrawn and modified after Gass and Smewing, 1973; reproduced by permission of Macmillan Press), correlated with oceanic seismic layers

produced by accretion, and the second by accretion and late fracturing of a magma chamber when the spreading had stopped (Desmet *et al.*, 1978). The Sheeted Dyke complex has a greenschist facies metamorphic overprint and the Lower Pillow Lavas a zeolite facies, from which a thermal gradient of $150°C$ km^{-1} has been calculated (Gass and Smewing, 1973).

The Upper Pillow Lavas are thought by Gass and Smewing (1973) to be unrelated to the Lower Lavas and are an off-axis sequence analogous to the seamounts and volcanic islands that are distant from the ridge in the modern ocean floor. Their undersaturated composition is in keeping with the idea that progressive undersaturation in silica takes place with increasing distance from a mid-oceanic rise. In contrast, Desmet (1976) produced geochemical data which support the idea that all the lavas were derived from a single parental magma, with progressively differentiated magmas feeding the successive lower and upper lavas.

The basal sediments (umbers) are closely comparable with Recent ferromanganiferous, trace element-enriched sediments along the East Pacific Rise and with their Tertiary 'rise' equivalents of the Atlantic (Horowitz and Cronan, 1976) and Pacific Oceans (Cronan,

1976). The metals (and REE enrichments) were picked up when circulating heated sea water percolated through hot layer 2 lavas; chloride complexes in reduced hydrothermal solutions were exhaled as thermal springs into aerobic sea water. Sulphide mineralization formed as chemical precipitates derived from metal-rich, sulphur-bearing exhalations derived from hydrothermally circulated hot seawater (Spooner and Bray, 1977; Parmentier and Spooner, 1978). The preservation of overlying radiolarites may be associated with silica derived from the thermal springs. The cherts contain Campanian (Upper Cretaceous) radiolaria and this agrees with the 76+ Ma age for the igneous rocks of the complex (Vine *et al.*, 1973).

There is a close correlation between the gravity and magnetic (Vine *et al.*, 1973) anomalies over the Troodos Massif. The positive magnetic anomalies coincide with the strongly and normally magnetized Lower Pillow Lavas (Fig. 20.5b). In the absence of reversely magnetized material, Vine *et al.* (1973) suggest that the lavas may have formed in the 10 Ma period just before 75–80 Ma ago when the earth's magnetic field was of a single normal polarity. It is interesting to note that the adjoining eastern Mediterranean is magnetically undisturbed; it may have formed in the same mid-late Cretaceous, constant, normal polarity field as the Troodos Complex (Vine *et al.*, 1973).

What about the tectonic environment of formation of the oceanic crust-mantle that gave rise to the Troodos Complex? The preferential chilling of the sheeted dykes suggests that the ophiolite was produced at a constructive margin that lay to the west of the present-day Complex (Kidd and Cann, 1974), but palaeomagnetic data (Moores and Vine, 1971) show that the complex has suffered 90° anticlockwise rotation in post-late Cretaceous time and its spreading axis originally was oriented E–W. From the geochemical data that show the rocks evolved from a hydrous magma (Pearce, 1980), and from the fact that the oceanic layers of the complex are much thinner than those in present-day major oceans, Gass (1980) concluded that the Troodos Complex originated in a back-arc marginal basin. In contrast, Robertson and Woodcock (1980) point out that all evidence for the contemporaneous island arc is lacking and thus conclude that it formed in an intra-continental rifted basin. There is evidence of island arc volcanism on Cyprus, but it post-dates the obduction of the ophiolite and is related to collision tectonics (Baroz, 1980).

Alpine and Flysch and Plate Collision

The term *flysch* has had problems of definition. For the present purpose flysch designates Cretaceous and Early Tertiary marine shaly formations in the Alps, characterized by the presence of regular intercalations of sandstone and/or impure limestone beds. These are detrital, deep-marine turbidites deposited by sediment gravity flow processes (Homewood and Caron, 1982).

It has been conventional to consider Alpine sediments in terms of pre-Alpine evaporites, carbonates and pelagics, synorogenic flysch and post-orogenic molasse. In plate tectonic terms flysch sedimentation took place in a variety of environments created as a result of different types of plate movements; they are synorogenic because they owe their origins to the movements of microcontinents. The flysch has a distinctive palaeotectonic setting because it did not come into existence in the early 'pre-Alpine' stage of continental fragmentation, but late when the microcontinental interactions resulted in rising cordilleras, island arcs and coastal ranges, with consequent erosion and transport of clastic terrigenous debris that largely constitutes this facies. The flysch is often deposited in the internal zones of the orogenic belt. The flysch troughs migrate systematically outward from the core of the belt and earlier flysch may be redeposited as it is deformed and incorporated into younger flysch in the front of advancing nappes (Smith and Woodcock, 1982).

The mode of formation of Alpine flysch, long misunderstood because of its variable occurrence, must be seen in terms of complex plate tectonic settings. There are at least three principal flysch environments:

1. Geophysical evidence suggests that the Balearic and Tyrrhenian basins in the western Mediterranean had an extensional origin and deep-sea drilling has confirmed a graben structure in the Balearic basin as well as the presence there, and in the Alboran and Valencia basins, of extensive turbidites of flysch type. In fact, seismic data indicate that these deep flysch-like basins in the western Mediterranean are floored by abyssal plains with Recent flysch-type sediments up to 1 km thick, many of which were transported there by bottom currents from the basin margins.
2. Recent flysch-like turbidites make up the submarine fan of the Nile delta. Many Tertiary turbidite sequences in the Tethyan region were deposited in deep-sea fan environments, e.g. Upper Eocene–Oligocene and Miocene in the northern Apennines and Oligocene–Lower Miocene on Rhodes, and the many Mesozoic and Tertiary fan deposits in this region were channelled down submarine canyons. The Eocene–Oligocene turbidites in the flysch trough of the western Carpathians were not supplied from 'internal' zones (uplifted cordilleras), as is commonly supposed, but were funnelled down submarine canyons in the continental shelf.

3. Flysch may be deposited in active trenches bordering island arcs if subduction rates are low or if the sediment supply rate is high (Dewey *et al.*, 1973). Quaternary compositionally-immature turbidites occur in the Hellenic trench formed by underthrusting of the Aegean (Hellenic) Arc (sometimes called the Ionia Basin) by the African plate. Modern flysch sedimentation in the Hellenic trench is reviewed by Stanley (1974).

The most extensive Alpine flysch deposits are of Cretaceous and Palaeogene age and occur in the external zone bordering the Alpine Front and extending from the Swiss Alps eastwards to the Carpathians and the Balkans. The early Cretaceous flysch in Rumania was deposited in a trench where the Greco-Italian microcontinent overrode the Tethys. Because Bavarian flysch sedimentation in the East Alps took place continuously for about 50 Ma from the early Cretaceous to the Palaeocene–Eocene, it may be difficult to imagine its accumulation in a trench where underthrusting would be expected to remove it quickly within a single tectonic–subduction cycle of 3–4 Ma. But it is probable that the relative motion between the plates at that time was largely lateral.

The classical model of Alpine flysch formation is in a compressional foredeep in front of a rising cordillera. If we reinterpret this in plate tectonic terms, the flysch is developed in a trench bordering an uplifted arc. Such is the Schlieren flysch of the western Alps deposited in an Eocene accretionary trench wedge (Homewood and Caron, 1982).

Late- to Post-orogenic Molasse

The accumulation of clastic wedges of the molasse facies in basins along cratonic margins (foredeeps or exogeosynclines) is a hallmark of the last stages of orogeny; it is initiated by uplift of the mountain belt following the last major tectonism. In the Alpine System the first molasse sediments were deposited before the last stages of nappe movement and the last were entirely post-tectonic (for distribution see Fig. 20.4). Van Houten (1974) gives a review of molasse deposits in the northern Alpine, Aquitaine and Ebro foredeeps.

The last main deformation phase in the European Alps began in the early Oligocene. By late Oligocene there was extensive uplift and transport of nappes (Ultrahelvetic and South Helvetic) and consequent erosion contributed 6 km of terrigenous detritus to the Oligocene–Miocene Molasse Basin of the Central Alps in Switzerland. Some proximal molasse deposits were overriden by nappes between the late Oligocene and the late Miocene (Helvetic nappes), and post-tectonic molasse continued to accumulate until the early Pliocene. Commonly, the autochthonous flysch sequences grade upward into molasse, which is dominated by a distinctive proximal non-marine facies consisting largely of alluvial fanconglomerates which may exceed 1 km in thickness (Van Houten, 1974).

During molasse accumulation sedimentation kept pace with subsidence with the result that a near-sea level surface fluctuated between non-marine and marine conditions; molasse may thus be intercalated with, or grade laterally into, marine sequences with lignite, coal, freshwater limestones and evaporites.

Each of the molasse foredeeps reviewed by Van Houten (1974) underwent a distal axis migration with time, the northern Alpine axis moving 60 km between the Oligocene and Pliocene at an average rate of 2 mm yr^{-1}. The molasse accumulation lasted for 25–35 Ma with a preservation rate of between 150 and 400 m Ma^{-1}. The average vertical stripping for the three regions varied from 3–4 km, suggesting that a maximum of 7–10 km must have been locally eroded from source areas. From geochronologic and heat flow data, Clark and Jäger (1969) calculated an erosion rate of 0.4–1 mm yr^{-1} in the Central Alps, from which they estimated that a maximum local stripping of 10–25 km took place during 25 Ma of molasse accumulation. The difference between this maximum value and that recorded by Van Houten reflects the proportion of eroded material that is not preserved in the molasse basins.

Late Miocene Evaporites

It is ironic that, having started its history with extensive evaporite deposition, the Tethys–Mediterranean should end the same way.

Deep-sea drilling has revealed the existence of substantial late Miocene (Messinian) evaporites beneath all the major basins of the Mediterranean, namely the Tyrrhenian, Ionian, Balearic and Levantine Basins (Van Couvering *et al.*, 1976; Cita, 1982) and they extend in interconnected basins as far as the Caspian Sea and the Yemen (Sonnenfeld, 1975). The evaporites are 1.5–2 km thick and formed between 6.5 and 5.0 Ma ago. They resulted from the precipitation of more than 10^6 km^3 of gypsum, halite and other salts from a volume of sea water estimated to be equivalent to 30 times that in the present Mediterranean basins. The western Tethys lost its connection with the Indo-Pacific ocean about 16.2 Ma ago and with the Atlantic 10 Ma later. This led to total isolation of the Mediterranean and its consequent desiccation. There is well-documented evidence of a major fall in the level of the world's oceans by 50–70 m in the late Eocene, and this was contemporaneous with an expansion of Antarctic glaciation. However, it is not certain whether the glaciation caused the lowering of global sea levels and this in turn caused the isolation of the Mediterranean and its eventual desiccation, or whether the extraction of 6% of the dissolved salts in the world ocean brought about by the desiccation could have induced the late Miocene glaciation by lowering high latitude salinity sufficiently to raise the freezing point of sea water (Adams *et al.*, 1977). The salinity crisis came to an end with marine incursions in the Pliocene.

Mineralization

There is a consistent and expected relationship between the kinds of Alpine mineral deposits, their host rocks, and their plate tectonic environment. The main types are as follows (see Fig. 20.7):

1. Those associated with early graben. Syngenetic lead–zinc deposits of the Bleiberg type occur in Middle Triassic limestones containing evaporites; barytes, celestine and anhydrite are found in the ores. This association may be analogous to the late Tertiary lead–zinc deposits in the Red Sea area which probably formed in early graben. A modern expression of this activity may be in the metal-rich brines and sediments of the Salton Sea (California) and the Red Sea.

2. Chromite in ophiolite ultramafics. Lenticular and lineated (podiform) chromite deposits occur in the lower serpentinized ultramafics of many ophiolite complexes in the eastern Alpine belt (Fig. 20.7) which formed at back-arc spreading centres (Pearce, 1980) and were subsequently emplaced tectonically (obducted) into or onto continental crust during arc–plate collision.

3. Massive cupriferous sulphide deposits (ochres) occur in many obducted ophiolite complexes, such as the Troodos Complex, Cyprus (Fig. 20.7) (Constantinou, 1976; Robertson, 1976; Spooner, 1977), and at Küre and Ergani–Maden in Turkey. The Cu sulphides are thought to have formed in fumarolic exhalations discharged into a highly reducing environment in depressions near the crest of a mid-oceanic ridge. Late- to post-volcanic Fe–Mn sediments (umbers) (Robertson, 1976b) are comparable with ferromanganoan concentrations from the East Pacific Rise. Mn-rich deposits in cherts and Fe–Cu–Zn sulphide deposits in basalts occur in Apennine ophiolites; they formed as a result of mobilization of metals from basalt during circulation of thermal waters near the Mesozoic spreading centre, following the model of Bonatti (1975).

4. Porphyry copper deposits occurring in the eastern part of the Alpine belt in calc-alkaline plutons, largely in the granodiorite–quartz monzonite range associated either with or without molybdenum concentrations. They formed, typically within 0.5–2 km of the surface, either on continental margins or in island arcs in relation to palaeo-Benioff zones and plate collision (Mitchell and Garson, 1981). One of the most prominent is the Sar Chesmeh deposit in Iran which lies in a belt of

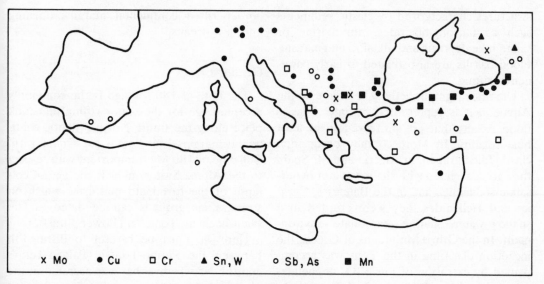

x Mo   • Cu   □ Cr   ▲ Sn, W   ○ Sb, As   ■ Mn

Fig. 20.7   Some important mineral deposits in the Alpine belt (some compiled from Kostov, 1978)

related copper occurrences over 160 km long (Fig. 20.7).

There is an interesting distribution of these ores along the Alpine belt. The ophiolitic chromite and sulphide ores and the porphyry copper deposits occur more in the east, whilst in the west there is remarkably little evidence of mineralization. To account for this phenomenon Evans (1975) proposed that in the west there was insufficient subduction of oceanic lithosphere and, further, following a suggestion of K. J. Hsü, that the Alpine orogenic belt did not develop at a convergent junction but in a region where the African and European plates moved laterally past each other. It is clear that transform motion is required to move South America westwards from its position south of western Europe in Permian times, according to the palaeomagnetic fit of Smith *et al.* (1981), but this motion would have been earlier than that required to produce the Alpine orogenic belt. I suggest that a more likely explanation lies in the wedge-shaped opening of Neo-Tethys as described by Sengör (1979a) and Sengör *et al.* (1980). In the west only Red Sea-type basins opened and, therefore, we find there only authigenic pitchblende and pyrite mineralization in sandstones belonging to a graben stage, Pb–Zn in Triassic carbonates with baryte and evaporites (Red Sea type), and Mn–Fe carbonates and silicates in marls, etc., comparable with deposits in the Afar Rift (Evans, 1975). The limited subduction of such narrow oceanic rifts was not sufficient to create significant calc-alkaline plutonism and associated Cu–Mo mineralization and the ophiolites there are of mid-ocean ridge type (Pearce, 1980). In contrast, further to the east, Neo-Tethys became a wide ocean fringed by island arcs and its subduction gave rise to appreciable porphyry Cu deposits, whilst obducted back-arc basin ophiolites carry important chromitite mineralization. Kostov (1978) shows that several Mn (chert), Mo and SnW (in granitic rocks) ore deposits occur in Anatolia and Greece but not further west—a distribution that is consistent with the above suggestion.

Tectonic Evolution

Having considered the development of the principal rock units we shall now review the tectonic evolution of the Alpine belt.

Early Rifting

The earliest expression of the rupture of a continent is the formation of graben-like rift structures characterized by clastic sediments such as fanconglomerates, non-marine or saline lake deposits and alkaline magmatism. Flood basalts are also related to early continental rifting.

The first magmatic activity belonging to the Alpine belt is represented by late Triassic flood basalts that are preserved in at least nine regions. In Morocco they occur in a clastic-filled graben, in Algeria and SE Spain they are associated with shallow-water hypersaline sediments, and in the Balearics, Carnics and Hellenides they were erupted in a shallow-water, marine, carbonate environment. In the Othris Mountains of Greece the inception of rifting in the Trias was accompanied by extrusion of low K_2O, nepheline-normative, light REE-enriched basalts. In Sicily alkali trachytes were erupted in the Toarcian of the early Jurassic and in the Venetian (southern) Alps there was rhyolitic–trachytic volcanism in the late Jurassic. Dewey *et al*. (1973) suggest that all this volcanic activity was related to early episodes of continental distension and rifting prior to the development of an ocean in the regions concerned. Permian–mid-Triassic volcanics in the southern Alps (*sensu stricto*) may mark a graben-like zone of rifting (Evans, 1975). The early Jurassic Tethys seaways opened in places parallel to, and elsewhere transverse to, the Triassic rifts.

A relevant question that may be asked at this stage is what was going on in the European platform when major plate movements were beginning to take place to the west and south between the late Trias and late Jurassic? In Britain during the Triassic a horst–graben topography controlled continental sedimentation. A similar tectonic regime continued to operate into the Jurassic when basement faults controlled depositional patterns during the formation of a north European epeiric sea. The North Sea graben (which controlled much oil accumulation) formed in the Callovian (Middle Jurassic) as a series of failed arms (Whiteman *et al*., 1975; Ziegler, 1975). The formation of all these structures is a reflection of the extensional tectonics that characterize shelf regions that border rifted continental margins opening into new oceans.

Oceanic Crust

The date of 180 Ma ago is the commonly accepted age for the major rifting that took place along the future North Atlantic continental margins—for details see Chapter 16. The date of 180 Ma is important with respect to the Alpine System as it is the age of collapse of the carbonate platforms which on stratigraphic grounds can be dated as late Pliensbachian–Toarcian (Lower Jurassic).

Here we must be careful to distinguish between the Permo-Triassic Palaeo-Tethys and the Neo-Tethys that was accreted in the period between the early Jurassic and Eocene.

The Palaeo-Tethys closed between the late Triassic and the mid-Jurassic. Sengör (1979a) and Bernoulli and Lemoine (1980) proposed that this closure was caused by the collision with Laurasia of a Cimmerian continent that was rifted away from northern Gondwanaland during the Triassic (Fig. 20.8). The Neo-Tethys (Argyriadis *et al*., 1980) may have opened in the mid-Jurassic, partly as a back-arc basin over a south-dipping Palaeo-Tethyan subduction zone. The suture of the Palaeo-Tethys is located in a series of early to mid-Mesozoic orogenic belts stretching from the South Rhodope orogen in N Greece, eastwards via N Turkey, N Iran and Afghanistan to central Tibet. The closure of Neo-Tethys gave rise to a southern belt of ophiolites extending from the coast of Yugoslavia, through the Othris zone of southern Greece, to southern Turkey (Fig. 20.4). This closure may have involved northward subduction (Smith and Woodcock, 1976).

The early basins of the western Neo-Tethys were narrow—Kelts (1981) has likened them to the present-day Gulf of California. Most models operate with one basin in the Alps (e.g. Dewey *et al*., 1973; Dietrich, 1976; Hsü, 1977). Frisch (1979), however, provides evidence of two basins in a tectonic progradation model. Tectonic retrogradation occurs in predominantly oceanic regimes with well-

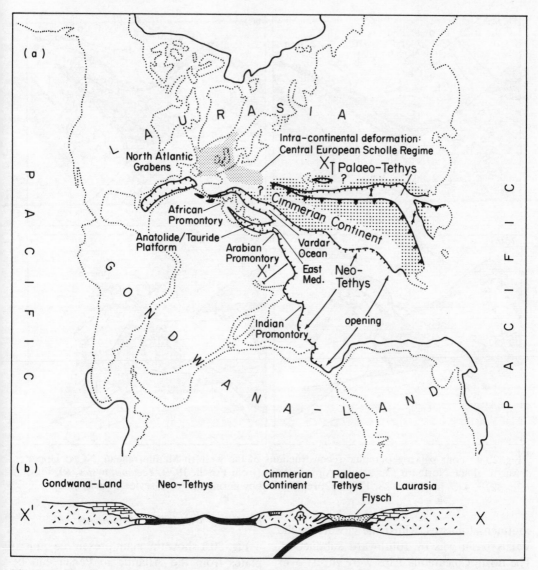

Fig. 20.8 Reconstruction of Pangaea for the latest Triassic–early Jurassic showing the growth of Neo-Tethys at the expense of Palaeo-Tethys. Lines with black triangles mostly on the south side of Palaeo-Tethys are subduction zones with the triangles on the upper plate. Lines with short hachures represent passive continental margins. XX′ indicates the line of the section (reprinted with permission from Sengor, 1979a, *Nature*, **279**, Fig. 2, p. 592. Copyright © 1979 Macmillan Journals Ltd)

developed magmatic arcs, and in this case a new ocean basin tends to develop as a back-arc basin. In contrast, tectonic progradation typically develops by rifting in an originally continental environment. The resulting oceanic basins are too narrow to give rise to magmatic arcs on subduction, and therefore a new ocean basin tends to develop not as a back-arc, but on the opposite side of the suture. In this way the sutures and tectonic

activity migrate from the internal (oceanic-side) to the external (continent-side) parts of the orogen. Fig. 20.9 presents four scenarios in the development of the Alps. The first Piedmont ocean basin developed about 140 Ma ago and was then subducted southwards. The new Valais ocean opened to the north of a narrow Briançonian plate 30 Ma later. The Piedmont ocean closed by 80 Ma ago with the collision of the Austroalpine and Briançonian

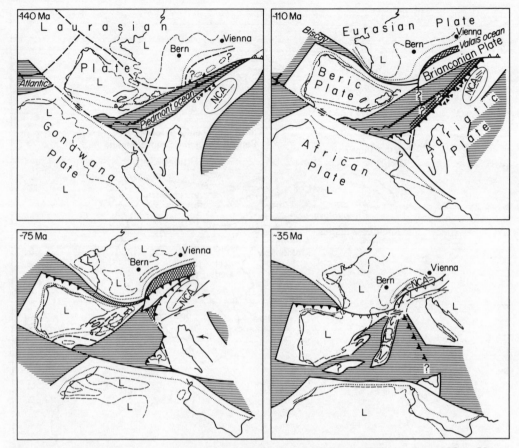

Fig. 20.9 Four palaeogeographic reconstructions of the western Mediterranean. NCA: Upper Austro-alpine Northern Calcareous Alps, L: land (from Frisch, 1979, *Tectonophysics*, **60**, Figs. 2, 3, 4, 5, pp. 126, 128, 130, 132; reproduced by permission of Elsevier, Amsterdam)

continental masses, whilst the closure of the Valais basin was by southward subduction. The north Calcareous Alps were thrust over the Briançonian microcontinent into the narrowed Valais trough in the late Eocene, and by the Oligocene all Mesozoic oceanic crust in the western Mediterranean area had been consumed so that there was a continuous orogenic belt.

Continued post-collisional shortening took place by adjustment in the geometry of the plates in response to boundary stresses at the plate margins. Hsü (1979) demonstrated that underthrusting of successive crustal wedges (not lithospheric plates), together with late décollement overthrusts of sedimentary covers, took place in the Helvetic Alps and the Jura in post-Eocene times.

Plate and Microplate Movements

Fig. 20.1 show the main present-day microplates from the Atlantic to Persia and between the major Eurasian and African plates, all plates being defined by the seismicity along their margins. Each of these plates is in motion although there is no unidirectional sense. The African plate is generally moving northwards; however, the net effect against the Eurasian plate is not simply compressional, because it is complicated by the variable interplay of movements by the intervening microplates. For example, a result of the northward movement of the Arabian plate is that the Turkish plate is being wedged sideways to the west, and therefore the Aegean plate is advancing southwestwards against the Ionian plate (Fig. 20.1). Fig. 20.10 shows that

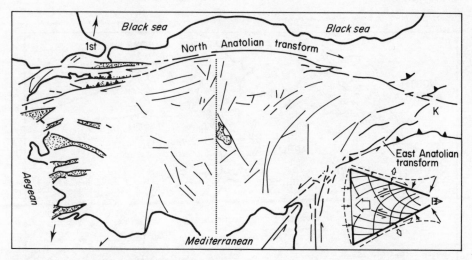

Fig. 20.10 Simplified map of the Turkish plate showing the bounding North and East Anatolian transform faults plus internal faults which are compared in the inset with the stress patterns in a modified Prandtl cell. The dotted line in the middle depicts the possible western boundary of the Prandtl cell analogy in central Anatolia (after Sengör, 1979b, *J. geol. Soc. Lond.*, **136**, Fig. 6, p. 279)

the Turkish plate contains an array of faults which Sengör (1979b) explained in terms of a modified Prandtl cell. The curved arrays of internal faults arise from the stress systems set up within a plate (or cell) that is indenting westwards, and the E–W rifts situated in the west are a further response to this type of indentation tectonics. Such rifts are analogous to the N–S rifts in Tibet caused by the northward indentation of India (Chapter 21).

So much for the present picture. The question now is were similar microplate movements responsible for the generation of the individual fold belts within the Alpine belt from the early Jurassic to the present? A. G. Smith (1971) suggested three main stages of movement between Africa and Eurasia, which were dependent upon the relative opening rates of the central and north Atlantic and which controlled all minor movements of intervening smaller plates. After the incipient rifting the main stages were as follows:

Period	Time (Ma)	Motion of Africa relative to Europe
1. Middle Jurassic to Upper Cretaceous	165–80	Eastward
2. Upper Cretaceous to late Eocene	80–40	Westward
3. Late Eocene to present	40	Northward

The stage 1 'eastward' movement occurred because the central Atlantic had begun to open but Europe was still joined to North America. The second stage resulted from the fact that the North Atlantic was now opening at a faster rate than the central Atlantic. Stage 3 occurred when the opening rates of both parts of the Atlantic were similar.

Fig. 20.11 shows the evolving plate boundaries for three important stages in Alpine history and, in particular, the decrease in size of the Tethyan plate, a modern remnant of which may be found in the Caspian and Black Seas.

Whilst the relative movements of some of the older microplates may be subject to reappraisal and discussion, the rotation of some of the younger microplates has been successfully determined by palaeomagnetic data especially Spain, Sardinia–Corsica and the Italian–Dinaride block.

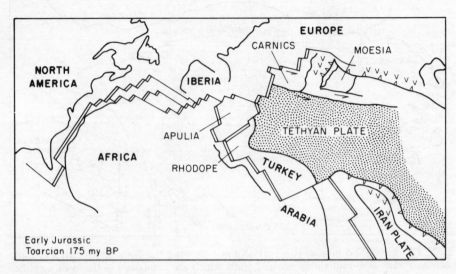

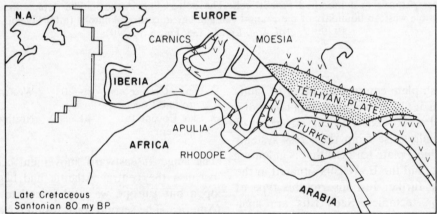

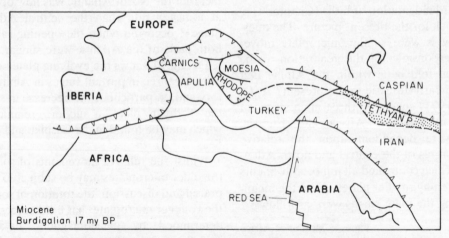

Fig. 20.11 Proposed plate boundary schemes in early Jurassic, late Cretaceous and late Tertiary times for the Alpine System (compiled from Dewey *et al*., 1973; reproduced by permission of The Geological Society of America)

Chapter 21

The Himalayas

The Himalayas is the world's youngest and highest orogenic belt. It consists of a collage of allochthonous terrains and completed its evolution by continent–continent collision. There is evidence here of the complete Wilson cycle, from the Mesozoic to the Eocene, followed by post-collisional deformation which is still active today. We shall reconsider the section of the belt that extends in an E–W direction from the Afghanistan–Pakistan border to the Burmese frontier and we must include Tibet which contains key data.

Topographically, the belt has many superlatives: the highest rate of uplift (nearly 1 cm yr^{-1} at Nanga Parbat), the highest relief (Nanga Parbat to the Indus), the greatest concentration of glaciers outside of the polar regions (the Karakorum), the highest plateau (Tibet) and the source of some of the greatest rivers (Indus, Zangbo-Brahmaputra, Ganges, Yangtse, Mekong, Salween, Huang Ho). For some of these reasons much of the belt is extremely well exposed, especially Tibet, Ladakh-Zanskar and Karakorum-Kohistan-Hindu Kush.

Geologically, the belt has several unique features. The central part provides an uncomplicated profile across a continent–continent collisional boundary; in other words, it does not have the complexities of many intervening oblique mini-plates as in the Alpine-Mediterranean belt. Along its 3000 km length there is a remarkable regular series of tectonic belts, although in the west, in Pakistan, the uplift has been very much higher than further east; thus there is an excellent opportunity to study variations along the length of a simple suture, tectonic belt, plutonic arc, thrust belt, etc. In the west, in Kohistan-Ladakh, there is a Cretaceous island arc that has been deformed and turned on end giving a complete profile from the top to the base of the arc, where there are granulites. The high uplift in the west has removed almost the whole Phanerozoic pile of sediments on the Indian plate so that here, and in the base of the adjacent arc, the lower crust of the Himalayan belt can be studied. Post-collisional deformation has given rise, firstly, in the south, to a unique intra-plate thrust belt with inverted metamorphic zones and lower crustal melt leucogranites and, in the north, to rifts and wrench-faults that follow a systematic pattern that can be explained by indentation slip-line theory. By extension of these slip-line faults to the southeast via Indonesia it is possible for the first time to interrelate the tectonics of continent–continent collision at the end of the Wilson Cycle with the tectonics of back-arc basins nearer the beginning of the cycle. For all these reasons, and many more, the Himalayas present for us an exciting display of geological phenomena.

Tectonic Evolution

Here it is useful to consider briefly the main large-scale aspects of the tectonic evolution of the Himalayas.

The Himalayan mountain range was created by the collision between continental plates (Gansser, 1964; Le Fort, 1975). The Palaeo-Tethys of late Palaeozoic age was

310

closed by the late Triassic by collision between the Eurasian plate and the Tibetan
Plateau which gave rise to the Kun Lun
range. According the Sengör (1979a), the
closure of Palaeo-Tethys was contemporaneous with and responsible for the opening of
Neo-Tethys in the Triassic, the closure of
which brought about the collision betwen the
Tibetan Plateau and the Indian Plate with
final production of the Indus-Zangbo Suture.
Subduction of Neo-Tethys gave rise to an
Andean-type margin in the east, which was
completed by the Eocene, and an island arc in
the west which went through two stages of

growth: early–mid-Cretaceous and Palaeocene–Eocene.

There are substantial palaeomagnetic data
on the India–Asia collision (Fig. 21.1),
although there is still disagreement about the
precise timing. According to Patriat *et al.*
(1982) subduction of the Indian plate under
Eurasia began 110 Ma ago (Aptian). The
Indian plate moved northwards at an average
rate of 14.9 ± 4.5 cm yr^{-1} from 70 Ma ago
until about 40 Ma ago, when it slowed to its
present rate of 5.2 ± 0.8 cm yr^{-1}, the time of
slowing corresponding to the time of
India–Asia collision in the late Eocene

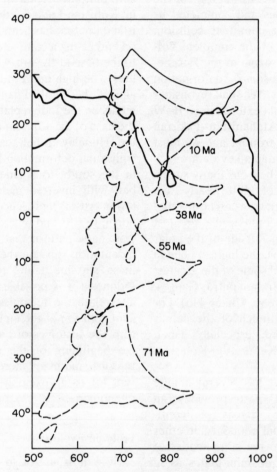

Fig. 21.1 The northward drift of India with
respect to Asia from 71 Ma ago to the present
determined from magnetic anomalies in the Indian
and Atlantic oceans (reproduced with permission
from Molnar and Tapponnier, 1975, *Science*, **189**,
Fig. 1, p. 240, copyright 1975 by the American
Association for the Advancement of Science)

(Peirce, 1978). In contrast, Klootwijk and Peirce (1979) suggested that initial collision between India and Asia occurred 55 Ma ago in the late Palaeocene–early Eocene, when India's northward motion reduced from 20–15 cm yr^{-1} to 6–4 cm yr^{-1}. However, according to Klootwijk and Radhakrishnamurty (1981), the initial collision between India and the Kohistan-Ladakh arc was in the late Palaeocene–early Eocene, and the rapid decrease in rate of northward movement of the Indian plate was at 40 Ma ago (as originally suggested by Molnar and Tapponnier in 1975) when the Indo-Pakistan island arc block collided with Asia. Thus there is good evidence for a major drop in movement rate of the Indian plate, but whether this occurred 55 or 40 Ma ago, and what it was that India collided with, is not certain.

There is general agreement, however, by the palaeomagnetists that the northward movement of the Indian plate at about 5 cm yr^{-1} has continued from about 40 Ma ago to the present. Therefore, it can be calculated that 2000 km (Molnar and Tapponnier, 1975) to 250–3500 km (Molnar and Chen, 1978) of post-collisional crustal shortening has taken place in the last 40 Ma, solely within the continental plates. This shortening may be expressed in four ways, relevant contributions of which are under much discussion.

1. Molnar and Tapponnier (1975) proposed that 500–1000 km of shortening took place along two arrays of transcurrent faults within the foreland of the indented Asian plate; this is consistent with palaeomagnetic evidence which suggests 1000 km of crustal convergence since the Palaeogene within Asia north of the Pamirs. A concomitant effect of this plastic-rigid indentation model is the formation of N–S graben and E–W extension in the 'dead zone' of Tibet (Tapponnier et al., 1981a,b).
2. Sengör and Kidd (1979) argued that crustal thickening took place in Tibet during Quaternary N–S shortening in the overriding plate. This is possibly expressed at a high level by the formation of extensive buckle folds with E–W trending axial surfaces and probably associated thrusts in Mesozoic sedimentary rocks; but these features have yet to be substantiated.
3. Extensive crustal shortening took place along the entire 2400 km long northern edge of the Indian plate, along a series of thrust planes which decrease in age southwards. The three principal ones are the Main Central Thrust (MCT) with a translation of more than 100 km (Andrieux et al., 1981), the Main Boundary Thrust (MBT) and the Main Frontal Thrust (MFT).
4. There may have been 200–350 km of underthrusting of continental lithosphere (Indian plate) under Tibet, along the Indus–Zangbo suture (Bingham and Klootwijk, 1980).

Tectonic Units

The following six units can be recognized across the Himalayan belt in a N–S direction (Figs 21.2 and 21.3):

1. *The Karakorum Range* in Pakistan lies along the southern border of the western part of the Tibetan Plateau and north of the 'Northern Suture' of Bard et al. (1980c) and Coward et al. (1982). It contains the calc-alkaline Karakorum batholith of late Cretaceous age.
2. *The Trans-Himalayas* occur along the southern border of the Tibetan Plateau, in India and China, and consist of Palaeozoic and younger beds intruded by Cretaceous to Palaeocene calc-alkaline batholiths; in the west there is an island arc complex in Kohistan (Pakistan) and Ladakh (India) separated from the Karakorum Range by the 'Northern Suture'.
3. *The Indus-Zangbo Suture Zone* separates the Indian plate from the Kohistan-Ladakh arc on the west and the Tibetan plateau on the east, and it contains belts of ophiolites, glaucophane schists, granulites, basic volcanics and molasse. The suture is termed the Main Mantle Thrust in the west (e.g. Tahirkheli et al., 1979; Andrieux et al., 1981).
4. *The Higher Himalayas* contain a basement

312

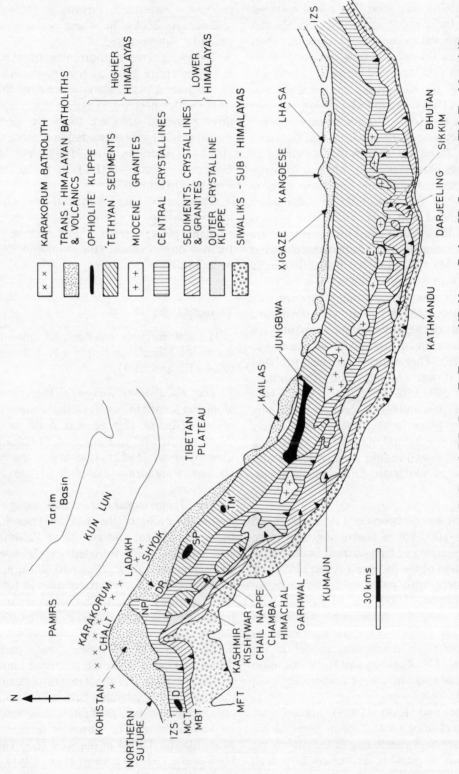

Fig. 21.2 Tectonic map of the Himalayas. D: Dargai, DR: Dras, E: Everest, NP: Nanga Parbat, SP: Spongtang, TM: Tso Morari (from Windley, 1983b)

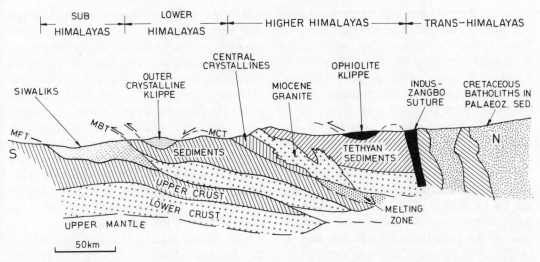

Fig. 21.3   Schematic cross-section through the Central Himalayas in Nepal and South Tibet. D: Dargai, DR: Dras, E: Everest, NP: Nanga Parbat, SP: Spongtang, TM: Tso Morari (from Windley, 1983b; modified from Andrieux *et al.*, 1977 and Mitchell, 1979)

of Proterozoic metamorphic rocks overlain by a conformable sequence of Cambrian-to-Eocene 'Tethyan' sediments (also referred to as the 'Tethys Himalayas': Gansser, 1980). This thrust belt was intruded by Cambrian granites, suffered prominent Miocene metamorphism and was intruded by Miocene leucogranites. The high-grade thrust slices are called the 'Central Crystallines' in India (Thakur, 1980). Ophiolite nappes were thrust southwards onto the Higher Himalayas from the Indus-Zangbo Suture in the early Eocene. The Higher Himalayas are separated from the Lower Himalayas by the Main Central Thrust, which formed not earlier than the early Miocene because beds of this age are the youngest exposed in tectonic windows (Fuchs, 1981), and which has a displacement of more than 100 km (Andrieux *et al.*, 1981).

5. *The Lower Himalayas* consist of sediments of Proterozoic, Palaeozoic and locally Mesozoic age metamorphosed mostly to a low greenschist grade and subdivided by thrusts with progressively older rocks to the north. This belt also contains thrust slabs of gneiss (the Lower Himalayan Crystallines) and major klippen of high-grade rocks termed the 'Outer Crystallines' which are thrust sheets derived from the Higher

Himalayas. The Lower Himalayas have been thrust southwards over the Siwaliks of the Sub-Himalayas on the Main Boundary Thrust of Plio-Pleistocene age.

6. *The Sub-Himalayas* contain the mid-Miocene–Pleistocene Siwalik molasse sediments and are commonly separated from the autochthonous Indo-Pakistan Shield to the south by the Main Frontal Thrust which has deformed Quaternary river terraces.

We are now in a position to consider each of these main tectonic units in detail.

The Karakorum

This region lies between the Northern Suture (Chalt–Shyok) and the Kun Lun Range in China and the Northern Pamirs of the USSR, but little is known of it except about its southern margin near the the Karakorum batholith at Hunza there is a progressive northward increase in metamorphism via four reaction isograds in pelites).

Between the Northern Suture at Chalt and the Karakorum batholith at Hunza there is a progressive northward increase in metamorphism via four reaction isograds in pelites (R. D. Broughton, pers. comm.). The isograds dip to the NE underneath the Karakorum batholith and are thus inverted. They formed in association with major southward thrusting on the margin of the Karakorum plate during post-collisional deformation in a late stage of indentation of the Indian plate (Coward *et al.*, in press).

The Karakorum Range is dominated by a complex array of dioritic-granodioritic-granitic plutons of the Karakorum batholith for which there is a U–Pb age on zircons of 95 ± 46 Ma, interpreted as an emplacement age by Le Fort *et al.* (in press). Later ages reflect Eocene regional metamorphism at the time of the Indo-Eurasian collision, and re-equilibration during Miocene rebound of the plate margin.

The Northern Suture This is the suture that lies between the Kohistan–Ladakh arc and the Karakorum belt to the north. It contains the Chalt mélange on the NW of the Nanga Parbat syntaxis and the Shyok mélange on the SE side.

The Chalt mélange comprises several mafic-to-ultramafic tectonic layers and lenses, each several kilometres long and tens of metres wide, in a matrix of chloritoid slate (Coward *et al.*, 1982, in press).

The Shyok mélange consists of highly deformed, partly metamorphosed, disrupted units of serpentinite, pyroxenite, greenschist, amphibolite, gabbro, phyllite, gneiss, marble, acid–basic volcanics, chert, greywacke and muscovite-chloritoid schist (Brookfield, 1981).

The Trans-Himalayan Plutonic Belt and the Kohistan–Ladakh Arc

The Trans-Himalayan plutons can be followed discontinuously for over 2500 km along the north side of the Indus–Zangbo Suture (Gansser, 1980). In the east the plutons were emplaced in an Andean-type margin, but in the west, where the Zangbo Suture has split into two, they occur between the two sutures and belong to the Kohistan–Ladakh island arc.

The Kangdese belt extends along the north side of the Zangbo Suture in Tibet (Xizang). The main rocks are slates, phyllites, schists, gneisses, amphibolites and migmatites, which are Ordovician to Cretaceous in age. Sub-horizontal, overlying andesites, dacites and rhyolites have an $^{40}Ar/^{39}Ar$ Palaeocene age of 60 Ma (Tapponnier *et al.*, 1981a). Most of the calc-alkaline plutons (diorites, granodiorites and granites) underlie these volcanics, and early basic bodies amongst them have $^{39}Ar/^{40}Ar$ ages of 90–100 Ma (Maluski *et al.*, 1982). Geological and isotopic evidence clearly indicates subduction in the Mid-Cretaceous and the Palaeocene.

The predominantly calc-alkaline rocks of the Kohistan–Ladakh arc (Fig. 21.4) were produced in two periods separated by a major deformation phase (Coward *et al.*, 1982). The rocks were first described as an island arc by Tahirkheli *et al.* (1979) and Bard *et al.* (1980c). The early suite comprises the following rock units (from N to S): the Yasin Group of intra-arc basin sediments, metavolcanics (schistose amphibolites, chlorite schists, hornblende tuffs) with prominent pillows in basic lavas with arc-type chemistry; foliated and gneissic tonalites and diorites, the Chilas Complex which is a stratiform cumulate body over 300 km long and 8+ km thick of dunites, norites and gabbros which formed in the magma chamber under the island arc; and the Kamila amphibolite belt of deformed arc volcanics (Coward *et al.*, 1982, in press). The minimum age for the growth of the arc is given by a zircon age of 103 Ma on granodiorites in Ladakh (Honegger *et al.*, 1982), and by mid-Cretaceous foraminifera in the Yasin Group. When this island arc underwent collision with the Karakorum plate, the rocks suffered two phases of isoclinal folding, which gave them a vertical attitude, and a regional metamorphism which ranges from greenschist grade in the metavolcanics in the north to granulite in the Chilas Complex (Coward *et al.*, 1982).

These deformed vertical rocks were then intruded by undeformed Palaeogene tonalites, diorites and pegmatites, and overlain by the Eocene Dir Group which contains pelites, minor limestones and a thick calc-alkaline

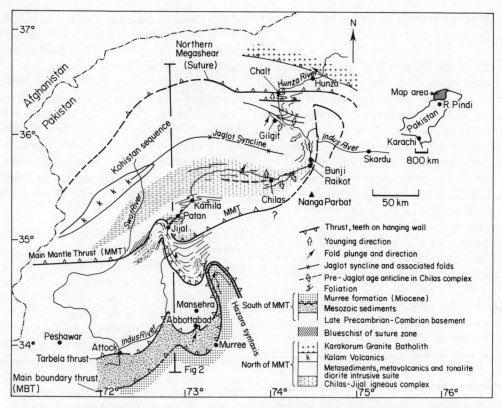

Fig. 21.4   Reginal map of the Himalayas of N Pakistan. The Kohistan island arc (sequence) is situated between the Northern Suture and the Main Mantle Thrust (or Indus Suture) (after Coward *et al.*, 1982, *J. geol. Soc. Lond.*, **139**, 299–308)

pile of andesites, dacites, rhyolites and ignimbritic tuffs (Tahirkheli, 1979). According to Brookfield and Reynolds (1981), this late calc-alkaline suite was formed when the Kohistan arc was attached to the Indian plate and thus acted as an advancing Andean-type margin with a subducting oceanic plate to the north; thus these rocks make up a second magmatic arc sequence superimposed on the first. However, Coward *et al.* (in press) show that the Andean-type plutons cut the structures associated with the closure of the Northern Suture and intrude to within a kilometre of that suture, and therefore argue that the Northern Suture closed before the Indus Suture and that the late calc-alkaline, Andean-type batholith was formed by renewed subduction on the Indus Suture after the arc had completed collision with the Karakorum plate. This conclusion is consis-

tent with that of Trommsdorf *et al.* (1982) for the Ladakh section.

East of Nanga Parbat a higher structural level is exposed than on the west, thus the volcanics of the first arc are better exposed there than the plutonics (Brookfield, 1981). The development of the second arc is well expressed by high-level granodiorites and quartz-diorites with dacites and andesites. $^{40}Ar/^{39}Ar$ ages on granodiorites range from 45 to 39 Ma (Brookfield and Reynolds, 1981), which is almost contemporaneous with the late Eocene–Oligocene Ladakh molasse.

The Indus–Zangbo Suture Zone

This zone contains several groups of rocks which provide information on the history of the suture.

Ophiolite   complexes   and   associated

glaucophane-bearing rocks occur as follows (east to west):

1. The Naga Hills in NE India contain an ophiolite belt with oceanic and volcanogenic sediments, basic-to-intermediate volcanics and mafic-to-ultramafic differentiates (Agrawal and Kacker, 1980). The ophiolite has spectacular narrow belts of glaucophane schists (Ghose and Singh, 1980).
2. The Xigaze ophiolite in Tibet (Nicholas et al., 1981) is interesting and unusual in having no gabbro zone, a sill rather than a dyke complex and a 6 km thick ultramafic unit of serpentinized harzburgites and dunites. Nicholas et al. (1981) interpret this ophiolite as the product of a slowly spreading ridge (<1 cm yr^{-1}), but the thin mafic section and the late intrusion of diabase into the ultramafic foundation led Sengör (1981) to suggest an origin close to a fracture zone.
3. In the Kailas region of NW India the suture zone contains a highly tectonized ophiolitic mélange with large ultramafic bodies imbricated with flysch-type sediments which are weakly metamorphosed and range in age from Triassic to Upper Cretaceous (Gansser, 1979).
4. Ophiolite melanges are prominent in the Indus Suture of the Ladakh Himalayas (Gupta and Kumar, 1979; Honegger et al., 1982). The suture zone is represented by basaltic and dacitic lavas (the Upper Jurassic–Upper Cretaceous Dras island arc volcanics), cherts, agglomerates, serpentinites and dunites which have been tectonically emplaced with radiolarites, limestones and ash beds in a complex order (Srikantia and Razdan, 1980). But in eastern Ladakh, Virdi (1981a) has distinguished the following four units (from north to south): an ophiolite mélange, serpentinite-peridotite zone, gabbro-diabase zone and pillow lavas with chert, jasper, tuffs and agglomerates. Evidence of high-pressure metamorphism is given by meta-gabbroic eclogite lenses within serpentinites (Gupta and Kumar, 1979), and by the presence of glaucophane in meta-basic rocks which contain key minerals in different assemblages (Virdi, 1981a). A post-early Eocene forma-

tion of the suture is suggested by the fact that the Spongtang Ophiolite was thrust on top of fossiliferous Palaeocene–early Eocene sediments (Baud et al., 1982).
5. In Pakistan there is a 3 km wide belt of blueschists with crossite; muscovite from such a blueschist has a K/Ar age of 84 ± 1.7 Ma (Shams, 1980), which suggests that the blueschist metamorphism did not take place during the continental collision and formation of the Indus-Zangbo Suture but formed in connection with the subduction zone of the earlier island arc.

In summary, whilst ophiolites and ophiolitic mélanges occur intermittently along the Indus Suture, glaucophane-bearing rocks are only present at the western and eastern extremities. This may imply less crustal uplift in the central part of the suture.

High-pressure garnet granulites are exposed in the Indus Suture in the Jijal Complex which is a 200 km^2 tectonic wedge on the south side of the Kohistan Arc in Pakistan (Jan and Howie, 1981). These rocks were equilibrated at 670–690°C and 12–14 kb at a deep level of the suture zone.

The Higher Himalayas

This tectonic unit has three main components:

1. A belt of high-grade metamorphic rocks—the 'Central Crystallines' or Tibetan Slab of Le Fort (1975).
2. An overlying sequence of Palaeozoic-to-Mesozoic (Tethyan) sediments to the north. On the south side of the crystalline Himalayas are the Kashmir and Chamba Synclinoria which contain rocks biostratigraphically similar to the Tethyan sediments.
3. Thrust klippen and ophiolites derived from the Indus Suture Zone.

The Central Crystallines Amphibolite facies gneisses form a principal 'crystalline' metamorphic belt up to c. 15 km thick along the entire length of the Himalayas. These are the Central Crystallines (Thakur, 1980;

Fuchs, 1981) which contain evidence of pre-Mesozoic protolith material:

1. A biotite-muscovite augen gneiss in the Kumaun Higher Himalayas has a 1830 Ma whole-rock isochron age with an initial Sr^{87}/Sr^{86} ratio of 0.725 (Bhanot et al., 1977).
2. Biotite gneisses from Himachal have Rb–Sr ages of 612 ± 100 Ma (Bhanot et al., 1975) and 581 ± 9 Ma (Mehta, 1977). These are deformed granites which are related to the Cambrian granites of the Outer Crystalline klippen in the Lower Himalayas.

Comparable biotite augen gneisses, which were derived from Cambrian porphyritic granites (Le Fort et al., 1980) and which underlie folded Mesozoic sediments, were metamorphosed during the Himalayan deformation and occur immediately south of the Indus Suture in Pakistan (Coward et al., in press).

Some gneisses in India which underlie Tethyan sediments contain fossiliferous Jurassic rocks metamorphosed to a kyanite grade (Powell and Conaghan, 1973). It has been widely thought that the crystalline gneisses are solely of Precambrian age and that they are overlain unconformably by the Tethyan sediments. However, Baud et al. (1982) point out that relations in eastern Zanskar in NW India show that the gneisses with meta-sedimentary Jurassic relicts are overlain by Infracambrian–Cambrian sediments at the base of the Tethyan succession and therefore that the gneisses and the Tethyan sediments must be in thrust contact. This is consistent with the presence in the Kashmir and Chamba synclinoria of Tethyan sediments that have been thrust southwestwards over the Central Crystallines.

Tourmaline-bearing, muscovite or two-mica leucogranites are prominent within and along the northern margin of the Central Crystallines of the Higher Himalayas (Fig. 21.2). The Manaslu granite in Nepal (Le Fort, 1981; Vidal et al., 1982) has a Rb–Sr whole-rock isochron age of 29 ± 1 Ma (Hamet and Allègre, 1976) and granites in Bhutan have an age of between 30 Ma and 12 Ma (Dietrich and Gansser, 1981). These granites have very high initial $^{87}Sr/^{86}Sr$ ratios: Manaslu, 0.7433 to 0.7874 (Hamet and Allègre, 1976); Bhutan, c. 0.77 (Dietrich and Gansser, 1981); Everest, 0.7630 to 0.773 (Kai, 1981). They owe their origin to partial melting of Palaeozoic or older crustal rocks at depths of 15–30 km (Dietrich and Gansser, 1981) or 30–40 km (Andrieux et al., 1977) below the Main Central Thrust as a result of crustal shortening by thrust–nappe stacking.

The 'Tethyan' Zone This is a 10 km thick pile of fossiliferous marine sediments ranging in age almost continuously from Cambro-Ordovician to mid-Eocene. The Mesozoic–Cenozoic sediments formed in a continental shelf environment (thick carbonates and shales on the northern border of the Indian plate) and they tend to show a deeper water facies to the north (pelites and flysch). The rocks have suffered a late Eocene greenschist-grade metamorphism in Tibet (Shackleton, 1981). Metamorphism of these sediments increases markedly in intensity towards the western and eastern ends of the Himalayan belt in Pakistan and near Burma (Gansser, 1980).

In Ladakh a structural dome has exposed crystalline gneisses overlain by Mesozoic sediments (Honegger et al., 1982). West of Nanga Parbat in Kohistan, the Indian plate south of the suture was uplifted at least 5 km in the period 30–20 Ma ago (Zeitler et al., 1982a), and the Tethyan sediments were largely eroded. A few relics were folded into isoclinal synclines in their crystalline basement and metamorphosed to a high grade with common marbles and graphite-rich mica schists (Coward et al., 1982). The highest rate of uplift in the Himalayas is in the Nanga Parbat region where it has been nearly 1 cm yr^{-1} during the Pleistocene (Zeitler et al., 1982b); for this reason the region consists of basement gneisses and migmatites.

In Nepal the Tethyan sediments are intruded by Miocene tourmaline-bearing leucogranites of Manaslu type (Stöcklin, 1980).

Ophiolite klippen from Indus Suture Zone Several ophiolite klippen have been thrust southwards for some 30 km (Ladakh) to 80 km (Tibet) as allochthonous sheets from the Indus–Zangbo Suture over the Tethyan sediments (Fig. 21.2): Jungbwa/ Kiogar/Amlang La in Nepal; Spongtang/Shilakong in Ladakh; and Dargai/ Malakhand in Pakistan. Typically, imbricated platform-type sediments have been overthrust by sheets of ophiolitic mélange, which are overlain by slices of ultramafics including harzburgites. The Spongtang/Shilakong klippe has ridden on a serpentine sole and coloured mélange over fossiliferous Palaeocene and early Eocene sediments (Fuchs, 1981; Baud *et al.*, 1982); thus it could be argued that the ocean could not have closed finally by the Maestrichtian, as suggested by Brookfield and Reynolds (1981) and Brookfield (1981). It consists largely of peridotite, has a total lack of cumulates and a thin sill complex (Reibel and Reuber, 1982).

The Lower Himalayas

The Lower Himalayas consist of a 20 km thick thrust pile of Proterozoic gneisses, thick Palaeozoic and thin Mesozoic sediments, and some Cambrian granites. The Main Central Thrust has brought crystalline gneissic rocks southwards from the Higher Himalayas as major klippen, commonly in synclines on the sediments; these are the 'Outer Crystallines', separating older sediments to the south from younger sediments to the north (Figs. 21.2 and 21.3).

The klippe of the Kathmandu Complex in Nepal contains two-mica tourmaline granites with high initial strontium isotope ratios and Rb–Sr ages of 26–22 Ma and which resemble the Manaslu-type granites of the Higher Himalayas with which they were once probably connected (Andrieux *et al.*, 1977).

In the Almora thrust slab SE of Kumaun, granite gneisses and metarhyolites have a Rb/Sr isochron of 1905 Ma using the Rb decay constant of 1.42×10^{-11} yr^{-1} (Frank *et al.*, 1977) and biotite gneisses have a Rb/Sr isochron of 1620 Ma with an initial $^{87}Sr/^{86}Sr$ ratio of 0.749 ± 0.0007 (Powell *et al.*, 1979).

In the Chail Nappe of Himachal Pradesh

(Virdi, 1981b) low-grade slates, quartzites, talc schists and phyllites are underlain by mylonitic gneisses which have a Rb–Sr whole-rock isochron age of 1430 ± 150 Ma, and they are intruded by large granitic-to-granodioritic plutons such as the Mandi granite which has Rb–Sr ages ranging from 456 ± 50 Ma to 545 ± 12 Ma. In the same nappe the Dhauladhar-Dalhousie batholith has a Rb–Sr age of 450 ± 50 Ma (Bhanot *et al.*, 1975) and consists of granodiorite, adamellite and biotite granite. These granites were commonly converted to gneisses by deformation associated with the greenschist-grade metamorphism of the whole Chail Nappe (Fuchs, 1981). The above relationships show that the Lower Himalayas Outer Crystalline thrust slabs contain evidence of Proterozoic protolith material and granite intrusion in the Cambrian and Miocene, in common with the Central Crystallines of the Higher Himalayas.

Inverted metamorphic zones below the Main Central Thrust (MCT) occur along the whole of the southern front of the Himalayas (Andrieux *et al.*, 1980; Lal *et al.*, 1981; Sinha-Roy, 1981, 1982; Arita, 1983). In places, above the MCT, there is an upward decrease in metamorphic grade, but elsewhere above the MCT there is an upward increase to a higher temperature zone, mostly sillimanite, followed upwards by a decrease in grade—the so-called divergent isograds (Thakur, 1977). Both problems are related in a broad way to the superimposition of the hot thrust slab of the Central Crystallines of the Higher Himalayas (and the Outer Crystallines of the Lower Himalayas) on the low-grade meta-sediments of the Lower Himalayas.

Of the many models to account for this reversed metamorphism, one of the most popular is that of Le Fort (1975), according to which the thrust speed exceeded that of thermal equilibration and so palaeo-isotherms were overturned into an S-form across the MCT. The folded kyanite–sillimanite boundary in Central Nepal was mapped by Bouchez and Pêcher (1981) and defines a divergent isograd pattern across a central migmatized domain in the Central Crystal-lines. The chlorite, biotite, garnet and kyanite isograds indicate increasing metamorphic grade upwards towards the MCT, which in this part of the Lower Himalayas is a 10 km wide shear zone. They proposed that the total thrust movement was accomplished in 2–3 Ma, corresponding to an average rate of 3–5 cm yr^{-1}. They also pointed out that at shallow depths, as in western Nepal, the isotherms are more closely spaced and the MCT is a shear zone and splits into discontinuities charac-teristic of deformation at lower temperatures.

There is some doubt as to whether the inverted isograds are of the same type and origin along the whole Himalayan front. Andrieux *et al*. (1980, 1981) suggested that in the NW Himalayas of India the inversion

does not result from the synkinematic over-turning of the isograds by the MCT, as in Nepal (Pêcher, 1979), but is due to later thrusting of high-grade over low-grade rocks. The overthrusting was also responsible for pronounced retrogression of high-grade assemblages (Andrieux *et al*., 1981). In many areas it is difficult to define the position of the MCT, particularly where it is close to the kyanite isograd with inverted zones below and above it in the Lower and Higher Himalayas, respectively (Andrieux *et al*., 1981). Sinha-Roy (1982) pointed out that this confusion could be avoided if the MCT were redefined to occur at the base of the nappe sequences, so that the inverted metamorphic rocks lie only above it, from the chlorite up to the sillimanite zones.

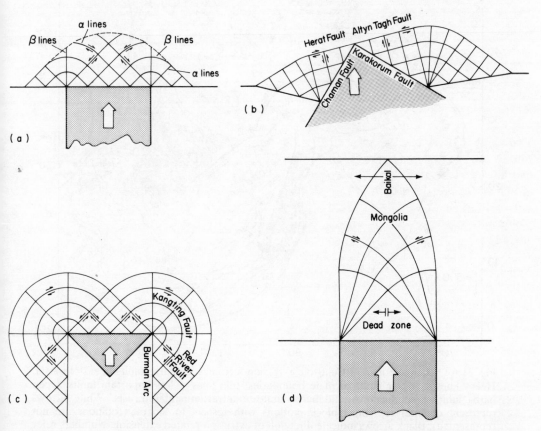

Fig. 21.5(a–d)   Plane indentation of a plastic medium by rigid dies of different shapes. Arrows indicate sense of shear along slip lines. Names of correlated faults are indicated (reprinted by permission from Tapponnier and Molnar, 1976, *Nature*, **264**: (a), (b), (c) Fig. 2, p. 321; (d) Fig. 4b, p. 322. Copyright © 1976 Macmillan Journals Ltd)

The Sub-Himalayas

The Siwalik rocks are unmetamorphosed arkoses, siltstones, shales and conglomerates. About 7 km of this molasse was deposited in the last 7 Ma.

Indentation Tectonics

It has long been a puzzle why there are so many earthquakes in central China, since it is not obvious that there is an active plate boundary in that region. Tapponnier and

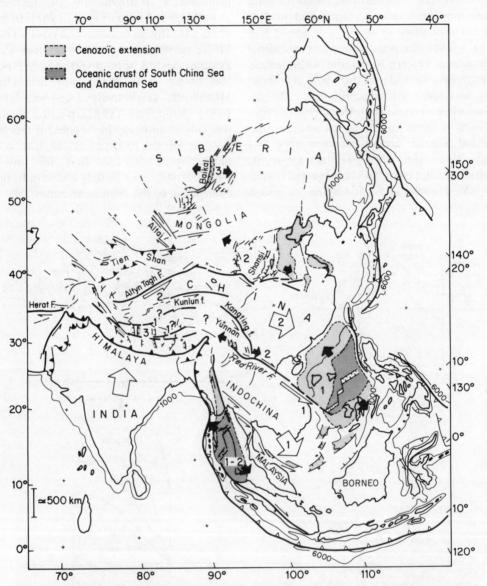

Fig. 21.6 Schematic map of Cenozoic extrusion tectonics and large faults in eastern Asia. Heavy lines = major faults or plate boundaries; thin lines = less important faults. Open barbs indicate subduction; solid barbs indicate intracontinental thrusts. White arrows represent qualitatively major block motions with respect to Siberia (rotations are not represented). Black arrows indicate direction of extrusion-related extension. Numbers refer to extrusion phases: 1 ≃ 50 to 20 Ma. B.P.; 2 ≃ 20 to 0 Ma. B.P.; 3 = most recent and future. Arrows on faults in western Malaysia, Gulf of Thailand and southwestern China Sea (earliest extrusion phase) do not correspond to present-day motions (after Tapponnier *et al.*, 1982, *Geology*, **10**, Fig. 1, p. 612, reproduced with permission)

Molnar (1976) made a major advancement in the understanding of intra-continental tectonics by proposing an analogy between the tectonics of Asia and deformation in a rigidly indented solid. According to this model, India behaves like a rigid die that indents northwards into Asia causing deformation over a large area that resembles the plane horizontal strain patterns calculated by mechanical engineers for the slip-line field theory. During indentation two sets of slip lines (α, β) are set up in the plastic indented medium (Fig. 21.5). These lines are maximum shear stress trajectories and correspond to linear and curved strike-slip faults in Asia. The pattern of the slip lines depends strongly on the shape of the indenter—Fig. 21.5 illustrates several possible patterns. In particular, the linear Herat and Altyn Tagh faults are similar in sense and trend to two symmetrical slip lines caused by wedge-shaped indentation, and the more curved Kun Lun and Kang-Ting faults to curved lines near the edge of a more planar indenter (Fig. 21.6).

Tension is also predicted by the theory in two regions in front of a flat indenter (Fig. 21.5). Firstly, at a high distance from the indenter, the Baikal Rift System and the Shansi Graben System (Fig. 21.6) may be examples of such structures. Secondly, immediately in front of the indenter there is a dead zone between the two sets of slip lines

and here a zone of tension exists that may be expressed by rifts oriented at a high angle to the indenter; the many young rifts in southern Tibet (Fig. 21.7) clearly formed by east–west extension and corroborate this model (Tapponnier et al., 1981b).

There is a final fascinating aspect to this story. Tapponnier et al. (1982) performed plane indentation experiments on unilaterally confined blocks of plasticine, i.e. with a free boundary on the 'eastern' side, which would be equivalent to the situation in China where a less rigid oceanic plate lies to the east in contrast to more rigid continental material in Eurasia to the west. The experiments suggested that the penetration of India into Asia has rotated (25°) and extruded (800 km) Indochina to the south-east, along the left-lateral Red River fault (Fig. 21.6) in the first 20–30 Ma of the collision. This accounts for the opening of the South China Sea before late Miocene time. Extrusion tectonics then migrated north, activating the Altyn Tagh fault as a second major left-lateral fault and moving South China hundreds of kilometres to the east. As this occurred, Indochina kept rotating clockwise (as much as 40°), but the sense of motion then reversed on the Red River fault. These extrusion tectonics may also account for the opening of the Andaman Sea up to the present. An important lesson provided by these experiments is that there

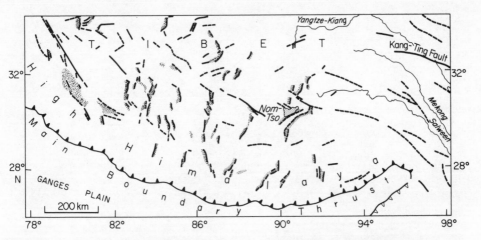

Fig. 21.7   N–S aligned rifts (stippled) in southern Tibet (modified after Tapponnier et al., 1981b, Nature, **294**, Fig. 1, p. 411, reproduced with permission)

may be a genetic relationship between continent–continent collision and indentation tectonics on the one hand, and the opening of marginal oceanic basins on the other. Thus the Wilson Cycle turns full circle.

Since Tapponnier and Molnar (1976), indentation fracture patterns have been recognized in the foreland of the Alpine belt of NW Europe (Sengör *et al*., 1978), of the Pan-African belts of Saudi Arabia (Fleck *et al*. in Al-Shanti, 1979, vol. 3) and the Hoggar (Ball, 1980) and of the early Proterozoic Wopman orogen in NW Canada (Hoffman, 1980).

Chapter 22

The Evolving Continents

In the previous chapters we have looked at the main features of Archaean, Proterozoic and Phanerozoic continental regions. From this survey two major points emerge.

1. Whether one considers Archaean greenstone belts, early Proterozoic dyke swarms, mid-Proterozoic anorthosites, rapakivi granites and alkaline complexes, late Proterozoic tillites or early–late Phanerozoic continental margins, the individual rock suites concerned and their associated structures on any one continent are essentially similar to those on another, although they may be diachronous. In fact the similarities are so striking that one need not hesitate to conclude that continental growth went through comparable stages in different places which means, in turn, that it is justifiable to talk in terms of a sequence of stages in continental evolution. What form these evolutionary stages took, viz. cyclic, unidirectional, repetitious, is widely debated.
2. Many rock groups are characteristic of particular periods of earth evolution, e.g. greenstone belts in the Archaean, massif anorthosites in the Proterozoic, granulites/charnockites and banded iron formations in the Precambrian. Conversely, red beds, tillites, massif anorthosites, dolomites, glaucophane-bearing rocks, eclogites, alkaline complexes, kimberlites and lead deposits are rare in the Archaean. Yet there are marked similarities between Archaean (greenstone belts) and Mesozoic–Cenozoic gold mineralization, volcanic rock units and chemistry and sedimentary facies. Having said this, the question arises as to the meaning of these occurrences in terms of continental evolution.

In this final chapter we shall bring together many of the features and relationships previously described in order to see what role they played in the evolution of the continents. We shall do this by considering the evolution of the sedimentary, magmatic, metamorphic and metallogenic record, and, finally, by reviewing some major factors that contributed to crustal evolution.

The Atmosphere and Hydrosphere

In earlier chapters we reviewed a great many sedimentary rocks but did not consider possible factors controlling their development, such as the evolution of the earth's atmosphere and oceans from the Archaean to the present.

A long-appreciated aspect of the terrestrial atmosphere is that the noble gases are depleted by factors of between 10^{-7} and 10^{-11} compared with their cosmic abundances. This led to the inference that the earliest atmospheric constituents were lost from the earth when it accreted and that the present atmosphere is of secondary degassing origin.

Degassing Models

There are two types of degassing model:

1. Catastrophic—the atmosphere was created instantaneously during a single major 'burp' event related to the process of core formation or impact melting.
2. Continuous—the atmosphere evolved in a continuous degassing process throughout geological time.

The Terrestrial Oxygen Budget

It is widely held that most of the gases of the secondary atmosphere were derived from volcanic exhalations, largely during the Archaean. Here one has to make an assumption: if the early degassing products were similar to the exhalations of modern volcanoes, water vapour and CO_2 would have made up the bulk of the volatiles, followed by H_2S, CO, H_2, N_2, CH_4, NH_3, HF, HCl, Ar, etc. However, there is negligible molecular oxygen in modern volcanic exhalations, and therefore it is commonly argued that the secondary atmosphere must have been anoxygenic in the Archaean, the free oxygen that we see today having evolved subsequently.

Because degassing could not supply free molecular oxygen to the atmosphere–ocean system, the oxygen must have been derived from dissociation of oxides like CO_2 or H_2O, solar radiation providing the energy source for two possible photochemical reactions, inorganic (photodissociation) and organic (photosynthesis) (Berkner and Marshall, 1967; Schidlowski, 1980).

Inorganic Photodissociation The ultraviolet-rich short wavelengths between 1500 and 2100 Å of the solar spectrum provide the energy to dissociate water vapour in the upper atmosphere, the hydrogen escaping preferentially from the earth's gravity field leaving the atmosphere enriched in free oxygen. However, the oxygen-producing capacity of this process is considerably smaller (about 10^{-3}) than that of photosynthesis and it is therefore this second process that must have produced the bulk of the oxygen now present in the atmosphere (Schidlowski 1980).

Organic Photosynthesis The low energy spectral range of visible light provided the energy for primitive organisms, particularly blue–green algae, to produce carbohydrates by photosynthesis from water and carbon dioxide, releasing oxygen as a by product:

$$CO_2 + H_2O \rightarrow CH_2O + O_2$$

The reaction is carried out expediently by green plants and certain blue–green algae (cyanophytes). There is reason to believe that this form of photosynthesis had commenced by the beginning of the geological record, some 3700–3800 Ma ago, particularly because of the widespread appearance in 3500 Ma sediments of stromatolites (Walter *et al* 1980; Orpen and Wilson, 1981), resulting from the matting of prokaryotic cyanobacteria. Also, there is a continuous record of reduced (organic) carbon (or kerogen) as CH_2O in sediments back to the 3800 Ma old Isua sediments. In fact, Archaean sediments have an 'astoundingly modern organic content', which is not surprising in view of the fact that contemporaneous prokaryotic systems sustained rates of production comparable to those of some modern agricultural crops. The storage in the oldest sediments of a reservoir of organic carbon of the order of 10^{22} g would imply the existence of an equivalent reservoir of photosynthetic oxidation products at that time. Schidlowski (in press) concludes that a formidable proportion, if not the bulk, of this reservoir consisted of molecular oxygen.

So we now have to ask the question: Does the early geological record provide evidence of substantial oxidative conditions, and if not, why not? There is interesting controversy currently surrounding this question, with two contrasting standpoints:

1. Cloud (1976b) suggested that the answer lay in early Precambrian banded iron formations. The basis of Cloud's argument is that free O_2 is a poison to organisms in the absence of oxygen-mediating enzymes and therefore the first ones to produce it were only able to survive by having an external oxygen acceptor. The Archaean iron formations may have acted as this acceptor. Ferrous iron was washed away from the continents as a result of an anoxygenic weathering cycle; the iron could only be transported in the ferrous state, ferrous salts being more soluble than ferric salts, and so the lack of oxygen in the seawater allowed extensive transport of the iron. However, the ferrous iron required the addition of oxygen in order to be precipi-

tated as ferric oxides or hydroxides, and it was the early organisms that provided this oxygen. The Archaean oceans were finally cleared of the large amount of dissolved ferrous iron when it was precipitated in the ferric state (in particular in the haematite–Fe_2O_3 facies) in the last and major period of banded iron formations about 2000 ± 200 Ma ago (Goldich, 1973). A decrease in the availability of the iron formations inevitably resulted in a rise in atmospheric oxygen which enabled the first continental red beds to form. Such red beds thus herald the start of an oxygenic weathering cycle on the continents and of the incipient build-up of an atmospheric oxygen reservoir. Thus the interval 2000 ± 200 Ma has come to be widely accepted as the most probable transition period from an anoxic to an oxygenic atmosphere (Margulis et al., 1976; Schidlowski, 1980).

Geological evidence can be cited in favour of an Archaean anoxygenic (but not necessarily reducing) atmosphere:

(a) There is a widespread weathered zone on the Archaean erosion surface beneath the basal unconformity of the 2300 Ma old Lower Huronian Supergroup in Canada (Frarey and Roscoe, 1970). The important feature here is that compared with modern soil profiles iron has been lost rather than accumulated and the ferric–ferrous ratio has been relatively decreased. These changes are taken to indicate that the ground water and atmosphere lacked free oxygen at the time.

(b) In the Lower Huronian sediments there are 'drab beds' that lack reddish coloration (Frarey and Roscoe, 1970). Comparable modern clastic sandstones and siltstones with a low haematite content and ferric/ferrous iron oxide ratio are known to form in an anoxygenic environment below the water table.

(c) The occurrence of detrital pyrite and uraninite in early Precambrian conglomerates, e.g. in the Witwatersrand and Transvaal Systems of South Africa, the Huronian Supergroup in Canada, the Bababudan greenstone belts in India and the Jacobina Series in Brazil. At present, these minerals do not survive weathering processes because they are readily oxidized (Schidlowski, 1980).

2. There is increasing geological evidence of the existence of oxygenic atmospheric and shallow-water hydrospheric conditions in the Archaean and early Proterozoic (Clemmey and Badham, 1982).

(a) Kimberley et al. (in press) report the discovery of a palaeosol on a palaeohill at the base of the early Proterozoic Elliot Lake Group in S Canada. This is an example of topographic control of weathering under oxidative conditions and it is interesting that the famous uraniferous conglomerates of the region lie only a few metres above the palaeosol.

(b) The pigment of red beds forms during oxidative diagenesis and there are many occurrences of red beds older than 2000 ± 200 Ma, e.g. the Loskop Formation of the Transvaal Supergroup (2150 Ma), the Huronian Supergroup in Canada (2350 Ma), Dharwar greenstone belts in India (2600 Ma), the Wawa greenstone belt in Canada (80 m thick red beds (2800 Ma); Shegelski, 1980), the Yengra Series of the Aldan Shield (3300 Ma; Kulish, 1979), the Fig Tree Group in the Barberton Greenstone belt (3450 Ma; Dimroth and Kimberley, 1976; M. Muir, personal communication).

(c) The iron of volcanic glass of pillow-bearing basalts in the 2800 Ma old Abitibi greenstone belt in Canada has been partially oxidized to goethite and haematite. The formation of such ferric oxide crusts, like those on modern pillow basalts, is clear evidence of late Archaean submarine diagenetic oxidation (Dimroth and Lichtblau, 1978).

(d) According to the Cloud (1968b) model, sedimentary sulphates should not have formed before 800 Ma ago. However, Schidlowski (in press) points out that because it is safe to assume that the anoxygenic forms of bacterial photosynthesis preceded the O_2-releasing photosynthesis reaction in the evolution of carbon fixation, oxidation products other than free oxygen, notably sulphate, might have been more abundant in the Archaean than in later times. This is interesting because in recent years there have been discoveries of bedded sulphate deposits in the 2750 Ma Yilgarn block of SW Australia (Golding and Walter, 1979), the 3450 Ma Barberton greenstone belt in South Africa (Reiner, 1980), the 3500 Ma Pilbara block of N Australia (Lambert *et al.*, 1978; Barley *et al.*, 1979), the pre-3000 Ma Sargur schists of S India (Radhakrishna and Vasudev, 1977) and the 3300 Ma Yengra Series of the Aldan Shield of the USSR (Kulish, 1979). The chemical processes probably involved the oxidation of sulphur by fixation of photosynthetic oxygen (Maisonneuve, 1982).

(e) There is a marked similarity in size and compositional range of detrital thorian uraninite (U : Th = 10 : 1) from the late Archaean Dominion Reef and Witwatersrand Supergroups and from the modern alluvium in the Indus and Hunza rivers in N Pakistan (Simpson and Bowles, 1977). There is evidence that the thorium content of uraninite increases its stability as a detrital mineral. The Indus–Hunza alluvium also contains appreciable detrital unoxidized pyrite (Tahirkheli, 1974). Simpson and Bowles (1981) suggested that the preservation of uraninite and pyrite would be favoured by catastrophic burial of the sediments. This suggestion is corroborated by the fact that the Indus and Hunza rivers drain off the Nanga Parbat massif which has been uplifted and eroded at nearly 1 cm yr^{-1} for the last 0.5 Ma (Zeitler *et al.*, 1982b), which is about 10 times the average Himalayan rate. Similar conditions could be expected in early Proterozoic river systems (Grandstaff, 1980) for two reasons: firstly, the very rapid rate of crustal growth and thickening in the late Archaean would have led to considerable uplift and emergence in the early Proterozoic, which gave rise to the very thick clastic sequences in contemporaneous sedimentary basins (Windley, 1977); Secondly, the late Archaean–early Proterozoic

was a period when large, relatively stable continental plates were assembled (as outlined in this book). Present-day average continental heights are closely related to continental areas (Harrison *et al.*, 1981), and thus in the early Proterozoic high relief would be accompanied by high rates of erosion, transport and deposition and catastrophic burial of detrital uraninite and pyrite without oxidation under extreme thicknesses of sediment.

The above considerations suggest that there were more oxidative conditions in the Archaean and early Proterozoic than has previously been realized. Most models for the evolution of the atmosphere have taken insufficient account of the tectonic environments of the early–mid Precambrian, which must have played a major role in controlling the development of, for example, shelf platforms, topographic relief and drainage patterns which, in turn, controlled the types of deposited sediments. As Kimberley and Dimroth (1976) remark, the lack of Archaean red beds is hardly surprising given the relative lack of preserved shelf facies where subaerial oxidation and shallow-water chemical precipitation could take place.

The Carbon Dioxide Balance

The input of CO_2 into the oceans and atmosphere, and its abundance relative to other constituents, have undoubtedly varied with geological time and this variation should be expressed in the composition and quantity of a variety of sediments, in particular the carbonates.

The appearance of biogenic oxygen in the early Archaean accelerated the oxidation of the juvenile CH_4 and CO, thus increasing the content of CO_2 in the atmosphere and its dissolution in the oceans, and an increase in the partial CO_2 pressure during the periods of major Archaean volcanism intensified weathering on land and made it possible for marine waters to carry larger amounts of carbonates in solution (Ronov, 1968). H. L. James

(1966) pointed out that, if the partial pressure of CO_2 in the mid-Precambrian atmosphere was one hundred times greater than that in the present atmosphere (i.e. 0.03 atm compared with 0.0003 atm), then the equilibrium pH of surface waters would be changed from the present weakly alkaline (8.17) to weakly acid (6.1). If, as would be expected, the total volume of water on the earth's surface was less than at present, the water would have been even more acidic. If such conditions did prevail, then surface waters would have had a much greater capacity for leaching iron and transporting it in solution than they have at present. Thus large quantities could have been moved in surface solution without accompanying detritus, such as Al_2O_3; Precambrian iron formations are characterized by an exceedingly low alumina content (Stanton, 1972).

The partial pressure of carbon dioxide in the Archaean atmosphere may have been many hundred times greater than at present (Kulish, 1979; Neruchev, 1979; Walker, 1983). Deposits of calcium carbonate are present in sedimentary rocks of all ages, back to the oldest yet discovered, and from this fact Walker (1976) and Sidorenko and Borshchevskiy (1979) concluded that the ocean has always been close to saturation with respect to calcium carbonate, as it is today. The enhanced CO_2 concentrations in the atmosphere would have created a greenhouse effect and the higher global temperatures could have been the sole mechanism for preventing glaciation on the Archaean earth at a time when there was reduced solar luminosity (Owen et al., 1979).

Isotope Chemistry of Sedimentary Rocks as a Function of Time

Oxygen, carbon and strontium isotope variations of sedimentary carbonate rocks can be used to determine the corresponding isotopic compositions of sea water. These measurements are limited to largely unaltered or unmetamorphosed carbonate rocks, but such rocks are well preserved as far back as the early–mid-Proterozoic and the Archaean; the

marbles in Archaean high-grade regions are unsuitable.

There is a notable increase in $\delta\,O^{18}$ of carbonate sediments throughout geological history; a likely explanation for this may be the rising proportion of dolomites in carbonate sequences with increasing age since dolomites are less susceptible to post-depositional equilibration with meteoric waters in comparison with limestones, and they should more easily retain their original heavy $\delta\,O^{18}$ (Veizer and Hoefs, 1976).

There is general agreement that C^{13}/C^{12} ratios of carbonate rocks do not show any definite age trend. Schidlowski et al. (1975) ascribed this constancy to the fact that total sedimentary carbon was almost always partitioned between organic carbon and carbonate carbon in the ratio of 20 : 80, as at present.

The trend of the isotopic variations of Sr during earth history is shown in Fig. 22.1. The lowest measured strontium isotope ratios in the sedimentary carbonates are taken as the best approximation for $^{87}Sr/^{86}Sr$ of coeval well-mixed seawater (Veizer and Compston, 1976). The sharp increase in values in the period 2500–2000 Ma ago and the decrease in the Phanerozoic parallel the K_2O/Na_2O trend in sedimentary and igneous rocks established by Engel et al. (1974); these trends probably define major fractionation stages of the earth's crust.

Fig. 22.2 shows the variations in the isotopic composition of sulphur (in evaporites), carbon (organic) and strontium (in limestones) throughout the Phanerozoic (Hoefs, 1981) compared with the global sea level curve of Vail et al. (1977). Global highstands of sea level throughout much of the Palaeozoic and Mesozoic are characterized by widespread shallow seas on continental shelves, whereas global lowstands, as in Permo–Triassic time during the formation of Pangaea, are characterized by reduction of the area of shallow seas and an increased rate of erosion. Highest sea level occurs at periods of maximum sea-floor spreading when mid-oceanic ridges have their greatest volume and elevation; such relations can explain the

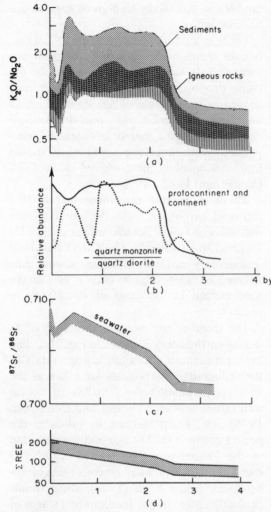

Fig. 22.1 (a) Variations in K_2O/Na_2O ratios of sedimentary and igneous rocks, relative proportion of continental crust and protocrust, and (b) quartz monzonite/quartz diorite ratio during geological history (after Engel *et al.*, 1974), and (c) $^{87}Sr/^{86}Sr$ of seawater with time (after Veizer and Compston, 1976; figure from Veizer, 1976a; reproduced by permission of J. Wiley). (d) The variation in REE with time in sedimentary rocks (after Taylor, 1979, Fig. 7d, in M. W. McElhinney (Ed.), *The Earth, its Origins, Structure and Evolution*; reproduced by permission of Academic Press, London)

isotopic variations. Periods of extended weathering introduce additional light continental sulphur into the ocean, which decreases the $\delta^{34}S$ of ocean sulphate (Brass, 1976). The carbon isotope composition would have been affected by an increase in

photosynthesis from Devonian towards Carboniferous time due to the appearance and increased productivity of land plants. The strontium isotope compositions are related to the increased mantle strontium, in the Ordovician–Devonian and Jurassic onwards, which is released by oceanic volcanism at ridge systems. The post-Cretaceous increase in $^{87}Sr/^{86}Sr$ may be explained by an increase in land area caused by continental plate collisions.

Atmospheric Evolution and the Development of Life Forms

There has been a close interdependence of atmospheric and biospheric activity throughout geological time. On the one hand, most of the free oxygen in the atmosphere resulted from biological activity through the photosynthesis reaction and, on the other, changes in atmospheric composition, in particular the progressive increase in the molecular oxygen content, triggered off major biological innovations which enabled life to advance and diversify (J. W. Schopf, 1974, 1975; Schidlowski, 1980).

The following milestones in biological evolution related to atmospheric conditions can be recorded, albeit tentatively (see Fig. 22.3). (For further aspects of Archaean and Proterozoic life forms see Chapters 3 and 8, respectively.)

1. In the absence of oxygen-mediating enzymes free O_2 is a poison to living cells. In the early Archaean free oxygen was produced by photosynthetic reaction in the first organisms and the banded iron formations conveniently acted as the oxygen acceptor, which they themselves needed for their formation (Cloud, 1968a). By keeping the oxygen levels down to a safe minimum the BIF gave the early life forms time to adapt to their oxygeneous waste product. These first (blue–green algae) organisms were procaryotes which were relatively resistant to ultraviolet radiation as the atmosphere in the early Archaean lacked a radiation-protective ozone screen. In the early to mid-

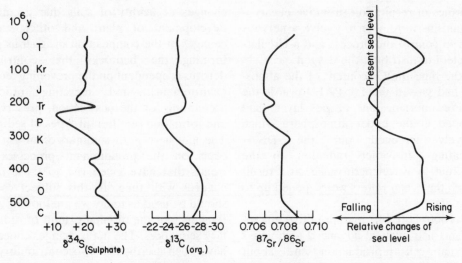

Fig. 22.2 Age curves of δ^{34}S, δ^{13}C and ^{87}Sr/^{86}Sr ratio together with the sea level curve of Vail *et al.* (1977) (from Hoefs, 1981, in R. J. O'Connell and W. S. Fyfe (Eds), *Evolution of the Earth*, Geodyn. Ser., vol. 5, Amer. Geophys. Un., Fig. 5, p. 116. Copyright by The American Geophysical Union)

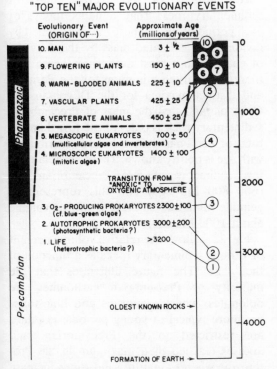

Fig. 22.3 The temporal distribution of important innovations that have occurred during the course of biological evolution (modified by Schopf after Schopf, 1974; reproduced by permission of D. Reidel Publishing Co.)

Precambrian oxygen levels were less than 10^{-2} to 10^{-6} PAL (present atmospheric level) (Grandstaff, 1980).

2. By 1300 Ma ago the oxygen content of the atmosphere had increased to such an extent that a primitive form of oxidative metabolism was possible; the first oxygen employing eucaryotes appeared at this time (Schopf, 1974). These were complex organisms with a nucleus capable of cell division enabling the genetic coding material (DNA) to be passed on to their descendants (Cloud, 1968a).

Louis Pasteur discovered that many modern microbes undergo a fundamental change in their type of metabolism from fermentation to respiration when the oxygen content reaches 1% of the present atmospheric level (PAL)—known as the Pasteur level. Berkner and Marshall (1967) identified the Precambrian–Cambrian boundary with an oxygen level of 1% PAL, but this Pasteur level must have been reached by 1300 Ma ago, the age of the oldest eucaryotes (Cloud, 1976a).

The rise of the oxygen pressure of the ancient atmosphere to 1% PAL enabled life to pass the Pasteur level and change from fermentation, an anaerobic process, to respiration, a highly advanced form of aerobic metabolism (Schidlowski, 1980). With the appearance of respiration primitive eucaryotic organisms were able to evolve a nervous system to control the process and a circulatory system to distribute the oxygen.

By the time the O_2 content of the atmosphere had passed the 1% PAL towards the late Precambrian, an ozone layer had developed in the upper atmosphere which effectively screened out the DNA-inactivating ultraviolet radiation in the 2400–2600 Å wavelength range, as a result of which the ocean waters were opened up to pelagic life.

3. By 1000–900 Ma ago primitive microorganisms had evolved advanced techniques of sexual cell reproduction and about 700–680 Ma ago the first Metazoa appeared–the Ediacara fauna (for details see Chapter 8). These are complex multicellular organisms that require oxygen for their growth—$<10^{-1}$ PAL according to Grandstaff (1980). The earliest fauna were soft bodied, i.e. jelly-fish, worms and sponges, etc., and they probably contained collagen, the main structural protein in Metazoan tissues, which requires molecular oxygen for its synthesis, although collagen is particularly concerned with the formation of hard skeletons and shells, evidence for which appeared in the fossil record about 570 Ma ago at the Precambrian–Cambrian boundary.

4. According to the Cloud (1968a,b; 1976a,b) model further increase in the oxygen content enabled the land surface to be protected from DNA-damaging ultraviolet radiation and this enabled land plants to evolve. At 50% PAL land animals appeared. The oxygen content in the atmosphere reached the present level by the Cretaceous.

Sedimentation

There are appreciable variations in the amounts, types and compositions of sedimentary rocks formed at different times in earth history (Fig. 22.12). A great many factors may have contributed to these variations, such as changes in the types of tectonic environment, atmospheric and hydrospheric evolution, erosion rates dependent on changes of acidity of soils due in turn to development of plant and other species, changes in the composition of igneous rocks forming the bedrocks that contributed detritus (dependent on the prevailing tectonic environment), and recycling processes accounting for the destruction, preservation and formation of different types of sediment. The aim here is to summarize current concepts on the predominant processes and trends that have given rise to these major changes with time. In this subject caution should be used as today we are looking at the remains, not the original extent, of sedimentary sequences. The main changes concerned have been ascribed to two contributory factors: recycling and evolutionary trends. But first we must look at the volume and type of preserved sedimentary rocks.

Sedimentary Deposits in Relation to Age

An estimate of the average mass of sedimentary rocks that remains today for each period of the Phanerozoic is obtained by dividing the estimated mass by the duration of each Period; most authorities agree that there is a marked minimum for the Permian and distinct maxima for the Devonian and Cenozoic and that the mass of Precambrian sedimentary rock is only a little less than the total of the Phanerozoic but its distribution with age is poorly known. There are relatively few sediments of late Precambrian age (600–800 Ma ago) and this represents a genuine minimum as there is a high volume about 1000 Ma ago.

Fig. 22.4 shows the relative volume percent of various sedimentary rocks as a function of their age. The figure illustrates that the majority of Precambrian carbonates are dolomites, whilst evaporites and limestones are more typical of young periods, jaspilites are restricted to the Precambrian, and arkoses reached their maximum in the Proterozoic whereas mature sandstones increase in younger periods and there are more Archaean greywackes than later. During the Precambrian lutites (argillaceous rocks) were fairly constant in amount.

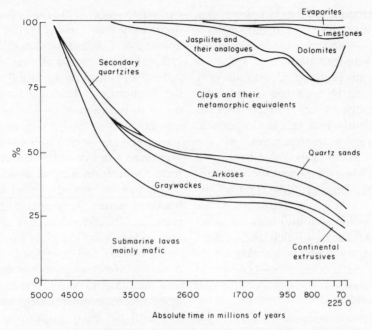

Fig. 22.4 Outline of the compositional evolution of sedimentary rocks (after Ronov, 1968; based on Ronov, 1964; reproduced by permission of Blackwell Scientific Publications Ltd.)

Recycling

The sedimentary recycling concept takes account of the fact that sediments have been eroded and redeposited (i.e. recycled) throughout time. The point is brought home when one considers that at the present rate of erosion half the currently exposed sediments in the world will be destroyed in about 100 million years. Some rocks such as lutites and volcanogenics are more resistant to recycling than, for example, limestones which may have been recycled at least five times in the last 3000 Ma. Thus the recycling of sediments has led to differences in the proportion of rock types originally deposited compared with those seen today (Veizer and Jansen, 1979).

Sedimentological Trends

Besides changes due to recycling there are several prominent long-term changes in the composition and relative abundance of sedimentary rocks:

Veizer (1982) summarized the main changes in sedimentary facies through the early–mid Precambrian:

1. Early- to mid-Archaean greenstone belts were built of mafic-to-felsic volcanogenic and volcaniclastic edifices with high relief, topped by chemical sediments such as cherts, Algoma-type iron formations, exhalative carbonates and barites (Lowe, 1980a). Similar rocks today occur near oceanic ridges and island arcs.

2. Late Archaean belts are characterized by a re-sedimented clastic assemblage of shallow-water turbidites deposited in submarine fans, and a non-marine facies of stratified conglomerates, sandstones and siltstones deposited on braided alluvial flood plains (Hyde, 1980).

3. These sediments pass laterally and vertically into cross-bedded sandstones and clast-supported conglomerates formed in coastal environments such as deltaic plains, tidal flats and marine shelves. Such sediments imply the existence of a stable hinterland and they represent a transition to the mature assemblages of the early Proterozoic.

4. Early Proterozoic sediments belong to a mature shelf assemblage of orthoconglomerate-quartzitic sandstone-shale-carbonate-Superior-type iron formation.

The Archaean–Proterozoic boundary is thus characterized by a change from immature to mature facies which resulted from the expanding epicontinental seas made possible by the progressive cratonization of the Archaean belts (as described in Chapter 4). The post-2000 Ma evolution of the sedimentary mass is typified by 65% cannibalistic recycling which therefore redispersed the more mafic (basaltic andesite) bulk compositions of the Archaean so that the present global sedimentary mass is imbalanced in favour of more mafic compositions compared with the more granodioritic upper continental crustal sources.

In Chapter 6 we considered the three types of supracrustal assemblages that are characteristic of the early to mid-Proterozoic (Condie, 1982a,d), viz. (1) the quartzite-carbonate-shale, (2) the bimodal-volcanic quartzite-arkose, and (3) the calc-alkaline volcanic (andesite)-greywacke. Condie (1982a) adds a fourth assemblage, the ophiolitic, typical of the late Proterozoic and the Phanerozoic. Table 22.1 shows the proposed distribution of these four supracrustal assemblages through geological time. Assemblage 3 formed in hot unstable rifts in Archaean greenstone belts, and continued to form in Proterozoic greenstone belts, after which time it formed in cool brittle rifts in back-arc basins. Assemblage 2 is found in cratonic rifts and aulacogens and so was not able to form until the early Proterozoic when stable cratons had developed. Assemblage 1 occurs in Archaean granulite–gneiss belts but was not able to survive the unstable permobile tectonic environment and thus it was not until the early Proterozoic that it survived, typically as a stable shelf assemblage on rifted continental margins.

Fig. 22.5 illustrates the percentage changes

Table 22.1 Summary of the distribution of supracrustal assemblages with geologic time from Condie, 1982a, reproduced by permission of Pergamon Press Ltd.

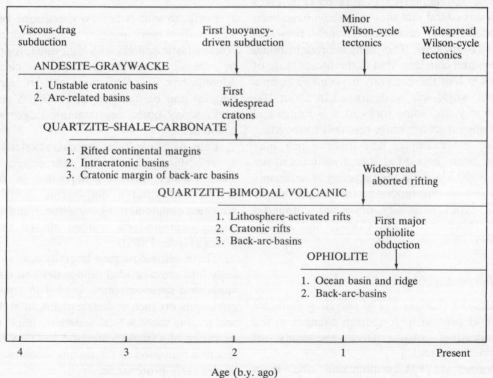

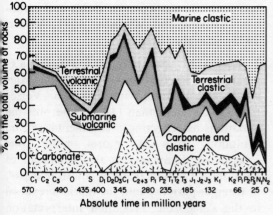

Fig. 22.5 Changes in the ratio of lithological associations during the Phanerozoic (from Ronov *et al.*, 1980, *Sed. Geol.*, **25**, Fig. 4, p. 319, reproduced by permission of Elsevier, Amsterdam)

in lithological associations throughout the Phanerozoic (Ronov *et al.*, 1980). Particularly marked are the mutual variations between the clastic and carbonate associations, the former almost substituted for the latter in the early Devonian, the late Permian–early Triassic and at the end of the Cenozoic. These periods correspond to the final stages of the Caledonian, Hercynian and Alpine–Himalayan orogeneses which were times of plate coalescence and consequent high land relief (Harrison *et al.*, 1981). The peaks of carbonate sedimentation coincide with periods of major transgressions, maximum sea-floor spreading activity and well-developed trailing continental margins. During the Phanerozoic there was a general

Table 22.2 Evolution of non-detrital sedimentary rocks in geological history (modified after Veizer, 1973; reproduced by permission of *Contrib. Mineral. Petrol.*)

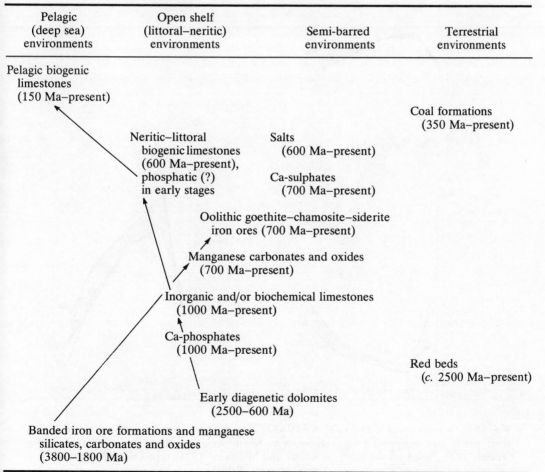

reduction in the sea-covered area within the continents, a marked increase in volume of sediments and in mean rates of subsidence. Table 22.2 depicts the evolution of major non-detrital sedimentary rocks throughout geological time. The depositional environment of carbonate sediments has spread with time from semi-barred to littoral sequences in the Precambrian into more open to pelagic types in the Phanerozoic.

Chemical Trends

There are several distinctive secular chemical variations in the composition of sedimentary rocks, in particular of shales and carbonates (Ronov and Migdisov, 1971; Veizer, 1973; Schwab, 1978).

1. K_2O/Al_2O_3, Na_2O/Al_2O_3 and K_2O/Na_2O variations in detrial rocks, especially shales which are illustrated in Fig. 22.6. Veizer (1973) and McLennan (1982) interpret the variations as reflecting a continental crustal evolution from a more mafic one (plagioclase-dominant) in the Archaean to more felsic (with increase in potash feldspar) in younger periods, until the marked reversal in the Palaeozoic. A similar interpretation may apply to the increase in K_2O/Al_2O_3 and

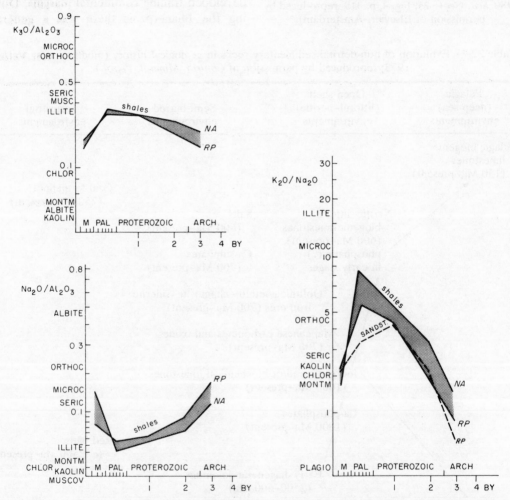

Fig. 22.6 Variations of K_2O/Al_2O_3, Na_2O/Al_2O_3 and K_2O/Na_2O ratios with age for shales and their partly metamorphosed equivalents. NA: North America, RP: Russian Platform (after Veizer, 1973, based on summary of Ronov and Migdisov, 1971; reproduced by permission of *Contr. Mineral. Petrol.*)

K/Rb ratios with age in carbonate rocks (Veizer and Garrett, 1978).

2. Increasing total Fe/Al_2O_3, MnO/Al_2O, $MnO/total\ Fe$, MgO/CaO, SiO_2/Al_2O_3 and Fe^{2+}/Fe^{3+} with age of the rocks. As the separation of Mn and Fe is caused, in particular, by the oxidation potential of Fe the trends may reflect the fact that the oxygen pressure was progressively increasing with time or that there was stronger volcanic degassing and more extensive submarine leaching of volcaniclastic sediments during early stages of the Earth's history (Veizer, 1978).

Another explanation for the decrease in the dolomite/calcite ratio with time reflects variations in the evolution of the ocean–atmosphere system, such as the input of CO_2 and Ca. As more CO_2 enters the oceans, so more Ca is deposited within $CaCO_3$; when more CO_2 is put into the system than Ca is released during weathering, as in the Archaean and Proterozoic, then Mg is also deposited as a carbonate phase in dolomites. It seems that the marked decrease in the dolomite/calcite ratio by the Phanerozoic was due to a decrease in the ratio of CO_2 supply to Mg and Ca demand, caused by a decrease in both the rate of juvenile degassing and of degassing of recycled CO_2 (Holland, 1976).

3. Engel et al. (1974) use the K_2O/Na_2O ratio of clastic sediments as an index of their increasing maturity, i.e. from the greywackes and subgreywackes in the Archaean to the argillaceous sandstones and shales characteristic of stable continental shelf assemblages in the Proterozoic. Fig. 22.1 shows a plot of this ratio increasing as a function of decreasing age, demonstrating that the sediments were progressively derived from more and more fractionated igneous rocks. This ratio, they suggest, may be used as a guide to thickness, composition and stability of the source region. There has been little change in the average major element chemistry of clastic sediments during post-Archaean time (McLennan, 1982).

Table 22.3 summarizes possible changes in chemical composition of sedimentary rocks by a variety of factors.

Magmatism

There are many who ascribe to the view that magmatic activity has been the major process contributing to the growth of the continents (Hamilton 1981; Wells, 1981). By 2500 Ma ago 85% of the present crustal mass had grown by the addition of mafic and calc-alkaline magmas at an average rate of 11.7 Pg yr^{-1}—1 Petagram $= 10^{15}$ g (Dewey and Windley, 1981). Alternative estimates are 65–75% (McLennan and Taylor, 1982) and nearly 100% (Armstrong, 1981). So we must consider, in particular, the contribution of this early Precambrian magmatism to the evolution of the continents.

There is an interesting current debate about two models for chemical evolution and

Table 22.3 Possible changes in the chemical composition of sedimentary rocks with increasing age caused by four factors (after Veizer, 1973; reproduced by permission of Contr. Mineral. Petrol.)

Process	Increasing with increasing age	Decreasing with increasing age
(a) Diagenetic and metamorphic effects	Al_2O_3, MgO, MgO/CaO, Fe^{2+}/Fe^{3+}, K_2O, K_2O/Na_2O, K_2O/Al_2O_3, Rb, Total Fe	CaO, Sr, Na_2O, Na_2O/Al_2O_3
(b) Evolution of sedimentary environments	MgO, MgO/CaO	CaO
(c) Changes in the chemical composition of atmosphere and hydrosphere	MnO, Total Fe, MnO/Total Fe, Fe^{2+}/Fe^{3+}, MgO, MgO/CaO, SiO_2	(CaO)
(d) Changes in chemical composition of the upper continental crust	Na_2O/K_2O, MgO/CaO, Fe^{2+}/Fe^{3+}, Al_2O_3/SiO_2, K/Rb, Sr, Mn, Ti, Mg, Fe, Al, Ca, P	K, Na, Si, Rb

magmatism of the continental crust:

1. From an analysis of Archaean sediments, Taylor (1977, 1979) concluded that the overall composition of the Archaean crust exposed to weathering was similar to that of average present-day island arc volcanic rocks as well as to that calculated for the present-day whole crust. It follows that if the bulk continental crust is andesitic, and if the present-day upper crust has a granodioritic composition, then the lower crust should have a mafic chemistry. This island arc model for continental growth thus provides a total crustal composition from which the upper granodioritic crust is derived by partial melting, leaving a lower depleted (mafic) crust.

2. Tarney *et al.* (1982) point out that tonalite is the dominant component of the Archaean continental crust in both low- and high-grade terrains, and Brown (1979) and Weaver and Tarney (1982) that calc-alkaline plutonism has been dominant over andesitic volcanism as the agent for crustal growth throughout geological time. According to these authors the chemistry of intra-oceanic island arcs is too basic to yield average crustal compositions, and thus the continental crust must have a composition more siliceous than andesite. Tarney and Windley (1979) produce a critique of the island arc model in reply to a critique by Taylor and McLennan of the tonalite model. Weaver and Tarney (1982) argue that Archaean granulites, which are our best representative of the lower crust, are not refractory mafic residues of intracrustal melting, but have a primary intermediate (tonalitic) composition, modified only slightly by granulite facies metamorphic processes. Fig. 22.7 shows that the high-level Archaean pluton tonalite from the Barberton greenstone belt has incompatible element abundances close to Tertiary Andean pluton tonalite, and that the lower crustal Lewisian tonalitic granulite is similar to these except for the low values in the heat-producing elements (Th, U, K, Rb) which were depleted during the granulite facies metamorphism. It is well known that tonalite is the dominant plutonic rock type in the Andean-type batholiths of North

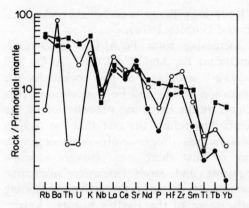

Fig. 22.7 Primordial mantle normalized trace elements in average Lewisian tonalitic granulite (○: Weaver and Tarney, 1980), Barberton 'ancient' tonalite (●: Glikson, 1976) and Tertiary Andean tonalite (■: Lopez-Escobar *et al.*, 1979) (modified with permission after Tarney *et al.*, 1982, *Rev. Bras. Geoc.*, **12**, Fig. 4. p. 56)

and South America and therefore the growth of the continental crust more likely occurred by vertical accretion of tonalitic plutons at some form of active continental margin rather than by andesitic volcanism and lateral accretion of island arcs. The differences between the Archaean and the modern Andean compositions probably reflect an increased component derived from hydrous partial melting of the subducted oceanic crust in the Archaean consistent with the higher heat flow (Weaver and Tarney, 1982). For more relevant details and ideas on this subject the reader is referred to Chapter 4.

Now let us consider Proterozoic plutonic magmatism which, compared with Archaean magmatism, was characterized by a dominance of high-potash minimum-melting granites over calc-alkaline silicic/intermediate igneous rocks, implying a dominance of crustal differentiation over crustal growth (Engel *et al.*, 1974; Dewey and Windley, 1981). This means that there has been a long-term secular trend over the entire history of the crust from more sodic to more potassic magmatism. For the Proterozoic this may be related to the following three processes.

1. In the mid-Proterozoic there were major

but abortive attempts to rift the continents (Chapter 7). Resultant magmatic products were characterized by a bimodal suite of anorthosites and alkaline-to-peralkaline granitic rocks with associated Sn mineralization. Geochemical studies suggest that a large part of the magmas was derived by partial melting of thickened continental crust (Simmons and Hanson, 1978).

2. Post-orogenic Andean-type high-K calc-alkaline complexes (E. Greenland, Wopmay Orogen, Wollaston Lake fold belt, Sveco-karelian belt: Chapter 6) were more likely derived from the volatile-fluxed mantle wedge rather than by direct melting of a downgoing slab (Tarney and Saunders, 1979). This may reflect the long-term declining importance of subducted oceanic lithosphere (in comparison with Archaean magmatism), and the increasing role of the overlying mantle wedge in magma generation (Brown, 1981).

3. The post-collisional stage of thrust-nappe stacking in Himalayan-type orogenic belts leads to partial melting of the tectonically thickened lower continental crust and the generation of S-type leucogranites and alkaline granites (Chapter 21). Examples are well documented in the early Proterozoic Wopmay Orogen in NW Canada (Hoffman, 1980) and the late Proterozoic Pan-African belts in NE Africa (Rogers *et al.*, 1978; Ries *et al.*, 1983).

None of the above three types of magma generation took place on any significant scale during the Archaean. They all reflect the thicker, larger and stronger plates of the Proterozoic, and thus they continued to form during the Phanerozoic where they can all be related to different stages of the Wilson Cycle.

Having so far considered plutonism, let us now turn our attention to the role of volcanism on earth through time (Burke and Kidd, 1980; Ronov *et al.*, 1980). An important group, virtually unique to the Archaean, is formed by the ultramafic komatiites (Nesbitt *et al.*, 1982). The extrusion temperature of the magmas was greater than 1600°C, or

some 300–350° above that of Phanerozoic magmas. Green (1981) concluded that such rocks were produced by processes similar to those operating in modern mid-oceanic ridges or back-arc basin spreading centres. Sleep and Windley (1980) used the difference in eruption temperature between present primitive mid-ocean ridge basalts and Archaean periodotitic komatiites to compute the differences between Archaean and present oceanic crustal thicknesses. The higher temperatures beneath Archaean ridges resulted in more partial melting which extended down to grea-

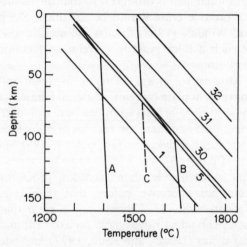

Fig. 22.8 A schematic diagram shows the total amount of melting in ascending mantle material as a function of depth and temperature. The slanting lines represent contours of the cumulative amount of melt produced. Dynamic melting occurs such that a small fraction of melt is retained in the source region. Rapid melting occurs between the 5% and 30% contours where the melting is considered to be nearly eutectic. Depleted residuum with a small fraction of retained melt continues to ascend above the 30% contour. The additional amount of melting is small. Curve A represents modern ridge basalts and curve B represents the komatiite studied by Arndt (1976, 1977). The depth interval over which melting increases from 5% to 30% is approximately the same for both curves. Komatiites, in small amounts, could also be produced from an adiabatically ascending residuum from previous melting events (curve C). The temperature below the source region is about 100°C lower for curve C than curve A (from Sleep and Windley, 1982, *J. Geol.*, **90**, Fig. 2, p. 366, copyright University of Chicago. Reproduced with permission)

ter depths (c. 100–70 km) than at present (Fig. 22.8), and the increased volume of magma produced therefore gave rise to an Archaean oceanic crust which was much thicker (> 20 km) than modern crust (c. 5 km). In Fig. 22.8 curve B represents material of the same composition as curve A but with a higher initial temperature. For B the bulk of the melting occurs below 70 km, before the lower melting phases are used up, and produces basalt which segregates to form oceanic crust. All material above 70 km melts by at least 30% to give 21 km of basalt, whilst between 70 and 100 km depth about 5 km of additional melt is produced so that the resulting oceanic crust should be 26 thick. Sleep and Windley (1982) discuss the implications of such a thick crust for Archaean tectonic evolution.

Modern volcanism occurs at divergent and convergent plate margins. Material created at these may be preserved in back-arc ophiolite complexes and there are records of ophiolites with sheeted dykes as far back as 1000 Ma ago (Chapter 9), which leads some to suppose that modern-style plate accretion processes were not operative before that date (e.g. Kröner, 1981a,b). However, it is clear that faster spreading ridges have no axial rift and less dykes (Dewey and Kidd, 1977) and one of the world's best ophiolites in Papua has no sheeted dykes, so we have to be flexible in our interpretation of the pre-1000 Ma oceanic record, in particular since there is much agreement that spreading rates in the early Precambrian were several times higher than those of today. Because the global heat flow is proportional to the square root of the global spreading rate, six times the current global spreading rate of 3 km^2 yr would be required if the heat flow were 2.5 times the present rate 2800 Ma ago (Bickle, 1978); thus normal development of sheeted dykes might not be expected in the Archaean. Nevertheless, De Wit et al. (1980) report the presence of a 1 km thick vertical sheeted dyke complex that feeds 2 km of flanking pillow lavas in the Onverwacht Group of the 3500 Ma old Barberton greenstone belt of South Africa. These are four well-known authorities

and a detailed report on this occurrence is awaited with much interest.

There are two extremes in current interpretations of the significance of Archaean ocean-floor material. First, greenstone belts are remnants of a globe-encircling volcanogenic ocean (Fryer et al., 1979). From one key standpoint this would seem unlikely because the same types of greenstone belt continued throughout the Proterozoic (see Chapter 10), and no one would suppose that these are remnants of a globe-encircling volcanogenic ocean. Second, greenstone belts of any age can best be interpreted as some form of marginal basin (Tarney and Windley, 1981). Some of the best geochemical work supports the idea of multiple back-arc basins, although above a thinner lithosphere and a shallower subduction zone than in modern environments (Drury, 1983); but as Drury points out, the marked similarities with modern basalts cannot justify a syllogistic conclusion that they occupied tectonic settings identical to those present today.

Thermal Regimes

Types of Metamorphism and Metamorphic Belts

Ignoring contact metamorphism, there are four factors about regional thermal regimes in space and time that have important implications for continental evolution.

1. Metamorphism occurs during the successive stages of the plate tectonic cycle in many different places within the plate framework, but they seem to be far more numerous and complex than is described in most papers and textbooks.

(a) Ocean-floor metamorphism at mid-oceanic ridges caused by hydrothermal circulation of hot seawater (Fyfe and Lonsdale, 1981), followed by low-temperature alteration caused by seawater–crust interaction (Honnorez, 1981). The effects of such metamorphism may be found in mid-oceanic ridges and in ophiolites.

(b) Contemporaneous paired metamorphic belts in island arcs (Miyashiro, 1981). A low-pressure, high-temperature type belt occurs in the depth of the volcanic arc and a high-pressure belt in the subduction zone. The latter consists of two types: low-temperature glaucophane-bearing blueschists and high-temperature garnet-granulites such as the Jijal Complex at the base of the Kohistan arc in Pakistan (Jan and Howie, 1981). The blueschists and granulites can be expected to occur close together on eventual collisional sutures, as in the Hercynian of NW Europe (Bard *et al.*, 1980a). Note, however, that the type area for paired metamorphic belts in Japan has recently been shown by Barber (1982) to comprise a group of allochthonous terrains, and thus the paired metamorphic belts concerned are not contemporaneous. Nevertheless, the contrasting thermal regimes in the arc and trench belts of island arcs are well known and should, in principle, give rise to the two types of metamorphic belts in question.

(c) In Andean-type continental margins the back-arc basin is easily susceptible to closure because of late compressional deformation at such margins. Folding and thrusting of the basin is associated with low-grade metamorphism in the 'rocas verdes' ophiolite complexes of southern Chile (Dalziel, 1981; De Wit and Stern, 1981). Low-pressure granulites were created during closure of the late Palaeozoic marginal basin of the Aracena mobile belt in S Spain (Bard, 1977).

(d) The metamorphic core complexes of the Cordilleran belt of N America developed in the back-arc area between the arc and the eastern thrust front (Chapter 19). In the décollement zone there is a late lineated cataclastic fabric with a steep metamorphic gradient that has been superimposed on earlier fabrics (Coney, 1979).

(e) During the accretion of 'suspect terrains' in Cordilleran-type orogenic belts, island arcs, for example, will be folded, thrusted and metamorphosed as they collide with the main continental mass in front of them, whilst an oceanic plate is sub-

ducted behind them. An example is the Kohistan arc in N Pakistan which was deformed and metamorphosed up to granulite grade when it collided with the Karakorum plate to the north before India had arrived (Coward *et al.*, in press).

(f) During the terminal continent–continent collision of Himalayan-type orogenic belts, the rocks at the front of the abutted belt will be deformed and metamorphosed. For example, the fore-arc basin sediments of the Xigaze Group in S Tibet are telescoped and cleaved (Tapponnier *et al.*, 1981a) and parts of the Kohistan arc in N. Pakistan were remetamorphosed (Coward *et al.*, in press) when the Indian plate collided. Where the incoming plate collides obliquely, strike–slip motion will take place along the suture, recrystallizing earlier fabrics.

(g) A result of post-collisional indentation may be intracontinental thrusting, nappe stacking and the production of inverted isograds, as on the foreland of the Indian plate (Chapter 21). Some thrusting may take place along earlier sutures, giving rise to inverted metamorphism as on the Northern Suture of the Karakorum Range in N Pakistan (Coward *et al.*, in press).

(h) Uplift of the various segments of Andean, Cordilleran and Himalayan belts will re-equilibrate earlier metamorphic assemblages during the creation of a PT trajectory that can be documented by geothermo-barometric methods. Each segment, suspect terrain and suture can be expected to have its own PT trajectory.

2. Certain types of metamorphic rocks seem to be time-restricted. For example, blueschists are only younger than late Precambrian, and granulite facies are most extensively developed in the Precambrian Shields, although granulites can form during every stage of the plate tectonic cycle (Windley, 1981). We shall discuss these two rock groups later.

3. There is a considerable difference in the depth of erosion of many Precambrian mobile belts and most Phanerozoic orogenic belts; this factor is related to the granulite problem.

4. It has been proposed that the geothermal gradients of various Phanerozoic orogenic belts were different, and that the high heat production in the early Precambrian gave rise to steeper geothermal gradients than today.

Phanerozoic Blueschists

The majority of glaucophane-bearing high-pressure belts are of Mesozoic–Cenozoic age having formed in island arcs, Cordilleran and Alpine belts related to plate subduction, obduction and collision since the break up of Pangaea (Ernst, 1971, 1972). There are some glaucophane rocks in Palaeozoic belts, especially in the early Japanese arcs, the Urals and Caledonides of Scotland and the Hercynian belt of NW Europe; those in the Caledonian Mona Complex of Wales and in Namibia are late Precambrian.

Ernst (1972) reviewed the occurrence and mineralogical evolution of blueschists with time. He pointed out that whilst the epidote-bearing greenschist–blueschist rocks occur throughout the Phanerozoic, lawsonite is rare in the Palaeozoic, and aragonite with jadeitic pyroxene and quartz are strictly confined to Mesozoic and Cenozoic terrains (Fig. 22.9). These index minerals indicate higher pressures in the formation of the younger blueschist rocks as indicated by experimental phase equilibrium studies.

Miyashiro (1981) suggested that blueschists may form on the upper surface of the descending slab, in the accretionary pile overlying the descending slab, and in both environments when a Benioff Zone shifts oceanwards. Thus successive stages of recrystallization in blueschists may result from successive shifts in the position of the subduction zone.

An interesting question concerns the absence of glaucophane in pre-1000 Ma rocks (Chapter 10), and this absence has been used as evidence against the operation of modern-type plate tectonics in these early times (Kröner, 1981a,b). It cannot be argued that the faster heat flow in the Archaean would have prevented the development of blueschists (Ernst, 1972) because it is most likely that the additional heat escaped by more rapid generation of oceanic crust rather than by high geothermal gradients in the continental plates (Bickle, 1978)—as Watson (1978) points out, the PT ranges of Archaean belts were little different from those of equivalent Phanerozoic belts. A major reason why there are no old blueschists is that on uplift they tend to increase in T before they decrease in T and P, and during this early T increase the glaucophane will be recrystallized upgrade to hornblende; thus blueschists are easily transformed to greenschists on uplift (England and Richardson, 1977). Another important fact is that a large number of present-day glaucophane schists occur in the orogenic

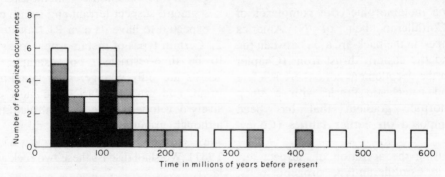

Fig. 22.9 Histogram showing incidence of blueschists of contrasting mineralogies with time. Open boxes represent epidote-bearing glaucophane schists, stippled pattern lawsonite (±epidote), and black boxes aragonite and/or jadeitic pyroxene and quartz (generally also lawsonite) (after Ernst, 1972; reproduced by permission of *American Journal of Science*)

belts surrounding the Pacific, but they cannot be compared with possible early Precambrian examples since these belts have not yet completed the continent–continent stage of the Wilson Cycle. If we look at the Himalayan belt there are only five localities of blueschists, each less than a few kilometres long, in a distance of over 3000 km, and with a few kilometres of uplift these will no doubt be eroded. So one must be careful about comparisons between the presence and absence of blueschists in modern and ancient belts.

Precambrian Granulites

Granulite facies rocks are found at the present surface level on every continent. Of course we would expect that the occurrence of granulites was once more widespread since many must have been obliterated by later retrograde metamorphism and orogenic activity and some are, no doubt, hidden below younger cover deposits.

Except for one small occurrence in West Greenland (Griffin *et al.*, 1980) there are no early Archaean granulites (>3600 Ma); this can be related to the fact that the early Archaean is widely regarded as a period of minor crustal growth.

In contrast, the main period of granulite formation was in the late Archaean (*c.* 3000–2600 Ma), which was the chief period of crustal growth in earth history (Moorbath, 1980). These are mostly medium- to high-pressure granulites, developed in the lower crust under moderate thermal gradients of 20–40°C km^{-1} and of great importance for the geochemical differentiation of the sialic crust. Wells (1981) proposed that they formed during prograde dehydration reactions during prograde PT conditions resulting from major magmatic crustal accretion and thickening mechanisms or from tectonic thickening, acting alone or with the magmatic thickening, provided that erosion was inhibited. A second environment for generation of granulites was the upper crust where temperatures were high (700–900°C) and thermal gradients were steep (*c.* 10°C km^{-1}); such low-pressure granulites formed by dehydration reactions and decompression melting in near-isothermal, falling-pressure regimes dominated by uplift and erosion (Wells, 1981). For further details on the characteristics and evolution of these granulites, see Chapter 2.

Chapter 19 gives details of the youngest (50 Ma) granulites in the world which occur in a Cordilleran continental margin with major calc-alkaline batholiths and large-scale thrust-nappe structures. They provide a modern analogue for the Archaean rocks but the granulites have only just begun to be exposed in the deep levels of this modern orogenic belt. Clearly one reason why we do not see many young granulites is because very few modern orogenic belts have been sufficiently uplifted and eroded to their lower crustal levels. A second example of such young granulites (*c.* 100 Ma old) is in the Kohistan arc of N Pakistan, described in Chapter 21, where again uplift has been very high.

Heat Production and Geothermal Gradients

There are close and interesting relationships between such factors as heat generation and loss, geothermal gradient, plate accretion rate, continental crustal volume and growth rate. What do the relevant data and thermal-tectonic models based on such interrelationships tell us about the early history of the earth?

All recent thermal history models indicate that radioactive heating does not balance heat flow, and that the earth has cooled by about 300°C over the last 3000 Ma (McKenzie and Weiss, 1980; Turcotte, 1980). Because the mantle was hotter, ultramafic komatiite magmas were produced which had eruption temperatures of *c.* 1630°C 2700 Ma ago and 1780°C 3300 Ma ago compared with 1400°C for present-day, primitive, mid-ocean ridge basalts (Sleep, 1979).

Hargraves (1978) suggested that the geothermal gradient beneath Archaean continents was much greater than at present because the large amounts of radioactive heat generation in the Archaean require high continental heat flow and a thin lithosphere.

However, this conclusion is unwarranted for the following reasons. Heat can escape by conduction and/or convection. If there were no convection and all heat escaped by conduction, this would have increased the geothermal gradient in the continental crust so much that in the lower crust the temperature would have been 2000°C (Bickle, 1978); this is obviously impossible because the lower crust would have melted if its temperature were raised by less than 50% to form granitic liquids, but the Archaean geological record shows a marked absence of granites. Burke and Kidd (1978) suggested therefore that the maximum temperature at the base of the Archaean crust was about 800°C. The geothermal gradient in the Archaean continental crust was apparently not dissimilar from that of today (Watson, 1978), and no evidence yet suggests that Archaean plate thicknesses were different from those at present (McKenzie *et al.*, 1980).

It is more likely that the additional heat available in the Archaean escaped by convection, and this means by more rapid generation and recycling of oceanic crust (Burke *et al.*, 1976a; Sleep, 1979). The present-day oceanic plate production rate is *c*. 3 km^2 yr^{-1}, and Bickle (1978) calculated from thermal constraints that plate production was at least 18 km^2 yr^{-1} 2800 Ma ago. This would have resulted in plates being subducted with a mean age of 20 Ma compared with 60 Ma at present. Sixty per cent of the heat loss from the earth today is used in oceanic plate creation and cooling (Sclater *et al.*, 1981). This is the only efficient means of using up heat, and it should have been so in the past in the absence of any other known means of heat loss. If there were oceanic plate creation in the Archaean, whatever the rate, then in a non-expanding earth there should have been a comparable amount of plate subduction. Also, most Archaean oceanic crust must have been subducted for the following geometrical argument: at the present rate 3 km^2 yr^{-1} of oceanic crust is created at ridges and an equivalent amount is subducted. Crustal area equal to the present continents (40% of the earth's area) is thus created in 70 Ma. Using

the present 5 km thickness of oceanic crust, a volume of oceanic crust equal to the volume of continent crust is subducted every 500 Ma. We certainly see no evidence today of such large amounts of Archaean oceanic crust, and it would be impossible geometrically to preserve such an amount in the present continents. Thus the bulk of the Archaean oceanic crust must have been subducted (Sleep and Windley, 1982). Baer (1977) contended that the higher temperatures in the Archaean would preclude subduction altogether by preventing the oceanic crust in the downgoing slab from converting to eclogite, but Sleep and Windley (1982) demonstrated the fallibility of this argument.

Davies (1979) accepted that most of the Archaean high heat flux would have been

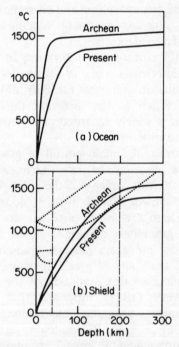

Fig. 22.10 Comparisons of Archean and present geotherms, assuming Archean deep mantle heat generation three times and lithospheric heat generation twice present values. (a) Oceanic geotherms; cooling age 3 Ma, surface heat flux 180 mW m^{-2}. (b) Shield geotherms; surface heat flux 64 mW m^{-2}. Granodiorite and peridotite wet and dry solidi included for comparison (from Davies, 1979, *Earth Planet. Sci. Lett.*, **44**, Fig. 4, p. 235, reproduced by permission of Elsevier, Amsterdam)

removed through faster sea-floor spreading, but pointed out that there would still have been a high heat flux at the base of the continental crust. To account for the absence of melting of Archaean lower crustal rock, and the remarkable intact preservation of such rocks today, Davies proposed that there must have existed beneath stable continental crust a root zone at least 200 cm thick which acted as a thermal buffer between the crust and the hot convecting mantle. This is comparable to the tectosphere root of Jordan (1978).

Fig. 22.10 gives possible present and Archaean oceanic and continental geotherms (Davies, 1979). The average near-surface oceanic geothermal gradient could be steeper in the Archaean in proportion to the higher heat flux. For the continental shield geotherms, the temperatures in the crust and lithosphere are increased by less than 200°C. Fig. 22.11 compares a variety of variables through geological time. Heat production

from breakdown of long-lived radiogenic isotopes was 2–3 times greater in the Archaean than at present, more than 75% of the present continental crust volume was created by the beginning of Proterozoic time, and in the Archaean the difference in the thickness of oceanic and continental plates was much smaller than today.

Reference at this point to Chapters 2–4 will show that the geological evidence is consistent with the above thermal constraints for high rates of plate creation and subduction in the Archaean. One point is worth repeating here: in order to account for the voluminous quantities of tonalitic material, a tectonic model is required that is capable of continuously replenishing over a sustained period the sites of continental growth with new mantle-derived basaltic and associated material. It was the partial melting of such material that created the prodigious quantities of calc-alkaline material which enabled the bulk of

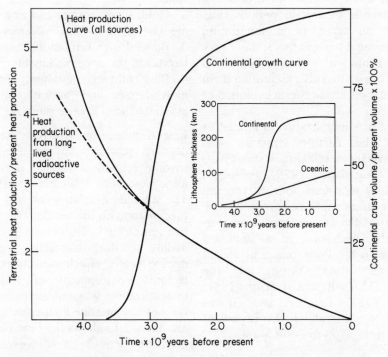

Fig. 22.11 Heat production versus continental growth. Inset shows changes in lithosphere thickness with time (from Brown, in press, in N. Snelling (Ed.), *Geochronology and Geological Record*, Geol. Soc. Lond. Sp. Publ.; inset: changes in lithosphere thickness with time)

the continental crust to be created by the end of the Archaean. But the partial melting of the basic roots of some down-sagged greenstone belts in some form of static tectonic model is obviously an inadequate source. Only a plate creation and subduction model is capable of providing such continuous and sustained replenishment. Nevertheless, it must be emphasized that other evidence suggests that modern-style plate tectonics began in the early Proterozoic, whilst some early variant of the plate tectonic process seems most likely to have been in operation in the Archaean.

Metallogeny

Throughout the previous chapters an attempt has been made to indicate the kinds of mineralization prominent in various environments at different times in earth history. Just as the rocks have undergone secular trends or variations, so too have their enclosed ores. The modes of occurrence of ore deposits have changed in sympathy with the prevailing environments, some of which were characteristic of certain periods (Fig. 22.12). Not all types of mineralization evolved at the same time; in fact it was a very diachronous evolution. The aim here is to review these ore occurrences by putting them into the context of the long-term evolution of the continents.

The record of mineralization is related to changes in the earth's structure and behaviour throughout 3800 Ma of time (Watson, 1973; Tugarinov, 1979; Folinsbee, 1982). The long-term variations in type and style of mineralization may be correlated with the evolution of the earth's crust from permobile in the Archaean, to a stabilized cratonic stage in the Proterozoic, to a final plate tectonic stage throughout the Phanerozoic (Mitchell and Garson 1981; Hutchinson, 1983). Lambert and Groves (1981) provide a useful lengthy review of metallogeny from 3800 to 2000 Ma ago.

The Archaean

During the Archaean period from 3800–2500 Ma ago there were two main tectonic environments, neither of which survived in an intact state. The few ores that exist in the high-grade regions occur within the gneisses in thin strips of metamorphosed volcanics, sediments and intruded layered complexes, i.e. iron formations at Isua (West Greenland), in NE China and many other regions (Prasad et al., 1982), Cu and Ni in meta-volcanic amphibolites at Pikwe (Botswana) in the Limpopo belt, and chromite in layered anorthositic complexes in West Greenland (Fiskenæsset), southern Africa (Limpopo belt–Messina Formation) and southern India (Sittampundi).

In contrast, the Archaean greenstone belts abound in ore deposits, a lot of which are economic. Summarizing the points brought out in Chapter 3 the main mineral deposits can be related to distinct rock groups in the stratigraphic pile and to the associated granites:

1. Chromite, nickel, asbestos, magnesite and talc in the lower ultramafic flows and intrusions.
2. Gold, silver, copper and zinc in the intermediate mafic-to-felsic volcanics.
3. Banded iron formations, manganese and barytes in the upper sediments.
4. Rare lithium, tantalum, beryllium, tin, molybdenum and bismuth in pegmatites associated with granite plutons.

Late Archaean massive base metal sulphide deposits (Cu–Zn–Au) in basalt–andesite–rhyolite sequences formed near centres of felsic volcanism (Hutchinson, 1980, 1982). The sulphide emplacement was by chemical precipitation on the sea floor through exhalative vents from hydrothermal fluids generated mainly by deep convective circulation of marine waters. Hutchinson considers that the tectonic environment of this Archaean mineralization was analogous to the ocean floor of a submarine Phanerozoic early island arc. Fig. 22.13 shows the time range of these primitive Zn–Cu volcanogenic sulphide deposits.

The earliest signs of porphyry-style Cu–Mo mineralization appeared 2700–2900 Ma ago in connection with subvolcanic plutons in

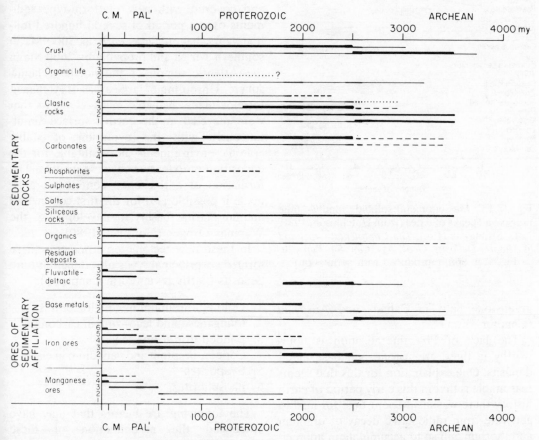

Fig. 22.12 Summary of variations in the composition of the crust, biosphere, sedimentary rocks and ores of sedimentary affiliation during geologic history (after Veizer, 1976b; reproduced by permission of Elsevier Scientific Publ.). Explanations:

Crust: 1. Archaean 'protocrust'. 2. Thick continental crust with 'granitic' upper layer.

Organic life: 1. Procaryota. 2. Eucaryota. 3. Metazoa. 4. Terrestrial life.

Clastic rocks: 1. Greywackes. 2. Shales and slates. 3. Arkoses. 4. Oligomictic conglomerates and ortho-quartzites. 5. Red beds.

Carbonates: 1. Early diagenetic dolomites. 2. Inorganic and/or biochemical (?) limestones. 3. Littoral–neritic organogenic and organodetrital limestones and late diagenetic dolomites. 4. Deep sea lime-stones.

Organics: 1. Bituminous shales. 2. Oil and gas. 3. Coal.

Fluviatile–deltaic deposits: 1. Conglomerate–U–Au–pyrite type. 2. Sandstone–U–V–Cu type. 3. Placers and palaeoplacers.

Base metal deposits: 1. Volcanogenic and volcano–sedimentary type. 2. Sedimentary basinal type. 3. Sedimentary red bed type. 4. Carbonate–Pb–Zn type.

Iron ores: 1. Algoma type. 2. Superior type. 3. Clinton type. 4. Bilbao type. 5. Minette type. 6. River bed, bog iron, laterite and similar types.

Manganese ores: 1. 'Jaspilitic' type. 2. Volcano–sedimentary and orthoquartzite–siliceous shale–Mn carbonate types. 3. Carbonate association. 4. Orthoquartzite–glauconite–clay association. 5. Marsh and lake deposits.

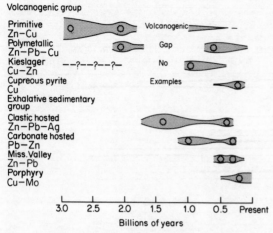

Fig. 22.13 Massive metal sulphide family—time ranges and peaks of types (from Hutchinson, 1980, in D. Strangway (Ed.), *The Continental Crust and its Mineral Deposits, Geol. Ass. Can. Sp. Pap.*, **20**, Fig. 1, p. 666, reproduced with permission)

greenstone belts in Canada and Australia (Chapter 3).

The lack of Pb mineralization is noteworthy in these greenstone belt base metal deposits. One explanation for this that seems reasonable is that in this early period of earth history there was insufficient time for lead to generate from radioactive decay of uranium and thorium and so to accumulate in appreciable quantities in the crust-by Proterozoic time lead makes its first appearance in volcanogenic sulphide deposits.

The mineralization that we see today in Archaean terrains occurs, or can potentially occur, in rocks spanning the period from 3800–2500 Ma ago, which is a third of geological time. What is striking is the remarkable degree of similarity in rock and ore development over such a long period. In the succeeding Proterozoic stage of earth history the types and styles of mineralization were substantially different from those of the Archaean.

The Proterozoic

There was a dramatic change in tectonic conditions between the Archaean and Proterozoic although the changeover was both transitional and diachronous. The Archaean crust cooled, was stabilized, uplifted, eroded and covered with new platform-type sediments over a period of several hundred million years, beginning as early as 3000 Ma in southern Africa and probably by 2600 Ma in the Atlantic region of the northern hemisphere. Unroofing of large areas of Archaean terrains made available many elements that were recycled at the earth's surface simultaneously with the evolution of stable cratonic environments more suitable for the deposition of extensive cover sequences. The formation of a thickened continental crust made it possible also for the first continental margin marine basins to develop, e.g. the Wopmay Orogen (Hoffman, 1980).

In these new tectonic environments there formed six principal types of sedimentary ore deposits (partly reviewed in Chapter 6):

1. Gold and uranium in conglomerates.
2. Manganese and lead–zinc in carbonates.
3. Banded iron formations.
4. Copper, uranium and vanadium in clastics.
5. Evaporites.
6. Phosphorites.

The following are factors that may have influenced the accumulation of these deposits:

1. Extensive erosion of the Archaean basement made available several suitable elements. In particular, the Au in the BIF were no doubt derived from the greenstone belts with their abundant basic volcanics. Note that the gold placer province in southern Africa is centrally located in the Kaapvaal craton and thus near to the uplifted greenstone belts.
2. It was suggested by Cloud (1976b) that a low oxygen content of the prevailing atmosphere may have contributed to the deposition of:

   (a) Detrital uraninite which readily oxidizes on weathering.

   (b) Banded iron formations which used photosynthetically derived oxygen for the oxidation of ferrous iron in the seawater. However, the timing of the development of the oxygenic atmosphere and its relationship to the ores is currently under much debate, and the reader should be cautious of these simple correlations—see earlier this chapter.

This time-bound character of uranium deposits has recently been much discussed (Robertson *et al.*, 1978; Dahlkamp, 1980; Derry, 1980).

3. The CO_2 content in the oceans had increased to such an extent by the early Proterozoic that thick dolomite sequences were deposited. The accumulation of Mn deposits in these carbonates was facilitated by the increasing partial pressure of oxygen because the separation of Mn from Fe is achieved by a higher oxidation of iron (Veizer, 1976b).

The dolomites typically belong to a shallow marine intertidal lagoonal-to-littoral facies, often with algal reefs that formed in a carbonate shelf-bank setting at rifted continental margins. Such dolomites provide the host for Cu–Pb–Zn mineralization at McArthur River and Mount Isa in narrow basins, aulacogens or at triple junctions; these are the first examples of this type of rift-controlled mineralization that became common in the late Phanerozoic (Table 22.4) (Sawkins, 1982).

Table 22.4 Relationship of various metal deposits to past rifting events from Sawkins, 1982, Table 3, in *Continental and Oceanic Rifts*, Geodynamic Ser., vol. 8, American Geophysical Union)

Rifting event	Deposit type	Examples
Breakup of Pangea	Stratiform Cu	Angola sed. Cu Kupferschiefer
	Cu breccia pipes	Messina, S Africa
	Carbonate hosted Pb–Zn	Benue Trough deposits Red Sea margins Pennines, England
	Crustal granite Sn	Jos Plateau tin district
Devonian rifting. Europe	Shale–hosted massive sulphide	Rammelsberg, W Germany Meggen, W Germany
	Volcanogenic massive sulphide	Rio Tinto district, Spain
Late Proterozoic rifting	Stratiform Cu	Zambian deposits Namibian deposits White Pine, USA Adelaidean deposits Seal Lake, Canada Coppermine River, Canada Redstone River, Canada
	Cu breccia pipes	Tribag, Ontario, Canada
	Magmatic Cu–Ni	Duluth Complex
	Shale–hosted massive sulphide	Ducktown, Tennessee Otjihase, Namibia Tsumeb, Namibia
	Carbonate–hosted Pb–Zn	
	Crustal granite Sn	Rhondonia deposits, Brazil
Mid-Proterozoic rifting	Shale–hosted massive sulphide	Mt. Isa, Australia McArthur River, Australia Sullivan, Canada Broken Hill, Australia
	Carbonate–hosted Pb–Zn	McArthur River area, Australia

There are also early Proterozoic Cu Pb Zn ores with a volcanogenic affiliation associated with intermediate-felsic calc-alkaline volcanics, such as at Errington and Vermilion Lake in the Sudbury basin and at Prescott, Arizona. These polymetallic pyrite–galena–sphalerite–chalcopyrite deposits (Fig. 22.13) have more Pb and Ag at the expense of Cu and Au than Archaean deposits. They are of the same type and have the same mode of occurrence as those of California and of the Kuroko district of Japan, all of which formed in active island arcs or continental margins (Hutchinson, 1980). Because these ores and felsic volcanics were derived from magmas that have been through a two-stage mantle fractionation process, the early Proterozoic examples may have formed in a similar plate tectonic environment.

Thus it seems possible that both types of early Proterozoic Cu–Pb–Zn deposits formed in the first active continental margins that are similar to those of the Mesozoic.

In the early to mid-Proterozoic major dyke swarms and layered complexes were intruded into either the Archaean basement itself, or the unconformable cover successions. Fractionation of the mantle-derived tholeiitic liquids gave rise to a distinctive group of metal concentrations, in particular nickel, chromium, platinum and copper (Chapter 5).

During mid-Proterozoic time there was a major abortive attempt to rift the continental plates that were extensive by this time (Chapter 7). Sn, Be, W mineralization occurs in rapakivigranites and andesine–labradorite anorthosites contain important ilmenite–titaniferous magnetic deposits. There is a distinct Ti metallogenic association with these mid-Proterozoic anorthosites on all continents and in this respect they differ from the commonly chromite-bearing Archaean anorthosites.

In the late Proterozoic 1300–600 Ma ago one or several large continents began to break-up and much igneous activity, sedimentation and related mineralization were localized in major rifts, aulacogens and along continental margins. Examples of such mineralization are as follows:

1. *Copper in volcanics and sediments*. Basalt flows were extensively extruded onto, or at the margins of, continents in this period with many located in rifts or failed arms. The most common ore in these volcanics is copper, as in the Coppermine River Group (1110 Ma) in NW Canada, the Keweenawan basalts (1100 Ma) in Michigan, and the slightly younger Bukoban lavas in East Africa (100–800 Ma) (Chapter 7).

Certain slightly younger clastic sediments contain enrichments in copper, but only some are economic. It is probable that the copper in these sediments was derived by recycling from that in the lavas (Watson, 1973). The Katanga System in Zambia has well-known economic deposits of this type, and the sediments of the Belt Basin in northwestern USA contain anomalously high values of copper in the range of 100 ppm over thousands of square miles (Chapter 8).

2. *Alkaline complexes*. These are often associated, both temporally and spatially, with carbonatites formed in conjunction with rift activity and they contain deposits of U, Th, Nb, Be and Zr (Ilimaussaq, South Greenland), Ba and rare earths (Mountain Pass, California) and nepheline (Blue Mountain and Bancroft, Ontario) (Chapter 7). During the formation of the Grenville, Dalslandian and Pan-African mobile belts in the late Proterozoic a wide range of mineral deposits were formed as part of the plate tectonic framework that controlled the formation of these belts. There has been little synthesis of the mineralization of these belts, except for the reviews of the metallogenesis of the Grenville Supergroup by Sangster and Bourne (1982), and of the Arabian–Nubian Shield by Al-Shanti and Roobol (in Al-Shanti, 1979, Vol. 1). The three volumes edited by Al-Shanti (1979) provide much useful detailed information.

Phanerozoic Plate Boundaries

Starting in the late Proterozoic a new type of tectonic pattern developed in the continents which eventually gave rise to Phanerozoic fold belts formed by continental drift. In the following discussion the formation of ores will be considered throughout Phanerozoic time in relation to the following principal tectonic environments; early domes, aulacogens and rifts in continents, accreting and consuming plate margins and transform faults (for summary see Fig. 22.14). For general reviews, see Mitchell and Garson (1981).

Domes, Aulacogens and Rifts We want to consider here the early structures and associated mineralization that developed in continental crust before separation took place and new oceanic crust was formed. These represent the incipient stages of continental fragmentation.

The first major structures to develop are topographic domes like one sees today in Africa and the earliest ores are tin deposits in anorogenic alkaline-to-peralkaline granites that lie in linear belts associated with the domes and whose formation stopped before the triple rift systems developed. Examples are the Younger Granites of Nigeria, and the granites of the Tibesti area of Chad and in Damaraland in South West Africa.

Although some rifts may open to give rise eventually to new oceans, some may not open and so remain as aulacogens which are failed riftarms that project from an oceanic margin into a continental plate (Burke and Dewey, 1973a). The main types of mineralization in such rift zones are:

1. Lead–Zinc. For details of this mineralization in the Benue Trough, the Oslo Graben and the Rio Grande Rift, see Chapter 16. These Pb–Zn deposits, often with Ag, are of the Sullivan type of Sawkins (1982) which form during the attempts to fragment continental plates; they make their first appearance during the early Proterozoic (Mount Isa) and later examples are given in Table 22.4.
2. Niobium and REE concentrations in carbonatites in Tanzania, Uganda, Zambia, Malawi and Oka (Quebec).

Accreting Plate Margins With further extension new oceanic lithosphere is created along the axial zone of the rifts giving rise to oceanic ridges which we see today preserved on a minor scale in the Red Sea and in back-arc marginal basins. Continental drift gives rise to wide oceans. Examination of material from these spreading-centre ridges is possible from dredges and drill cores and from onland ophiolite complexes. Fig. 20.6 gives an idealized cross-section through the oceanic crust showing the stratigraphic location of the principal mineral deposits. Evidence of the metal concentrations in the topmost sediments comes from currently active ridges like the East Pacific Rise and Red Sea as well as from correlated examples capping ophiolite sequences; the mineralization in the volcanic–plutonic part of the cross-section, however, is known only from 'fossil' ophiolites.

The oldest Phanerozoic ores formed at constructive plate boundaries, whether in a main ocean or a back-arc basin, are in the ophiolite complexes of western Newfoundland which were obducted in Ordovician time; asbestos, chromite and nickel ores occur in the lower plutonics (Strong, 1974).

Podiform chromite deposits are common in the lower serpentinized ultramafics of many ophiolite complexes in the eastern part of the Alpine fold belt (Chapter 20) and in many other parts of the world (Panayiotou, 1980).

Cupreous pyrite deposits (ochres) with minor zinc (but no lead) occur in or above the basaltic pillow lavas of several Phanerozoic ophiolite successions (Table 22.5).

The topmost sediments in ophiolites are associated with Mn–Fe rich sediments free of sulphides which lie unconformably in depressions on the basalt lava surface: these are the umbers of Cyprus (Robertson, 1976).

Layer 1 sediments on the flanks of the East Pacific Rise contain anomalously high concentrations of Mn and Fe, and lesser quantities of Cu, Zn, Pb, B, As, U and Hg; similar metal-enrichments occur elsewhere on active oceanic ridges. High concentrations of Fe, Mn and a variety of non-ferrous metals also occur in hot brines in the Red Sea and beneath the Salton Sea, California. In the Red Sea there are sulphide-rich muds with sphalerite and lesser chalcopyrite, pyrite and marcasite overlain by layers enriched in Fe

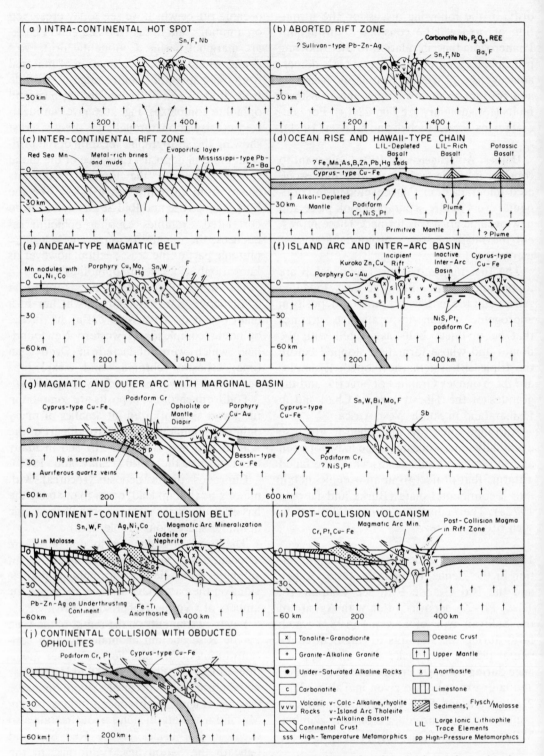

Fig. 22.14   Schematic cross-sections through plate boundary-related tectonic settings showing principal types of mineralization (after Mitchell and Garson, 1976; reproduced by permission of *Minerals, Science and Engineering*)

Table 22.5 Examples of ophiolites with cupreous pyrite deposits

Period	Locality
Early Palaeozoic	Whalesback, Tilt Cove, Notre Dame Bay, Betts Cove, Bay of Island complexes, etc. (Newfoundland)
Late Palaeozoic	150+ deposits, Sanbagawa zones, etc. (Japan) 100+ deposits (Urals, USSR)
Mesozoic	Troodos complex (Cyprus) Küre (Turkey) Island Mtn (Franciscan, Calif.)
Cenozoic	Ergani Maden (Turkey) Phillipines

and Mn minerals; compaction of these metalliferous muds would give rise to deposits similar in several ways to the umbers in ophiolites. Magnesite vein deposits are prominent in the ultramafic rocks of the ophiolites in Greece (Dabitzias, 1980).

Consuming Plate Margins When an oceanic lithospheric plate is carried down a subduction zone, new magmas are generated at successive pressure/temperature levels which arise through the overlying plate. These magmas give rise to plutonic and volcanic rocks, as well as mineral deposits, which change in composition and type with increasing distance from the trench. The metals are emplaced as components of magmas that produced tholeiites nearest the ocean, calc-alkaline granitic batholiths and andesitic–rhyolitic lavas in the main arc, and alkaline rocks furthest from the ocean. The zonal distribution and types of mineral deposits across island arcs and Andean-type continental margins are substantially similar (Mitchell, 1976; Frutos, 1982). The general sequences across the American Andean margin and the arcs of the western Pacific (omiting extensional marginal basins) are (Sillitoe, 1976):

Oceanic side
Eastern Pacific
Fe, Cu–(Au–Mo), Cu–Pb–Zn–Ag, Sn–(W–Ag–Bi)

Western Pacific
Au–Cu–Mo, Au–Pb–Zn, Sn–W–Sb–Hg

There are various types of base metal sulphide ores at Phanerozoic active plate boundaries. The volcanogenic zinc–copper type (Fig. 22.13), is present in Palaeozoic belts (West Shasta district in California; Fleur de Lys Supergroup, W Newfoundland; Bathurst in Brisbane–New England eugeosyncline, NSW, Australia), but it did not survive into post-Palaeozoic times, whereas the lead–zinc–copper–silver volcanogenic type reached a maximum in early Palaeozoic orogenic belts (New Brunswick; Newfoundland; NSW, Australia) and continued into younger belts (Triassic in East Shasta district of California; Jurassic in Foothill Copper Zinc belt of California; Tertiary in Kuroko district of Japan). Hutchinson (1980, 1982) suggests that the first type developed in early stages of subduction along continental margins, and the second in later, more felsic, calc-alkaline volcanics.

Transform Faults Mitchell and Garson (1976) suggest that the following types of mineralization occur along transform faults, either in the oceanic crust where they offset ridges or along their extensions in continental fractures.

The location of the metalliferous brine pools in the Red Sea is controlled by transform faults (Garson and Krs, 1976). Also Zn, Pb, Cu and Mn deposits of mainly Miocene age occur near the Red Sea in continental fractures which extend onshore from presumed former transform faults.

Serpentinized peridotites, gabbros and alkaline plutonic rocks have been derived from transform faults in the equatorial mid-Atlantic ridge. Some of the peridotites contain high Ni, Co, Ti and Cu values.

Alkaline complexes and carbonatites are aligned along marked lineaments in Angola, Brazil, SW Africa and Uruguay, the lineaments lying along transform faults centred on the Cretaceous poles of rotation for the South Atlantic. Diamond-bearing kimberlite pipes are centred on these transform lineaments, especially along the Lucapa graben in Angola.

Onland transcurrent faults that may be extensions of transform faults have localized porphyry copper deposits in the Philippines, porphyries with sulphide mineralization and Kuroko-type Cu–Pb–Zn deposits in the Red Sea region, and tin-bearing granites and lepidolite–tin pegmatites in Thailand.

In the Troodos ophiolite complex there is a fossil (Arakapas) oceanic fracture zone, along which there are Fe-rich mudstones precipitated chemically from hydrothermal solutions, and minor Cu sulphide mineralization derived by high temperature leaching (Robertson, 1978).

Crustal Evolution

During the 3800 Ma dealt with in these pages the continents evolved to their present form; they passed through three distinct stages of growth which roughly correspond to the Archaean, Proterozoic and Phanerozoic. The tectonic patterns during these periods are a reflection of the conditions of aggregation, stabilization and fragmentation, respectively, that contributed to the growth of the continents. For an alternative viewpoint see Salop (1983). Here we shall look at some of the growth models for the continental crust, followed by a summary of the main evolutionary trends through the Precambrian.

Continental Growth Models

Three basic models can be considered to explain the volumetric growth rate of the continental crust—Fig. 22.15 (Veizer, 1976a):

1. The rate of continental crust formation increases exponentially towards the present.
2. Rapid growth at an early stage of earth history and afterwards recycling of crustal material through the mantle or within the crust.
3. Linear crustal growth is combined with recycling through the crust if the rate of recycling is faster than the rate of linear growth.

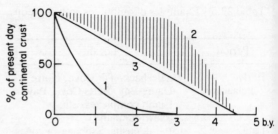

Fig. 22.15 Basic models of the volumetric growth of the continental crust during geological history. The % on vertical axis are cumulative. The lined area includes the majority of recent estimates of crustal growth rates indicated in the text (modified after Veizer 1976a, in B. F. Windley (Ed.), *The Early History of the Earth*, Wiley, London, Fig. 5, p. 573)

Of course models 1 and 3 are limiting cases and much recent and current research is devoted to finding out where the real growth rate took place between these extremes.

Model 1 predicts that a little more than half of the present continental crust formed in Phanerozoic time. However, the geological map of the world (Fig. 1.1) shows that rather more than half of the present area formed in Precambrian time, and it is well known that all orogenic belts of any age contain reworked older material; for example, most of the Himalayas must be underlain by Precambrian basement. Thus the present areal distribution of continental crust with respect to its age (Fig. 22.16) does not support this model, which is therefore an unlikely possibility.

Model 2 has been expounded in particular by Armstrong (1981) who concluded that rates of crustal growth and recycling reached a near-steady state for the first 1000 Ma of earth history and that there was negligible crustal growth after 2900 Ma ago. He placed emphasis on two points. Firstly, the ingenious freeboard model produced by Wise (1974) according to which the relative elevation of continents with respect to sea level has stayed the same throughout the Proterozoic and Phanerozoic. This means that the continents have had at least 90% of their present thickness for a minimum of 2500 Ma. Secondly, the demonstration by Condie (1973) that the seismically determined thicknesses of

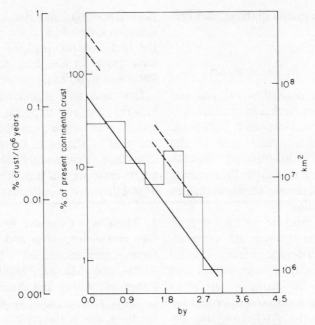

Fig. 22.16 Present-day areal distribution of the continental basement as a function of its geological 'age' (based on compilation of Hurley and Rand, 1969, after Veizer, 1976a; reproduced by permission of J. Wiley)

cratonic areas for the period 2500 Ma to the present vary only between 38 and 40 km and are independent of age. A further important result is that of McCulloch and Wasserburg (1978) from their study of Sm–Nd and Rb–Sr chronology—namely that the Superior, Slave and Churchill Provinces of Canada were all formed in the period 2700–2500 Ma ago, with no crustal growth since then. In the models of Fyfe (1978) and Hargraves (1976, 1978) the bulk of the continental crust is formed early, some is reworked and some is partly preserved, and this assumes a degree of recycling of continental crust.

According to model 3, which involves slow crustal growth, new material has been added to the continents throughout geological time. The isotopic interpretation of Moorbath (1980), that the continental crust has grown episodically and irreversibly during geological time, is close to this model. From isotopic constraints DePaolo (1980) concluded that since 3600 Ma ago there has been semi-continuous crustal growth with zero or small amounts of recycling.

It is interesting at this point to summarize the main conclusions of recent authors who have made percentage estimates of continental crust created by the end of the Archaean (see also Windley, 1977):

60–80%	Cook and Turcotte (1981)
50–60%	Moorbath and Taylor (1981)
70–85%	McLennan and Taylor (1982)
85%	Dewey and Windley (1981)
80%	Brown (in press)

Other estimates are that 40–50% of the continental crust formed between 3000 and 2700 Ma ago (Condie, 1980), and 65–75% in the period 3200–2500 Ma (McLennan and Taylor, 1982). Fig. 22.15 indicates the range of estimates by authors quoted in this section for crustal growth models which lie between models 2 and 3.

Many of the above growth rate estimates were based on isotopic constraints. The geological evidence, as outlined in Chapters 2–4, is consistent with these conclusions and so it is appropriate here to summarize the main geological features which tell us about how the continents evolved in the critical Precambrian period.

Cratonization During the Precambrian

The stratigraphic, petrochemical and tectonic evidence for late Archaean crustal thickening in greenstone belts and high-grade regions provide an adequate clue to the marked increase in lithospheric stability which took place by the early Proterozoic. This process of progressive stabilization can be termed cratonization.

During the first third of geological time (roughly 1300 million years) in the Archaean there was such a high degree of tectonic activity that no rocks escaped deformation and metamorphism. But it would be wrong to think that permobile conditions operated for long or throughout the Archaean; the isochron age peaks are indicative of several short-lived, but widespread events.

So far it has not proved possible to define any well-structured continental–oceanic margins or platform-basement contacts in the Archaean. Admittedly there are a few unconformities between greenstone belts and higher grade gneissose/granitic rocks (Windley, 1973; Bickle *et al.*, 1975), but these are only local erosional remnants and are not at the borders of well-defined cratons. Nevertheless, geochemical differentiation had progressed sufficiently, even by 3700–3800 Ma ago, to produce the well-known bimodal granite–greenstone association. There was certainly a marked change in crustal conditions by the late Archaean in southern Africa (somewhat ahead of other continents—Windley, 1973) because the formation of the greenstone belts in Rhodesia (2900 Ma for the Bulawayan—overlapped with the deposition of platform-type cover sediments of the Pongola (*c.* 2850 Ma—Hunter, 1974b) and Dominion Reef (*c.* 2800 Ma) sequences.

There are only a few places in the world where pre-3200 Ma old rocks have been defined isotopically: Minnesota River Valley in the USA, Godthaab–Isua in West Greenland, northern Norway, Barberton–Swaziland–Rhodesia, and the Labrador coast of Canada. One of the main reasons for this is the fact that the peak of accretion activity took place in the late Archaean between 3100 and 2700 Ma.

Evidence and arguments supporting the idea that the most important period of crustal growth was in the permobile late Archaean and that by the early Proterozoic the continents had consequently achieved roughly or nearly their present size, thickness and degree of stability, are as follows:

1. There was extensive formation of granulites and charnockites and high amphibolite facies metamorphism in the period 3100–2800 Ma, *viz.* Scotland, Greenland, Labrador, Aldan and Anabar Shields, Limpopo, W Africa, India, W Australia, etc., and the thickness of the crust in these areas must have been more than that of modern continental crust for it to be able to support such high temperature metamorphism in its lower part.
2. The majority of greenstone belts formed between 3000 and 2700 Ma. Some continued to form in the early Proterozoic.
3. In some places such as W Greenland, and probably in others like Scotland, S India and the Limpopo, there was such an intensive intrusion of late Archaean tonalites that they make up the greater part of the present surface area. The low initial strontium isotope ratios of about 0.701 of the tonalites and greenstone belt volcanics suggest that the isochron ages are close to the time of derivation and accretion of material from the mantle in the late Archaean (Moorbath, 1980).
4. At least twice, maybe even three times, as much heat was generated by radioactive decay in the Archaean than today (Fig. 22.11). Burke *et al.* (1976a) suggested that if this heat were the sole source of formation of the continents through a two-stage fractionation process, then about two-thirds of the continental crust would have been produced by 2500 Ma ago; but with probable additional heat sources to promote mantle fractionation in the early Archaean (short-lived isotopes, core formation and meteorite impacts), it is

likely that much, if not nearly all, of the present continental mass was produced during this permobile era.

5. By comparing the potash contents of modern and Archaean calc-alkaline volcanics, Condie (1973) inferred that the crust of North America approached or reached present-day thicknesses by about 2700 Ma ago, whilst Wise (1974) showed that there is a constant relationship between the thickness, and the volume of the continents.

6. The presence of thin strips of marble–orthoquartzite–K pelite in the Archaean (high-grade regions) suggests that shallow-water continental margin/platform conditions were in existence, although very poorly developed, in this period. The Dharwars of India are the only greenstone belts to have a prominent marble–orthoquartzite association. Conditions obviously changed drastically between about 3000 and 2500 Ma because by the early Proterozoic kilometre-thick, carbonate–quartzite sequences appear in, for example, the Witwatersrand–Transvaal series and the Coronation and Mount Isa geosynclines. From their world survey Engel et al. (1974) calculated that the ratio of Archaean carbonate plus orthoquartzite to other Archaean sediments was about 1:100 whereas in the Proterozoic the ratio was 1:5. This can be regarded as an approximate estimate of the growth of stable shelf envionments at this time.

7. There was an appreciable increase in amount of mature quartz sediments about 2500 ± 200 Ma ago, which was complementary to an abrupt decrease in relative volume of immature wackes, conglomerates and arkoses. The K_2O/Na_2O index of clastic sediments, roughly reflecting their degree of maturity, increased from less than 1.0 in Archaean greenstone belts to 1.5–5.0 in Proterozoic sequences (Engel et al., 1974). Also the unstable tectonic environments of the Archaean would have been unsuitable hosts for deposition of mature sediments, whereas the broad platforms from the early Proterozoic onwards were ideal. These sedimentary relationships are reflected by the increasing areal extent of basins with time in southern Africa, considered by Anhaeusser (1973) to be directly proportional to the crustal stability (Fig. 6.1).

8. In the Archaean two different proto-plate tectonic regimes can be distinguished: back-arc marginal basins (greenstone belts) and continental margin areas (now in a high-grade state). The first recognizable sedimentary sequences comparable to those along modern continental margins occur in the Wopmay Orogen which developed 2100–1750 Ma ago (Hoffman, 1980). Its depositional and structural evolution so faithfully mimics that of modern Andean–Himalayan orogenic belts that it is difficult to escape the conclusion (bearing in mind that such structures are unknown in the Archaean) that plate tectonics at modern-type continental margins began in the early Proterozoic.

9. Aulacogens are sediment-filled failed rift arms that extend from an oceanic margin into a continent (Burke and Dewey, 1973a). They develop at triple junctions related to domal uplifts and are a key indication of the existence of stable continents and plate tectonics. No Archaean aulacogens are known; they first appear in the early Proterozoic in connection with the Wopmay Orogen and are increasingly common in the Proterozoic such as the Adelaide and Amadeus basins (Rutland, 1973b; Milanovsky, 1983) and the Phanerozoic.

10. Stable continental crust must have existed by 2800 Ma ago in southern Africa and by about 2500 Ma ago elsewhere for the enormously thick supracrustal piles to be deposited against or on them. Examples are:

Sequence	Age (Ma)	Thickness (km)
Pongola	c. 2850	10.7
Dominion Reef, Witwatersrand	2700–2800	11.9
Ventersdorp	>2300	5.0
Huronian	2300	c. 12.0
Labrador Trough	c. 2000	17.0
Coronation	2100	10.7
MacArthur	c. 1700	8.5
Mount Isa	c. 1700	22.2

11. The intrusion of extensive basic dyke swarms implies the existence of widespread

areas of continental crust. The Ameralik dykes in West Greenland and the Saglek dykes in Labrador belong to the only notable swarms in the Archaean (McGregor, 1973; Bridgewater *et al.*, 1978); there are a few other thin amphibolite dykes elsewhere but they amount to very little. In contrast, the Proterozoic abounds in major dyke swarms, particularly in the periods 2500–2000 Ma and 1300–600 Ma ago, reviewed in Chapters 5 and 8. The most recent age determinations given in these chapters demonstrate that there was a broad bimodal time distribution with some, but fewer, dykes intruded in the intervening period. The times of dyke intrusion in general coincide with 'lows' in the histogram of isochron ages of rocks from mobile belts and thus correlate with periods of crustal stability. They also seem to coincide with the hairpin bends of some polar wander curves which relate to large changes in the direction of movement of continental masses.

The Proterozoic dykes broadly reflect the various occasions when different parts of the continents had become stable and began to fracture, perhaps during an early abortive attempt at break-up; it was a diachronous stabilization. The late Proterozoic dykes were related in some way to various initial stages of fragmentation of the continents (or of a super-continent) independently indicated by the formation of many aulacogens at this time (Burke and Dewey, 1973a; Milanovsky, 1981).

12. Intrusive alkaline complexes, carbonatites and kimberlites are characteristic of stable cratons. They are all unknown in the Archaean but appear in the Proterozoic.

13. Ultramafic komatiites are most restricted to the Archaean. They reflect higher mantle temperatures and higher crustal growth rates then than in later geological times.

The above points, and there are no doubt more, all lead to the conclusion that there was a dramatic change in the stability of the continents in the early Precambrian. Table 22.6 summarizes in a comparative way the main features of Archaean and Proterozoic ter-

rains. The probably thin and unstable early Archaean sialic crust was transformed by particularly rapid growth in the late Archaean into a thick crust by the early Proterozoic and therefore the boundary between these periods represents the most important tectonic turning point in earth history (Salop and Scheinmann, 1969; Ronov, 1972; Hoffman, 1980; Engel *et al.*, 1974).

Let us now consider possible reasons for this changeover in crustal conditions from the Archaean to the Proterozoic.

The Archaean–Proterozoic Boundary

Realizing that this boundary separates the permobile from the platformal stage of earth evolution, Salop and Scheinmann (1969) proposed that it was a tectonic reflection of a dramatic change in the energy regimes of the crust and upper mantle. Brown (in press) reasons that in pre-Archaean/early Archaean times earth heat flow was $c.$ 300 mWm^{-2}, which would correspond to a lithosphere thickness of only 10–12 km (Fig. 22.11), but by early Proterozoic times the flow had decreased to $c.$ 120–150 mWm^{-2}, corresponding to an average lithospheric thickness approaching 25 km. If, as Davies (1979) proposed, the continental lithosphere had a 200 km root by this time, this would have acted as a thermal buffer with the result that most of the earth's heat loss had become concentrated, as today, into the oceanic areas, implying that sea-floor spreading by high heat flow plate accretion processes must have been active at that time. Thus, with a high mantle temperature and high heat flow the Archaean can be expected to have been a period of rapid continental growth, but whether by hot spot tectonics (Fyfe, 1978) or some early variant of the plate tectonic process (Sleep and Windley, 1982) is still under debate. By whatever means, the Archaean was dominated by the formation and aggregation of many proto/mini continents or platelets which coalesced into one or several large continental masses by early Proterozoic.

The Archaean–Proterozoic boundary was certainly the most important threshold in the geothermal/tectonic evolution of the conti-

Table 22.6 Comparison of major features of Archaean and Proterozoic terrains (from Lambert and Groves (1981), Early earth evolution and metallogeny in K. H. Wolf (Ed.), *Handbook of Strata Bound and Stratiform Ore Deposits*, Elsevier, Amsterdam, p. 352)

	Archaean	Proterozoic
Regional setting	Greenstone belts and extensive granitoid complexes. Several ages of greenstone and granitoid formation result in variable rock relationships	Extensive epicratonic sequences plus intercratonic mobile belts
Main structures and metamorphism	Synclinal or nappe-like greenstone belts; strike–slip faults common; mainly low to moderate metamorphism (low pressure). Complex structure and moderate to high grade metamorphism (low to high pressure) in gneissic terrains	Epicratonic sequences little deformed or metamorphosed; mobile belts strongly folded, faulted and metamorphosed up to granulite facies (low to high pressure)
Volcanic rocks	Ultramafic–mafic komatiites, low-K tholeiites, Na-rich dacites and rhyolites. Andesites locally important	Tholeiitic basalts, predominantly normal-K; acid volcanics mainly K-rich. Andesites locally abundant. Minor komatiite-like rocks
Intrusive rocks	In some terrains, early granitoids mainly tonalite, trondjhemite and granodiorite, whereas late granitoids mainly granodiorite, adamellite and granite (s.s.); temporal overlap between subvolcanic ultramafic, mafic and felsic intrusives in greenstone belts. Dolerite dyke swarms	Large, layered mafic–ultramafic intrusives, dolerite dyke swarms. Granodiorite, adamellite, granite (including rapakivi types). Subvolcanic felsic porphyries. Na-rich granitoids uncommon
Sediments	Chert, BIF, conglomerate, volcaniclastic greywacke and siltstone; minor carbonate. Quartz-rich sandstone, aluminous shale are uncommon, except in some late Archaean sequences	Quartz-rich sandstone (quartzite) arkose, polymictic conglomerate, carbonate (mainly dolomite), chert, BIF, shale, greywacke
Biological activity	Photosynthesis probably established before 3.5 Gyr; putative microfossils from ~3.5 Gyr; sporadic stromatolite occurrences. Bacterial sulphate reduction not of widespread significance	Proliferation of microorganisms: common stromatolites, sulphate-reducing bacteria. First eucaryotic micro-organisms and megascopic algae
Important mineralization	Ni–Cu, Cr and asbestos in ultramafic volcanics and intrusives; Fe and Au in BIF; Au in altered greenstones; Cu–Zn (–Pb) in mafic–felsic volcanics. Minor Ba in sediments and Cu–Mo in felsic porphyries	Cr, Ni–Cu, Pt in large layered mafic–ultramafic intrusives: Fe in BIF; Au–U in quartz-rich clastics; U associated with unconformities; Pb–Zn–Ag in shales; Cu in 'red bed' sandstone–dolomitic–shale sequences

nents, as is reflected in the substantial evidence for the appearance of modern-style crustal conditions in the early Proterozoic, documented in Chapters 6 and 10, in comparison with those permobile crustal features and conditions outlined in Chapters 2–4. However, this statement must be qualified as follows. At the time of the completion of the 1st edition of this volume in 1976, it was generally thought that greenstone belts were not only characteristic of, but were confined to, the Archaean; but as summarized in Chapter 10 and in Tarney and Windley (1981), it is now clear from well-documented examples that they continued to form, albeit in lesser abundance, through the Proterozoic. Likewise the granulite–gneiss

358

belts, so characteristic of the Archaean, continued to form throughout Proterozoic time. In this important respect the Archaean–Proterozoic boundary is less important than previously thought. This conclusion is further illustrated by two features of sedimentation.

Firstly, there have been several recent remarkable discoveries of surprisingly thick and extensive evaporites in the Archaean, as well as several excellent examples in early Proterozoic basins. Secondly, it has long been thought there are only two types of Precambrian banded iron formation (BIF)—the Algoma type in Archaean greenstone belts, and the Superior type in Proterozoic shelf sequences. However, Prasad et al. (1982) demonstrated that BIF are present in most Archaean granulite–gneiss belts where they are associated with quartzite–marble–mica schist sequences. These represent a proto-Superior type of BIF in thin unstable shelf sequences which did not survive the permobile magmatic and tectonic conditions in the Archaean. Particularly impressive are the surprisingly abundant BIF (with associated quartzites) in the Archaean granulites and gneisses in NE China which are so thick and extensive that they are the foundation of the major steel industry in that region, centred on the city of Anshan (my personal observations; and Li Jiliang et al., 1979). Moreover, Algoma-type BIF continued to form throughout the Proterozoic in the greenstone belts of that period (Windley, 1983a). So we now perceive that BIF formed in the two types of Archaean terrain and that the same two types continued in the Proterozoic, but with opposite degrees of abundance.

Finally, the Archaean–Proterozoic boundary is well demonstrated by the marked changes in elemental values in sediments and igneous rocks, shown by Fig. 22.1. Veizer et al. (1982) pointed out that the Archaean calcites and dolomites, if compared with their Phanerozoic counterparts, are enriched in Sr^{2+}, Ba^{2+}, Mn^{2+}, depleted in ^{18}O, Na^+, and contain mantle-like $^{87}Sr/^{86}Sr$ and $^{34}S/^{32}S$. To explain these differences they suggested that in the Archaean there was massive seawater pumping through the coeval basaltic oceanic crust but, starting in the Proterozoic, the con-

trolling factor of seawater composition was river discharge off the stable continental shelves that had developed by then.

A Summary

Here let us consider some of the main conclusions arrived at in this volume, in particular those relating to secular changes that may have contributed to the evolution of the continents.

1. During the first third of the geological record from 3800 to about 2500 Ma ago the earth evolved in a broadly similar manner giving rise to the granite–greenstone/granulite–gneiss association. In this permobile period an overall high radiogenic heat flow was expressed by a rapid growth and consequent thickening of continental crust. Addition of mantle-derived magmas to the continents gave rise to suites of volcanic and plutonic rocks whose chemical parameters are so remarkably similar to those of Mesozoic–Cenozoic times that it is unlikely that early continental growth did not take place by some form of subduction-controlled proto-plate collision mechanism. The consistently low K_2O/Na_2O ratios of Archaean volcanic and plutonic rocks are suggestive of an arc–Andean margin–back-arc basin type of crustal growth. In other words, continental aggregation in the Archaean and dispersal in the Phanerozoic both depended on subduction mechanisms and thus both gave rise to similar magma products. Aggregation of many small proto-continental plates created a few large, more stable continental masses by the early Proterozoic. A great variety of geological, geochemical (major and trace element) and isotopic evidence plus a consideration of the constancy of freeboard of the continents with time all corroborate the idea that the continental lithosphere attained a degree of rigidity 2500 Ma ago that began to approximate that of today.

2. A major result of this newly attained stability was that modern-style plate tectonic processes, as we understand them today, were able to evolve from the early Proterozoic as also did aulacogens, transcontinental basic

dyke swarms, rift systems with alkaline complexes, and extensive, thick, sedimentary basins in stable continental interiors. For the first time in earth history orogenic belts were able to develop along the margins of stable continental plates and the full array of active plate margin and plate collisional structures are seen from the early Proterozoic. However, greenstone belts and granulite-gneiss belts continued to evolve throughout the Proterozoic, but their tectonic relationships with other structures (such as Andean-type batholithic belts, island arcs or aulacogens) are not yet well understood.

3. In the mid-Proterozoic there were widespread attempts to fragment what had become large continental plates or even perhaps a supercontinent, and the magmatic products are comparable to anorogenic, rift-related, mid-Phanerozoic rocks in W Africa. Late Proterozoic orogenic belts have been through a complete Wilson Cycle, but early Proterozoic belts are only just beginning to disclose their origins. Future studies should attempt to define the differences between early Proterozoic and Mesozoic–Cenozoic orogenic belts and account for them in terms of differences in definable variables such as heat budget, thickness of crust and lithosphere, accretion rates, etc.

4. It is still widely considered that the Archaean atmosphere was anoxygenic and that it was not until 2000 Ma ago that the atmospheric oxygen content had increased sufficiently for banded iron formations to give way to red beds; however, this idea is now being strongly challenged on several fronts, as evidence is accumulated of early palaeosols, red beds, evaporites, oxidized pillow basalts, together with a better understanding of the conditions of formation of detrital uraninite. By 1300 Ma ago the atmospheric oxygen content had reached about 1% of the present level, allowing the appearance of oxygen-employing eucaryotes and, eventually, of complex multicellular organisms.

The deposition of carbonate sediments increased drastically between the Archaean and the early Proterozoic and there was a marked decrease in their dolomite–calcite ratio between the Precambrian and the Phanerozoic. These and other chemical variations in sedimentary rocks reflect secular changes in the evolution of the ocean–atmospheric system.

5. Beginning about 1000 Ma ago a new stage of continental fragmentation, drift and collision began which led to the formation of the Appalachian–Caledonian–Hercynian–Uralian belts and the creation of a Pangaea landmass which itself broke up into continental segments that drifted throughout the Mesozoic–Cenozoic into their present positions. The different types of sedimentary, igneous and metamorphic rocks that developed along plate boundaries during the growth and closure of oceans provide a major key to the unravelling of the complex tectonic relationships found in modern and most ancient fold belts.

Finally, what about uniformitarianism? Similar physicochemical principles must have controlled the processes of sedimentation, magmatism and metamorphism throughout earth history. But these were influenced by the major differences between the Archaean and the Proterozoic tectonic regimes which were responsible for a significant change in development of the sedimentary, igneous, metamorphic and metallogenic record. Bearing these constraints in mind, the principles of uniformitarianism have applied throughout the evolution of the continents.

We should remember that uniformitarianism does not mean that the past and the present are identical but rather that the present provides a model for understanding the past. Uniformitarianism cannot provide the sole means of interpreting the distant past because many features of earlier history are predictably the result of the greater heat production at that time. Thus, in order to understand the early Precambrian, we must combine knowledge of the present with predictions based on thermal constraints.

References

Aalto, K. R. (1971). Glacial marine sedimentation and stratigraphy of the Toby conglomerate (Upper Proterozoic), southeastern British Columbia, North West Idaho and northeastern Washington. *Can. J. Earth Sci.*, **8**, 753–787.

Acharya, H. (1981). Volcanism and aseismic slip in subduction zones. *J. Geophys. Res.*, **86**, B1, 335–344.

Adams, C. G., Benson, R. H., Kidd, R. B., Ryan, W. B. F. and Wright, R. C. (1977). The Messinian salinity crisis and evidence of late Miocene eustatic changes in the world ocean. *Nature*, **269**, 383–386.

Ager, D. V. (1976). The nature of the fossil record, *Proc. geol. Ass.*, **87**, 131–160.

Agrawal, O. P. and Kacker, R. N. (1980). Nagaland ophiolites, India. A subduction zone ophiolite complex in a Tethyan orogenic belt. In A. Panayiotou (Ed.), *Ophiolites, Cyprus Geol. Surv.*, 454–461.

Aldiss, D. T. (1981). Plagiogranites from the ocean crust and ophiolites. *Nature*, **289**, 577–578.

Allaart, J. H. (1978). The pre-3760 my old supracrustal rocks of the Isua area, central West Greenland, and the associated occurrence of quartz-banded ironstone. In B. F. Windley (Ed.), *The Early History of the Earth*, Wiley–Interscience, London, pp. 177–189.

Allard, G. O. (1970). The Dore Lake Complex, Chibougamau, Quebec, a metamorphosed Bushveld-type layered intrusion. *Geol. Soc. South Africa Sp. Publ.*, **1**, 477–491.

Al-Shanti, A. M. S. (Ed.) (1979). *Evolution and Mineralisation of the Arabian-Nubian Shield*, vols 1–3, Pergamon, Oxford.

Alvarez, C. W., Alvarez, W., Asaro, F. and Michel, H. V. (1980). Extraterrestrial cause for the Cretaceous–Tertiary extinctions. *Science*, **208**, 1095–1108.

Anderson, C. A., Blacet, P. M., Silver, L. T. and Stern, T. W. (1971). Revision of Precambrian stratigraphy in the Prescott–Jerome area, Yavapai County, Arizona. *U.S. Geol. Surv. Bull.*, **1324C**, 1–16.

Anderson, J. L. (1980). Mineral equilibria and crystallization conditions in the late Precambrian Wolf River rapakivi massif, Wisconsin. *Amer. J. Sci.*, **280**, 289–332.

Anderson, J. L., Cullers, R. L. and van Schmus, W. R. (1980). Anorogenic metaluminous and peraluminous granite plutonism in the mid-Proterozoic of Wisconsin, USA. *Contrib. Mineral. Petrol.*, **74**, 311–328.

Anderson, J. M. and Schwyzer, R. U. (1977). The biostratography of the Permian and Triassic pt 4. Palaeomagnetic evidence for large-scale intra-Gondwana plate movements during the Carboniferous to Jurassic. *Trans. geol. Soc. S. Afr.*. **80**, 211–234.

Anderson, P. (1977). Proterozoic convergent plate tectonics, *25th Int. Geol. Cong. Sydney, Aust.*, Vol. 1, 73.

Anderson, R. N. (1975). Heat flow in the Mariana marginal basin. *J. Geophys. Res.*, **80**, 4043–4048.

Anderton, R. (1980). Did Iapetus start to open during the Cambrian? *Nature*, **286**, 706–708.

Andrieux, J., Brunel, M. and Hamet, J. (1977). Metamorphism, granitisations and relations with the Main Central Thrust in Central Nepal: $^{87}Rb/^{87}St$ age determination and discussion. In *Himalaya. Sci. de la Terre. Coll. Int. de CNRS*, **No. 268**, 31–40, Paris.

Andrieux, J., Arthaud, F., Brunel, M. and Sauniac, S. (1980). Le caractère postmétamorphique des grands chevauchements himalayens dans l'Himalaya du Nord-Ouest. Aperçu sur les relations possible avec le métamorphism inverse du Népal. *C.R. Acad. Sci. Paris*, **291**, 525–528.

Andrieux, J., Arthaud, F., Brunel, M. and Sauniac, D. (1981). Gëomëotrie et cinématique des chevauchements, en Himalaya du Nort-Ouest. *Bull. Soc. géol. France*, **23**, 651–661.

Anhaeusser, C. R. (1971a). The Barberton Mountain Land, South Africa—a guide to the understanding of the Archaean geology of western Australia. *Geol. Soc. Australia Sp. Publ.*, No. 3, 103–120.

Anhaeusser, C. R. (1971b). Cyclic volcanicity and sedimentation in the evolutionary development of Archaean greenstone belts of Shield areas. *Geol. Soc. Australia Sp. Publ.*, No. 3, 57–70.

Anhaeusser, C. R. (1973). The evolution of the early Precambrian crust of Southern Africa. *Phil. Trans. R. Soc. Lond.*, **A273**, 359–388.

Anhaeusser, C. R. (1976). The nature and distribution of Archaean gold mineralisation in southern Africa. *Minerals Sci. Engng.*, **8**, 46–84.

Anhaeusser, C. R. (1978). The geological evolution of the primitive earth—evidence from the Barberton Mountain Land. In D. H. Tarling (Ed.), *Evolution of the Earth's Crust*, Academic Press, London, pp. 72–106.

Anhaeusser, C. R. (1981). Geotectonic evolution of the Archaean successions in the Barberton Mountain Land, South Africa. In A. Kröner (Ed.), *Precambrian Plate Tectonics*, Elsevier, Amsterdam, pp. 137–160.

Anhaeusser, C. R. (1982). Archaean greenstone terranes: geological evolution and metallogenesis. *Rev. Bras. Geoc.*, **12**, 1–6.

Anhaeusser, C. R., Mason, R., Viljoen, M. J. and Viljoen, R. P. (1969). A reappraisal of some aspects of Precambrian Shield geology. *Bull. geol. Soc. Amer.*, **80**, 2175–2200.

Anhaeusser, C. R., and Wilson, J. F. (1981). Southern Africa: the granitic–gneiss greenstone shield. In D. R. Hunter (Ed.), *Precambrian of the Southern Hemisphere*, Elsevier, Amsterdam, pp. 423–499.

Annells, R. N. (1974). Keweenawan volcanic rocks of Michipicoten island, Lake Superior, Ontario: an eruptive centre of Proterozoic age. *Bull. geol. Surv. Canada*, **218**, 141 pp.

Appel, P. W. U. (1978). Stratabound galena and sulphides of Ni–Sb–Cd–Ag in the early Precambrian Isua supracrustal belt, W. Greenland. *Int. Mineral Assoc. XI Meeting, Novosibirsk*, pp. 140–148.

Appel, P. W. U. (1979). Stratabound copper sulfides in a banded iron formation and in basaltic tuffs in the early Precambrian Isua supracrustal belt, W. Greenland. *Econ. Geol.*, **74**, 45–52.

Appel, P. W. U. (1980). On the early Precambrian Isua iron formation, West Greenland. *Precamb. Res.*, **11**, 73–87.

Archibald, N. J. and Bettenay, L. F. (1977). Indirect evidence for tectonic reactivation of a pre-greenstone sialic basement in W Australia. *Earth Planet. Sci. Lett.*, **33**, 370–378.

Archibald, N. J., Bettenay, L. F., Binns, R. A., Groves, D. I. and Ginthrope, R. J. (1978). The evolution of Archaean greenstone terrains, eastern Goldfields Province, W. Australia. *Precamb. Res.*, **6**, 103–132.

Argyriadis, I., De Graciznsky, P. C., Marcoux, J. and Ricou, L. E. (1980). The opening of the Mesozoic Tethys between Eurasia and Arabia–Africa. In *Geology of the Alpine chains born of the Tethys. 26th Int. Geol. Cong.* (Mem. BRGM, **115**), 119–214.

Arita, K. (1983). Origin of the inverted metamorphism of the Lower Himalayas, central Nepal. *Tectonophysics*, **95**, 43–60.

Armstrong, R. L. (1978). Pre-Cenozoic Phanerozoic time scale—computer file of critical dates and consequences of new and in-progress decay-constant revisions. In G. V. Cohee, M. F. Glaessner, and H. D. Hedberg (Eds), Contributors to the *Geologic Time Scale, Amer. Ass. Petrol. Geol. Studies in Geology*, No. 6, pp. 73–91.

Armstrong, R. L. (1981). Radiogenic isotopes: the case for crustal recycling on a near-steady-state no-continental-growth Earth. *Phil. Trans. R. Soc. Lond.*, **A301**, 443–472.

Armstrong, R. L. (1982). Cordilleran metamorphic core complexes—from Arizona to southern Canada. *Ann. Rev. Earth Planet. Sci.*, **10**, 129–154.

Armstrong, R. L. and Runkie, D. (1979). Rb–Sr geochronometry of the Ecstall, Kitkiata, and Quottoon plutons and their country rocks, Prince Rupert region, Coast Plutonic Complex, British Columbia. *Can. J. Earth Sci.*, **16**, 387–399.

Arndt, N. T. (1977). Ultrabasic magmas and high degree melting of the mantle. *Contrib. Mineral. Petrol.*, **64**, 205–221.

Arndt, N. T., Francis, D. and Hynes, A. J. (1979). The field characteristics and petrology of Archaean and Proterozoic komatiites. *Can. Mineral.*, **17**, 147–163.

Arndt, N. T. and Nisbet, E. G. (Eds) (1982). *Komatiites*, Allen and Unwin, London, 544 pp.

Arthaud, F. and Matte, P. (1977). Late Palaeozoic strike-slip faulting in southern Europe and northern Africa: result of a right-lateral shear zone between the Appalachians and the Urals. *Bull. geol. Soc. Amer.*, **88**, 1305–1320.

Arthur, M. A. and Jenkyns, H. C. (1981). Phosphorites and paleoceanography. *Oceanol. Acta, Proc. 26th Int. Geol. Cong. Geol. of Oceans*, Paris, pp. 83–96.

Asmus, H. E. and Ponte, F. C. (1973). The Brazilian marginal basins. In A. E. M. Nairn and F. G. Stehli (Eds), *The South Atlantic*, vol. 1, Plenum Press, New York, 87–134.

Atwater, T. and Macdonald, K. C. (1977). Are spreading centers perpendicular to their transform faults? *Nature*, **270**, 715–719.

Autran, A. (1978). Synthèse provisoire des évenements orogenique caledoniens en France. *Geol. Surv. Can. Pap.*, **78–13**, 159–175.

Awramik, S. M. (1971). Precambrian columnar stromatolite diversity: reflection of Metazoan appearance. *Science*, **174**, 825–826.

Awramik, S. M. and Barghoorn, E. S. (1977). The Gunflint microbiota. *Precamb. Res.*, **5**, 121–142.

Ayres, L. D., Averill, S. A. and Wolfe, W. J. (1982). An Archaean molybdenite occurrence of possible porphyry type at Setting Net Lake, NW Ontario, Canada. *Econ. Geol.*, **77**, 1105–1119.

Baadsgaard, H. (1973). U–Th–Pb dates on zircons from the early Precambrian Amîtsoq gneisses, Godthaab district, West Greenland. *Earth Planet. Sci. Lett.*, **19**, 22–28.

Baadsgaard, H., Collerson, K. D. and Bridgwater, D. (1979). The Archaean gneiss complex of N. Labrador. I. Preliminary U–Th–Pb geochronology. *Can. J. Earth Sci.*, **16**, 951–961.

Baadsgaard, H. and McGregor, V. R. (1981). The U–Th–Pb systematics of zircons from the type Nûk gneisses, Godthåbsfjord, W. Greenland. *Geochim. Cosmochim. Acta*, **45**, 1099–1109.

Bachtadse, V., Heller, F. and Kroner, A. (1983). Palaeomagnetic investigations in the Hercynian mountain belt of central Europe. *Tectonophysics*, **91**, 285–299.

Badham, J. P. N. (1978). The early history and tectonic signifi-

cance of the East Arm graben, Great Slave Lake, Canada. *Tectonophysics*, **45**, 201–215.

Badham, J. P. N. (1982). Strike-slip orogens—an explanation for the Hercynides. *J. geol. Soc. Lond.*, **139**, 493–504.

Badham, J. P. N. and Stanworth, C. W. (1977). Evaporites from the lower Proterozoic of the East Arm, Great Slave Lake. *Nature*, **268**, 516–518.

Baer, A. J. (1967). The Grenville Province in Helikian times: a possible model of evolution. *Phil. Trans. R. Soc. Lond.*, **A280**, 499–515.

Baer, A. J. (1977). The Grenville Province as a shear zone. *Nature*, **267**, 337–338.

Baer, A. J. (1978). Speculations on the evolution of the lithosphere. *Precamb. Res.*, **5**, 249–260.

Baer, A. J. (1981a). Two orogenies in the Grenville belt? *Nature*, **290**, 129–131.

Baer, A. J. (1981b). A Grenville model of Proterozoic plate tectonic. In A. Kröner (Ed.), *Precambrian Plate Tectonics*, Elsevier, Amsterdam, pp. 353–385.

Baer, A. J., Emslie, R. F., Irving, E. and Tanner, J. C. (1974). Grenville geology and plate tectonics. *Geoscience, Canada*, **1**, 54–61.

Bahat, D. (1979). On the African rift system, theoretical and experimental study. *Earth Planet. Sci. Lett.*, **45**, 445–452.

Bailey, D. K. (1974). Continental rifting and alkaline magmatism. In H. Sørensen (Ed.), *The Alkaline Rocks*, Wiley, London, pp. 148–159.

Bakor, A. R. and Gass, I. G. (1976). Jabal al Wask, NW Saudi Arabia; an Eocambrian back arc ophiolite. *Earth Planet. Sci. Lett.*, **30**, 1–9.

Baker, M. C. W. (1977). Geochronology of Upper Tertiary volcanic activity in the Andes of North Chile. *Geol. Rund.*, **66**, 455–465.

Baker, M. C. W. and Francis, P. W. (1978). Upper Cenozoic volcanism in the central Andes—ages and volumes. *Earth Planet. Sci. Lett.*, **41**, 175–187.

Ball, E. (1980). An example of very consistent brittle deformation over a wide intercontinental area; the late Pan-African fracture system of the Tuareg and Nigerian Shields. *Tectonophysics*, **61**, 363–379.

Ballard, R. D. and Uchupi, E. (1975). Triassic rift structure in Gulf of Maine. *Amer. Ass. Petrol. Geol. Bull.*, **59**, 1041–1072.

Bambach, R. K., Scotese, C. R. and Ziegler, A. M. (1980). Before Pangaea: the geography of the Paleozoic world. *Amer. J. Sci.*, **68**, 26–38.

Bamford, D., Nunn, K., Prodehl, C. and Jacob, B. (1978). Crustal structure of northern Britain. *Geophys. J. R. astra. Soc.*, **54**, 43–60.

Baragar, W. R. A. (1977). Volcanism of the stable crust. In W. R. A. Baragar *et al.* (Eds), *Volcanic Regimes in Canada*, Geol. Ass. Canada. Sp. Pap., **16**, 377–405.

Baragar, W. R. A. and Scoates, R. F. J. (1981). The Circum-Superior belt: a Proterozoic plate margin? In A. Kröner (Ed.), *Precambrian Plate Tectonics*, Elsevier, Amsterdam, pp. 261–296.

Barazangi, M. and Dorman, J. (1969). World seismicity map of ESSA coast and geodetic survey epicenter data for 1961–7. *Bull. Seismol. Soc. Amer.*, **59**, 369–380.

Barber, A. J. (1982). Interpretations of the tectonic evolution of southwest Japan. *Proc. Geol. Ass.*, **93**, 131–145.

Barber, A. J. and Max, M. D. (1979). A new look at the Mona Complex (Anglesey, N. Wales). *J. geol. Soc. Lond.*, **136**, 407–432.

Barberi, F., Bonatti, E., Marinelli, G. and Varet, J. (1974). Transverse tectonics during the split of a continent; data from the Afar rift. *Tectonophysics*, **23**, 17–29.

Barbier, M. J. (1974). Continental weathering as a possible origin of vein-type uranium deposits. *Mineral Deposita.*, **9**, 271–288.

Bard, J. P. (1977). Signification tectonique des métatholéites d'affinité abyssale de la ceinture métamorphique de basse presion d'Aracena (Huelva, Espagne), *Bull. Soc. géol. France*, **19**, 385–393.

Bard, J. P., Burg, J. P., Matte, P. L. and Ribeiro, A. (1980). La chaine hercynienne d'Europe occidentale en terms de tectoni-

362

que des plaques. In J. Cogné and M. Slansky (Eds), *Geology of Europe*, BRGM, Orléans, France, 233–246.

Bard, J. P. *et al.* (1980b). Le métamorphisme en France. In *26th Int. Geol. Cong. Paris. Coll, C7*, Geologie de la France, pp. 162–189.

Bard, J. P., Maluski, H., Matte, P. L. and Proust, F. (1980c). The Kohistan sequence: crust and mantle of an obducted island arc. *Univ. Peshawar. Sp. Issue Geol. Bull.*, **13**, 87–94.

Barghoorn, E. S., Knoll, A. H., Dembicki, H. and Meinschein, W. G. (1977). Variation in stable carbon isotopes in organic matter from the Gunflint Iron Formation. *Geochim. Cosmochim. Acta*, **41**, 425–430.

Barker, F. and Peterman, Z. E. (1974). Bimodal tholeiitic-dacitic magmatism and the early Precambrian crust. *Precamb. Res.*, **1**, 1–12.

Barker, F., Wones, D. R., Sharp, W. N. and Desborough, G. A. (1975). The Pikes Peak batholith, Colorado Front Range, and a model for the origin of the gabbro–anorthosite–syenite–potassic granite suite. *Precamb. Res.*, **2**, 97–160.

Barker, P. F. (1972). A spreading centre in the East Scotia Sea. *Earth Planet. Sci. Lett.*, **15**, 123–132.

Barley, M. E. (1982). Porphyry-style mineralization associated with early Archaean calc–alkaline igneous activity, E. Pilbara, W. Australia. *Econ. Geol.*, **77**, 1230–1236.

Barley, M. E., Dunlop, J. S. R., Glover, J. E. and Groves, D. I. (1979). Sedimentary evidence from an Archean shallow water volcanic–sedimentary facies, eastern Pilbara block, W. Australia. *Earth Planet. Sci. Lett.*, **43**, 74–84.

Barnes, S. J. and Sawyer, E. W. (1980). An alternative model for the Damara mobile belt: ocean crust subduction and continental convergence. *Precamb. Res.*, **13**, 297–336.

Baroz, F. (1980). Volcanism and continental island arc collision in the Penta Dakfyles Range, Cyprus. In A. Panayiotou (Ed.), *Ophiolites, Cyprus Geol. Surv.*, 73–85.

Barton, J. M. Jr., Fripp, R. E. P., Horrocks, P. and McLean, N. (1977). The geology, age and tectonic setting of the Messina layered intrusion, Limpopo mobile belt. In I. F. Ermanovics, R. M. Key and G. McEwen (Eds), The Proc. of a Seminar Pertaining to the Limpopo Mobile Belt. *Geol. Surv. Botswana Bull*, **12**, 75–82.

Barton, J. M. Jr., Fripp, R. E. P., Horrocks, P. and McLean, N. (1979). The geology, age and tectonic setting of the Messina layered intrusion, Limpopo mobile belt, Southern Africa. *Amer. J. Sci.*, **279**, 1108–1134.

Barton, J. M. Jr., Hunter, D. R., Jackson, M. P. A. and Wilson, A. C. (1980). Rb–Sr age and source of the bimodal suite of the Ancient Gneiss Complex, Swaziland, *Nature*, **283**, 756–758.

Barton, J. M. Jr. and Key, R. M. (1981). The tectonic development of the Limpopo mobile belt and the evolution of the Archaean cratons of Southern Africa. In A. Kröner (Ed.), *Precambrian Plate Tectonics*, Elsevier, Amsterdam, pp. 185–212.

Bateman, P. L. (1980). Geologic and geophysical constraints on models for the origin of the Sierre Nevada batholith, California. In W. G. Ernst (Ed.), *The Geotectonic Development of California*, Prentice-Hall, London, pp. 71–86.

Baud, A. *et al.* (1982). Le contact Gondwana—péri-Gondwana dans le Zanskar oriental (Ladakh, Himalaya). *Bull. Soc. géol. France*, **7**, 341–361.

Bayer, R. and Matte, P. L. (1979). Is the mafic/ultramafic massif of Cabo Ortegal (NW. Spain) a nappe emplaced during a Variscan obduction? A new gravity interpretation. *Tectonophysics*, **57**, 9–18.

Beckinsale, R. D., Bowles, J. F. W., Pankhurst, R. J. and Wells, M. K. (1977). Rubidium–strontium age studies and geochemistry of acid veins in the Freetown complex, Sierra Leone. *Mineral. Mag.*, **41**, 501–511.

Beckinsale, R. D. and Thorpe, R. S. (1979). Rubidium–strontium whole rock isochrons: evidence for the age of metamorphism and magmatism in the Mona Complex of Anglesey. *J. geol. Soc. Lond.*, **136**, 433–439.

Beckinsale, R. D., Drury, S. A. and Holt, R. W. (1980). 3360 myr old gneisses from the South Indian Craton. *Nature*, **283**, 469–470.

Behr, H. J., Walliser, O. H. and Weber, K. (1980). The development of the Rheno-Hercynian and Saxo-Thuringian zones of the mid-European Variscides. *26th Int. Geol. Cong. Paris*, Coll. C6, *Geology of Europe*, pp. 77–89.

Behr, H. J., Engel, W. and Franke, W. (1982). Variscan wild-flysch and nappe tectonics in the Saxo-Thuringian zone (NE Bavaria, W Germany). *Amer. J. Sci.*, **282**, 1438–1470.

Bell, K., Blenkinsop, J. and Moore, J. M. (1975). Evidence for a Proterozoic greenstone belt from Snow Lake, Manitoba. *Nature*, **258**, 698–701.

Bell, R. T. and Jackson, G. D. (1974). Aphebian halite and sulphate indications in the Belcher Group, Northwest Territories. *Can. J. Earth Sci.*, **11**, 722–728.

Ben-Avraham, Z. and Nur, A. (1980). The elevation of volcanoes and their edifice heights at subduction zones. *J. Geophys. Res.*, **85**, B8, 4325–4335.

Ben-Avraham, Z., Nur, A., Jones, D. and Cox, A. (1981). Continental accretion from oceanic plateaus to alloththonous terrains. *Science*, **213**, 47–54.

Ben-Avraham, Z., Nur, A. and Jones, D. (1982). The emplacement of ophiolites by collision. *J. Geophys. Res.*, **87**, B5, 3861–3867.

Berg, J. H. (1977). Regional geobarometry in the contact aureole of the anorthositic Nain complex, Labrador. *J. Petrol.*, **18**, 339–430.

Berger, G. W., York, D. and Dunlop, D. J. (1979). Calibration of Grenvillian palaeopoles by $^{40}Ar/^{39}Ar$ dating. *Nature*, **277**, 46–47.

Berkner, L. V. and Marshall, L. C. (1967). The rise of oxygen in the Earth's atmosphere with notes on the Martian atmosphere. *Adv. Geophys.*, **12**, 309–331.

Bernoulli, D. and Jenkins, H. C. (1974). Alpine, Mediterranean and Central Atlantic Mesozoic facies in relation to the early evolution of the Tethys. In R. H. Dott Jr and R. H. Shaver (Eds), *Modern and Ancient Geosynclinal Sedimentation, Soc. Econ. Pal. Mineral. Sp. Publ.*, **19**, 129–160.

Bernoulli, D. and Lemoine, M. (1980). Birth and early evolution of the Tethys: the overall situation. In *Geology of the Alpine Chains born of the Tethys. 26th Int. Geol. Cong. Paris* (Mem. BRGN, **115**), 168–179.

Berthelsen, A. (1978). Himalayan and Sveconorwegian tectonics—a comparison. In P. S. Saklani (Ed.), *Tectonic Geology of the Himalayas*, pp. 287–294.

Berthelsen, A. (1980). Towards a palinspastic tectonic analysis of the Baltic Shield. In J. Cogné and M. Slansky (Eds), *Geology of Europe*, BRGM, Orléans, France, pp. 5–21.

Bertrand, J. M. L. and Caby, R. (1978). Geodynamic evolution of the Pan-African orogenic belt: a new interpretation of the Hoggar Shield (Algerian Sahara). *Geol. Rund.*, **67**, 357–388.

Besairie, H. *et al.* (1951). Le Graphite à Madagascar. *Madagascar Bur. Geol. Tr.*, No. 27, 94 pp.

Bhanot, V. B., Geol, A. K., Singh, V. P. and Kwatra, S. K. (1975). Rb–Sr radiometric studies for Dalhousie and Rohtang areas, Himachal Pradesh. *Curr. Sci.*, **44**, 219.

Bhanot, V. B., Singh, V. P., Kansal, A. K. and Thakur, V. C. (1977). Early Proterozoic Rb–Sr whole rock age for central crystalline gneiss of Higher Himalaya, Kumaun. *J. geol. Soc. India*, **18**, 90–91.

Bibikova, Y. V., Makarov, V. A., Gracheva, T. V. and Seslaviskiy, K. B. (1978). Age of oldest rocks of the Omolon block. *Dokl. Earth Sci. Lett.*, **241**, 43–45.

Bichan, R. (1970). The evolution and structural setting of the Great Dyke, Rhodesia. In T. N. Clifford and I. G. Gass (Eds), *African Magmatism and Tectonics*, Oliver and Boyd, Edinburgh, 51–71.

Bickford, M. E. and Mose, D. G. (1975). Geochronology of Precambrian rocks in the St. Francois Mountains, SE Missouri. *Geol. Soc. Amer. Sp. Pap.*, **165**, 48 pp.

Bickle, M. J. (1978). Heat loss from the earth: a constraint on Archaean tectonics from the relation between geothermal gradients and the rate of plate production. *Earth Planet. Sci. Lett.*, **40**, 301–315.

Bickle, M. J., Martin, A. and Nisbet, E. G. (1975). Basaltic and peridotitic komatiites and stromatolites above a basal unconformity in the Belingwe greenstone belt, Rhodesia. *Earth Planet. Sci. Lett.*, **27**, 155–162.

Bickle, M. J., Bettenay, L. F., Boulter, C. A., Groves, D. L. and

Morant, P. (1980). Horizontal tectonic interaction of an Archaean gneiss belt and greenstones, Pilbara block, W Australia. *Geology*, **8**, 525–529.

Bickle, M. J. and Eriksson, K. A. (1982). Evolution and subsidence of early Precambrian sedimentary basins. *Phil. Trans. R. Soc. Lond.*, **A305**, 225–247.

Bignell, R. D. (1975). Timing, distribution and origin of submarine mineralization in the Red Sea. *Trans. Inst. Min. Metall.*, **84**, 1–6.

Bingham, D. K. and Klootwijk, C. J. (1980). Palaeomagnetic constraints on Greater India's underthrusting of the Tibetan plateau. *Nature*, **284**, 336–338.

Bird, J. M. and Dewey, J. F. (1970). Lithosphere plate: continental margin tectonics and the evolution of the Appalachian orogen. *Bull. geol. Soc. Amer.*, **81**, 1031–1059.

Bird, P. (1979). Continental delamination and the Colorado plateau. *J. Geophys. Res.*, **84**, 7561–7571.

Bishop, F. C., Smith, J. V. and Windley, B. F. (1980). The Fiskenaesset Complex, W Greenland, part 4, Chemistry of sulphide minerals. *Bull. Grønlands geol. Unders*, **137**, 35 pp. (the last of 4 monographs).

Black, R., Ba, H., Bertrand, J. M. L., Bouillier, A. M., Caby, R., Davison, I., Fabre, J., Leblanc, M. and Wright, L. I. (1979a). Outline of the Pan-African geology of Adrar des Iforas (Republic of Mali). *Geol. Rund.*, **68**, 543–564.

Black, R., Caby, R., Moussine-Pouchkine, A., Bayer, R., Bertrand, J. M., Bouillier, A. M., Fabre, J and Lesquer, A. (1979b). Evidence for late Precambrian plate tectonics in West Africa, *Nature*, **278**, 223–227.

Blackburn, C. E. (1980). Towards a mobilist tectonic model for part of the Archaean of NW Ontario. *Geosci. Can.*, **7**, 64–72.

Blais, S., Auvray, B., Capdevila, R. and Hameurt, I. (1977). Les Séries komatiitique et tholéiitique des ceintures archéennes de roches vertes de Finlande orientale. *Bull. Soc. géol. France*, **19**, 965–970.

Blake, M. C. *et al.* (1978). Neogene basin formation in relation to plate-tectonic evolution of San Andreas fault system, California. *Amer. Ass. Petrol Geol. Bull.*, **62**, 344–372.

Bluck, B. J. (1978). Geology of a continental margin: The Ballantrae Complex. In D. R. Bowes and B. E. Leake (Eds), *Crustal Evolution in NW Britain and Adjacent Regions*, Seel House Press, Liverpool, pp. 150–162.

Bluck, B. J., Halliday, A. N., Aftalion, M. and MacIntyre, R. M. (1980). Age and origin of Ballantrae ophiolite and its significance to the Caledonian orogeny and Ordovician time scale. *Geology*, **8**, 492–495.

Boak, J. L. and Dymek, R. F. (1982). Metamorphism of the *ca* 3800 Ma supracrustal rocks at Isua, W Greenland: implications for early Archaean crustal evolution. *Earth Planet. Sci. Lett.*, **59**, 155–176.

Bogatikov, O. A. (1974). *Anorthosites of the USSR*, Acad. Nauk, Moscow, 122 pp (in Russian).

Bogomolov, M. A. (1981). Gradients of parameters of Archaean metamorphism of the Anabar Shield. *Dokl. Akad. Nauk., SSR*, **256**, 95–97 (English translation).

Bonatti, E. (1973). Origin of offsets of the Mid-Atlantic ridge in fracture zones. *J. Geol.*, **81**, 144–156.

Bonatti, E. (1975). Metallogenesis at oceanic spreading centres. *Ann. Rev. Earth Planet. Sci.*, **3**, 401–431.

Bonatti, E. (1978). Vertical tectonism in oceanic fracture zones. *Earth Planet. Sci. Lett.*, **37**, 369–379.

Bonatti, E. (1981). Metal deposits in the oceanic lithosphere. In C. Emiliani (Ed.), *The Oceanic Lithosphere*, Wiley, New York, pp. 639–686.

Bonatti, E. and Honnorez, J. (1976). Sections of the earth's crust in the equatorial Atlantic. *J. Geophys. Res.*, **81**, 4104–4116.

Bonatti, E., Honnorez-Guerstein, M. B., Honnorez, J. and Stern, C. (1976). Hydrothermal pyrite concretions from the Romanche trench (equatorial Atlantic): metallogenesis in oceanic fracture zones. *Earth Planet. Sci. Lett.*, **32**, 1–10.

Bonatti, E., Harrison, C. G. A., Fisher, D. E., Honnorez, J., Schilling, J. G., Stipp, J. J. and Zentilli, M. (1977). Easter volcanic chain (Southeast Pacific): a mantle hot line. *J. Geophys. Res.*, **82**, 2457–2478.

Bondesen, E. (1970). The stratigraphy and deformation of the Precambrian rocks of the Graenseland area, SW Greenland. *Meddr. Grønland*, **185**, 125 pp.

Bondesen, E., Pedersen, K. R. and Jørgensen, O. (1967). Precambrian organisms and the isotopic composition of organic remains in the Ketilidian of south-west Greenland. *Meddr. Grønland*, **164**, 4, 41 pp.

Bose, M. K. and Chakraborti, M. K. (1981). Fossil marginal basin from the Indian shield: a model for the evolution of Singhbhum Precambrian belt, E India. *Geol. Rund.*, **70**, 504–518.

Bosellini, A. and Winterer, E. L. (1975). Pelagic limestone and radiolarite of the Tethyan Mesozoic: a genetic model. *Geology*, **3**, 279–282.

Bouchez, J. L. and Pêcher, A. (1981). The Himalayan Main Central Thrust pile and its quartz-rich tectonites in Central Nepal. *Tectonophysics*, **78**, 23–50.

Bowden, P. and Turner, D. C. (1974). Peralkaline and associated ring complexes in the Nigeria-Niger province, West Africa. In H. Sørensen (Ed.), *The Alkaline Rocks*, Wiley, London, pp. 330–351.

Bowden, P. and Kinnaird, J. A. (1978). Younger granites of Nigeria—a zinc-rich tin province. *Trans. Inst. Min. Metall. Sect. B*, **87**, 66–69.

Bowes, D. R. (1976). Archaean crustal history in the Baltic Shield. In B. F. Windley (Ed.), *The Early History of the Earth*, Wiley–Interscience, London, pp. 481–488.

Bowes, D. R. (1978). Shield formation in early Precambrian times: the Lewisian complex. In D. R. Bowes and B. E. Leake (Eds), *Crustal Evolution in N.W. Britain and Adjacent Regions*, Seel House Press, Liverpool, pp. 39–80.

Bowes, D. R. (1980). Correlation in the Svecokarelides and a crustal model. In F. P. Mitranov (Ed.), *Principles and Criteria of Subdivision of Precambrian in Mobile Zones*, Nauka, Leningrad, pp. 294–303.

Boyle, R. W. (1976). Mineralization processes in Archaean greenstone and sedimentary belts. *Geol. Surv. Can. Pap.*, **75**—15, 45 pp.

Brass, G. W. (1976). The variation of the marine $^{87}Sr/^{86}Sr$ ratio diring Phanerozoic time: interpretation using a flux model. *Geochim. Cosmochim. Acta*, **40**, 721–730.

Brewer, M. S., Brook, M. and Powell, D. (1979). Dating of the tectonometamorphic history of the SW Moines, Scotland. In A. L. Harris, C. H. Holland and B. E. Leake (Eds), *The Caledonides of the British Isles—reviewed*, Geol. Soc. Lond., Scottish Academic Press, Edinburgh, pp. 129–137.

Briden, J. C. (1976). Application of palaeomagnetism to Proterozoic tectonics. *Phil. Trans. R. Soc. Lond.*, **A280**, 405–416.

Briden, J. C. and Irving, E. (1964). Palaeolatitude spectra of sedimentary palaeoclimatic indicators. In A. E. M. Nairn (Ed.), *Problems in Palaeoclimatology*, Interscience, New York, pp. 199–224.

Briden, J. C., Morris, W. A. and Piper, J. D. A. (1973). Regional and global implications. In *Palaeomagnetic studies in the British Caledonides, Geophys. J. R. Astr. Soc.*, **34**, 107–134.

Bridgwater, D. and Windley, B. F. (1973). Anorthosites, post-orogenic granites, acid volcanic rocks, and crustal development in the North Atlantic Shield during the mid-Proterozoic. In L. A. Lister (Ed.), *Symposium on Granites, Gneisses and Related Rocks, Geol. Soc. S. Afr. Sp. Publ.*, **3**, 307–318.

Bridgwater, D., Watson, J. and Windley, B. F. (1973). The Archaean craton of the North Atlantic region. *Phil. Trans. R. Soc. Lond.*, **A273**, 493–512.

Bridgwater, D., McGregor, V. R. and Myers, J. S. (1974a). A horizontal tectonic regime in the Archaean of Greenland and its implications for early crustal thickening. *Precamb. Res.*, **1**, 179–198.

Bridgwater, D., Sutton, J. and Watterson, J. (1974b). Crustal down-folding associated with igneous activity. *Tectonophysics*, **21**, 57–77.

Bridgwater, D. and Collerson, K. (1976). The major petrological and geochemical characters of the 3600 m.y. Uivak gneisses from Labrador. *Contrib. Mineral. Petrol.*, **54**, 43–60.

Bridgwater, D., Collerson, K. D. and Myers, J. S. (1978). The development of Archaean gneiss complex of the North Atlan-

364

tic region. In D. H. Tarling (Ed.), *Evolution of the Earth's Crust*, Academic Press, London, pp. 19–69.

Bridgwater, D. *et al*. (1981). Microfossil-like objects from the Archaean of Greenland: a cautionary note. *Nature*, **289**, 51–53.

Brocoum, S. J. and Dalziel, I. W. D. (1974). The Sudbury basin, the Southern Province, the Grenville and Penokean orogeny. *Bull. geol. Soc. Amer.*, **85**, 1571–1580.

Bromley, A. V. (1976). A new interpretation of the Lizard complex, S Cornwall in the light of the ocean crust model. *J. geol. Soc. Lond.*, **132**, 114 (abstract).

Brook, M., Powell, D. and Brewer, M. S. (1977). Grenville events in the Moine rocks of the Northern Highlands, Scotland. *J. geol. Soc. Lond.*, **133**, 489–496.

Brookfield, M. E. (1981). Metamorphic distributions and events in the Ladakh range, Indus suture zone and Karakorum mountains. In P. S. Saklani (Ed.), *Metamorphic Tectonites of the Himalaya*, Today and Tomorrow's Publ., New Delhi, pp. 1–14.

Brookfield, M. E. and Reynolds, P. H. (1981). Late Cretaceous emplacement of the Indnus suture zone ophiolitic mélanges and an Eocene-Oligocene magmatic arc on the northern edge of the Indian plate. *Earth Planet. Sci. Lett.*, **55**, 157–162.

Brooks, C., Ludden, J., Pigeon, Y. and Hubregtse, J. J. M. W. (1982). Volcanism of shoshonite to high-K andesite affinity in an Archaean arc environment. *Can. J. Earth Sci.*, **19**, 55–67.

Brown, D. A. (1968). Some problems of distribution of late Palaeozoic and Triassic terrestrial vertebrates. *Australian J. Sci.*, **30**, 439–445.

Brown, D. A., Campbell, K. S. W. and Crook, K. A. W. (1968). *The Geological Evolution of Australia and New Zealand*, Commonwealth and Int. Lib., Pergamon Press, 409 pp.

Brown, G. C. (1977). Mantle origin of Cordilleran granites. *Nature*, **265**, 21–24.

Brown, G. C. (1979). The changing patterns of batholith emplacement during earth history. In M. P. Atherton and J. Tarney (Eds), *Origin of Granite Batholiths: Geochemical Evidence*, Shiva Publications, Kent, pp. 106–115.

Brown, G. C. (1981). Space and time in granite plutonism. *Phil. Trans. R. Soc. Lond.*, **A301**, 321–336.

Brown, G. C. (1982). Calc-alkaline intrusive rocks: their diversity, evolution, and relation to volcanic arcs. In R. S. Thorpe (Ed.), *Andesites*, Wiley, London, pp. 437–459.

Brown, G. C. (in press). Processes and problems in the continental lithosphere: geological history and physical implications. In N. Snelling (Ed.), *Geochronology and Geological Record*, Geol. Soc. Lond. Sp. Publ., pp. 000–000.

Brown, M., Friend, C. R. L., McGregor, V. R. and Perkins, W. T. (1981). The late Archaean Qôrqut granite complex of SW Greenland. *J. Geophys. Res.*, **86 B11**, 10617–10632.

Brown, P. A. (1976). Ophiolites in SW Newfoundland. *Nature*, **264**, 712–715.

Brown, R. L. (1978). Structural evolution of the SE Canadian Cordillera: a new hypothesis. *Tectonophysics*, **48**, 133–151.

Brown, R. L., Chappell, J. F., Moore, J. M. and Thompson, P. H. (1975). An ensimatic island arc and ocean closure in the Grenville Province of SE Ontario, Canada. *Geosci. Can.*, **2**, 141–144.

Brun, J. P. and Burg, J. P. (1982). Combined thrusting and wrenching in the Ibero–Armorican arc: a corner effect during continental collision. *Earth Planet. Sci. Lett.*, **61**, 319–332.

Buffetaut, E. and Taquet, P. (1979). An early Creteceous terrestrial crocodile and the opening of the South Atlantic. *Nature*, **280**, 486–487.

Bukharev, V. P., Stekolinikov, A. V. and Polyanskiy, V. D. (1973). Tectonics and deep structure of anorthosite massifs in the northwest of the Ukrainian Shield. *Geotectonics*, **4**, 207–210.

Bullard, E., Everett, J. E. and Smith, A. G. (1965). The fit of the continents around the Atlantic. *Phil. Trans. R. Soc. Lond.*, **A258**, 41–51.

Burg, J. P. and Matte, P. L. (1978). A cross-section through the French Massif Central and the scope of its Variscan geodynamic evolution. *Zeit. Deutsch geol. Ges.*, **129**, 429–460.

Burg, J. P., Iglesias, M., Laurent, P. L. and Ribeiro, A. (1981). Variscan intracontinental deformation: the Coimbra-Cordoba

shear zone (SW Iberian Peninsula). *Tectonophysics*, **78**, 161–177.

Burke, K. (1975). Atlantic evaporites formed by evaporation of water spilled from Pacific, Tethyan and southern oceans. *Geology*, **3**, 613–616.

Burke, K. (1976). Development of graben associated with the initial ruptures of the Atlantic ocean. *Tectonophysics*, **36**, 83–112.

Burke, K. (1977). Aulacogens and continental break-up. *Ann. Rev. Earth Planet. Sci.*, **5**, 371–396.

Burke, K. and Dewey, J. F. (1973a). Plume-generated triple junctions: key indicators in applying plate tectonics to old rocks. *J. Geol.*, **81**, 406–433.

Burke, K. and Dewey, J. F. (1973b). An outline of Precambrian plate development. In D. H. Tarling and S. K. Runcorn (Eds), *Implications of Continental Drift to the Earth Sciences*, vol. 2, Academic Press, London, 1035–1045.

Burke, K., Dewey, J. F. and Kidd, W. S. F. (1976a). Dominance of horizontal movements, arc and microcontinental collisions during the later permobile regime. In B. F. Windley (Ed.), *The Early History of the Earth*, Wiley–Interscience, London, pp. 113–129.

Burke, K., Dewey, J. F. and Kidd, W. S. F. (1976b). Precambrian palaeomagnetic results compatible with contemporary operation of the Wilson cycle. *Tectonophysics*, **33**, 287–299.

Burke, K. and Kidd, W. S. F. (1978). Were Archaean continental geothermal gradients much steeper than those of today? *Nature*, **272**, 240–241.

Burke, D. and Kidd, W. S. F. (1980). Volcanism on earth through time. In D. W. Strangway (Ed.), *The Continental Crust and its Mineral Deposits*, Geol. Ass. Can. Sp. Pap., **20**, 503–522.

Button, A. (1976a). Iron-formation as an end-member in carbonate sedimentary cycles in the Transvaal Supergroup, South Africa. *Econ. Geol.*, **71**, 193–201.

Button, A. (1976b). Transvaal and Hammersley basins—review of basin development and mineral deposits. *Mineral Sci. Engng.*, **8**, 262–293.

Button, A. *et al*. (1981). Southern Africa: the cratonic environment. In D. R. Hunter (Ed.), *Precambrian of the Southern Hemisphere*, Elsevier, Amsterdam, pp. 501–639.

Caby, R., Bertrand, J. M. L. and Black, R. (1981). Pan-African ocean closure and continental collision in the Hoggar-Iforas segment, central Sahara. In A. Kröner (Ed.), *Precambrian Plate Tectonics*, Elsevier, Amsterdam, pp. 407–434.

Caia, J. (1976). Palaeogeographical and sedimentological controls of copper, lead and zinc mineralizations in the lower Creteceous sandstones of Africa. *Econ. Geol.*, **71**, 409–422.

Camfield, P. A. and Gough, D. I. (1977). A possible Proterozoic plate boundary in North America. *Can. J. Earth Sci.*, **14**, 1229–1238.

Campbell, D. S. (1980). Structural and metamorphic development of migmatites in the Svecokarelides near Tampere, Finland. *Trans. R. Soc. Edin. Earth Sci.*, **71**, 185–200.

Campbell, I. H. (1977). A study of macro-rhythmic layering and cumulate processes in the Jimberlana intrusion, W Australia. I. The upper layered series. *J. Petrol.*, **18**, 183–215.

Cann, J. R. (1981). Basalts from the ocean floor. In C. Emiliani (Ed.), *The Oceanic Lithosphere*, Wiley, New York, 363–390.

Capdevila, R., Goodwin, A. M., Ujike, O. and Gorton, M. P. (1982). Trace element geochemistry of Archaean volcanic rocks and crustal growth in SW Abitibi belt, Canada. *Geology*, **10**, 418–422.

Carey, S. W. (1958). A tectonic approach to continental drift. In S. W. Carey (Ed.), *Continental Drift*, A symposium, University of Tasmania, Hobart, Tasmania, pp. 177–355.

Carpenter, M. S. N. and Civetta, L. (1976). Hercynian high pressure/low temperature metamorphism in the Île de Groix blueschists. *Nature*, **262**, 276–277.

Cathles, L. M., Cuber, A. L., Lenagh, T. C., Dudas, F. and Horikoshi, E. (1980). Kuroko massive sulfide deposits in Japan: products of an attempt to form a new marginal basin? *Geol. Soc. Amer. Abst. with Prog.*, **12**, 400.

365

Cavanaugh, M. D. and Seyfert, C. K. (1977). Apparent polar wander paths and the joining of the Superior and Slave Provinces during early Proterozoic times. *Geology*, **5**, 207–211.

Cawthorn, R. G. (1977). Pyroxene compositions, reaction relations, and the lack of silica enrichment in the eastern Bushveld Complex. *Trans. geol. Soc. S. Afr.*, **80**, 139–144.

Cesbron, F. and Bariand, P. (1975). The uranium–vanadium deposit of Mounana, Gabon. *Min. Record*, **6**, 237–249.

Chadwick, B. (1981). Field relations, petrography and geochemistry of Archaean amphibolite dykes and Malene supracrustal amphibolites, N. W. Buksefjorden, SW Greenland. *Precamb. Res.*, **14**, 221–259.

Chadwick, B. and Coe, K. (1976). New evidence relating to Archaean events in southern West Greenland. In B. F. Windley (Ed.), *The Early History of the Earth*, Wiley–Interscience, London, pp. 203–212.

Chadwick, B., Ramakrishnan, M., Viswanatha, M. V. and Srinivasa Murthy, V. (1978). Structural studies in the Archaean Sargur and Dharwar supracrustal rocks of the Karnataka craton. *J. geol. Soc. India*, **19**, 531–549.

Chadwick, B., Ramakrishnan, M. and Viswanatha, M. N. (1981a). The stratigraphy and structure of the Chitradurga region: an illustration of cover-basement interaction in the late Archaean evolution of the Karnataka craton, S India. *Precamb. Res.*, **16**, 31–54.

Chadwick, B., Ramakrishnan, M. and Viswanatha, M. N. (1981b). Structural and metamorphic relations between Sargur and Dharwar supracrustal rocks and Peninsular Gneiss in central Karnataka. *J. geol. Soc. India*, **22**, 557–569.

Chandler, S. W. and Schwarz, E. J. (1980). Tectonics of the Richmond Gulf area, northern Quebec—a hypothesis. In *Geol. Surv. Can. Pap.*, **80–1C**, 59–68.

Chapman, H. J. (1979). 2390 My Rb–Sr whole rock for the Scourie dykes of north western Scotland. *Nature*, **277**, 642–643.

Chapman, H. J. and Moorbath, S. (1977). Lead isotope measurements from the oldest recognised Lewisian gneisses of NW Scotland. *Nature*, **268**, 41–42.

Chappell, B. W. and White, A. J. R. (1974). Two contrasting granite types. *Pacific Geol.*, **8**, 173–174.

Chappell, J. F., Brown, R. L. and Moore, J. M. (1975). Subduction and continental collision in the Grenville province of SE Ontario. *Geol. Soc. Amer. Abst. with Prog.*, **7**, 733–734.

Chauhan, D. S. (1979). Phosphate-bearing stromatolites of the Precambrian Aravalli phosphorite deposits of the Udaipur region, their environmental significance and genesis of phosphorite. *Precamb. Res.*, **8**, 95–126.

Cheng Yu-Chi, Chao Yih-Ming and Lu Song-Nien (1978). Main type groups of iron deposits of China. *Acta Geologica Sinica* No. 4 (Chinese with English abnstract, English translation by *Geol. Soc. China*, 1976, 1–30).

Cheng Yugi, Bai Jin, Sun Dazhong (1982). The Lower Precambrian of China. *Rev. Bras. Geol.*, **12**, 65–73.

Choubert, G. and Faure-Muret, A. (1980). The Precambrian in North Peri-Atlantic and South Mediterranean mobile zones: general results. *Earth Sci. Rev.*, **16**, 85–219.

Choudhuri, A. (1980). The early Proterozoic greenstone belts of the northern Guyana Shield, S America. *Precamb. Res.*, **13**, 363–374.

Christensen, N. I. and Salisbury, M. H. (1975). Structure and constitution of the lower oceanic crust. *Rev. Geophys. Space Phys.*, **13**, 57–86.

Churkhrov, F. V., Vinogradov, V. I. and Ermilova, L. P. (1970). On the isotopic sulfur composition of some Precambrian strata. *Mineral Deposita*, **5**, 209–222.

Chumakov, N. M. (1981). Upper Proterozoic glaciogenic rocks and their stratigraphic significance. *Precamb. Res.*, **15**, 373–395.

Churkin, M. Jr. (1974). Palaeozoic marginal ocean basin-volcanic arc systems in the Cordilleran fold belt. In R. H. Dott Jr and R. H. Shaver (Eds), *Modern and Ancient Geosynclinal Sedimentation*, *Soc. Econ. Pal. Mineral. Sp. Publ.*, **19**, 174–192.

Churkin, M. and Eberlein, G. D. (1977). Ancient borderland terranes of the North American Cordillera: correlation and microplate tectonics. *Bull. geol. Soc. Amer.*, **88**, 769–786.

Cita, M. B. (1982). The Messinian salinity crisis in the Mediterranean (review). In H. Berckhemer and K. Hsü (Eds), *Alpine–Mediterranean Geodynamics*, Amer. Geophys. Un., pp. 113–140.

Clague, D. A. and Jarrard, R. D. (1973). Tertiary Pacific plate motion deduced from the Hawaiian-Emperor chain. *Bull. geol. Soc. Amer.*, **84**, 1135–1154.

Clark, A. H. *et al.* (1976). Longitudinal variations in the metallogenetic evolution of the central Andes: a progress report. *Geol. Ass. Canada Sp. Pap.*, **14**, 24–58.

Clark, G. S., Bald, R. and Ayres, L. D. (1981). Geochronology of orthogneiss adjacent to the Archaean Lake of the Woods greenstone belts, NW Ontario: a possible basement complex. *Can. J. Earth Sci.*, **18**, 94–102.

Clark, S. P. and Jäger, E. (1969). Denudation rate in the Alps from geochronologic heat flow data. *Amer. J. Sci.*, **267**, 1143–1160.

Clemmey, H. (1981). Some aspects of the genesis of heavy mineral assemblages in Lower Proterozoic uranium–gold conglomerates. *Mineral. Mag.*, **44**, 399–408.

Clemmey, H. and Badham, N. (1982). Oxygen in the Precambrian atmosphere: an evaluation of the geological evidence. *Geology*, **10**, 141–146.

Cloud, P. (1968a). Pre-Metazoan evolution and the origins of the Metazoa. In E. T. Drake (Ed.), *Evolution and Environment*, Yale University Press, New York, pp. 1–72.

Cloud, P. (1968b). Atmospheric and hydrospheric evolution on the primitive earth. *Science*, **160**, 729–736.

Cloud, P. (1973). Palaeoecological significance of the banded iron formation. *Econ. Geol.*, **68**, 1135–1143.

Cloud, P. (1976a). Beginnings of biospheric evolution and their biogeochemical consequences. *Paleobiology*, **2**, 351–387.

Cloud, P. (1976b). Major features of crustal evolution. *Geol. Soc. S. Afr. Annex.*, **79**, 1–33.

Cloud, P. and Semikhatov, M. A. (1969). Proterozoic stromatolite zonation. *Amer. J. Sci.*, **267**, 1017–1061.

Cobbing, E. J. (1976). The geosynclinal pair at the continental margin of Peru. *Tectonophysics*, **36**, 157–165.

Cobbing, E. J. (1982). The segmented coastal batholith of Peru: its relationship to volcanicity and metallogenesis. *Earth Sci. Rev.*, **18**, 241–251.

Cocks, C. R. M. and Fortey, R. A. (1982). Faunal evidence for oceanic separations in the Palaeozoic of Britain. *J. geol. Soc. Lond.*, **139**, 465–478.

Cocks, L. R. M. and Toghill, P. (1973). The biostratigraphy of the Silurian rocks of the Girvan district, Scotland. *J. geol. Soc. Lond.*, **129**, 209–243.

Coertze, F. J., Burger, A. J., Walraven, F., Marlow, A. G. and MacCaskie, D. R. (1978). Field relations and age determinations in the Bushveld complex. *Trans. geol. Soc. S. Afr.*, **81**, 1–11.

Cogné, J. (1977). La chaîne hercynienne ouest-Européene correspond-elle à un orogène par collision? Propositions pour une interpretation géodynamique globale. In *Himalaya: Sciences de la Terre*, CNRS No. 268, Paris, pp. 111–130.

Cogné, J. and Wright, J. B. (1980). L'orogéne cadomien. In J. Cogné and M. Slansky (Eds), *Geology of Europe*, BRGM, Orléans, France, pp. 29–55.

Colbert, E. H. (1973). Continental drift and the distribution of fossil reptiles. In D. H. Tarling and S. K. Runcorn (Eds), *Implications of Continental Drift to the Earth Sciences*, vol. 1, Academic Press, London, pp. 395–412.

Coleman, R. G. (1977). *Ophiolites*. Springer Verlag, Berlin, 229 pp.

Collerson, K. D. (1983). The Archaean gneiss complex of northern Labrador. 2. Mineral ages, secondary isochrons, and diffusions of strontium during polymetamorphism of the Uivak gneisses. *Can. J. Earth Sci.*, **20**, 707–718.

Collerson, K. D., Jesseau, C. W. and Bridgwater, D. (1976). Crustal development of the Archaean gneiss complex, eastern Labrador. In B. F. Windley (Ed.), *The Early History of the Earth*, Wiley–Interscience, London, pp. 237–253.

Collerson, K. D. and Fryer, B. J. (1978). The role of fluids in the formation and subsequent development of early continental crust. *Contr. Mineral. Petrol.*, **67**, 151–167.

Collerson, K. D. and Bridgwater, D. (1979). Metamorphic development of early Archaean tonalitic and trondhjemitic gneisses: Saglek area, Labrador. In F. Barker (Ed.), *Trondhjemites, Dacites and Related Rocks*, Elsevier, Amsterdam, pp. 205–273.

Collerson, K. D., Brooks, C., Ryan, A. B. and Compston, W. (1982). A reappraisal of the Rb–Sr systematics of early Archaean gneisses from Hebron, Labrador. *Earth Planet. Sci. Lett.*, **60**, 325–336.

Collette, B. J. (1974). Thermal contraction joints in a spreading sea-floor as origin of fracture zones. *Nature*, **251**, 299–300.

Colman-Sadd, S. P. (1980). Geology of south-central Newfoundland and evolution of the eastern margin of Iapetus. *Amer. J. Sci.*, **280**, 991–1017.

Compton, P. (1978). Rare earth evidence for the origin of the Nûk gneisses, Buksefjorden region, SW Greenland. *Contr. Mineral Petrol.*, **66**, 283–293.

Condie, K. C. (1973). Archaean magmatism and crustal thickening. *Bull. geol. Soc. Amer.*, **84**, 2981–2992.

Condie, K. C. (1976). The Wyoming Archean Province in the western United States. In B. F. Windley (Ed.), *The Early History of the Earth*, Wiley–Interscience, London, pp. 419–424.

Condie, K. C. (1980). Origin and early development of the earth's crust. *Precamb. Res.*, **11**, 183–197.

Condie, K. C. (1981). *Archean Greenstone Belts*, Elsevier, Amsterdam, 434 pp.

Condie, K. C. (1982a). *Plate Tectonics and Crustal Evolution*, 2nd edn, Pergamon Press, New York, 301 pp.

Condie, K. C. (1982b). Archaean andesites. In R. S. Thorpe (Ed.), *Andesites*, Wiley, London, pp. 575–590.

Condie, K. C. (1982c). Plate tectonic models for Proterozoic continental accretion in the SW United States. *Geology*, **10**, 37–42.

Condie, K. C. (1982d). Early and middle Proterozoic supracrustal successions and their tectonic settings. *Amer. J. Sci.*, **282**, 341–357.

Condie, K. C. and Harrison, N. M. (1976). Geochemistry of the Archaean Bulawayan Group, Midlands greenstone belt, Rhodesia. *Precamb. Res.*, **3**, 253–271.

Condie, D. C. and Hunter, D. R. (1976). Trace element geochemistry of Archaean granitic rocks from the Barberton region, South Africa. *Earth Planet. Sci. Lett.*, **29**, 389–400.

Condie, K. C. and Moore, J. M. (1977). Geochemistry of Proterozoic volcanic rocks from the Grenville Province, eastern Ontario. In W. R. A. Baragar, L. C. Coleman and J. M. Hall (Eds), *Volcanic Regimes in Canada*, Geol. Soc. Can. Sp. Pap., **16**, 149–168.

Condie, K. C. and Nuter, J. A. (1981). Geochemistry of the Dubois greenstone succession: an early Proterozoic bimodal volcanic association in west-central Colorado. *Precamb. Res.*, **15**, 131–155.

Condie, K. C., Allen, P. and Narayana, B. L. (1982). Geochemistry of the Archean low- to high-grade transition zone, S India. *Contr. Mineral. Petrol.*, **81**, 157–167.

Coney, P. J. (1972). Cordilleran tectonics and North America plate motion. *Amer. J. Sci.*, **272**, 603–628.

Coney, P. J. (1979). Tertiary evolution of Cordilleran metamorphic core complexes. In J. M. Armentrout, M. R. Cole and H. Terbest (Eds), *Cenozoic Paleogeography of the Western United States*, Soc. Econ. Pal. Mineral., 15–28.

Coney, P. J., Jones, D. L. and Monger, J. W. H. (1980). Cordilleran suspect terranes. *Nature*, **288**, 329–333.

Constantinou, G. (1976). Genesis of the conglomerate structure, porosity and collomorphic textures of the massive sulphide ores of Cyprus. In D. F. Strong (Ed.), *Metallogeny and Plate Tectonics, Geol. Ass. Can. Sp. Pap.*, **14**, 187–210.

Cook, F. A., Albaugh, D. S., Brown, L. D., Kaufman, S., Oliver, J. E. and Hatcher, R. D. (1979). Thin skinned tectonics in the crystalline southern Appalachians: COCORP seismic-reflection profiling of the Blue Ridge and Piedmont. *Geology*, **7**, 563–567.

Cook, F. A. and Oliver, F. E. (1981). The late Precambrian-early Palaeozoic continental edge in the Appalachian orogen. *Amer. J. Sci.*, **281**, 993–1008.

Cook, F. A. and Turcotte, D. L. (1981). Parametrized convection and the thermal evolution of the earth. *Tectonophysics*, **75**, 1–17.

Cook, F. A., Brown, L. D., Kaufman, S., Oliver, J. E. and Peterson, T. A. (1981). COCORP seismic profiling of the southern Appalachian orogen beneath the coastal plain of Georgia. *Bull. geol. Soc. Amer.*, part 1, **92**, 738–748.

Cook, P. J. and McElhinny, M. W. (1979). A reevaluation of the spatial and temporal distribution of sedimentary phosphate deposits in the light of plate tectonics. *Econ. Geol.*, **74**, 315–330.

Cooper, J. A., Nesbitt, R. W., Platt, J. P. and Mortimer, G. E. (1978). Crustal development in the Agnew region, W Australia, as shown by Rb/Sr isotopic and geochemical studies. *Precamb. Res.*, **7**, 31–59.

Corliss, J. B. et al. (1979). Submarine thermal springs on the Galapagos rift. *Science*, **203**, 1073–1083.

Corliss, J. B., Baross, J. A. and Hoffman, S. E. (1981). A hypothesis concerning the relationship between submarine hot springs and the origin of life on earth. *Oceanologica Acta. Sp. Publ.*, **80–7**, 59–69.

Coulon, C. and Thorpe, R. S. (1981). Role of continental crust in petrogenesis of orogenic volcanic associations. *Tectonophysics*, **77**, 79–93.

Cowan, D. S. (1978). Origin of blueschist-bearing chaotic rocks in the Franciscan complex, San Simeon, California. *Bull. geol. Soc. Amer.*, **89**, 1415–1423.

Coward, M. P. and James, P. R. (1974). The deformation patterns of two Archaean greenstone belts in Rhodesia and Botswana. *Precamb. Res.*, **1**, 235–258.

Coward, M. P., Lintern, B. C. and Wright, L. I. (1976). The pre-cleavage deformation of the sediments and gneisses of the northern part of the Limpopo belt. In B. F. Windley (Ed.), *The Early History of the Earth*, Wiley–Interscience, London, pp. 323–330.

Coward, M. P., Jan, M. Q., Rex, D., Tarney, J., Thirwall, M. and Windley, B. F. (1982). Geotectonic framework of the Himalaya of N Pakistan. *J. geol. Soc. Lond.*, **139**, 299–308.

Coward, M. P. and McClay, K. R. (1983). Thrust tectonics of S Devon. *J. geol. Soc. Lond.*, **140**, 215–228.

Coward, M. P., Windley, B. F., Pudsey, C., Broughton, R. D., Petterson, M. P., Luff, I. and Khan, M. A. (in press). Collision tectonics in the NW Himalaya. In J. G. Ramsay, M. P. Coward and A. Ries (Eds), *Collision Tectonics*, Geol. Soc. Lond.

Cowie, J. W. (1974). The Cambrian of Spitzbergen and Scotland. In C. H. Holland (Ed.), *Cambrian of the British Isles, Norden and Spitzbergen. Lower Palaeozoic rocks of the World*, vol. 2, Wiley, London, pp. 123–156.

Cox, C. B. (1974). Vertebrate palaeodistributional patterns and continental drift. *J. Biogeogr.*, **1**, 75–94.

Crawford, A. R. (1979). The myth of a vast oceanic Tethys, the India–Asia problem and earth expansion. *J. Petrol. Geol.*, **2**, 3–9.

Crawford, A. R. and Daily, B. (1971). Probable non-synchroneity of late Precambrian glaciations. *Nature*, **230**, 111–112.

Crittenden, M. D., Coney, P. J. and Davis, G. H. (Eds) (1980). Cordilleran metamorphic core complexes. *Geol. Soc. Amer. Mem.*, **153**, 490 pp.

Cronan, D. S. (1976). Basal metalliferous sediments from the eastern Pacific. *Bull. geol. Soc. Amer.*, **87**, 928–934.

Cross, T. A. and Pilger, R. H. (1978). Constraints on absolute motion and plate interactions inferred from Cenozoic igneous activity in the Western United States. *Amer. J. Sci.*, **278**, 865–902.

Cross, T. A. and Pilger, R. H. (1982). Controls of subduction geometry, location of magmatic arcs, and tectonics of arc and back-arc regions. *Bull. geol. Soc. Amer.*, **93**, 545–562.

Crough, S. T., Morgan, W. J. and Hargraves, R. B. (1980). Kimberlites: their relation to mantle hotspots. *Earth Planet. Sci. Lett.*, **50**, 260–274.

Crouse, R. A., Cerny, P., Trueman, D. L. and Burt, R. O. (1979). The Tranco pegmatite, SE Manitoba. *Can. Min. Metall. Bull.*, **72**, 142–151.

Crowell, J. C. (1978). Gondwana glaciation, cyclotherms, continental positioning and climate change. *Amer. J. Sci.*, **278**, 1345–1372.

Dabitzias, S. G. (1980). Petrology and genesis of the Vavdos crypto-crystalline magnesite deposits, Chalkidiki peninsula, N Greece. *Econ. Geol.*, **75**, 1138–1151.

Dagelayskaya, I. N. and Moshkin, V. N. (1979). Petrology of the anorthositization zones in the Precambrian of the Aldan-Stanovoy Shield. *Int. Geol. Rev.*, **21**, 281–290.

Dahlkamp, F. J. (1980). The time-related occurrence of uranium deposits. *Mineral. Deposita*, **15**, 67–79.

Dallmeyer, R. D. (1975). The Pallisades sill: a Jurassic intrusion? Evidence from ^{40}Ar/^{39}Ar incremental release ages, *Geology*, **3**, 243–245.

Dalrymple, G. B., Grommé, C. S. and White, R. W. (1975). Potassium–argon age and palaeomagnetism of diabase dikes in Liberia: initiation of central Atlantic rifting. *Bull. geol. Soc. Amer.*, **86**, 399–411.

Dalziel, I. W. D. (1981). Back arc extension in the southern Andes: a review and critical appraisal. *Phil. Trans. R. Soc. Lond.*, **A300**, 319–335.

Davidson, A., Culshaw, N. G. and Nadeau, L. (1982). A tectonic metamorphic framework for part of the Grenville Province, Parry Sound region, Ontario. *Geol. Surv. Canada, Pap.*, **82–1A**, 175–190.

Davidson, D. (1973). Plate tectonic model for the Musgrave Block, Amadeus Basin complex of Central Australia. *Nature,*, *Phys. Sci.*, **245**, 21–33.

Davies, G. F. (1979). Thickness and thermal history of continental crust and root zones. *Earth Planet. Sci. Lett.*, **44**, 231–238.

Davies, J. F. and Luhta, L. E. (1978). An Archaean 'porphyry-type' disseminated copper deposit, Timmins, Ontario. *Econ. Geol.*, **73**, 383–369.

Davis, G. H. (1983). Shear zone model for the origin of metamorphic core complexes. *Geology*, **11**, 342–347.

Davis, G. H. and Coney, P. J. (1979). Geologic development of the Cordilleran metamorphic core complexes. *Geology*, **7**, 120–124.

Dawson, J. B. (1980). *Kimberlites and Their Xenoliths*, Springer, Berlin, 280 pp.

de Brodtkorb, M. K. and Brodtkorb, A. (1977). Strata-bound scheelite deposits in the Precambrian basement of San Luis (Argentina). In D. D. Klemm and H. J. Schneider (Eds), *Time and Strata-bound Ore Deposits*, Springer, Berlin, pp. 141–149.

De Jong, K. A. and Scholten, R. (Eds) (1973). *Gravity and Tectonics*, Wiley, New York, 502 pp.

De Laeter, J. R., Fletcher, I. R., Rosman, K. J. R., Williams, I. R., Gee, R. D. and Libby, W. G. (1981). Early Archaean gneisses from the Yilgarn block, W Australia. *Nature*, **292**, 322–323.

Delfour, J. (1976). Volcanisme et gîtes minéraux du Bouclier arabo-nubien. *Mém. h. Sér. Soc. géol. France*, no. 7, 137–142.

DeLong, S. E., Dewey, J. F. and Fox, P. J. (1977). Displacement history of oceanic fracture zones. *Geology*, **5**, 199–201.

DeLong, S. E., Dewey, J. F. and Fox, P. J. (1979). Topographic and geologic evolution of fracture zones. *J. geol. Soc. Lond.*, **136**, 303–310.

Demaiffe, D. and Javoy, M. (1980). ^{18}O/^{16}O ratios of anorthosites and related rocks from the Rogaland complex (SW Norway). *Contrib. Mineral. Petrol.*, **72**, 311–317.

DePaolo, D. J. (1980). Crustal growth and mantle evolution: inferences from models of element transport and Nb and Sr isotopes. *Geochim. Cosmochim. Acta*, **44**, 1185–1196.

DePaolo, D. J. and Wasserburg, G. J. (1979). Sm–Nd age of the Stillwater complex and the mantle evolution curve for neodymium. *Geochim. Cosmochim. Acta*, **43**, 999–1008.

Derry, D. R. (1980). Uranium deposits through time. In D. W. Strangway (Ed.), *The Continental Crust and its Mineral Deposits, Geol. Ass. Can. Sp. Pap.*, **20**, 625–632.

Deruelle, B. (1978). Calc-alkaline and shoshonitic lavas from five Archean volcanoes (between latitudes 21° 45' and 24° 30' S) and the distribution of the Plio-Quaternary volcanism of the south-central and southern Andes. *J. Volcanol. Geotherm. Res.*, **3**, 281–298.

Desmet, A. P. (1976). Evidence of co-genesis of the Troodos lavas, Cyprus. *Geol. Mag.*, **113**, 165–168.

Desmet, A., Lapierre, H., Roccl, G., Gagny, C. L., Parrot, J. F. and Delaloye, M. (1978). Constitution and significance of the Troodos sheeted complex. *Nature*, **273**, 527–530.

Dewey, J. F. (1969). Evolution of the Appalachian/Caledonian orogen. *Nature*, **222**, 124–129.

Dewey, J. F. (1972). Plate tectonics. *Sci. Amer.*, **226**, 56–68.

Dewey, J. F. (1980). Episodicity, sequence and style at convergent plate boundaries. In D. W. Strangway (Ed.), *The Continental Crust and its Mineral Deposits, Geol. Ass. Can. Sp. Pap.*, **20**, 553–573.

Dewey, J. F. and Burke, K. C. A. (1973). Tibetan, Variscan and Precambrian basement reactivation: products of continental collision, *J. Geol.*, **81**, 683–692.

Dewey, J. F., Pitman, W. C. III, Ryan, W. B. F. and Bonnin, J. (1973). Plate tectonics and the evolution of the Alpine System, *Bull. geol. Soc. Amer.*, **84**, 3137–3180.

Dewey, J. F. and Burke, K. (1974). Hot spots and continental break-up: implications for collisional orogeny. *Geology*, **2**, 57–60.

Dewey, J. F. and Kidd, W. S. F. (1974). Continental collisions in the Appalachian-Caledonian orogenic belt: variations related to complete and incomplete suturing. *Geology*, **2**, 543–546.

Dewey, J. F. and Kidd, W. S. F. (1977). Geometry of plate accretion. *Bull. geol. Soc. Amer.*, **88**, 960–968.

Dewey, J. F. and Windley, B. F. (1981). Growth and differentiation of the continental crust. *Phil. Trans. R. Soc. Lond.*, **A301**, 189–206.

De Wit, M. J. (1982). Sliding and overthrust nappe tectonics in the Barberton greenstone belt. *J. Struct. Geol.*, **4**, 117–136.

De Wit, M., Hart, R., Stern, C. and Barton, C. M. (1980). Metallogenesis related to seawater interaction with 3.5 by oceanic crust. *Eos*, **61**, 386.

De Wit, M. J. and Stern, C. R. (1981). Variations in the degree of crustal extension during formation of a back arc basin. *Tectonophysics*, **72**, 229–260.

De Wit, M. J., Hart, R., Martin, A. and Abbottt, P. (1982). Archaean abiogenic and probable biogenic structures associated with mineralized hydrothermal vent systems and regional metamorphism with implications for greenstone belt studies. *Econ. Geol.*, **77**, 1783–1802.

Dick, H. J. B. (1982). The petrology of two back arc basins of the northern Philippine sea. *Amer. J. Sci.*, **282**, 644–700.

Dickinson, W. R. (1973a). Widths of modern arc-trench gaps proportional to past duration of igneous activity in associated magmatic arcs. *J. Geophys. Res.*, **78**, 3376–3389.

Dickinson, W. R. (1973b). Reconstruction of past arc-trench systems from petrotectonic assemblages in the island arcs of the western Pacific. In P. J. Coleman (Ed.), *The Western Pacific Island Arcs, Marginal Seas, Geochemistry*, University of W Australia Press, Nedlands, pp. 569–601.

Dickinson, W. R. (1974a). Plate tectonics and sedimentation. In W. R. Dickinson (Ed.), *Tectonics and Sedimentation, Soc. Econ. Pal. Mineral. Sp. Publ.*, **22**, 1–27.

Dickinson, W. R. (1974b). Sedimentation within and beside ancient and modern magmatic arcs. In R. H. Dott Jr and R. H. Shaver (Eds), *Modern and Ancient Geosynclinal Sedimentation, Soc. Econ. Pal. Mineral. Sp. Publ.*, **19**, 230–239.

Dickinson, W. R. (1976). Sedimentary basins developed during evolution of the Mesozoic–Cenozoic. *Can. J. Earth Sci.*, **13**, 1268–1288.

Dickinson, W. R. (1981). Plate tectonics through geological time. *Phil. Trans. R. Soc. Lond.*, **A301**, 207–215.

Dickinson, W. R. (1982). Compositions of sandstones in circum-Pacific subduction complexes and fore-arc basins. *Amer. Ass. Petrol Geol. Bull.*, **66**, 121–137.

Dickinson, W. R. and Seely, D. R. (1979). Stratigraphy and structure of forearc regions. *Amer. Ass. Petrol Geol. Bull.*, **63**, 2–31.

Dietrich, V. J. (1976). Plattentektonik in den Ostalpen: Eine Arbeitshypothese. *Geotektonische Forsch.*, **50**, 1–84.

Dietrich, V. and Gansser, A. (1981). The leucogranites of the Bhutan Himalaya. *Schweiz. mineral. petrogr. Mitt.*, **61**, 177–202.

368

Dietz, R. S. (1964). Sudbury structure as an astrobleme. *J. Geol.*, **72**, 412–434.

Dilley, F. C. (1973). Large foraminifera and seas through time. In N. F. Hughes (Ed.), *Organisms and Continents through Time, Pal. Ass. Sp. Pap. in Pal.*, **12**, 155–168.

Dimroth, E. (1977). Facies models. 5. Models of physical sedimentation of iron formations. *Geosci. Can.*,m **4**, 23–30.

Dimroth, E. (1979). Significance of diagenesis for the origin of Witwatersrand-type uraniferous conglomerates. *Phil. Trans. R. Soc. Lond.*, **A291**, 277–287.

Dimroth, E. (1981). Labrador Geosyncline: type example of early Proterozoic cratonic reactivation. In A. Kröner (Ed.), *Precambrian Plate Tectonics*, Elsevier, Amsterdam, pp. 331–352.

Dimroth, E. and Kimberley, M. M. (1976). Precambrian atmospheric oxygen: evidence in the sedimentary distributions of carbon, sulfur, uranium and iron. *Can. J. Earth Sci.*, **13**, 1161–1185.

Dimroth, E. and Lichtblau, A. P. (1978). Oxygen in the Archaean ocean: comparison of ferric oxide crusts on Archaean and Cainozoic pillow basalts. *Neues Jahrb. Min. Abhdl.*, **133**, 1–22.

Dimroth, E., Imreh, L., Rocheleau, M. and Goulet, N. (1982). Evolution of the south central part of the Abitibi Belt, Quebec, Pt. 1. Stratigraphy and paleogeographic model. *Can. J. Earth Sci.*, **19**, 1729–1758.

Donn, W. L. and Shaw, D. M. (1977). Model of climate evolution based on continental drift and polar wandering. *Bull. geol. Soc. Amer.*, **88**, 390–396.

Dostal, J., Caby, R. and Dupuy, C. (1979). Metamorphosed alkaline intrusions and dyke complexes within the Pan-African belt of Western Hoggar (Algeria): geology and geochemistry. *Precamb. Res.*, **10**, 1–20.

Dougan, T. W. (1976). Origin of trondhjemitic biotite–quartz–oligoclase gneisses from the Venezuelan Guyana Shield. *Precamb. Res.*, **3**, 317–342.

Douglas R. J. W. (1980). Proposals for time classification and correlation of Precambrian rocks and events in Canada and adjacent areas of the Canadian Shield. *Geol. Surv. Can. Pap.*, **80-24**, 19 pp.

Doumani, G. A. and Long, W. E. (1962). The ancient life of the Antarctic. *Sci. Amer.*, **207**, 169–184.

Downing, K. N. and Coward, M. P. (1981). The Okahandha lineament and its significance for Damaran tectonics in Namibia. *Geol. Rund.*, **70**, 972–1000.

Drake, R., Vergar, M., Munizaga, F. and Vicente, J. C. (1982). Geochronology of Mesozoic–Cenozoic magmatism in Central Chile, Lat. 31°–36° S. *Earth Sci. Rev.*, **18**, 353–363.

Drewry, G. E., Ramsay, A. T. S. and Smith. A. G. (1974). Climatically controlled sediments, the geomagnetic field and trade wind belts in Phanerozoic time. *J. Geol.*, **82**, 531–533.

Drury, S. A. (1977). Structures induced by granite diapirs in the Archaean greenstone belt at Yellowknife, Canada: implications for Archaean geotectonics. *J. Geol.*, **85**, 345–358.

Drury, S. A. (1978). REE distributions in a high grade Archaean gneiss complex in Scotland: implications for the genesis of ancient stable crust. *Precamb. Res.*, **7**, 237–257.

Drury, S. A. (1983). The petrogenesis and setting of Archaean metavolcanics from Karnataka State, South India. *Geochim. Cosmochim. Acta*, **47**, 317–329.

Drury, S. A. and Holt, R. W. (1980). The tectonic framework of the South Indian craton: a reconnaisance involving LANDSAT imagery. *Tectonophysics*, **65**, 71–75.

Dudenko, I. N. and Zhdanov, V. V. (1979). Chemical evolution of meta-anorthosite assemblage in the NE Kola peninsula. *Int. Geol. Rev.*, **21**, 57–62.

Dunlop, J. S. R., Muir, M. D., Milne, V. A. and Groves, D. I. (1978). A new microfossil assemblage from the Archaean of W Australia. *Nature*, **274**, 676–678.

Dunlop, D. J., York, D., Berger, G. W., Buchan, K. L. and Stirling, J. M. (1980). The Grenville Province: a paleomagnetic case study of Precambrian continental drift. In D. W. Strangway (Ed.), *The Continental Crust and its Mineral Deposits, Geol. Ass. Can. Sp. Pap.*, **20**, 487–502.

Dunn, P. R., Thomson, B. P. and Rankama, K. (1971). Late Precambrian glaciation in Australia as a stratigraphic boundary. *Nature*, **231**, 498–502.

Dupré, B. and Allègre, C. J. (1980). Pb–Sr–Nd isotopic correlations and the chemistry of the North Atlantic mantle. *Nature*, **286**, 17–22.

Edhorn, A. S. (1977). Early Cambrian algae croppers. *Can. J. Earth Sci.*, **14**, 1014–1020.

Eguchi, T., Uyeda, S. and Maki, T. (1980). Seismotectonics and tectonic history of the Andaman Sea. In M. N. Toksoz, S. Uyeda and J. Francheteau (Eds), *Oceanic Ridges and Arcs: Geodynamic Processes*, Elsevier, Amsterdam, pp. 425–441.

Eisbacher, G. H. (1974). Evolution of successor basins in the Canadian Cordillera. In R. H. Dott Jr and R. H. Shaver (Eds), *Modern and Ancient Geosynclinal Sedimentation, Soc. Econ. Pal. Mineral. Sp. Publ.*, **19**, 274–291.

Elbers, F. J. (1976). Calc-alkaline plutonism, volcanism and related hydrothermal mineralization in the Superior Province of NE Manitoba. *Can. Min. Metall. Bull.*, **69**, 83–95.

Ellis, R. A. (1977). Disseminated copper in a Caledonian calc-alkaline intrusion, Argyllshire, Scotland. *Trans. Inst. Min. Metall.*, **86B**, 52–54.

Elthon, D. (1981). Metamorphism in oceanic spreading centres. In C. Emiliani (Ed.), *The Oceanic Lithosphere*, Wiley, New York, pp. 285–303.

Embleton, B. J. J. and Schmidt, P. W. (1979). Recognition of common Precambrian polar wandering reveals a conflict with plate tectonics. *Nature*, **282**, 705–707.

Emeleus, C. H. and Upton, B. G. J. (1976). The Gardar Province in South Greenland. In A. Escher and W. S. Watt (Eds), *Geology of Greenland, Geol. Surv. Greenland*, Copenhagen, pp. 152–181.

Emiliani, C., Kraus, E. R. and Shoemaker, E. M. (1981). Sudden death at the end of the Mesozoic. *Earth Planet. Sci. Lett.*, **55**, 317–334.

Emslie, R. F. (1970). The geology of the Michikamau intrusion Labrador. *Geol. Surv. Can. Pap.*, **68–57**, 88 pp.

Emslie, R. F. (1975). Nature and origin of anorthositic rocks. *Geosci. Can.*, **2**, 99–104.

Emslie, R. F. (1978a). Anorthosite massifs, rapakivi granites and late Proterozoic rifting of North America. *Precamb. Res.*, **7**, 61–98.

Emslie, R. F. (1978b). Elsonian magmatism in Labrador: age, characteristics and tectonic setting. *Can. J. Earth Sci.*, **15**, 438–453.

Engel, A. E. J. and Kelm, D. L. (1972). Pre-Permian global tectonics: a tectonic test. *Bull. geol. Soc. Amer.*, **83**, 2325–2340.

Engel, A. E. J., Ibson, S. P., Engel, C. G., Stickney, D. M. and Cray, E. J. Jr (1974). Crustal evolution and global tectonics: a petrogenic view. *Bull. geol. Soc. Amer.*, **85**, 843–858.

Engel, A. E. J., Dixon, T. H. and Stern, R. J. (1980). Late Precambrian evolution of Afro-Arabian crust from ocean arc to craton. *Bull. geol. Soc. Amer.*, part 1, **91**, 699–706.

England, P. C. and Richardson, S. W. (1977). The influence of erosion upon the mineral facies of rocks from different tectonic environments. *J. geol. Soc. Lond.*, **134**, 2£1–213.

Eriksson, K. A. (1979). Marginal marine depositional processes from the Archaean Moodies Group, Barberton Mountain Land, South Africa: evidence and significance. *Precamb. Res.*, **8**, 153–182.

Eriksson, K. A. (1980). Transitional sedimentation styles in the Moodies and Fig Tree Groups, Barberton Mountain Land, South Africa: evidence favouring an Archaean continental margin. *Precamb. Res.*, **12**, 141–160.

Eriksson, K. A., Truswell, J. F. and Button, A. (1976). Palaeoenvironmental and geochemical models from an Early Proterozoic carbonate succession in South Africa. In M. R. Walter (Ed.), *Stromatolites*, Elsevier, Amsterdam, pp. 635–644.

Eriksson, K. A. and Truswell, J. F. (1978). Geological processes and atmospheric evolution in the Precambrian. In D. H. Tarling (Ed.), *Evolution of the Earth's Crust*, Academic Press, London, pp. 219–238.

Ermanovics, I. F., Key, R. M. and McEwen, G. (Eds) (1977).

The Proceedings of a Seminar pertaining to the Limpopo mobile belt. *Botswana Geol. Surv. Bull.*, **12**, 247 pp.

Ernst, W.A. G. (1971). Metamorphic zonations on presumably subducted lithospheric plates from Japan, California and the Alps. *Contr. Mineral. Petrol.*, **34**, 43–59.

Ernst, W. G. (1972). Occurrence and mineralogic evolution of blueschist belts with time. *Amer. J. Sci.*, **272**, 657–668.

Escher, A., Escher, J. C. and Watterson, J. (1975). The reorientation of the Kangâmiut dike swarm, West Greenland. *Can. J. Earth Sci.*, **12**, 158–173.

Escher, J. C. and Myers, J. S. (1975). New evidence concerning the original relationship of early Precambrian volcanics and anorthosite in the Fiskenasset region, southern Greenland, *Rap. Grønlands geol. Unders.*, **75**, 72–76.

Escher, A., Jack, S. and Watterson, J. (1976). Tectonics of the North Atlantic Proterozoic dyke swarm. *Phil. Trans. R. Soc. Lond.*, **A280**, 529–539.

Evans, A. M. (1975). Mineralisation in geosynclines—the Alpine enigma. *Mineral. Deposita*, **10**, 254–260.

Evans, A. M. (1980). *An Introduction to Ore Geology*, Blackwell, Oxford, 231 pp.

Evans, A. M. (1977). Copper–molybdenum mineralization in the Ballachulish granite, Argyllshire, Scotland. *Trans. Inst. Min. Metall.*, **86B**, 152–153.

Evans, R. (1978). Origin and significance of evaporites in basins around Atlantic margin. *Amer. Ass. Petrol Geol. Bull.*, **62**, 223–234.

Evarts, R. C. and Schiffman, P. (1983). Submarine hydrothermal metamorphism of the Del Puerto ophiolite, California. *Amer. J. Sci.*, **283**, 289–340.

Ewing, J. and Ewing, M. (1970). Seismic reflection. In A. E. Maxwell (Ed.), *The Sea*, vol. 4, part 1, Wiley–Interscience, New York, pp. 1–51.

Exley, C. S., Stone, M. and Floyd, P. A. (1983). Composition and petrogenesis of the Cornubian granite batholith and post-orogenic volcanic rocks in SW England. In P. L. Hancock (Ed.), *The Variscan Fold Belt in the British Isles*, Adam Hilger, Bristol, pp. 153–177.

Fahrig, W. F., Irving, E. and Jackson, G. D. (1971). Palaeomagnetism of the Franklin diabases, *Can. J. Earth Sci.*, **8**, 455–467.

Fairhead, J. DS. (1980). The structure of the cross-cutting volcanic chain of Northern Tanzania and its relation to the east African rift system. *Tectonophysics*, **65**, 193–208.

Falkum, T. and Petersen, J. S. (1980). The Sveconorwegian orogenic belt, a case of late Proterozoic plate collision. *Geol. Rund.*, **69**, 622–647.

Fallaw, W. C. (1979). Trans-North Atlantic similarity among Mesozoic and Cenozoic invertebrates correlated with widening of the ocean basin. *Geology*, **7**, 389–400.

Fallaw, W. C. and Dromgoole, E. I. (1980). Faunal similarities across the South Atlantic among Mesozoic and Cenozoic invertebrates correlated with widening of the ocean basin. *J. Geol.*, **88**, 723–727.

Farrar, E. H., Clark, S. J., Haynes, G. S., Quirt, H., Conn, T. and Zentilli, M. (1970). K–Ar evidence for the post-Paleozoic migration of granite intrusion foci in the Andes of northern Chile. *Earth Planet. Sci. Lett.*, **10**, 60–66.

Fedorovskij, V. S. and Lejtes, A. M. (1980). Aldanian Shield: development stages of crystalline complexes and the age of the continental earth's crust. *Krystalinikum*, **15**, 7–32.

Finlow-Bates, T. and Stumpfl, E. F. (1979). The copper and lead-zinc-silver ore bodies of Mt. Isa Mine, Queensland: products of ore hydrothermal system. *Ann. Soc. geol. Belg.*, **102**, 497–517.

Fitton, J. G. (1980). The Benue Trough and Cameroon Line—a migrating rift system in West Africa. *Earth Planet. Sci. Lett.*, **51**, 132–138.

Flores, G. (1970). Suggested origin of the Mozambique Channel. *Trans. geol. Soc. S. Afr.*, **73**, 1–16.

Flower, M. F. J. (1981). Thermal and kinematic control on ocean ridge magma fractionation: contrasts between Atlantic and Pacific spreading axes. *J. geol. Soc. Lond.*, **138**, 695–712.

Floyd, P. A., Exley, C. S. and Stone, M. (1983). Variscan magmatism in SW England—discussion and synthesis. In P. L. Hancock (Ed.), *The Variscan Fold Belt in the British Isles*, Adam Hilger, Bristol, pp. 178–185.

Foland, K. A. and Faul, H. (1977). Ages of the White Mountain intrusions in New Hampshire, Vermont and Maine, USA. *Amer. J. Sci.*, **277**, 888–904.

Folinsbee, R. E. (1982). Variations in the distribution of mineral deposits with time. In H. D. Holland and M. Schidlowski (Eds), *Mineral Deposits and the Evolution of the Biosphere*, Springer Verlag, Berlin, 219–236.

Förster, R. (1978). Evidence for an open seaway between northern and southern proto-Atlantic in Albian times. *Nature*, **272**, 158–159.

Fox, P. J., Schreiber, E., Rowlett, H. and McCamy, K. (1976). The geology of the Oceanographer fracture zone: a model for fracture zones. *J. Geophys. Res.*, **81**, 4117–4128.

Fox, P. J. and Stroup, J. B. (1981). The plutonic foundation of the oceanic crust. In C. Emiliani (Ed.), *The Oceanic Lithosphere*, Wiley, New York, pp. 119–218.

Frakes, L. A. (1979). *Climates Throughout Geologic Time*, Elsevier, Amsterdam, 310 pp.

Francheteau, J., Choukroune, P., Hekinian, R., Le Pichon, X. and Needham, H. D. (1976). Oceanic fracture zones do not provide deep sections in the crust. *Can. J. Earth Sci.*, **13**, 1223–1235.

Francheteau, J. *et al.* (1979). Massive deep sea sulphide ore deposits discovered on the East Pacific Rise. *Nature*, **277**, 523–528.

Francis, E. H. (1978). The Midland Valley as a rift, seen in connection with the late Palaeozoic European rift system. In I. B. Ramberg and E. R. Neumann (Eds), *Tectonics and Geophysics of Continental Rifts*, Reidel, Dordrecht, Holland, pp. 133–147.

Frank, W., Thöni, M. and Purtscheller, F. (1977). Geology and petrology of Kulu—S. Lahul area. In *Himalaya. Sciences de la Terre*, Coll. Int. CNRS No. **268**, Paris, pp. 147–172.

Frarey, M. J. and Roscoe, S. M. (1970). The Huronian Supergroup north of Lake Huron. In A. J. Baer (Ed.), *Symp. on Basins and Geosynclines of the Canadian Shield*, Geol. Surv. Can. Pap., **70–40**, 143–157.

Fraser, J. A., Hoffman, P. F. and Irvine, T. N. (1972). The Bear Province. In *Variations in Tectonic Styles in Canada*, Geol. Ass. Can. Sp. Pap., **11**, 453–504.

French, B. M. (1972). Shock-metamorphic features in the Sudbury structure, Ontario: a review. *Geol. Ass. Canad. Sp. Pap.*, **10**, 19–28.

Friend, C. R. L., Hall, R. P. and Hughes, D. H. (1981). The geochemistry of the Malene (mid-Archaean) ultramafic-mafic amphibolite suite, SW Greenland. In J. E. Glover and D. I. Groves (Eds), *Archaean Geology, Geol. Soc. Australia, Sp. Publ.*, **7**, 301–312.

Fripp, R. E. P. (1976). Gold metallogeny in the Archaean of Rhodesia. In B. F. Windley (Ed.), *The Early History of the Earth*, Wiley–Interscience, London, pp. 455–466.

Fripp, R. E. P. (1981). The ancient Sand River gneisses, Limpopo mobile belt, South Africa. In J. E. Glover and D. I. Groves (Eds), *Archaean Geology, Geol. Soc. Australia, Sp. Publ.*, **7**, 330–335.

Fripp, R. E. P., van Nierop, D. A., Callow, M. J., Lilly, P. A. and du Plessis, L. U. (1980). Deformation in part of the Archaean Kaapvaal craton, South Africa. *Precamb. Res.*, **13**, 241–251.

Frisch, W. (1979). Tectonic progradation and plate tectonic evolution of the Alps. *Tectonophysics*, **60**, 121–139.

Frisch, W. and Al-Shanti, A. (1977). Ophiolite belts and the collision of island arcs in the Arabian shield. *Tectonophysics*, **43**, 293–306.

Frith, R. A. and Doig, R. (1975). Pre-Kenoran tonalitic gneisses in the Grenville Province. *Can. J. Earth Sci.*, **12**, 844–849.

Frutos, J. (1982). Archaean metallogeny related to the tectonic and petrologic evolution of the Cordillera: some remarkable points. In G. S. Amstutz *et al.* (Eds), *Ore Genesis: the State of the Art*, Springer Verlag, Berlin, pp. 493–507.

Fryer, B. J., Fyfe, W. S. and Kerrich, R. (1979). Archaean volcanogenic oceans. *Chem. Geol.*, **24**, 25–35.

370

Fuchs, G. (1981). Outline of the geology of the Himalaya. *Mitt. Österr. geol. Ges.*, **74/75**, 101–127.

Furlong, K. P. and Chapman, D. S. (1982). Thermal modelling of the geometry of the tectonics of the overriding plate. *J. Geophys. Res.*, **87**, B3, 1786–1802.

Fyfe, W. S. (1973). The granulite facies, partial melting and the Archaean crust. *Phil. Trans. R. Soc. Lond.*, **A273**, 457–461.

Fyfe, W. S. (1978). The evolution of the earth's crust: modern plate tectonics to ancient hot spot tectonics? *Chem. Geol.*, **23**, 89–114.

Fyfe, W. S. (1981). How do we recognise plate tectonics in very old rocks? In A. Kröner (Ed.), *Precambrian Plate Tectonics*, Elsevier, Amsterdam, pp. 549–560.

Fyfe, W. S. and Lonsdale, P. (1981). Ocean floor hydrothermal activity. In C. Emiliani (Ed.), *The Oceanic Lithosphere*, Wiley, New York, pp. 589–638.

Gaal, G. (1982). Proterozoic tectonic evolution and late Svecokarelian plate deformation of the central Baltic Shield. *Geol. Rund.*, **71**, 158–170.

Gabelman, J. W. (1976). Orogenic and taphrogenic mineralisation belts at continental margins. In D. F. Strong (Ed.), *Metallogeny and Plate Tectonics, Geol. Ass. Can. Sp. Pap.*, **14**, 273–300.

Gancarz, A. J. and Wasserburg, G. J. (1977). Initial Pb of the Amîtsoq gneiss, W Greenland and implications for the age of the earth. *Geochim. Cosmochim. Acta*, **41**, 1283–1301.

Gansser, A. (1964). *Geology of the Himalayas*, Wiley, London, 289 pp.

Gansser, A. (1979). The ophiolitic suture zones of the Ladakh and the Kailas regions—a comparison. *J. geol. Soc. India*, **20**, 277–281.

Gansser, A. (1980). The significance of the Himalayan suture zone. *Tectonophysics*, **62**, 37–52.

Garcia, M. O. (1978). Criteria for the identification of ancient volcanic arcs. *Earth Sci. Rev.*, **14**, 147–165.

Garrison, J. R. (1981). Coral Creek serpentinite, Llano uplift, Texas; a fragment of an incomplete Precambrian ophiolite. *Geology*, **9**, 225–230.

Garson, M. S. and Krs, M. (1976). Geophysical and geological evidence of the relationship of Red Sea transverse tectonics to ancient fractures. *Bull. geol. Soc. Amer.*, **87**, 169–181.

Gartner, S. and Keany, J. (1978). The terminal Cretaceous event: a geologic problem with an oceanographic solution. *Geology*, **6**, 708–712.

Gass, I. G. (1975). Magmatic and tectonic processes in the development of the Afro-Arabian domes. In A. Pilger and A. Rosler (Eds), *Afar Depression of Ethiopia*, Schweizerbartsche Verlag, Stuttgart, pp. 10–18.

Gass, I. G. (1980). The Troodos massif; its role in the unravelling of the ophiolite problem and its significance in the understanding of constructive plate margin processes. In A. Panayiotou (Ed.), *Ophiolites, Cyprus Geol. Surv.*, pp. 23–35.

Gass, I. G. (1981). Pan-African (Upper Proterozoic) plate tectonics of the Arabian-Nubian Shield. In A. Kröner (Ed.), *Precambrian Plate Tectonics*, Elsevier, Amsterdam, pp. 388–405.

Gass, I. G., Mallick, D. I. J. and Cox, K. G. (1973). Volcanic islands of the Red Sea. *J. geol. Soc. Lond.*, **129**, 275–310.

Gass, I. G. and Smewing, J. D. (1973). Intrusion, extrusion and metamorphism at constructive margins: evidence from the Troodos Massif, Cyprus. *Nature*, **242**, 26–29.

Gay, A. L. and Grandstaff, D. E. (1980). Chemistry and mineralogy of Precambrian palaeosols at Elliot Lake, Ontario, Canada. *Precamb. Res.*, **12**, 349–373.

Gee, R. D., Baxter, J. L., Wilde, S. A. and Williams, I. R. (1981). Crustal development in the Archaean Yilgarn block, W Australia. In J. E. Glover and D. I. Groves (Eds), *Archaean Geology, Geol. Soc. Australia, Sp. Publ.*, **7**, 43–56.

Gélinas, L., Brooks, C., Perrault, G., Carignan, J., Trudel, P. and Grasso, F. (1977). Chemo-stratigraphic divisions within the Abitibi volcanic belt, Rouyn-Noranda district, Quebec. In W. R. A. Baragar, L. C. Coleman and J. M. Hall (Eds), *Volcanic Regimes in Canada. Geol. Ass. Can. Sp. Pap.*, **16**, 265–295.

Ghisler, M. and Windley, B. F. (1967). The chromite deposits of the Fiskenasset region, West Greenland. *Rapp. Grønlands geol. Unders.*, **12**, 1–39.

Ghose, N. C. and Singh, R. N. (1980). Occurrence of blueschist facies in the ophiolite belt of Naga hills E of Kiphire, NE India. *Geol. Rdsch.*, **69**, 41–48.

Gibb, R. A. (1978). Slave-Churchill collision tectonics. *Nature*, **271**, 50–52.

Gibb, R. A. and Walcott, R. I. (1971). A Precambrian suture in the Canadian Shield. *Earth Planet. Sci. Lett.*, **10**, 417–422.

Gibb, R. A. and Halliday, D. W. (1974). Gravity measurements in southern districts of Keewatin and SE districts of Mackenzie. *Grav. Map. Sci. Earth Phys. Br. Can.*, nos. 124–131, 1–36.

Gibb, R. A. and Thomas, M. D. (1976). Gravity signature of fossil plate boundaries in the Canadian Shield. *Nature*, **262**, 199–200.

Gibb, R. A. and Thomas, M. D. (1977). The Thelon Front: a cryptic suture in the Canadian Shield. *Tectonophysics*, **38**, 211–222.

Gibbins, W. A. and McNutt, R. H. (1975). The age of the Sudbury Nickel Irruptive and the Murray Granite. *Can. J. Earth Sci.*, **12**, 1970–1989.

Gibbons, W. (1983). Stratigraphy, subduction and strike–slip faulting in the Mona Complex of North Wales—a review. *Proc. Geol. Ass.*, **94**, 147–163.

Gibbons, W. and Mann, A. (1983). Pre-Mesozoic lawsonite in Anglesey, N Wales: preservation of ancient blueschists. *Geology*, **11**, 3–6.

Gidskehaug, A., Creer, K. M. and Mitchell, J. G. (1975). Palaeomagnetism and K–Ar ages of the south west African basalts and their bearing on the time of initial rifting of the South Atlantic ocean. *Geophys. J. R. astra. Soc.*, **42**, 1–20.

Gill, J. B. (1982). Mountain building and volcanism. In K. J. Hsü (Ed.), *Mountain Building Processes*, Academic Press, London, pp. 13–18.

Gill, R. C. O. (1979). Comparative petrogenesis of Archaean and modern low-K tholeiites: a critical review of some geochemical aspects. In L. H. Ahrens (Ed.), *Origin and Distribution of the Elements*, Pergamon, Oxford, Vol. 11, pp. 431–447.

Gill, R. C. O., Bridgwater, D. and Allaart, J. H. (1981). The geochemistry of the earliest known basic metavolcanic rocks at Isua, W Greenland: a preliminary investigation. In J. E. Glover and D. I. Groves (Eds), *Archaean Geology, Geol. Soc. Australia, Sp. Publ.*, **7**, 313–325.

Gilluly, J. (1973). Steady plate motion and episodic orogeny and magmatism. *Bull. geol. Soc. Amer.*, **84**, 499–514.

Girdler, R. W. (1983). Processes of planetary rifting as seen in the rifting and breakup of Africa. *Tectonophysics*, **94**, 241–252.

Glaessner, M. F. (1971). Geographic distribution and time range of the Ediacara Precambrian fauna. *Bull. geol. Soc. Amer.*, **82**, 509–514.

Glikson, A. Y. (1972). Early Precambrian evidence of a primitive ocean crust and island nuclei of sodic granite. *Bull. geol. Soc. Amer.*, **83**, 3323–3344.

Glikson, A. Y. (1976a). Stratigraphy and evolution of primary and secondary greenstones: significance of data from Shields of the southern hemisphere. In B. F. Windley (Ed.), *The Early History of the Earth*, Wiley–Interscience, London, pp. 257–277.

Glikson, A. Y. (1976b). Trace element geochemistry and origin of early Precambrian igneous series, Barberton Mountain Land, Transvaal. *Geochim. Cosmochim. Acta*, **40**, 1261–1280.

Glikson, A. Y. (1979). Early Precambrian tonalite-trondhjemite sialic nuclei. *Earth Sci. Rev.*, **15**, 1–73.

Glikson, A. Y. (1981). Uniformitarian assumptions, plate tectonics and the Precambrian earth. In A. Kröner (Ed.), *Precambrian Plate Tectonics*, Elsevier, Amsterdam, pp. 91–104.

Glikson, A. Y. (1982). The early Precambrian crust with reference to the Indian Shield: an essay. *J. geol. Soc. India*, **23**, 581–603.

Glikson, A. Y. and Lambert, I. B. (1976). Vertical zonation and petrogenesis of the early Precambrian crust in western Australia. *Tectonophysics*, **30**, 55–89.

Goldich, S. S. (1973). Ages of Precambrian banded iron formations. *Econ. Geol.*, **68**, 1126–1134.

Goldich, S. S., Hedge, C. E. and Stern, T. W. (1970). Age of the Morton and Montevideo gneisses and related rocks, southwestern Minnesota. *Bull. geol. Soc. Amer.*, **81**, 3671–3696.

Golding, L. Y. and Walter, M. R. (1979). Evidence of evaporite minerals in the Archaean Black Flag beds, Kalgoorlie, W Australia. *BMR. J. Austral. Geol. and Geophys.*, **4**, 67–71.

Gole, M. J. and Klein, C. (1981). High-grade metamorphic Archaean banded iron formations; assemblages with coexisting pyroxenes ± fayalite. *Amer. Mineral.*, **66**, 87–99.

Goodwin, A. M. (1973). Archaean iron-formations and tectonic basins of the Canadian Shield. *Econ. Geol.*, **68**, 915–933.

Goodwin, A. M. (1974). The most ancient continental margins. In C. A. Burke and C. L. Drake (Eds), *The Geology of Continental Margins*, Springer, Berlin, pp. 767–780.

Goodwin, A. M. (1977a). Archaean volcanism in Superior Province, Canadian Shield. In W. R. A. Baragar, L. C. Coleman and J. M. Hall (Eds), *Volcanic Regimes in Canada, Geol. Ass. Can. Sp. Pap.*, **16**, 205–241.

Goodwin, A. M. (1977b). Archaean basin-craton complexes and the growth of Precambrian Shields. *Can. J. Earth Sci.*, **14**, 2737–2759.

Goodwin, A. M. (1981). Archaean plates and greenstone belts. In A. Kröner (Ed.), *Precambrian Plate Tectonics*, Elsevier, Amsterdam, pp. 105–135.

Goodwin, A. M. (1982a). Archaean volcanoes in SW Abitiba belt, Ontario and Quebec: form, composition and development. *Can. J. Earth Sci.*, **19**, 1140–1155.

Goodwin, A. M. (1982b). Distribution and origin of Precambrian banded iron formations. *Rev. Bras. Geoc.*, **12**, 457–462.

Goodwin, A. M. and Ridler, R. H. (1970). The Abitibi orogenic belt. *Geol. Surv. Can. Pap.*, **70–40**, 1–24.

Goodwin, A. M. and Smith, I. E. M. (1980). Chemical discontinuities in Archaean metavolcanic terrains and the development of Archaean crust. *Precamb. Res.*, **10**, 301–311.

Goosens, P. J. (1982). Graphite deposits of the Precambrian and their mining development. In *The Development Potential of Precambrian Mineral Deposits*, UN. Nat. Res. Energy Div.-Pergamon Press, New York, pp. 123–156.

Gordon, W. A. (1975). Distribution by latitude of Phanerozoic evaporite deposits. *J. Geol.*, **83**, 671–684.

Gorokhov, I. M. (1964). Whole rock Rb–Sr ages of the Koresten granites, Dnieper migmatites, and metamorphosed mafic rocks of the Ukraine. *Geochem. for 1964*, 738–746.

Grachev, A. F. and Fedorovsky, V. S. (1981). On the nature of greenstone belts in the Precambrian. *Tectonophysics*, **73**, 195–212.

Grandstaff, D. E. (1980). Origin of uraniferous conglomerates Elliot Lake, Canada and Witwatersrand, South Africa: implications for oxygen in the Precambrian atmosphere. *Precamb. Res.*, **13**, 1–26.

Green, D. H. (1981). Petrogenesis of Archaean ultramafic magmas and implications for Archaean tectonics. In A. Kröner (Ed.), *Precambrian Plate Tectonics*, Elsevier, Amsterdam, pp. 469–489.

Green, D. H., Nicholls, I. A., Viljoen, M. and Viljoen, R. (1975). Experimental demonstration of the existence of peridotitic liquids in earliest Archaean magmatism, *Geology*, **3**, 11–14.

Green, J. C. (1977). Keweenawan plateau volcanism in the Lake Superior region. In W. R. A. Baragar, L. C. Coleman and J. H. Hall (Eds), *Volcanic Regimes in Canada. Geol. Ass. Can. Sp. Pap.*, **16**, 407–422.

Green, J. C. (1983). Geological and geochemical evidence for the nature and development of the Middle Proterozoic (Keweenawan) mid-Continent rift of North America. *Tectonophysics*, **94**, 413–438.

Green, T. H. (1980). Island arc and continent building magmatism—a review of petrogenic models based on experimental petrology and geochemistry. *Tectonophysics*, **63**, 367–385.

Green, T. H. and Ringwood, A. E. (1968). Genesis of the calc-alkaline igneous rock suite. *Contrib. Mineral. Petrol.*, **18**, 105–162.

Greenwood, W. R., Hadley, D. G., Anderson, R. E., Fleck, R. J. and Schmidt, D. L. (1976). Proterozoic cratonization in SW Saudi Arabia. *Phil. Trans. R. Soc. Lond.*, **A280**, 517–527.

Grellet, C. and Dubois, J. (1982). The depth of trenches as a function of the subduction rate and age of the lithosphere. *Tectonophysics*, **82**, 45–56.

Griffin, W. L., McGregor, V. R., Nutman, A., Taylor, P. N. and Bridgwater, D. (1980). Early Archaean granulite-facies metamorphism south of Ameralik, W Greenland. *Earth Planet. Sci. Lett.*, **50**, 59–74.

Gross, G. A. (1980). A classification of iron formations based on depositional environments. *Can. Mineral.*, **18**, 215–222.

Groves, D. I., Archibald, N. J., Bettenay, L. F. and Binns, R. A. (1978). Greenstone belts in ancient marginal basins or ensialic rift zones. *Nature*, **273**, 460–461.

Gulson, B. L. and Krogh, T. E. (1975). Evidence of multiple intrusion, possible resetting of U–Pb ages, and new crystallization of zircons in post-tectonic intrusions (in rapakivi granites) and gneisses from South Greenland. *Geochim. Cosmochim. Acta*, **39**, 65–82.

Gupta, V. J. and Kumar, S. (1979). Ophiolites in the tectonic framework of Ladakh. In P. S. Saklani (Ed.), *Structural Geology of he Himalaya*, Today and Tomorrow's Publ., New Delhi, pp. 265–272.

Gupta, A., Basu, A. and Ghosh, P. K. (1980). The Proterozoic ultramafic and mafic lavas and tuffs of the Dalma greenstone belt, Singhbhum, E India. *Can. J. Earth Sci.*, **17**, 210–231.

Haapala, I. (1977). The controls of tin and related mineralizations in the rapakivi-granite areas of SE Fennoscandia. *Geol. Foren. Stockholm, Förh.*, **99**, 130–142.

Habicht, J. K. A. (1979). Paleoclimate, paleomagnetism and continental drift. *Amer. Ass. Petrol. Geol. Studies in Geology*, No. 9, 31 pp.

Hackett, J. P. and Bischoff, J. L. (1973). New data on the stratigraphy, extent and geologic history of the Red Sea geothermal deposits. *Econ. Geol.*, **68**, 533–564.

Hailwood, E. A. and Tarling, D. H. (1973). Palaeomagnetic evidence for a proto-Atlantic ocean. In D. H. Tarling and S. K. Runcorn (Eds), *Implications of Continental Drift to the Earth Sciences*, vol. 1, Academic Press, London, pp. 37–46.

Hallam, A. (1969). Faunal realms and facies in the Jurassic. *Palaeontology*, **12**, 1–18.

Hallam, A. (1979). The end of the Cretaceous. *Nature*, **281**, 430–431.

Hallam, A. (1980). A reassessment of the fit of Pangaea components and the time of their breakup. In D. W. Strangway (Ed.), *The Continental Crust and its Mineral Deposits, Geol. Ass. Can. Sp. Pap.*, **20**, 375–387.

Hallbauer, D. K. (1975). The plant origin of the Witwatersrand 'carbon', *Minerals Sci. Engng.*, **7**, 111–130.

Hallbauer, D. K. and Utter, T. (1977). Geochemical and morphological characteristics of gold particles from Recent river deposits and the fossil placers of the Witwatersrand. *Mineral Deposita*, **12**, 293–306.

Hallberg, J. A. and Glikson, A. Y. (1981). Archaean granite–greenstone terrains of W. Australia. In D. R. Hunter (Ed.), *Precambrian of the Southern Hemisphere*, Elsevier, Amsterdam, pp. 33–103.

Halls, H. C. (1978). The late Precambrian central North American rift system—a survey of recent geological and geophysical investigations. In I. B. Ramberg and E. R. Neumann (Eds), *Tectonics and Geophysics of Continental Rifts*, Reidel, Dordrecht, Holland, pp. 111–123.

Halls, H. C. (1982). The importance and potential of mafic dyke swarms in studies of geodynamic processes. *Geosci. Can.*, **9**, 145–154.

Hambrey, M. J. (1983). Correlation of late Proterozoic tillites in the North Atlantic region and Europe. *Geol. Mag.*, **120**, 209–232.

Hambrey, M. J. and Harland, W. B. (eds) (1981). *Earth's Pre-Pleistocene Glacial Record*, Cambridge University Press, Cambridge, 1004 pp.

Hamet, J. and Allègre, C. J. (1976). Rb–Sr systematics in granite from central Nepal (Manaslu): significance of the Oligocene age and high $^{87}Sr/^{86}Sr$ ratio in Himalayan orogeny. *Geology*, **4**, 470–472.

Hamilton, J. (1977). Isotope and trace element studies of the Great Dyke and Bushveld mafic phase and their relation to early Proterozoic magma genesis in Southern Africa. *J. Petrol.*, **18**, 24–52.

Hamilton, P. J., O'Nions, R. K. and Evensen, N. M. (1977). Sm–Nd dating of Archaean basic and ultrabasic volcanics. *Earth Planet. Sci. Lett.*, **36**, 263–268.

Hamilton, P. J., O'Nions, R. K., Evensen, N. M., Bridgwater, D. and Allaart, J. H. (1978). Sm–Nd isotopic investigations of Isua supracrustals and implications for mantle evolution. *Nature*, **272**, 41–43.

Hamilton, P. J., Evensen, N. M., O'Nions, R. K. and Tarney, J. (1979a). Sm–Nd systematics of Lewisian gneisses: implications for the origin of granulites. *Nature*, **277**, 25–28.

Hamilton, P. J., Evensen, N. M., O'Nions, R. K., Smith, H. S. and Erlank, A. J. (1969b). Sm–Nd dating of Onverwacht Group volcanics, S Africa. *Nature*, **279**, 298–300.

Hamilton, W. (1978). Mesozoic tectonics of the western United States. In D. G. Howell and K. A. McDougall (Eds), *Mesozoic Paleogeography of the Eastern United States, Soc. Econ. Pal. Mineral.*, 33–70.

Hamilton, W. (1981). Crustal evolution by arc magmatism. *Phil. Trans. R. Soc. Lond.*, **A301**, 279–291.

Hanmer, S. K. (1977). Age and tectonic implications of the Baie d'Audierne basic-ultrabasic complex. *Nature*, **270**, 336–338.

Hargraves, R. B. (1976). Precambrian geologic history. *Science*, **193**, 363–371.

Hargraves, R. B. (1978). Punctuated evolution of tectonic style. *Nature*, **276**, 459–461.

Harland, W. B. (1964). Evidence of late Precambrian glaciation and its significance. In A. E. M. Nairn (Ed.), *Problems in Palaeoclimatology*, Interscience, New York, pp. 119–149.

Harland, W. B. (1972). The Ordovician Ice Age. *Geol. Mag.*, **109**, 451–456.

Harland, W. B. (1974). The Precambrian-Cambrian boundary. In C. H. Holland (Ed.), *Lower Palaeozoic Rocks of the World*, vol. 2, *Cambrian of the British Isles, Norden and Spitzbergen*, Wiley, London, pp. 15–42.

Harland, W. B. (1975). The two geological time scales. *Nature*, **253**, 503–507.

Harland, W. B. and Rudwick, M. J. S. (1964). The great Infra-Cambrian ice age. *Sci. Amer.*, **211**, 28–36.

Harland, W. B. and Herod, K. N. (1975). Glaciations through time. In A. E. Wright and F. Moseley (Eds), *Ice Ages: Ancient and Modern*, Seel House Press, Liverpool, pp. 189–216.

Harland, W. B., Cox, A. V., Llewellyn, P. G., Pickton, C. A. G., Smith, A. G. and Walters, R. (1982). *A Geologic Time Scale*, Cambridge University Press, Cambridge, 131 pp.

Harris, A. L., Baldwin, C. T., Bradbury, H. J., Johnson, H. D. and Smith, R. A. (1978). Ensialic basin sedimentation: the Dalradian Supergroup. In D. R. Bowes and B. E. Leake (Eds), *Crustal Evolution in NW Britain and Adjacent Regions*, Seel House Press, Liverpool, pp. 115–138.

Harris, A. L., Holland, C. H. and Leake, E. E. (Eds) (1979). *The Caledonides of the British Isles—reviewed*, Geol. Soc. Lond., Scottish Academic Press, London, 768 pp.

Harris, L. D. and Bayer, K. C. (1979). Sequential development of the Appalachian orogen about a master décollement—a hypothesis. *Geology*, 7, 568–572.

Harris, N. B. W., Holt, R. W. and Drury, S. A. (1982). Geobarometry, geothermometry and late Archaean geotherms from the granulite facies terrain of S India. *J. Geol.*, **90**, 509–527.

Harrison, C. G. A. and Bonatti, E. (1981). The oceanic lithosphere. In C. Emiliani (Ed.), *The Oceanic Lithosphere*, Wiley, New York, pp. 21–48.

Harrison, C. G. A., Brass, G. W., Saltzman, E., Sloan, J., Southam, J. and Whitman, J. M. (1981). Sea level variation, global sedimentation rates and the hypsographic curve. *Earth Planet. Sci. Lett.*, **54**, 1–16.

Harrison, J. E. (1972). Precambrian Belt basin of northwestern United States: its geometry, sedimentation and copper occurrences. *Bull. geol. Soc. Amer.*, **83**, 1215–1240.

Harrison, J. E. and Peterman, Z. E. (1980). North American Commission on Stratigraphic Nomenclature. Note 52 A preliminary proposal for a chronometric time scale for the Pre-cambrian of the United States and Mexico. *Bull. geol. Soc. Amer.*, **91**, 377–380.

Harrison, T. M., Armstrong, R. L., Naeser, C. W. and Harakal, J. E. (1979). Geochronology and thermal history of the Coast plutonic complex, near Prince Rupert, British Columbia. *Can. J. Earth Sci.*, **16**, 400–410.

Hatcher, R. D. (1978). Tectonics of the western Piedmont and Blue Ridge, southern Appalachians: review and speculation. *Amer. J. Sci.*, **278**, 276–304.

Hatcher, R. D. and Odom, A. L. (1980). Timing of thrusting in the southern Appalachians, USA: model for orogeny? *J. geol. Soc. Lond.*, **137**, 321–328.

Hawkesworth, C. J., Moorbath, S. and O'Nions, R. K. (1975). Age relationships between greenstone belts and 'granites' in the Rhodesian Archaean craton. *Earth Planet. Sci. Lett.*, **25**, 251–262.

Hawkins, J. W. (1977). Petrologic and geochemical characteristics of marginal basalts. In M. Talwani and W. C. Pitman (Eds), *Island Arcs, Deep Sea Trenches and Back Arc Basins, Amer. Geophys. Un.*, Maurice Ewing Ser. 1, pp. 355–365.

Hay, W. W. (1981). Sedimentological and geochemical trends resulting from the breakup of Pangaea. *26th Int. Geol. Cong. Coll. 4 Geology of Oceans*, 135–147 (*Oceanologica Acta*, **4**, Suppl.).

Hay, W. W., Barron, E. J., Sloan, J. L. and Southam (1981). Continental drift and the global pattern of sedimentation. *Geol. Rund.*, **70**, 302–315.

Hayes, D. E. (1976). Nature and implications of asymmetric sea-floor spreading—'different rates for different plates'. *Bull. geol. Soc. Amer.*, **87**, 994–1002.

Heirtzler, J. R. (1968). Sea floor spreading. *Sci. Amer.*, **219**, 60–70.

Heirtzler, J. R., Le Pichon, X. and Baron, J. G. (1966). Magnetic anomalies over the Reykyanes Ridge. *Deep-Sea Res.*, **12**, 427–443.

Henderson, J. B. (1975). Archean stromatolites in the northern Slave Province, North West Territories, Canada. *Can. J. Earth Sci.*, **12**, 1619–1630.

Henderson, J. B. (1981). Archaean basin evolution in the Slave Province. In A. Kröner (Ed.), *Precambrian Plate Tectonics*, Elsevier, Amsterdam, pp. 213–236.

Henderson, J. F. (1975). Sedimentology of the Archaean Yellowknife Supergroup of the Yellowknife District of Mackenzie. *Bull. geol. Surv. Can.*, **246**, 62 pp.

Henderson, W. G. and Robertson, A. H. G. (1982). The Highland Border rocks and their relations to marginal basin development in the Scottish Caledonides. *J. geol. Soc. Lond.*, **139**, 433–452.

Herman, Y. (1981). Causes of massive biotic extinctions and explosive evolutionary diversification throughout Phanerozoic time. *Geology*, **9**, 104–108.

Herz, N. (1969). Anorthosite belts, continental drift and the anorthosite event. *Science*, **164**, 944–947.

Herz, N. (1976). Titanium deposits in anorthosite massifs. *U.S. geol. Surv. Prof. Pap.*, **959D**, 1–6.

Hibbard, J. P. and Williams, H. (1979). The regional setting of the Dunnage mélange in the Newfoundland Appalachians. *Amer. J. Sci.*, **279**, 993–1021.

Hickman, A. H. (1981). Crustal evolution of the Pilbara block, W Australia. In J. E. Glover and D. I. Groves (Eds), *Archaean Geology, Geol. Soc. Australia, Sp. Publ.*, 7, 57–69.

Hietanen, A. (1975). Generation of potassium-poor magmas in the northern Sierra Nevada and the Svecofennian of Finland. *J. Res. U.S. geol. Surv.*, **3**, 631–645.

Hills, F. A. and Armstrong, R. L. (1974). Geochronology od Precambrian rocks in the Laramie Range and implications for the tectonic framework of Precambrian, southern Wyoming. *Precamb. Res.*, **1**, 213–225.

Hills, F. A., Houston, R. S. and Subbarayudu, G. V. (1975). Possible Proterozoic plate boundary in southern Wyoming. *Geol. Soc. Amer. Abst. with Prog.*, **7**, 614.

Hoefs, J. (1981). Isotopic composition of the ocean-atmospheric system in the geological past. In R. J. O'Connell and W. S. Fyfe (Eds), *Evolution of the Earth*, Geodyn. Ser., vol. 5, Amer. Geophys. Un., 110–119.

Hoffman, P. F. (1973). Evolution of an early Proterozoic continental margin: the Coronation Geosyncline and associated aulacogens of the northwestern Canadian Shield. *Phil. Trans. R. Soc. Lond.*, **A273**, 547–581.

Hoffman, P. F. (1980). Wopmay orogen: a Wilson-cycle of early Proterozoic age in the northwest of the Canadian Shield. In *Geol. Assoc. Can. Sp. Pap.*, **20**, 523–549.

Hoffman, P. F. (1981). Early Proterozoic magmatic arcs, subduction and collision zones in Wopmay Orogen, NW Canadian Shield: a review of the evidence. In *Proterozoic Symposium*, Madison, Wisconsin. Abst.

Hoffman, P., Dewey, J. F. and Burke, K. (1974). Aulacogens and their genetic relation to geosynclines, with a Proterozoic example from Great Slave Lake, Canada. In R. H. Dott Jr and R. H. Shaver (Eds), *Modern and Ancient Geosynclinal Sedimentation, Soc. Econ. Pal. Mineral. Sp. Publ.*, **19**, 38–55.

Hoffman, P. F. and McGlynn, J. C. (1977). Great Bear batholith: a volcano-plutonic depression. In *Geol. Ass. Can. Sp. Pap.*, **16**, 169–192.

Hogarth, D. D. and Griffin, W. L. (1978). Lapis Lazuli from Baffin Island—Precambrian meta-evaporite. *Lithos*, **11**, 37–60.

Holland, C. H. and Hughes, C. P. (1979). Evolving life of the developing Caledonides. In A. L. Harris, C. H. Holland and B. E. Leake (Eds), *The Caledonides of the British Isles—reviewed*, Geol. Soc. Lond., Scottish Academic Press, Edinburgh, pp. 387–403.

Holland, H. D. (1976). The evolution of sea water. In B. F. Windley (Ed.), *The Early History of the Earth*, Wiley–Interscience, London, pp. 559–567.

Holland, J. G. and Lambert, R. St J. (1975). The chemistry and origin of the Lewisian gneisses of the Scottish mainland: the Scourie and Inver assemblages and sub-crustal accretion. *Precamb. Res.*, **2**, 161–188.

Hollister, L. S. (1975). Granulite facies metamorphism in the Coast Range crystalline belt. *Can. J. Earth Sci.*, **12**, 1953–1955.

Hollister, L. S. (1979). Metamorphism and crustal displacements: new insights. *Episodes*, **1979**, 3–8.

Hollister, L. S. (1982). Metamorphic evidence for rapid (2 mm/yr) uplift of a portion of the Central Gneiss Complex, Coast Mountains, BC. *Can. Mineral.*, **20**, 319–332.

Hollister, V. F., Potter, R. R. and Barker, A. L. (1974). Porphyry-type deposits of the Appalachian orogen. *Econ. Geol.*, **69**, 618–630.

Homewood, P. and Caron, C. (1982). Flysch of the western Alps. In K. J. Hsü (Ed.), *Mountain Building Processes*, Academic Press, London, pp. 157–168.

Honegger, K., Dietrich, V., Frank W., Gansser, A., Thöni, M. and Trommsdorff, V. (1982). Magmatism and metamorphism in the Ladakh Himalayas (the Indus–Tsangpo suture zone). *Earth Planet. Sci. Lett.*, **60**, 253–292.

Honnorez, J. (1981). The aging of the oceanic crust at low temperature. In C. Emiliani (Ed.), *The Oceanic Lithosphere*, Wiley, New York, pp. 525–587.

Horowitz, A. and Cronan, D. S. (1976). The geochemistry of basal sediments from the North Atlantic Ocean. *Mar. Geol.*, **20**, 205–228.

Hsü, K. J. (1977). Tectonic evolution of the Mediterranean basins. In E. M. Nairn, W. H. Kanes and F. G. Stehli (Eds), *The Ocean Basins and Margins, 4A The Eastern Mediterranean*, Plenum Press, New York, 29–75.

Hsü, K. J. (1979). Thin-skinned plate tectonics during Neo-Alpine orogenesis. *Amer. J. Sci.*, **279**, 353–366.

Hsü, K. J. (1980). Terrestrial catastrophe caused by cometary impact at the end of the Cretaceous. *Nature*, **285**, 201–203.

Hsü, K. J. (1981). Origin of geochemical anomalies at Cretaceous–Tertiary boundary. Asteroid or cometary impact? *Oceanol. Acta. Proc. 26th Int. Geol. Cong. Geol. of Oceans*, Paris, 129–133.

Hsui, A. T. and Toksöz, M. N. (1981). Back-arc spreading: trench migration, continental pull or induced convection? *Tectonophysics*, **74**, 89–98.

Hubbard, F. H. and Whitley, J. E. (1978). Rapakivi granite, anorthosite and charmockiti plutonism. *Nature*, **271**, 439–440.

Hubregtse, J. J. M. W. (1980). The Archaean Pikwitonei granu-

lite domain and its position at the margin of the northwest Superior Province (central Manitoba). *Geol. Surv. Can. Pap. G. P.*, **80-3**, 16 pp.

Hughes, C. J. (1976). Parental magma of the Great Dyke of Rhodesia—voluminous late Archaean high magnesium basalt. *Trans. geol. Soc. S. Afr.*, **79**, 179–182.

Humphries, F. J. and Cliff, R. A. (1982). Sm–Nd dating and cooling history of Scourian granites, Sutherland. *Nature*, **295**, 515–517.

Hunter, D. R. (1974a). Crustal development in the Kaapvaal Craton. 1. The Archaean. *Precamb. Res.*, **1**, 259–294.

Hunter, D. R. (1974b). Crustal development in the Kaapvaal Craton. 2. The Proterozoic. *Precamb. Res.*, **1**, 295–326.

Hunter, D. R. and Hamilton, P. J. (1978). The Bushveld complex. In D. H. Tarling (Ed.), *Evolution of the Earth's Crust*, Academic Press, London, pp. 107–173.

Hunter, D. R., Barker, F. and Millard, H. T. Jr. (1978). The geochemical nature of the Archean ancient gneiss complex and granodiorite suite, Swaziland: a preliminary study. *Precamb. Res.*, **7**, 105–127.

Hurst, R. W. (1978). Sr evolution in the West Greenland—Labrador craton: a model for early Rb depletion in the mantle. *Geochim. Cosmochim. Acta*, **42**, 39–44.

Hurst, R. W., Bridgwater, D., Collerson, K. D. and Wetherill, G. W. (1975). 3600 m.y. Rb–Sr ages from the very early Archaean gneisses from Saglek Bay, Labrador. *Earth Planet. Sci. Lett.*, **27**, 393–403.

Hurst, R. W. and Farhat, J. (1977). Geochronologic investigations of the Sudbury Nickel Irruptive and the Superior Province granites north of Sudbury. *Geochim. Cosmochim. Acta*, **41**, 1803–1815.

Husch, J. M. and Moreau, C. (1982). Geology and major element geochemistry of anorthositic rocks associated with Paleozoic hyperbyssal ring complexes, Air massif, Niger, West Africa. *J. Volc. Geotherm. Res.*, **14**, 47–66.

Hutchinson, C. S. (1983). *Economic Deposits and their Tectonic Setting*, Wiley, New York, 328 pp.

Hutchinson, R. W. (1980). Massive base metal sulphide deposits as guides to tectonic evolution. In D. W. Strangway (Ed.), *The Continental Crust and its Mineral Deposits, Geol. Ass. Can. Sp. Pap.*, **20**, 659–684.

Hutchinson, R. W. (1981). Metallogenic evolution and Precambrian tectonics. In A. Kröner (Ed.), *Precambrian Plate Tectonics*, Elsevier, Amsterdam, pp. 733–759.

Hutchinson, R. W. (1982). Syn-depositional hydrothermal processes and Precambrian sulphide deposits. In R. W. Hutchinson, C. D. Spence and J. M. Franklin (Eds), *Precambrian Sulphide Deposits. Geol. Ass. Can. Sp. Pap.*, **25**, 761–791.

Hutchinson, R. W., Ridler, R. H. and Suffel, G. G. (1971). Metallogenic relationships in the Abitibi belt, Canada. A model for Archaean metallogeny. *Trans. Can. Inst. Min.*, **74**, 106–115.

Hutchinson, W. W. (1982). Geology of the Prince Rupert-Skeena map area, British Columbia. *Geol. Surv. Canada. Mem.*, **394**, 116 pp.

Hyde, R. S. (1980). Sedimentary facies in the Archean Timiskaming Group, and their tectonic implications, Abitibi greenstone belt, NE Ontario, Canada. *Precamb. Res.*, **12**, 161–195.

Hynes, A. (1982). Stability of the oceanic tectosphere—a model for early Proterozoic intercratonic orogeny. *Earth Planet. Sci. Lett.*, **61**, 333–345.

Hynes, A. and Francis, D. M. (1982). A transect of the early Proterozoic Cape Smith fold belt, New Quebec. *Tectonophysics*, **88**, 23–59.

Ilyin, A. V. (1980a). A reevaluation of the spatial and temporal distribution of sedimentary phosphate deposits in the light of plate tectonics—a discussion. *Econ. Geol.*, **75**, 771–777.

Ilyin, A. V. (1980b). Sea floor spreading in the Atlantic and accumulation of phosphate. *Doklady Acak. Nauk, SSSR*, **240**, 108–110.

Irvine, T. N. and Baragar, W. R. A. (1972). Muskox intrusion and Coppermine River lavas, Northwest Territories, Canada. *24th Int. Geol. Cong.*, Montreal, Field Excursion A29 Guidebook, 70 pp.

374

Irving, E. (1964). *Palaeomagnetism and Its Application to Geological and Geophysical Problems*, Wiley, New York.

Irving, E. (1977). Drift of the major continental blocks since the Devonian. *Nature*, **270**, 304–309.

Irving, E. (1979). Paleopoles and paleolatitudes of North America and speculations about displaced terrains. *Can. J. Earth Sci.*, **16**, 669–694.

Irving, E. (1981). Phanerozoic continental drift. *Phys. Earth Planet. Int.*, **24**, 197–204.

Irving, E. and Brown, D. A. (1966). Reply to Stehli's discussion of labyrinthodont abundance and diversity. *Amer. J. Sci.*, **264**, 488–496.

Irving, E. and Park, J. K. (1972). Hairpins and super-intervals. *Can. J. Earth Sci.*, **9**, 1318–1324.

Irving, E., North, F. K. and Couillard, R. (1974a). Oil, climate and tectonics. *Can. J. Earth Sci.*, **11**, 1–17.

Irving, E., Emslie, R. F. and Ueno, H. (1974b). Upper Proterozoic palaeomagnetic poles from Laurentia and the history of the Grenville structural province. *J. Geophys. Res.*, **79**, 5491–5502.

Irving, E. and Lapointe, P. L. (1975). Palaeomagnetism of Precambrian rocks of Laurentia. *Geoscience, Canada*, **2**, 90–98.

Irving, E. and McGlynn, J. C. (1976). Proterozoic magnetostratigraphy and the tectonic evolution of Laurentia. *Phil. Trans. R. Soc. Lond.*, **A280**, 433–468.

Irving, E. and McGlynn, J. C. (1979). Palaeomagnetism in the Coronation Geosyncline and arrangement of continents in the middle Proterozoic. *Geophys. J. R. astra. Soc.*, **58**, 309–336.

Irving, E. and McGlynn, J. C. (1981). On the coherence, relation and palaeolatitudes of Laurentia in the Proterozoic. In A. Kröner (Ed.), *Precambrian Plate Tectonics*, Elsevier, Amsterdam, pp. 561–598.

Isacks, B. L. and Barazangi, M. (1977). Geometry of Benioff zones: lateral segmentation and downwards bending of the subducted lithosphere. In M. Talwani and W. C. Pitman (Eds), *Island Arcs, Deep Sea Trenches and Back Arc Basins, Amer. Geophys. Un.*, Maurice Ewing Ser. 1, pp. 99–114.

Ishiara, S. (1978). Metallogenesis in the Japanese island arc system. *J. geol. Soc. Lond.*, **135**, 389–406.

Ivanov, S. N., Perfilien, A. S., Efimov, A. A., Smirnov, G. A., Necheukhin, V. M. and Fershtater, G. B. (1975). Fundamental features in the structure and evolution of the Urals. *Amer. J. Sci.*, **275A**, 107–130.

Iverson, W. P. and Smithson, S. B. (1982). Master décollement root zone beneath the southern Appalachians and crustal balance. *Geology*, **10**, 241–245.

Jackson, N. J. and Ramsay, C. R. (1980). What is the 'Pan-African'? A consensus is needed. *Geology*, **8**, 210–210.

Jacobsen, J. B. E. (1975). Copper deposits in time and space. *Minerals Sci. Engng.*, **7**, 337–371.

Jahn, B. M. and Sun, S. S. (1979). Trace element distribution and isotopic composition of Archaean greenstones. In L. H. Ahrens (Ed.), *Origin and Distribution of the Elements*, Pergamon, Oxford, vol. 11, pp. 597–618.

Jahn, B. M., Vidal, P. and Tilton, G. R. (1980). Archaean mantle heterogeneity; evidence from chemical and isotopic abundances in Archaean igneous rocks. *Phil. Trans. R. Soc. Lond.*, **A297**, 353–364.

Jahn, B. M., Glikson, A. Y., Peucat, J. J. and Hickman, A. H. (1981). REE geochemistry and isotopic data of Archean silicic volcanics and granitoids from the Pilbara block, W Australia: implications for early crustal evolution. *Geochim. Cosmochim. Acta*, **45**, 1633–1652.

James, H. L. (1954). Sedimentary facies of iron formation. *Econ. Geol.*, **49**, 235–293.

James, H. L. (1966). Chemistry of the iron-rich sedimentary rocks. *U.S. geol. Surv. Prof. Pap.*, **440**, Chap. W.

James, H. L. (1972). Subdivision of the Precambrian: an interim scheme to be used by U.S. Geological Survey. *Amer. Ass. Petrol. Geol. Bull.*, **56**, 1026–1030.

James, H. L. (1978). Subdivision of the Precambrian—a brief review and a report on recent decisions by the Subcommission of Precambrian Stratigraphy. *Precamb. Res.*, **7**, 193–204.

Jan. M. Q. and Howie, R. A. (1981). The mineralogy and geochemistry of the metamorphosed basic and ultrabasic rocks of the Jijal Complex, Kohistan, NW Pakistan. *J. Petrol.*, **22**, 85–126.

Janardhan, A. S., Srikantappa, C. and Ramachandra, H. M. (1978). The Sargur schist complex—an Archean high grade terrain in S India. In B. F. Windley and S. M. Naqvi (Eds), *Archaean Geochemistry*, Elsevier, Amsterdam, pp. 127–150.

Jansa, L. F. and Wade, J. A. (1975). Geology of the continental margin off Nova Scotia and Newfoundland. In *Offshore Geology of Eastern Canada*, vol. 2, *Geol. Surv. Can. Pap.*, **74–30**, 51–105.

Jenkyns, H. C. (1980). Tethys: past and present. *Proc. Geol. Ass.*, **91**, 107–118.

Jenner, G. A., Fryer, B. J. and McLennan, S. M. (1981). Geochemistry of the Archean Yellowknife Supergroup. *Geochim. Cosmochim. Acta*, **45**, 1111–1129.

Johnson, G. A. L. (1978). European plate movement during the Carboniferous. In D. H. Tarling (Ed.), *Evolution of the Earth's Crust*, Academic Press, London, pp. 342–360.

Johnson, G. A. L. (1980). Carboniferous geography and terrestrial migration routes. In A. L. Panchen (Ed.), *The Terrestrial Environment and the Origin of Land Vertebrates*, Academic Press, London, pp. 39–54.

Johnson, G. A. L. (1981). Geographical evolution from Laurasia to Pangaea. *Proc. York. Geol. Soc.*, **43**, 221–252.

Johnson, G. A. L. (1982). Geographical change in Britain during the Carboniferous period. *Proc. York. Geol. Soc.*, **44**, 181–203.

Johnstone, G. S. (1975). The Moine Succession. In A. L. Harris *et al.* (Eds), *A Correlation of the Archaean Rocks in the British Isles, Geol. Soc. London, Sp. Rep.*, **6**, 30–45.

Joliffe, A. W. (1966). Stratigraphy of the Steeprock group, Steeprock Lake, Ontario. *Precambrian Symp. geol. Ass. Can. Sp. Pap.*, **3**, 75–98.

Jones, D. L. and McElhinny, M. W. (1966). Palaeomagnetic correlation of basic intrusions in the Precambrian of southern Africa. *J. Geophys. Res.*, **71**, 543–552.

Jones, D. L., Irwin, W. P. and Ovenshine, A. T. (1972). Southeastern Alaska—a displaced continental fragment? In *Geol. Surv. Research 1972: U.S. Geol. Surv. Prof. Pap.*, **800B**, 211–217.

Jones, D. L., Silberling, N. J. and Hillhouse, J. (1977). Wrangellia—a displaced terrane of northwestern North America. *Can. J. Earth Sci.*, **14**, 2565–257u.

Jones, D. L., Silberling, N. J., Gilbert, W. and Coney, P. (1982). Character, distribution and tectonic significance of accretionary terranes in the central Alaska range. *J. Geophys. Res.*, **87**, 3709–3717.

Jordan, T. E., Isacks, B. L., Allmendinger, R. W., Brewer, J. A., Ramos, V. A. and Ando, C. J. (1983). Archaean tectonics related to geometry of subducted Nazca plate. *Bull. geol. Soc. Amer.*, **94**, 341–361.

Jordan, T. H. (1978). Composition and development of the continental tectosphere. *Nature*, **275**, 544–548.

Jurdy, D. M. (1979). Relative plate motions and the formation of marginal basins. *J. Geophys. Res.*, **84**, B12, 6796–6801.

Kai, K. (1981). Rb–Sr geochronology of the rocks of the Himalayas, E. Nepal. Part II. The age and origin of the granite of the Higher Himalayas. *Mem. Fac. Sci. Kyoto Univ. Ser. Geol. and Mineral.*, **47/2**, 149–157.

Kanamori, H. (1977). Seismic and aseismic slip along subduction zones and their tectonic implications. In M. Talwani and W. C. Pitman (Eds), *Island Arcs, Deep Sea Trenches and Back Arc Basins, Amer. Geophys. Un.*, Maurice Ewing Ser. 1, pp. 163–174.

Kanasevich, E. R. (1968). Precambrian rift: genesis of stratabound ore deposits. *Science*, **161**, 1002–1005.

Karamata, S. (1980). Tethyan ophiolites: a short review and the main problems. In A. Panayiotou (Ed.), *Ophiolites, Cyprus Geol. Surv.*, 257–260.

Karig, D. E. (1970). Ridges and basins of the Tongo-Kermadec island arc system. *J. Geophys. Res.*, **75**, 239–254.

Karig, D. E. (1972). Remnant arcs. *Bull. geol. Soc. Amer.*, **83**, 1057–1068.

Karig, D. E. (1974). Evolution of arc systems in the western Pacific. *Ann. Rev. Earth Planet. Sci.*, **2**, 51–75.

Karig, D. E. and Moore, G. F. (1975). Tectonically controlled sedimentation in marginal basins. *Earth Planet. Sci. Lett.*, **26**, 233–238.

Karig, D. E. and Sharman, G. F. (1975). Subduction and accretion in trenches. *Bull. geol. Soc. Amer.*, **86**, 377–389.

Karig, D. E., Caldwell, J. G. and Parmentier, E. M. (1976). Effects of accretion on the geometry of the descending lithosphere. *J. Geophys. Res.*, **81B**, 6281–6291.

Karig, D. E. and Kay, R. W. (1981). Fate of sediments on the descending plate at converging margins. *Phil. Trans. R. Soc. Lond.*, **A301**, 233–251.

Karson, J. and Dewey, J. F. (1978). Coastal complex, western Newfoundland: an early Ordovician oceanic fracture zone. *Bull. geol. Soc. Amer.*, **89**, 1037–1049.

Katz, M. B. (1981). Note on the layered anorthosite complex of the Harts Range ruby deposit in the Arunta block, central Australia. *J. geol. Soc. Australia*, **28**, 247–248.

Kay, M. (1976). Dunnage mélange and subduction of the Protacadic ocean. NE Newfoundland. *Geol. Soc. Amer. Sp. Pap.*, **175**, 49 pp.

Kazansky, V. I. (1982). Metallogenic processes in the early history of the earth. *Rev. Bras. Geoc.*, **12**, 476–483.

Kazansky, V. I. and Moralev, V. M. (1981). Archaean geology and metallogeny of the Aldan Shield, USSR. In J. E. Glover and D. I. Groves (Eds), *Archaean Geology, Geol. Soc. Australia, Sp. Publ.*, **7**, 110–120.

Kazmin, V. (1980). Geodynamic control of rift volcanism. *Geol. Rund.*, **69**, 757–769.

Kazmin, V., Shifferaw, A. and Balchia, T. (1978). The Ethiopian basement: stratigraphy and possible manner of evolution. *Geol. Rund.*, **67**, 531–546.

Kean, B. F. and Strong, D. F. (1975). Geochemical evolution of an Ordovician island arc of the central Newfoundland Appalachians. *Amer. J. Sci.*, **275**, 97–118.

Keith, S. B. (1978). Paleosubduction geometries inferred from Cretaceous and Tertiary magmatic patterns in southwestern North America. *Geology*, **6**, 516–521.

Keith, S. B. (1982). Paleoconvergence rates determined from K_2O/SiO_2 ratios in magmatic rocks and their application to Cretaceous and Tertiary tectonic patterns in southwestern North America. *Bull. geol. Soc. Amer.*, **93**, 524–532.

Keller, B. M. (1979). Precambrian stratigraphic scale of the USSR. *Geol. Mag.*, **116**, 419–429.

Kelly, W. C. and Rye, R. O. (1979). Geology, fluid inclusions and stable isotope studies of the tin-tungsten deposits of Panasqueira, Portugal. *Econ. Geol.*, **74**, 1721–1822.

Kelts, K. (1981). A comparison of some aspects of sedimentation and tectonics from the Gulf of California and the Mesozoic Tethys, northern Penninic margin. *Eclogae geol. Helv.*, **74/2**, 317–338.

Kennedy, W. J. and Cooper, M. R. (1975). Cretaceous ammonite distributions and the opening of the South Atlantic. *J. geol. Soc. Lond.*, **131**, 283–288.

Kent, P. E. (1976). Major synchronous events in continental shelves. *Tectonophysics*, **36**, 87–91.

Keppie, J. D. (1977). Plate tectonic interpretation of Palaeozoic world maps. *Nova Scotia Dept of Mines. Pap.*, **77-3**, 1–30.

Kerrich, R. and Fyfe, W. S. (1981). The gold–carbonate association: source of CO_2, and CO_2 fixation reactions in Archaean deposits. *Chem. Geol.*, **33**, 265–294.

Key, R. M. (1977). The geological history of the Limpopo mobile belt based on the field mapping of the Botswana Geological Survey. *Geol. Surv. Botswana Bull.*, **12**, 41–59.

Key, R. M., Litherland, M. and Hepworth, J. W. (1976). The evolution of the Archaean crust of northeast Botswana. *Precamb. Res.*, **3**, 375–413.

Khan, M. A. (1975). The Afro-Arabian rift system. *Sci. Prog.*, **62**, 207–236.

Kidd, R. G. W. and Cann, J. R. (1974). Chilling statistics indicate an ocean floor spreading origin for the Troodos complex, Cyprus. *Earth Planet. Sci. Lett.*, **24**, 151–155.

Kidd, W. S. F. (1977). The Baie Verte lineament, Newfoundland: ophiolite complex floor and mafic volcanic fill of a small Ordovician marginal basin. In M. Talwani and W. C. Pitman (Eds), *Island Arcs, Deep Sea Trenches and Back Arc Basins, Amer. Geophys. Un.*, Maurice Ewing Ser. 1, pp. 407–418.

Kidd, W. S. F., Dewey, J. F. and Bird, J. M. (1978). The Mings Bight ophiolite complex, Newfoundland: Appalachian oceanic crust and mantle. *Can. J. Earth Sci.*, **15**, 781–804.

Kimberley, M. M. (1978). Paleoenvironmental classification of iron formations. *Econ. Geol.*, **73**, 215–229.

Kimberley, M. M. and Dimroth, E. (1976). Basic similarity of Archaean to subsequent atmospheric and hydrospheric compositions as evidence in the distributions of sedimentary carbon, sulphur, uranium and iron. In B. F. Windley (Ed.), *The Early History of the Earth*, Wiley–Interscience, London, pp. 579–585.

Kimberley, M. M., Grandstaff, D. E. and Tanaka, R. T. (in press). Topographic control on Precambrian weathering in the Elliot Lake uranium district, Canada. *J. geol. Soc. Lond.*, **00**, 000–000.

Kindle, E. D. (1972). Classification and description of copper deposits, Coppermine River area, district of Mackenzie, *Bull. Geol. Surv. Canada*, **214**, 109 pp.

Kinsman, D. J. J. (1975). Salt floors to geosynclines, *Nature*, **5**, 375–378.

Kirby, G. A. (1979). The Lizard complex as an ophiolite, *Nature*, **282**, 58–61.

Kistler, R. W. (1974). Phanerozoic batholiths in western North America. *Ann. Rev. Earth Planet. Sci.*, **2**, 403–418.

Kisvarsanyi, E. B. (1980). Granitic ring complexes and Precambrian hot spot activity in the St Francois terrance, midcontinent region, United States. *Geology*, **8**, 43–47.

Klein, C. and Bricker, O. P. (1977). Some aspects of the sedimentary and diagenetic environment of Proterozoic banded iron-formation. *Econ. Geol.*, **77**, 1457–1470.

Klemm, D. D. (1979). A biogenetic model of the formation of the banded iron formations in the Transvaal Supergroup, South Africa. *Mineral. Deposita*, **14**, 381–385.

Klevtsova, A. A. (1979). Late Riphean stage of development of the Russian plate. *Int. Geol. Rev.*, **21**, 167–178.

Klootwijk, C. T. and Peirce, J. W. (1979). India's and Australia's pole path since the late Mesozoic and the India-Asia collision. *Nature*, **282**, 605–607.

Klootwijk, C. J. and Radhakrishnamurty, C. (1981). Phanerozoic paleomagnetism of the Indian plate and the India-Asia collision. In M. W. McElhinny and D. A. Valencio (Eds), *Paleoreconstruction of the Continents, Geodynamic Ser.*, vol. 2, Amer. Geophys. Un., pp. 93–105.

Knoll, A. H. and Barghoorn, E. S. (1975). Precambrian eukaryotic organisms: a reassessment of the evidence. *Science*, **190**, 52–54.

Knoll, A. H. and Barghoorn, E. S. (1977). Archean microfossils showing cell divisions from the Swaziland System of South Africa. *Science*, **198**, 396–398.

Kornfält, K. A. (1976). Petrology of the Ragunda rapakivi massif, central Sweden. *Sver. Geol. Unders.*, **70**, Ser. C, 111 pp.

Kostov, I. (1978). Crustal pattern and mineralization of the Mediterranean frameland. *Proc. Geol. Ass.*, **89**, 101–123.

Kratz, K. O., Gerling, E. K. and Lobach-Zhuchenko, S. B. (1968). The isotope geology of the Precambrian of the Baltic Shield. *Can. J. Earth Sci.*, **5**, 657–660.

Kratz, K. and Mitrofanov, F. (1980). Main type reference sequences of the early Precambrian in the USSR. *Earth Sci. Rev.*, **16**, 295–301.

Krebs, W. (1977). The tectonic evolution of Variscan Meso-Europe. In D. V. Ager and M. Brooks (Eds), *Europe from Crust to Core*, Wiley, London, pp. 119–142.

Krogh, T. E., McNutt, R. H. and Davis, G. L. (1982). Two high precision U–Pb Zircon ages from the Sudbury nickel irruptive. *Can. J. Earth Sci.*, **19**, 723–728.

Kröner, A. (1976a). Proterozoic crustal evolution in parts of southern Africa and evidence for extensive sialic crust since the end of the Archaean. *Phil. Trans. R. Soc. Lond.*, **A280**, 541–554.

Kröner, A. (1976b). Non-synchroneity of late Precambrian glaciations in Africa. *J. Geol.*, **85**, 289–300.

Kröner, A. (1977a). Precambrian mobile belts of southern and

376

western Africa—ancient sutures or sites of ensialic mobility? A case for crustal evolution towards plate tectonics. *Tectonophysics*, **40**, 101–135.

Kröner, A. (1977b). The Precambrian geotectonic evolution of Africa plate accretion versus plate destruction. *Precamb. Res.*, **4**, 163–213.

Kröner, A. (1979). Pan-African plate tectonics and its repercussions on the crust of NE Africa. *Geol. Rund.*, **68**, 565–583.

Kröner, A. (1980). Pan-African crustal evolution. *Episodes*, **1980**, 3–8.

Kröner, A. (1981a). Precambrian crustal evolution and continental drift. *Geol. Rund.*, **70**, 412–428.

Kröner, A. (1981b). Precambrian plate tectonics. In A. Kröner (Ed.), *Precambrian Plate Tectonics*, Elsevier, Amsterdam, pp. 57–90.

Kroonenberg, S. B. (1982). A Grenvillian granulite belt in the Columbian Andes and its relation to the Guiana Shield. *Geol. en Mijnb.*, **61**, 325–333.

Kulish, E. A. (1979). The Archaean atmosphere of the Earth (Aldan time). In L. H. Ahrens (Ed.), *Origin and Distribution of the Elements, Physics Chemistry of the Earth*, Vol. 11, Pergamon, Oxford, pp. 141–148.

Kurtén, B. (1969). Continental drift and evolution. *Sci. Amer.*, **220**, pp. 54–64.

Lal, R. K., Mukerji, S. and Ackermand, D. (1981). Deformation and Barrovian metamorphism at Takdah, Darjeeling (Eastern Himalaya). In P. S. Saklani (Ed.), *Metamorphic Tectonites of the Himalaya*, Today and Tomorrow's Publ, New Delhi, pp. 231–278.

Lambert, I. B. (1976). The McArthur zinc–lead–silver deposits: features, metallogenesis and comparison with some other stratiform ores. In K. H. Wolf (Ed.), *Handbook of Stratabound and Stratiform Ore Deposits*, Elsevier, Amsterdam, pp. 535–585.

Lambert, I. B., Donnelly, T. H., Dunlop, J. S. R. and Groves, D. I. (1978). Stable isotopic compositions of early Archaean sulphate deposits of probable evaporite and volcanogenic origins. *Nature*, **276**, 808–811.

Lambert, I. B. and Groves, D. I. (1981). Early earth evolution and metallogeny. In K. H. Wolf (Ed.), *Handbook of Stratabound and Stratiform Ore Deposits*, Elsevier, Amsterdam, pp. 339–447.

Lambert, R. St J., Chamberlain, V. E. and Holland, J. G. (1976). The geochemistry of Archaean rocks. In B. F. Windley (Ed.), *The Early History of the Earth*, Wiley–Interscience, London, pp. 377–387.

Lambert, R. St J. and Holland, J. G. (1976). Amîtsoq gneiss geochemistry: preliminary observations. In B. F. Windley (Ed.), *The Early History of the Earth*, Wiley–Interscience, London, pp. 191–201.

Lambert, R. St J. and McKerrow, W. S. (1976). The Grampian orogeny. *Scott. J. Geol.*, **12**, 271–292.

Langford, F. F. and Morin, J. A. (1976). The development of the Superior Province of NW Ontario by merging island arcs. *Amer. J. Sci.*, **276**, 1023–1034.

Lappin, A. R. and Hollister, L. S. (1980). Partial melting in the Central Gneiss Complex near Prince Rupert, British Columbia. *Amer. J. Sci.*, **280**, 518–545.

Large, R. R. (1977). Chemical evolution and zonation of massive sulfide deposits in volcanic terrains. *Econ. Geol.*, **72**, 549–572.

Larsen, H. C. (1978). Offshore continuation of East Greenland dyke swarm and North Atlantic ocean formation, *Nature*, **274**, 220–223.

Larson, R. L. and Ladd, J. W., III (1973). Evidence for the opening of the South Atlantic in the early Cretaceous. *Nature*, **246**, 209.

Larue, D. K. and Sloss, L. L. (1980). Early Proterozoic sedimentary basins of the Lake Superior region: summary. *Bull. geol. Soc. Amer.*, pt. 1, **91**, 450–452.

Laubscher, H. and Bernoulli, D. (1977). Mediterranean and Tethys. In A. E. M. Nairn, W. H. Kanes and F. G. Stehli (Eds), *The Ocean Basins and Margins*, vol. 4A, *The Eastern Mediterranean*, Plenum Press, New York, pp. 1–28.

Laubscher, H. and Bernoulli, D. (1982). History and deformation of The Alps. In K. J. Hsü (Ed.), *Mountain Building Processes*, Academic Press, Lond., 263 pp.

Laurén, L. (1970). An interpretation of the negative gravity anomalies associated with the rapakivi granites and Jotnian sandstone in southern Sweden. *Geol. Foren. Stock. Forh.*, **92,1**, 21–34.

Lawver, L. A., Hawkins, J. W. and Sclater, J. G. (1976). Magnetic anomalies and crustal dilation in the Lau Basin. *Earth Planet. Sci. Lett.*, **33**, 27–35.

Lawver, L. A. and Hawkins, J. W. (1978). Diffuse magnetic anomalies in marginal basins: their possible tectonic and petrologic significance. *Tectonophysics*, **45**, 323–339.

Leake, B. E., Farrow, C. M. and Townend, R. (1979). A pre-2000 Myr old granulite facies metamorphosed evaporite from Caraiba, Brazil? *Nature*, **277**, 49–50.

Le Bas, M. J. (1980a). The East African Cenozoic magmatic province. In Geodynamic Evolution of the Afro-Arabian Rift System. *Accad. Naz. dei Lincei*, **47**, 113–122.

Le Bas, M. J. (1980b). Alkaline magmatism and uplift of continental crust. *Proc. Geol. Ass.*, **91**, 33–38.

Le Bas, M. J. (1982). Quaternary to Recent volcanicity in Japan. *Proc. Geol. Ass.*, **93**, 179–194.

Leblanc, M. (1981). The late Proterozoic ophiolites of Bou Azzer (Morocco): evidence for Pan-African plate tectonics. In A. Kröner (Ed.), *Precambrian Plate Tectonics*, Elsevier, Amsterdam, pp. 435–451.

Le Douaran, S., Needham, H. D. and Francheteau, J. (1982). Pattern of opening rates along the axis of the Mid-Atlantic Ridge. *Nature*, **300**, 254–257.

Leeder, M. R. (1982). Upper Palaeozoic basins of the British Isles—Caledonide inheritance versus Hercynian plate margin processes. *J. geol. Soc. Lond.*, **139**, 481–494.

Lefort, J. P. (1979). Iberian-Armorican arc and Hercynian orogeny in western Europe. *Geology*, **7**, 384–388.

Lefort, J. P. and van der Voo, R. (1981). A kinematic model for the collision and complete suturing between Gondwanaland and Laurasia in the Carboniferous. *J. Geol.*, **89**, 537–550.

Lefort, J. P., Audren, C. L. and Max, M. D. (1982). The southern part of the Armorican orogeny: a result of crustal shortening related to reactivation of a pre-Hercynian mafic belt during Carboniferous time. *Tectonophysics*, **89**, 359–377.

Le Fort, P. (1975). Himalaya: the collided range. Present knowledge of the continental arc. *Amer. J. Sci.*, **75**, 1–44.

Le Fort, P. (1981). Manaslu leucogranite: a collision signature of the Himalaya. A model for its genesis and emplacement. *J. Geophys. Res.*, **86**, **B11**, 10545–10568.

Le Fort, P., Debon, F. and Sonet, J. (1980). The 'Lesser Himalayan' cordierite granite belt: typology and age of the pluton of Manserah (Pakistan). *Univ. Peshawar Sp. Issue, Geol. Bull.*, **13**, 51–61.

Le Fort, P., Michard, A., Sonet, J. and Zimmermann, J. L. (in press). Petrography, geochemistry and geochronology of some samples from the Karakorum axial batholith, N. Pakistan. In A. Desio (Ed.), *The Granite Rocks of N Pakistan*, Lahore University Press, Lahore, Pakistan.

Leggett, J. K. (Ed.) (1982). *Trench-Forearc Geology: Sedimentation and Tectonics in Modern and Ancient Active Plate Margins*, Geol. Soc. London, Blackwell, Oxford, 576 pp.

Leggett, J. K., McKerrow, W. S. and Eales, M. H. (1979). The Southern Uplands of Scotland: a lower Palaeozoic accretionary prism. *J. geol. Soc. Lond.*, **136**, 755–770.

Leonardos, O. H. and Fyfe, W. S. (1974). Ultrametamorphism and melting of a continental margin, Brazil. *Contrib. Mineral. Petrol.*, **46**, 201–214.

Le Pichon, X. (1968). Sea-floor spreading and continental drift. *J. Geophys. Res.*, **73**, 3661–3697.

Le Pichon, X. and Francheteau, J. (1978). A plate tectonic analysis of the Red Sea-Gulf of Aden area. *Tectonophysics*, **46**, 369–406.

Le Roy, J. (1978). The Margnac and Fanay uranium deposits of the La Crouzille district (western Massif Central, France): geologic and fluid inclusion studies. *Econ. Geol.*, **73**, 1611–1634.

Lewis, A. D. and Bloxam, T. W. (1977). Petrotectonic environment of the Girvan-Ballantrae lavas from rare-earth element distributions. *Scott. J. Geol.*, **13**, 211–222.

Lewry, J. F. (1981). Lower Proterozoic arc-microcontinent collisional tectonic in the western Churchill Province. *Nature*, **294**, 69–72.

Lewry, J. F. and Sibbald, T. H. (1980). Thermotectonic evolution of the Churchill Province in northern Saskatchewan. *Tectonophysics*, **68**, 45–82.

Lewry, J. F., Stauffer, M. R. and Fumerton, S. (1981). A Cordilleran-type batholithic belt in the Churchill Province in northern Saskatchewan. *Precamb. Res.*, **14**, 277–313.

Light, M. P. R. (1982). The Limpopo mobile belt: a resuit of continental collision. *Tectonics*, **1**, 325–342.

Li Jiliang, Cong Bolin and Zhang Wenhua (1979). A preliminary study of the tectonic evolution of the late Archean iron formation in the SW of North China fault block. *Scientia Sinica*, **22**, 1430–1442.

Li Jiliang, Cong Bolin and Zhang Ruyuan (1982). Petrogenesis of early Archaean greenstone of Taipingzhai, Lianxi, E. Hebei Province. *Research on Geol., Inst. Geology., Acad. Sin., Beijing*, 64–65 (English Summary).

Li Shunguang (1982). Geochemical model for the genesis of the Gongchangling rich magnetite deposit in China. *Geochimica*, **2**, 113–121. (Chinese with English abstract).

Loberg, B. E. H. (1980). A Proterozoic subduction zone in southern Sweden. *Earth Planet. Sci. Lett.*, **46**, 287–294.

Lonsdale, P. (1979). A deep-sea hydrothermal site on a strike-slip fault. *Nature*, **281**, 531–534.

Losert, J. (1977). Sillimanite fibrolitique du Moldanubicum du Massif de Bohème et ses analogies avec celle des massifs cristallins varisiques de l'Europe occidentale. In *La Chaine Varisque d'Europe Moyenne et Occidentale*, Editions du CNRS, no. 243, Paris, pp. 329–340.

Lowe, D. R. (1980a). Archaean sedimentation. *Ann. Rev. Earth Planet. Sci.*, **8**, 145–167.

Lowe, D. R. (1980b). Stromatolites 3400 myr old from the Archean of Western Australia. *Nature*, **284**, 441–443.

Lowe, D. R. (1982). Comparative sedimentology of the principal volcanic sequences of Archean greenstone belts in South Africa, W Australia, and Canada: implications for crustal evolution. *Precamb. Res.*, **17**, 1–29.

Lowe, D. R. and Knauth, L. P. (1977). Sedimentology of the Onverwacht Group (3.4 billion years), Transvaal, South Africa, and its bearing on the characteristics and evolution of the early earth. *J. Geol.*, **85**, 699–723.

Lundberg, N. (1983). Development of forearcs of intraoceanic subduction zones. *Tectonics*, **2**, 51–61.

Luyendyk, B. P. (1970). Dips of downgoing lithospheric plates beneath island arcs. *Bull. geol. Soc. Amer.*, **81**, 3411–3416.

Luyendyk, B. P., Forsyth, D. and Phillips, J. D. (1972). Experimental approach to the palaeocirculation of the oceanic surface waters. *Bull. geol. Soc. Amer.*, **83**, 2649–2664.

Ma Xingyuan and Wu Zhengwen (1981). Early tectonic evolution of China. *Precamb. Res.*, **14**, 185–202.

Maaløe, S. (1982). Petrogenesis of Archaean tonalities. *Geol. Rund.*, **71**, 328–346.

McCall, G. J. H. (1971). Some ultrabasic and basic igneous rock occurrences in the Archaean of western Australia. *Geol. Soc. Australia Sp. Publ.*, **3**, 429–442.

McCall, G. J. H. (1981). Progress in research into the early history of the earth: a review, 1970–1980. *Sp. Publ. Geol. Soc. Australia*, **7**, 3–18.

McCall, G. J. H. and Peers, R. (1971). Geology of the Binneringie dyke, western Australia. *Geol. Rund.*, **40**, 1174–1263.

McClay, K. R. and Campbell, I. H. (1976). The structure and shape of the Jimberlana intrusion, western Australia, as indicated by an investigation of the bronzite complex. *Geol. Mag.*, **113**, 129–139.

McClay, K. R. and Carlile, D. G. (1978). Mid-Proterozoic sulphate evaporites at Mount Isa mine, Queensland, Australia. *Nature*, **274**, 240–241.

McConnell, R. B. (1977). East African Rift System dynamics in view of Mesozoic apparent polar wander. *J. geol. Soc. Lond.*, **134**, 33–39.

McCulloch, M. T. and Wasserburg, G. J. (1978). Sm–Nd and Rb–Sr chronology of continental crust formations. *Science*, **200**, 1003–1011.

McCulloch, M. T. and Wasserburg, G. J. (1980). Early Archaean Sm–Nd model ages from a tonalitic gneiss. N Michigan. In G. D. Morey and G. N. Hanson (Eds), *Selected Studies of Archaean Gneisses and Lower Proterozoic Rocks, Southern Canadian Shield, Geol. Soc. Amer. Sp. Pap.*, **182**, 135–138.

McCulloch, M. T. and Compston, W. (1981). Sm–Nd age of Kambalda and Kanowna greenstones and heterogeneity in the Archaean mantle. *Nature*, **294**, 322–326.

McCurry, P. and Wright, J. B. (1977). Geochemistry of late alkaline volcanics in NW Nigeria and a possible Pan-African suture zone. *Earth Planet. Sci. Lett.*, **37**, 90–96.

MacDonald, K. C. (1982). Mid-ocean ridges; fine scale tectonic, volcanic and hydrothermal processes within the plate boundary zone. *Ann. Rev. Earth Planet. Sci.*, **10**, 155–190.

McDougall, I. and Duncan, R. A. (1980). Linear volcanic chain recording plate motions? *Tectonophysics*, **63**, 275–295.

McElhinny, M. W. (1973). *Palaeomagnetism and Plate Tectonics*, Cambridge University Press, Cambridge, 358 pp.

McElhinny, M. W. and Briden, J. C. (1971). Continental drift during the Palaeozoic. *Earth Planet. Sci. Lett.*, **10**, 407–416.

McElhinny, M. W. and Embleton, B. J. J. (1976). Precambrian and early Palaeozoic palaeomagnetism in Australia. *Phil. Trans. R. Soc. Lond.*, **A280**, 417–432.

McElhinny, M. W., Giddings, J. W. and Embleton, B. J. J. (1974). Palaeomagnetic results and late Precambrian glaciations. *Nature*, **248**, 557–561.

MacGeehan, P. J. and MacLean, W. H. (1980a). Tholeiite basalt–rhyolite magmatism and massive sulphide deposits at Matagami, Quebec. *Nature*, **283**, 153–156.

MacGeehan, P. J. and MacLean, W. H. (1980b). A Precambrian subseafloor geothermal system, calc-alkali trends, and massive sulphide genesis. *Nature*, **286**, 767–771.

McGlynn, J. C. and Irving, E. (1981). Horizontal motions and rotations in the Canadian Shield during the early Proterozoic. In F. H. A. Campbell (Ed.), *Proterozoic Basins of Canada, Geol. Surv. Can. Pap.*, **81-10**, 183–190.

McGlynn, J. C., Irving, E., Bell, K. and Pullaiah, G. (1975). Palaeomagnetic poles and a Proterozoic supercontinent. *Nature*, **255**, 318–319.

McGregor, V. R. (1973). The early Precambrian gneisses of the Godthaab district, West Greenland. *Phil. Trans. R. Soc. Lond.*, **A273**, 343–358.

McGregor, V. R. (1979). Archaean gray gneisses and the origin of the continental crust: evidence from the Godthaab region, W Greenland. In F. Barker (Ed.), *Trondhjemites, Dacites and Related Rocks*, Elsevier, Amsterdam, pp. 169–204.

McGregor, V. R. and Mason, B. (1977). Petrogenesis and geochemistry of metabasaltic and metasedimentary enclaves in the Amîtsoq gneisses, W Greenland. *Amer. Mineral.*, **62**, 887–904.

McKenzie, D. P. (1977). The initiation of trenches: a finite amplitude instability. In M. Talwani and W. C. Pitman (Eds), *Island Arcs, Deep Sea Trenches and Back Arc Basins, Amer. Geophys. Un.*, Maurice Ewing Ser. 1, 57–62.

McKenzie, D. (1978). Some remarks on the development of sedimentary basins. *Earth Planet. Sci. Lett.*, **40**, 25–32.

McKenzie, D. and Weiss, N. (1980). The thermal history of the earth. In D. W. Strangway (Ed.), *The Continental Crust and its Mineral Deposits, Geol. Ass. Can. Sp. Pap.*, **20**, 575–590.

McKenzie, D., Nisbet, E. and Sclater, J. G. (1980). Sedimentary basin development in the Archaean. *Earth Planet. Sci. Lett.*, **48**, 35–41.

McKerrow, W. S. and Cocks, L. R. M. (1976). Progressive faunal migration across the Iapetus Ocean. *Nature*, **263**, 304–306.

McLaren, D. (1970). Time, life and boundaries. *J. Palaeont.*, **44**, 801–815.

McLaren, D. (197q). The Silurian–Devonian boundary committee. In A. Martinsson (Ed.), *The Silurian–Devonian Boundary*, IUGS Ser. A, No. 5, Schweizerbart'sche Verlag, Stuttgart, pp. 1–34.

McLennan, S. M. (1982). On the geochemical evolution of sedimentary rocks. *Chem. Geol.*, **37**, 335–350.

378

McLennan, S. M. and Taylor, S. R. (1982). Geochemical constraints on the growth of the continental crust. *J. Geol.*, **90**, 347–361.

MacLeod, W. N., Turner, D. C. and Wright, E. P. (1971). The geology of the Jos Plateau: vol. 1, General Geology. *Geol. Surv. Niger. Bull.*, **32**, 110 pp.

McWilliams, M. O. and Dunlop, D. J. (1978). Grenville palaeomagnetism and tectonics. *Can. J. Earth Sci.*, **15**, 687–695.

McWilliams, M. O. and McElhinny, M. W. (1980). Late Precambrian palaeomagnetism of Australia: the Adelaide Geosyncline. *J. Geol.*, **88**, 1–26.

McWilliams, M. O. and Kröner, A. (1981). Palaeomagnetism and tectonic evolution of the Pan-African Damara belt, southern Africa. *J. Geophys. Res.*, **86**, 156, 5147–5162.

Maisonneuve, J. (1982). The composition of the Precambrian ocean waters. *Sed. Geol.*, **31**, 1–11.

Mäkelä, K. (1980). Geochemistry and origin of Haveri and Kiipu Proterozoic stratabound volcanogenic gold–copper and zinc mineralization from SW Finland. *Bull. Geol. Surv. Finland*, **310**, 79 pp.

Malm, O. A. and Ormaasen, D. E. (1978). Mangerite-charnockite intrusives in the Lofoten-Vesteraalen area, N Norway: petrography, chemistry and petrology. *Norges Geol. Unders.*, **338**, 83–114.

Maltman, A. J. (1977). Serpentinites and related rocks of Anglesey. *Geol. J.*, **12**, 113–128.

Maluski, H., Proust, F. and Xiao, X. C. (1982). $^{39}Ar/^{40}Ar$ dating of the trans-Himalayan calc-alkaline magmatism of southern Tibet. *Nature*, **298**, 152–154.

Margulis, L., Walker, J. C. G. and Rambler, M. (1976). Reassessment of roles of oxygen and ultraviolet light in Precambrian evolution. *Nature*, **264**, 620–624.

Marsh, B. D. and Carmichael, I. S. E. (1974). Benioff zone magmatism. *J. Geophys. Res.*, **79**, 1196–1206.

Martin, A., Nisbet, E. G. and Bickle, M. J. (1980). Archaean stromatolites of the Belingwe greenstone belt, Zimbabwe (Rhodesia). *Precamb. Res.*, **13**, 337–362.

Martin, H. (1981). The late Palaeozoic Gondwana glaciation. *Geol. Rund.*, **70**, 480–498.

Martin, H. and Porada, H. (1977). The intracratonic branch of the Damara orogen in South West Africa. *Precamb. Res.*, **I, 5**, 311–318, **II, 5**, 339–357.

Mascle, J. (1977). Le Golfe de Guinée (Atlantique Sud): un exemple d'évolutions de marges atlantiques en cisaillement. *Mém. Soc. Geol. France*, **45**. *Mém.*, **128**, 1–104.

Mason, T. R. and von Brunn, V. (1977). 3-Gyr-old stromatolites from South Africa. *Nature*, **266**, 47–49.

Matsuda, T. and Uyeda, S. (1971). On the Pacific-type orogeny and its model-extension of the paired belts concept and possible origin of marginal seas. *Tectonophysics*, **11**, 5–27.

Matte, P. L. and Ribeiro, A. (1975). Forme et orientation de la virgation hercynienne de Galice. Relations avec le plissement et hypothèses sur la genèse de l'arc Ibero-Armoricain. *Compte r. Seances Acad. Sci. Paris.*, **280D**, 2825–2828.

Matte, P. L. and Burg, J. P. (1981). Sutures, thrusts and nappes in the Variscan arc of Western Europe: plate tectonic implications. In K. R. McClay and N. J. Price (Eds), *Thrust and Nappe Tectonics*, Geol. Soc. London, pp. 353–362.

Max, M. D. (1979). Extent and disposition of Grenville tectonism in the Precambrian continental crust adjacent to the North Atlantic. *Geology*, **7**, 76–78.

May, P. R. (1971). Pattern of Triassic-Jurassic diabase dikes around the North Atlantic in the context of predrift position of the continents. *Bull. geol. Soc. Amer.*, **82**, 1285–1292.

Mehta, P. K. (1977). Rb–Sr geochronology of the Kulu-Mandi belt: its implications for Himalayan tectonogenesis. *Geol. Rund.*, **66**, 156–175.

Meissner, R., Bartelsen, H. and Murawski, H. (1981). Thin-skinned tectonics in the Northern Rhenish Massif, Germany. *Nature*, **290**, 399–401.

Mel'nik, Y. P. (1982). *Precambrian Banded iron-formations*, Elsevier, Amsterdam, 310 pp.

Menard, H. W. (1969). Elevation and subsidence of the oceanic crust. *Earth Planet. Sci. Lett.*, **6**, 275–284.

Mendelsohn, F. (Ed.) (1961). *The Geology of the Northern Rhodesian Copper Belt*, Roan Antelope Copper Mines Ltd, and MacDonald Press, London.

Menzies, M., Blanchard, D. and Xenophontos, C. (1980). Genesis of the Smartville arc-ophiolite, Sierra Nevada foothills, California. *Amer. J. Sci.*, **280A**, 329–344.

Michard-Vitrac, A., Lancelot, J., Allègre, C. J. and Moorbath, S. (1977). U–Pb ages on simple zircons from the early Precambrian rocks of W Greenland and the Minnesota River Valley. *Earth Planet. Sci. Lett.*, **35**, 449–453.

Mikkola, A. (1980). The metallogeny of Finland. *Geol. Surv. Finland, Bull.*, **305**, 22 pp.

Milanovsky, E. E. (1981). Aulacogens on ancient platforms: problems of their origin and tectonic development. *Tectonophysics*, **73**, 213–248.

Milanovsky, E. E. (1983). Major stages of rifting evolution in the earth's history. *Tectonophysics*, **94**, 599–607.

Miller, C. F. and Bradfish, L. J. (1980). An inner Cordilleran belt of muscovite-bearing plutons. *Geology*, **8**, 412–416.

Misra, K. C. and Keller, F. B. (1978). Ultramafic bodies in the southern Appalachians: a review. *Amer. J. Sci.*, **278**, 389–418.

Mitchell, A. H. G. (1974). Southwest England granites: magmatism and tin mineralization in a post-collision tectonic setting. *Trans. Inst. Min. Metall. Sect. B*, **83**, 95–97.

Mitchell, A. H. G. (1976). Tectonic settings for emplacement of subduction-related magmas and associated mineral deposits. In D. F. Strong (Ed.), *Metallogeny and Plate Tectonics, Geol. Ass. Can. Sp. Pap.*, **14**, 3–22.

Mitchell, A. H. G. (1979). Guides to metal provinces in the central Himalaya collision belt: the value of regional stratigraphic correlations and tectonic analogies. *Mem. Geol. Soc. China*, No. 3, 167–194.

Mitchell, A. H. G. and Garson, M. S. (1976). Mineralization at plate boundaries. *Minerals Sci. Engng.*, **8**, 129–169.

Mitchell, A. H. G. and Garson, M. S. (1981). *Mineral Deposits and Global Tectonic Settings*, Academic Press, London, 405 pp.

Mitchell, A. H. G. and McKerrow, W. S. (1975). Analogous evolution of the Burma Orogen and the Scottish Caledonides. *Bull. geol. Soc. Amer.*, **86**, 305–315.

Miyashiro, A. (1972). Metamorphism and related magmatism in plate tectonics. *Amer. J. Sci.*, **272**, 629–656.

Miyashiro, A. (1981). Metamorphism and plate convergence. In D. W. Strangway (Ed.), *The Continental Crust and its Mineral Deposits, Geol. Ass. Can. Sp. Pap.*, **20**, 591–605.

Mobus, G. (1977). Charactéristiques de l'évolution de la zone Saxo-Thuringienne dans l'orogenèse hercynienne en Europe central. In *La Chaine Varisque d'Europe Moyenne et Occidentale*, Editions du CNRS, no. 243, Paris, pp. 391–404.

Mohr, P. (1978). Afar. *Ann. Rev. Earth Planet. Sci.*, **6**, 145–172.

Molnar, P. and Atwater, T. (1973). Relative motion of hot spots in the mantle. *Nature*, **246**, 288–291.

Molnar, P. and Tapponnier, P. (1975). Cainozoic tectonics of Asia: the effects of a continental collision. *Science*, **189**, 419–426.

Molnar, P. and Atwater, T. (1978). Interarc spreading and Cordilleran tectonics as alternates related to the age of subducted oceanic lithosphere. *Earth Planet. Sci. Lett.*, **41**, 330–340.

Molnar, P. and Chen, W. P. (1978). Evidence of large Cainozoic crustal shortening of Asia. *Nature*, **273**, 328–220.

Monger, J. W. H., Price, R. A. and Tempelman-Kluit, D. J. (1982). Tectonic accretion and the origin of the two major metamorphic and plutonic belts in the Canadian Cordillera. *Geology*, **10**, 70–75.

Monger, J. W. H., Souther, J. G. and Gabrielse, H. (1972). Evolution of the Canadian Cordillera: a plate tectonic model. *Amer. J. Sci.*, **272**, 577–602.

Montgomery, C. W. (1979). Uranium–lead geochronology of the Archaean Imataca Series, Venezuelan Guayana Shield. *Contrib. Mineral. Petrol.*, **69**, 167–176.

Moorbath, S. (1969). Evidence for the age of deposition of the Torridonian sediments of north-west Scotland. *Scott. J. Geol.*, **5**, 154–170.

Moorbath, S. (1977). Age, isotopes and evolution of Precambrian continental crust. *Chem. Geol.*, **20**, 151–187.

Moorbath, S. (1978). Age and isotope evidence for the evolution of continental crust. *Phil. Trans. R. Soc. Lond.*, **A288**, 401–413.

Moorbath, S. (1980). Aspects of the chronology of ancient rocks related to continental evolution. In D. W. Strangway (Ed.), *The Continental Crust and its Mineral Deposits, Geol. Ass. Can. Sp. Pap.*, **20**, 89–116.

Moorbath, S., Wilson, J. F. and Cotterill, P. (1976). Early Archaean age for the Sebakwian group at Selukwe, Rhodesia. *Nature*, **264**, 536–538.

Moorbath, S., Allaart, J. H., Bridgwater, D. and McGregor, V. R. (1977). Rb–Sr ages of early Archaean supracrustal rocks and Amîtsoq gneisses at Isua. *Nature*, **270**, 43–45.

Moorbath, S. and Taylor, P. N. (1981). Isotopic evidence for continental growth in the Precambrian. In A. Kröner (Ed.), *Precambrian Plate Tectonics*, Elsevier, Amsterdam, pp. 491–526.

Moorbath, S., Taylor, P. N. and Goodwin, R. (1981). Origin of granite magma by crustal remobilisation: Rb–Sr and Pb–Pb geochronology and isotope geochemistry of the late Archaean Qôrqut granite complex of SW Greenland. *Geochim. Cosmochim. Acta*, **45**, 1051–1060.

Moore, C. F., Curray, J. R. and Emmel, F. J. (1982). Sedimentation in the Sunda trench and forearc region. In J. K. Leggett (Ed.), *Trench-Forearc Geology: Sedimentation and Tectonics in Modern and Ancient Active Plate Margins*, Geol. Soc. London, Blackwell, Oxford, pp. 245–258.

Moore, J. C. and Wheeler, R. L. (1978). Structural fabric of a mélange, Kodiak Islands, Alaska. *Amer. J. Sci.*, **278**, 739–765.

Moore, J. C. *et al.* (14 authors) (1979). Progressive accretion in the middle America trench, southern Mexico. *Nature*, **281**, 638–642.

Moore, J. G. (1959). The quartz diorite boundary line in the western United States. *J. Geol.*, **67**, 198–210.

Moore, J. M. (1977). Orogenic volcanism in the Proterozoic of Canada. In W. R. A. Baragar *et al.* (Eds), *Volcanic Regimes in Canada, Geol. Ass. Can. Sp. Pap.*, **16**, 127–148.

Moores, E. M. (1973). Geotectonic significance of ultramafic rocks. *Earth Sci. Rev.*, **9**, 241–258.

Moores, E. M. (1979). Anatomy of a Jurassic arc–arc collision, N Sierra Nevada, California: possible model for Archaean tectonics. *Geol. Ass. Can. Prog. with Abstr.*, **4**, 68.

Moores, E. M. and Vine, F. J. (1971). The Troodos Massif, Cyprus, and other ophiolites as oceanic crust: evaluations and implications. *Phil. Trans. R. Soc. Lond.*, **A268**, 443–466.

Moralev, V. M. (1981). Tectonics and petrogenesis of early Precambrian complexes of the Aldan Shield, Siberia. In A. Kröner (Ed.), *Precambrian Plate Tectonics*, Elsevier, Amsterdam, pp. 237–260.

Morel, P. and Irving, E. (1977). Tentative maps of continental position for 1150 my to 200 my. *IX Coll. Afr. Geol. Gottingen, Abstract*, 62.

Morel, P. and Irving, E. (1981). Palaeomagnetism and the evolution of Pangaea. *J. Geophys. Res.*, **86B**, 1858–1872.

Morgan, G. E. and Briden, J. C. (1981). Aspects of Precambrian palaeomagnetism with new data from the Limpopo mobile belt and Kaapvaal craton in southern Africa. *Phys. Earth. Planet. Int.*, **24**, 142–168.

Morgan, W. J. (1968). Rises, trenches, great faults, and crustal blocks. *J. Geophys. Res.*, **73**, 1959–1982.

Morgan, W. J. (1972). Deep mantle convection plumes and plate motions. *Amer. Ass. Petrol. Geol. Bull.*, **56**, 203–213.

Morgan, W. J. (1973). Plate motions and deep mantle convection. *Geol. Soc. Amer. Mem.*, **132**, 7–22.

Morgan, W. J. (1981). Hot spot tracks and the opening of the Atlantic and Indian oceans. In C. Emiliani (Ed.), *The Oceanic Lithosphere*, Wiley, New York, pp. 443–487.

Morgan, W. R. (1982). A layered ultramafic intrusion in Archaean granulites near Lake Kondinim, W Australia. *J. Roy. Soc. Austral.*, **65**, 69–85.

Morris, W. A. and Roy, J. W. (1977). Discovery of the Hadrynian polar track and further study of the Grenville problem. *Nature*, **266**, 689–692.

Morse, S. A. (1982). A partisan review of Proterozoic anorthosites. *Amer. Mineral.*, **67**, 1087–1100.

Morton, R. D., Goble, R. J. and Friz, P. (1974). The mineralogy,

sulfur-isotope composition and origin of some copper deposits in the Belt Supergroup, SW Alberta, Canada. *Mineral. Deposita*, **9**, 223–241.

Moseley, F. (1977). Caledonian plate tectonics and the place of the English Lake District. *Bull. geol. Soc. Amer.*, **88**, 764–768.

Moshkin, V. N. and Dageldiskaya, I. N. (1972). The Precambrian anorthosites of the USSR. *24th Int. geol. Congr. Montreal Sect. 2*, 329–333.

Mottl, M. J. (1983). Metabasalts, axial springs, and the structure of hydrothermal systems at mid-ocean ridges. *Bull. geol. Soc. Amer.*, **94**, 161–180.

Mueller, P. A. and Wooden, J. L. (1976). Rb–Sr whole rock age of the contact aureole of the Stillwater igneous complex, Montana. *Earth Planet. Sci. Lett.*, **29**, 384–388.

Muir, M. D. and Grant, P. R. (1976). Micropalaeontological evidence from the Onverwacht Group, South Africa. In B. F. Windley (Ed.), *The Early History of the Earth*, Wiley–Interscience, London, pp. 595–604.

Murray, G. E., Kaczor, M. J. and McArthur, R. E. (1980). Indigenous Precambrian petroleum revisited. *Amer. Ass. Petrol. Geol. Bull.*, **64**, 168–170.

Myers, J. S. (1975). Cauldron subsidence and fluidization: mechanisms of intrusion of the Coastal Batholith of Peru into its own volcanic ejecta. *Bull. geol. Soc. Amer.*, **86**, 1209–1220.

Myers, J. S. (1976). Erosion surfaces and ignimbrite eruption, measures of Archaean uplift in northern Peru. *Geol. J.*, **11**, 29–44.

Myers, J. S. (1980). Structure of the coastal dyke swarm and associated plutonic intrusions of East Greenland. *Earth Planet. Sci. Lett.*, **46**, 407–418.

Myers, J. S. (1981). The Fiskenaesset anorthosite complex—a stratigraphic key to the tectonic evolution of the W Greenland gneiss complex 3000–2800 my ago. In J. E. Glover and D. I. Groves (Eds), *Archaean Geology, Geol. Soc. Australia, Sp. Publ.*, **7**, 351–360.

Nairn, A. E. M. and Ressetar, R. (1978). Palaeomagnetism of the Peri-Atlantic Precambrian. *Ann. Rev. Earth Planet. Sci.*, **6**, 75–91.

Nairn, A. E. M. and Stehli, F. C. (1973). A model for the South Atlantic. In A. E. M. Nairn and F. G. Stehli (Eds), *The Ocean Basins and Margins*, vol. 1, *The South Atlantic*, Plenum Press, New York, pp. 1–24.

Nakazawa, K. and Runnegar, B. (1973). The Permian-Triassic boundary: a crisis for bivalves? *Can. Soc. Petrol. Geol. Mem.*, **2**, 608–621.

Naqvi, S. M., Viswanathan, S. and Viswanatha, M. N. (1978). Geology and geochemistry of the Holenarasipur schist belt and its place in the evolutionary history of the Indian peninsula. In B. F. Windley (Ed.), *The Early History of the Earth*, Wiley–Interscience, London, pp. 109–126.

Naqvi, S. M., Govil, P. K. and Rogers, J. J. W. (1981). Chemical sedimentation in Archaean–early Proterozoic greenschist belts of the Dharwar Craton, India. In J. E. Glover and D. I. Groves (Eds), *Archaean Geology, Geol. Soc. Australia, Sp. Publ.*, **7**, 245–253.

Neary, C. R., Gass, I. G. and Cavanagh, B. J. (1976). Granite association of NE Sudan. *Bull. geol. Soc. Amer.*, **87**, 1501–1512.

Neill, W. M. (1973). Possible continental rifting in Brazil and Angola related to the opening of the South Atlantic. *Nature Phys. Sci.*, **245**, 104–107.

Neruchev, S. G. (1979). An attempt at a quantitative estimate of the parameters of the ancient atmospheres of the earth. *Int. Geol. Rev.*, **21**, 373–384.

Nesbitt, R. W., Sun, S. S. and Purvis, A. C. (1979). Komatiites: geochemistry and genesis. *Can. Mineral*, **17**, 165–186.

Nesbitt, H. W. and Young, G. M. (1982). Early Proterozoic climates and plate motions inferred from major element chemistry of lutites. *Nature*, **299**, 715–171.

Nesbitt, R. W., Jahn, B. M. and Purvis, A. C. (1982). Komatiites: an early Precambrian phenomenon. *J. Volc. Geochem. Res.*, **14**, 31–45.

Neudert, M. K. and Russell, R. E. (1981). Shallow water and

380

hypersaline features from the middle Proterozoic Mt. Isa sequence. *Nature*, **293**, 284–286.

Newton, R. C. (1978). Experimental and thermodynamic evidence for the operation of high pressures in Archaean metamorphism. In B. F. Windley (Ed.), *The Early History of the Earth*, Wiley–Interscience, London, pp. 221–240.

Newton, R. C., Smith, J. V. and Windley, B. F. (1980). Carbonic metamorphism, granulites and crustal growth. *Nature*, **288**, 45–50.

Neymark, L. A., Iskanderova, A. D., Chukhonin, A. P., Mironyuk, Ye. P. and Romina, Ye, Ye. (1981). U–Pb data on the Archaean age of the metamorphic rocks in the Stanovoy Range. *Geochem. Intern.*, **18**, 55–64.

Nicholas, A. *et al.* (8 authors) (1981). The Xigaze ophiolite (Tibet): a peculiar oceanic lithosphere. *Nature*, **294**, 414–417.

Nisbet, E. G., Bickle, M. J. and Martin, A. (1977). The mafic and ultramafic lavas of the Belingwe greenstone belt, Rhodesia. *J. Petrol.*, **18**, 521–566.

Nisbet, E. G., Wilson, J. F. and Bickle, M. J. (1981). The evolution of the Rhodesian craton and adjacent Archaean terrain: tectonic models. In A. Kröner (Ed.), *Precambrian Plate Tectonics*, Elsevier, Amsterdam, pp. 161–183.

Noble, D. C., McKee, E. H., Farrar, E. and Petersen, U. (1974). Episodic Cenozoic volcanism and tectonism in the Andes of Peru, *Earth Planet. Sci. Lett.*, **21**, 213–220.

Noble, J. A. (1976). Metallogenic provinces of the Cordillera of western North and South America. *Mineral. Deposita.*, **11**, 219–233.

Noltimier, H. C. (1974). The geophysics of the North Atlantic Basin. In A. E. M. Nairn and F. G. Stehli (Eds), *The Ocean Basins and Margins*, vol. 2, Plenum Press, New York, pp. 539–588.

Normark, W. R., Morton, J. L., Koski, A., Clague, D. A. and Delaney, J. R. (1983). Active hydrothermal vents and sulfide deposits on the southern Juan de Fuca Ridge. *Geology*, **11**, 158–163.

Norton, I. O. and Sclater, J. G. (1979). A model for the evolution of the Indian Ocean and the break-up of Gondwanaland. *J. Geophys. Res.*, **84**, 6803–6830.

Nunes, P. D. (1981). The age of the Stillwater complex—a comparison of U–Pb zircon and Sm–Nd isochron systematics. *Geochim. Cosmochim. Acta*, **45**, 1961–1963.

Nur, A. and Ben-Avraham, Z. (1981). Volcanic gaps and the consumption of aseismic ridges in South America. *Geol. Soc. Amer. Mem.*, **154**, 729–740.

Nur, A. and Ben-Avraham, Z. (1982a). Displaced terranes and mountain building. In K. J. Hsü (Ed.), *Mountain Building Processes*, Academic Press, London, pp. 73–84.

Nur, A. and Ben-Avraham, Z. (1982b). Oceanic plateaus, the fragmentation of continents and mountain building. *J. Geophys. Res.*, **87**, 3644–3661.

Nyström, J. O. (1982). Post-Svecokarelian Andinotype evolution in central Sweden. *Geol. Rund.*, **71**, 114–157.

Obradovich, J. D. and Peterman, Z. E. (1968). Geochronology of the Belt Series, Montana. *Can. J. Earth Sci.*, **5**, 737–747.

Ocola, L. C. and Meyer, R. P. (1973). Central North America rift system. 1. Structure of the axial zone from seismic and gravimetric data. *J. Geophys. Res.*, **78**, 5173–5194.

Oehler, J. H. and Logan, R. G. (1977). Microfossils, cherts and associated mineralization in the Proterozoic McArthur (H.Y.C.) lead–zinc–silver deposit. *Econ. Geol.*, **72**, 1393–1409.

Oftedahl, C. (1978a). Main geological features of the Oslo graben. In I. B. Ramberg and E. R. Neumann (Eds), *Tectonics and Geophysics of Continental Rifts*, Reidel, Dordrecht, Holland, pp. 149–165.

Oftedahl, C. (1978b). Cauldrons of the Permian Oslo rift. *J. Volc. Geotherm. Res.*, **3**, 343–371.

O'Hara, M. J. (1961). Petrology of the Scourie dyke, Sutherland. *Mineral. Mag.*, **32**, 848–865.

Olade, M. A. (Ø978). Early Cretaceous basalt volcanism and initial rifting in Benue Trough, Nigeria. *Nature*, **273**, 458–459.

Olade, M. A. (1980). Plate tectonics and metallogeny of intra-continental rifts and aulacogens in Africa—a review. In *Proc.*

Fifth Quad. IAGOD Symp., Schweizerbart'sche Verlag, Stuttgart, **91**, 111.

O'Nions, R. K., Evensen, N. W. and Hamiltion, P. J. (1979). Geochemical modelling of mantle differentiation and crustal growth. *J. Geophys. Res.*, **84**, 6091–6101.

Orpen, J. Z. and Wilson, J. F. (1981). Stromatolites at 3500 Myr and a greenstone-granite unconformity in the Zimbabwean Archaean. *Nature*, **291**, 218–220.

Oudin, E., Picot, P. and Pouit, G. (1981). Comparison of sulphide deposits from the East Pacific Rise and Cyprus. *Nature*, **291**, 404–407.

Owen, T., Cess, R. D. and Ramanathan, V. (1979). Enhanced CO_2 greenhouse to compensate for reduced solar luminosity on early Earth. *Nature*, **277**, 640–642.

Packham, G. H. and Falvey, D. A. (1971). An hypothesis for the formation of marginal seas in the western Pacific. *Tectonophysics*, **11**, 79–109.

Page, N. J. (1977). Stillwater complex, Montana: rock succession, metamorphism and structure of the complex and adjacent rocks. *U.S. geol Surv. Prof. Pap.*, **999**, 79 pp.

Palmer, H. C. and Carmichael, K. M. (1973). Palaeomagnetism of some Grenville Province rocks. *Can. J. Earth Sci.*, **10**, 1175–1190.

Panayiotou, A. (Ed.) (1980). *Ophiolites. Proc. Int. Ophiolite Symp. Cyprus Geol. Surv.*, 781 pp.

Pankhurst, R. J. (1974). Rb–Sr whole rock chronology of Caledonian events in NE Scotland. *Bull. geol. Soc. Amer.*, **85**, 345–350.

Pannella, G. (1972). Precambrian stromatolites as palaeontological clocks. *24th Int. geol. Cong. Montreal, Sect. 1*, 50–57.

Park, R. G. (1973). The Laxfordian belts of the Scottish mainland. In R. G. Park and J. Tarney (Eds), *The Early Precambrian of Scotland and Related Rocks of Greenland*, University of Keele, pp. 65–76.

Park, R. G. and Ermanovics, I. F. (1978). Tectonic evolution of two greenstone belts from the Superior Province in Manitoba. *Can. J. Earth Sci.*, **15**, 1808–1816.

Park, A. F. and Bowes, D. R. (1981). Metamorphosed and deformed pillow lavas from Losomäki: evidence of subaqueous volcanism in the Outokumpu association, E Finland. *Bull. geol. Soc. Finland*, **53**, 135–145.

Parmentier, E. M. and Spooner, E. T. C. (1978). A theoretical study of hydrothermal convection and the origin of the ophiolite sulphide ore deposits of Cyprus. *Earth Planet. Sci. Lett.*, **40**, 33–44.

Patchett, P. J., Bylund, G. and Upton, B. G. J. (1978). Palaeomagnetism and the Grenville orogeny; new Rb–Sr ages from dolerites in Canada and Greenland. *Earth Planet. Sci. Lett.*, **40**, 349–364.

Patriat, P. *et al.* (8 authors) (1982). Les mouvements relatifs de l'Inde, de l'Afrique et de l'Eurasie. *Bull. Soc. géol. France*, **24**, 363–373.

Pavlovskiy, Ye. V. (1980). Problem of the lower Precambrian 'greenstone belts'. *Int. Geol. Rev.*, **22**, 1314–1326.

Pearce, J. A. (1980). Geochemical evidence for the genesis and eruptive setting of lavas from Tethyan ophiolites. In A. Panayiotou (Ed.), *Ophiolites, Cyprus Geol. Surv.*, 261–272.

Pêcher, A. (1979). Les inclusions fluides des quartz d'exsudation de la zone du MCT himalayan au Népal central: données sur la phase fluide dans une grande zone de cisaillement crustal. *Bull. Mineral.*, **102**, 537–554.

Pedersen, F. D. (1981). Polyphase deformation of the massive sulphide ore of the Black Angel mine, central west Greenland. *Mineral. Deposita.*, **16**, 157–176.

Pedersen, K. R. and Lam, J. (1970). Precambrian organic compounds from the Ketilidian of south-west Greenland, parts 1–3. *Meddr. Grønland*, **185**, 5–7, 57 pp.

Pedreira, A. J. (1979). Possible evidence of a Precambrian continental collision in the Riv Pardo basin of eastern Brazil. *Geology*, **7**, 445–448.

Peirce, J. W. (1978). The northward motion of India since the late Cretaceous. *R. Astron. Soc. Geophys. J.*, **52**, 277–311.

Perfit, M. R., Gust, D. A., Bence, A. E., Arculus, R. J. and Taylor, S. R. (1980). Chemical characteristics of island-arc basalts: implications for mantle sources. *Chem. Geol.*, **30**, 227–256.

Perry, E. C. Jr., Ahmad, S. N. and Swulius, T. M. (1978). The oxygen isotope composition of 3800 my old metamorphosed chert and iron formation from Isukasia, W Greenland. *J. Geol.*, **86**, 223–239.

Peterman, Z. E., Zartman, R. E. and Sims, P. K. (1980). Tonalitic gneiss of early Archaean age from N Michigan. In G. D. Morey and G. N. Hanson (Eds), *Selected Studies of Archaean Gneisses and Lower Proterozoic rocks, Southern Shield, Geol. Soc. Am. Sp. Pap.*, **182**, 125–134.

Petters, W. S. (1978). Stratigraphic evolution of the Benue Trough and its implications for the Upper Cretaceous palaeogeography of W Africa. *J. Geol.*, **86**, 311–322.

Pettingill, H. S. and Patchett, P. J. (1981). Lu–Hf total-rock age for the Amîtsoq gneisses, W Greenland. *Earth Planet. Sci. Lett.*, **55**, 150–156.

Phillips, J. D. and Forsyth, O. (1972). Plate tectonics, palaeomagnetism and the opening of the Atlantic. *Bull. geol. Soc. Amer.*, **83**, 1579–1600.

Phillips, W. E. A., Stillman, C. J. and Murphy, T. A. (1976). Caledonian plate tectonic model. *J. geol. Soc. Lond.*, **132**, 579–609.

Philpotts, A. R. (1981). A model for the generation of massif-type anorthosites. *Can. Mineral.*, **19**, 233–253.

Piasecki, M. A. J. and van Breemen, O. (1979). A Morarian age for the 'younger Moines' of central and western Scotland. *Nature*, **278**, 734–736.

Pichamuthu, C. S. (1982). Schist-gneiss relationship in the Archaean of Dharwar craton. *Current Science*, **51**, 118–124.

Pidgeon, R. T. (1978). 3450 my old volcanics in the Archaean layered greenstone succession of the Pilbara block, W Australia. *Earth Planet. Sci. Lett.*, **37**, 421–428.

Pin, C. and Vielzeuf, D. (1983). Granulites and related rocks in a dualistic interpretation. *Tectonophysics*, **93**, 47–74.

Piper, J. D. A. (1973a). Latitudinal extent of late Precambrian glaciations. *Nature*, **244**, 342–344.

Piper, J. D. A. (1973b). Geological interpretation of palaeomagnetic results from the African Precambrian. In D. H. Tarling and S. K. Runcorn (Eds), *Implications of Continental Drift to the Earth Sciences*, vol. 1, Academic Press, London, pp. 19–32.

Piper, J. D. A. (1978). Geological and geophysical evidence relating to continental growth and dynamics and the hydrosphere in Precambrian times: a review and analysis. In P. Brosche and J. Sündermann (Eds), *Tidal Friction and the Earth's Rotation*, Springer, Berlin, pp. 197–241.

Piper, J. D. A. (1979). Aspects of Caledonian palaeomagnetism and their tectonic implications. *Earth Planet. Sci. Lett.*, **44**, 176–192.

Piper, J. D. A. (1980a). Palaeomagnetic study of the Swedish rapakivi suite: Proterozoic tectonics of the Baltic Shield. *Earth Planet. Sci. Lett.*, **46**, 443–461.

Piper, J. D. A. (1980b). Analogous Upper Proterozoic apparent polar wander loops. *Nature*, **283**, 845–847.

Piper, J. D. A. (1982). The Precambrian palaeomagnetic record: the case for the Proterozoic supercontinent. *Earth Planet. Sci. Lett.*, **59**, 61–89.

Pitcher, W. S. (1975). On the rate of emplacement of batholiths. *J. geol. Soc. Lond.*, **131**, 587–591.

Pitcher, W. S. (1978). The anatomy of a batholith. *J. geol. Soc. Lond.*, **135**, 157–182.

Pitcher, W. S. (1982). Granite type and tectonic environment. In K. J. Hsü (Ed.), *Mountain Building Processes*, Academic Press, London, pp. 19–40.

Plotnick, R. E. (1980). Relationship between biological extinctions and geomagnetic reversals. *Geology*, **8**, 578–581.

Plumstead, E. P. (1973). The late Plaeozoic Glossopteris flora. In A. Hallam (Ed.), *Atlas of Palaeobiogeography*, Elsevier, Amsterdam, pp. 187–206.

Porada, H. (1979). The Damara–Ribeira orogen of the Pan-African/Braziliano cycle in Namibia (southwest Africa) and Brazil as interpreted in terms of continental collision. *Tectonophysics*, **57**, 237–265.

Poulsen, K. H., Borradaile, G. J. and Kehlenbeck, M. M. (1980). An inverted Archaean succession at Rainy Lake, Ontario. *Can. J. Earth Sci.*, **17**, 1358–1369.

Powell, C. Mc. A. and Conaghan, P. J. (1973). Polyphase deformation in Phanerozoic rocks of the central Himalayan gneiss, NW india. *J. geol.*, **81**, 127–143.

Powell, C. Mc. A., Crawford, A. R., Armstrong, R. L., Prakash, R. and Wynne-Edwards, H. R. (1979). Reconnaissance Rb–Sr dates for the Himalayan Central Gneiss, NW India. *Ind. J. Earth Sci.*, **6**, 139–51.

Powell, C. Mc. A., Johnson, B. D. and Veevers, J. J. (1980). A revised fit of east and west Gondwanaland. *Tectonophysics*, **63**, 13–29.

Prasad, C. V. R. K., Subba Reddy, N. and Windley, B. F. (1982). Iron formations in Archaean granulite gneiss belts with special reference to southern India. *J. geol. Soc. India.* **23**, 112–122.

Preiss, W. V. (1977). The biostratigraphic potential of Precambrian stromatolites. *Precamb. Res.*, **5**, 207–219.

Pretorius, D. A. (1976). The nature of the Witwatersrand gold–uranium deposits. In K. H. Wolf (Ed.), *Handbook of Stratabound and Stratiform Ore Deposits*, Elsevier, Amsterdam, vol. 7, pp. 29–88.

Pretorius, D. A. (1981). Gold and uranium in quartz-pebble conglomerates. *Econ. Geol. 75th Anniv. Vol.*, 117–138.

Price, R. A., Monger, J. W. H. and Muller, J. E. (1981). Cordilleran cross-section—Calgary to Victoria. *Field Guides to Geology and Mineral Deposits, Calgary 81 GAC, MAC, CGU*, 261–269.

Price, C. and Moore, J. M. (1983). Petrogenesis of the Elsevier batholith and related trondhjemitic intrusions in the Grenville Province of eastern Ontario, Canada, *Contrib. Mineral. Petrol.*, **82**, 187–194.

Puscharovsky, Yu. M. (1973). Tectonics of the Pacific segment of the earth. In P. J. Coleman (Ed.), *The Western Pacific*, University of Western Australian Press, pp. 21–30.

Quinguis, H. and Choukroune, P. (1981). Les schistes bleus de l'île de Groix dans la chaîne hercynienne: implications cinématiques. *Bull. Soc. géol. France*, **23**, 409–418.

Rackley, R. I. (1976). Origin of Western States-type uranium mineralization. In K. H. Wolf, (Ed.), *Handbook of Stratabound and Stratiform Ore Deposits*, vol. 7, Elsevier, Amsterdam, pp. 89–156.

Radain, A. A. M., Fyfe, L. I. S. and Kerrich,. R. (1981). Origin of peralkaline granites of Saudi Arabia. *Contrib. Mineral. Petrol.*, **78**, 358–366.

Radhakrishna, B. P. and Sreenivasaiah, S. (1974). Bedded barytes from the Precambrian of Karnataka. *J. geol. Soc. India.*, **15**, 314–315.

Radhakrishna, B. P. (1976). Mineralisation episodes in the Dharwar craton of peninsular India. *J. geol. Soc. India*, **17**, 79–88.

Radhakrishna, B. P. and Vasudev, V. N. (1977). The early Precambrian of the southern Indian Shield. *J. geol. Soc. India.*, **18**, 525–541.

Raith, M., Ackermand, D. and Lal, R. K. (1982a). The Archaean craton of S India: metamorphic evolution and PT conditions. *Geol. Rund.*, **71**, 280–290.

Raith, M., Raase, P. and Hörmann, P. K. (1982b). The Precambrian of Finnish Lapland: evolution and regime of metamorphism. *Geol. Rund.*, **71**, 230–244.

Rajagopalan, P. T., Jayaram, S. and Venkatasubramaniam, V. S. (1980). Rb–Sr isochron age for gneisses in the western region of the Dharwar craton. *J. geol. Soc. India.*, **21**, 540–56.

Ramakrishnan, M., Viswanatha, M. N. and Swami Nath, J. (1976). Basement-cover relationships of Peninsular gneiss with high-grade schists and greenstone belts of southern Karnataka. *J. geol. Soc. India*, **17**, 9e–110.

Ramakrishnan, M., Viswanatha, M. N., Chayapathi, N. and Narayanan Kutty, T. R. (1978). Geology and geochemistry of anorthosites of Karnataka craton and their tectonic significance. *J. geol. Soc. India*, **19**, 115–134.

Ramiengar, A. S., Devadu, G. R., Viswanatha, M. N., Chayapathi, N. and Ramakrishnan, M. (1978). Banded chromite-fuchsite quartzite in the older supracrustal sequence of Karnataka. *J. geol. Soc. India*, **19**, 577–582.

Ramsay, J. G. (1963). Structural investigations in the Barberton Mountain Land, eastern Transvaal. *Trans. geol. Soc. S. Afr.*, **66**, 353–398.

Rast, N. (1983). Variscan orogeny. In P. L. Hancock (Ed.), *The Variscan Fold Belt in the British Isles*, Adam Hilger, Bristol, pp. 1–19.

Raü, A. and Tongiorgi, M. (1981). Some problems regarding the Palaeozoic palaeogeography in Mediterranean western Europe. *J. Geol.*, **89**, 663–673.

Ray, G. E. and Wanless, R. K. (1980). The age and geological history of the Wollaston, Peter Lake and Rottenstone domains in northern Saskatchewan. *Can. J. Earth Sci.*, **17**, 333–347.

Reeves, C. V. (1978). A failed Gondwana spreading axis in southern Africa. *Nature*, **273**, 222.

Reibel, G. and Reuber, I. (1982). La klippe ophiolitique de Spongtang–Photoksar (Himalaya du Ladakh): une ophiolite sans cumulats. *C.R. Seances Acad. Sci. Paris*, **294**, sér. II, 557–562.

Reimer, T. O. (1980). Archaean sedimentary baryte deposits of the Swaziland Supergroup (Barberton Mountain Land, South Africa). *Precamb. Res.*, **12**, 393–410.

Rentzsch, J. (1974). The 'Kupferschiefer' in comparison with the deposits of the Zambian Copper belt. In P. Bartholomé (Ed.), *Gisements Stratiformes et Provinces Cuprifères, Soc. geol. Belg., Liège*, 235–254.

Reyment, R. A. (1980). Palaeo-oceanology and palaeobiogeography of the Cretaceous South Atlantic Ocean. *Oceanologica Acta*, **3**, 127–133.

Reyment, R. A. and Mörner, N. A. (1977). Cretaceous transgressions and regressions exemplified by the South Atlantic. *Palaeont. Soc. Japan, Sp. Pap.*, **21**, 247–261.

Richter, F. M. (1973). Convection and large-scale circulation of the mantle. *J. Geophys. Res.*, **78**, 8735–8745.

Rickard, M. J. and Belbin, L. (1980). A new continental assembly for Pangaea. *Tectonophysics*, **63**, 1–12.

Ries, A. C., Shackleton, R. M., Graham, R. H. and Fitches, W. R. (1983). Pan-African structures, ophiolites and mélange in the eastern desert of Egypt: a traverse at 26°N. *J. geol. Soc. Lond.*, **140**, 75–95.

Ringwood, A. E. (1974). The petrological evolution of island arc systems. *J. geol. Soc. Lond.*, **130**, 183–204.

Ringwood, A. E. (1977). Petrogenesis in island arc systems. In M. Talwani and W. C. Pitman (Eds), *Island Arcs, Deep Sea Trenches and Back Arc Basins, Amer. Geophys. Un.*, Maurice Ewing Ser. 1, pp. 311–324.

Rivalenti, G., Garuti, G., Rossi, A., Siena, F. and Sinigoi, S. (1981). Existence of different peridotite types and of a layered igneous complex in the Ivrea Zone of the western Alps. *J. Petrol.*, **22**, 127–153.

Roberts, D. G. (1975). Evaporite deposition in the Aptian South Atlantic Ocean. *Marine Geol.*, **18**, M65–71.

Roberts, D. and Gale, G. H. (1978). The Caledonian-Appalachian Iapetus Ocean. In D. H. Tarling (Ed.), *Evolution of the Earth's Crust*, Academic Press, London, pp. 255–342.

Robertson, A. H. F. (1976). Origin of ochres and umbers: evidence from Skouriotissa, Troodos massif, Cyprus. *Trans. Inst. Min. Metall.*, **85**, B245–251.

Robertson, A. H. F. (1978). Metallogenesis along a fossil oceanic fracture: Arakapas fault belt, Troodos massif, Cyprus. *Earth Planet. Sci. Lett.*, **41**, 317–329.

Robertson, A. H. F. and Woodcock, N. H. (1980). Tectonic setting of the Troodos massif in the east Mediterranean. In A. Panayiotou (Ed.), *Ophiolites, Cyprus Geol. Surv.*, 34–46.

Robertson, D. S., Tilsley, J. E. and Hogg, G. M. (1978). The timebound character of uranium deposits. *Econ. Geol.*, **73**, 1409–1419.

Robertson, I. D. M. and du Toit, M. C. (1981). The Limpopo belt. In D. R. Hunter (Ed.), *Precambrian of the Southern Hemisphere*, Elsevier, Amsterdam, pp. 641–671.

Robertson, J. M. (1981). Bimodal volcanism in the early Proterozoic Pecos greenstone belt, S Sangre de Cristo Mountains, New Mexico. *Geol. Soc. Amer. Abst. with Prog.*, **103**.

Robinson, A. and Spooner, E. T. C. (1982). Source of the detrital components of uraniferous conglomerates, Quirke ore zone, Elliot Lake, Ontario, Canada. *Nature*, **299**, 622–624.

Robinson, P. L. (1973). Palaeoclimatology and continental drift. In D. H. Tarling and S. K. Runcorn (Eds), *Implications of Continental Drift to the Earth Sciences*, vol. 1, Academic Press, London, pp. 451–476.

Rocci, G. *et al.* (8 authors) (1980). The Mediterranean ophiolites and their related Mesozoic volcano-sedimentary sequences. In A. Panayiotou (Ed.), *Ophiolites, Cyprus geol. Surv.*, 273–283.

Rodgers, J. (1970). *The Tectonics of the Appalachians*, Wiley, New York.

Roedder, E. (1981). Are the 3800 Myr old Isua objects microfossils, limonite-stained fluid inclusions, or neither? *Nature*, **298**, 459–462.

Roeder, D. (1978). Three central Mediterranean orogens—a geodynamic synthesis. In H. Cloops, D. Roeder and K. Schmidt (Eds), *Alps, Apennines and Hellenides*, Schweizerbart'sche Verlag, Stuttgart, pp. 589–620.

Rogers, J. J. W., Ghuma, M. A., Nagy, R. M., Greenberg, J. K. and Fullagar, P. D. (1978). Plutonism in Pan-African belts and the geologic evolution of NE Africa. *Earth Planet. Sci. Lett.*, **39**, 109–117.

Rogers, J. J. W., Hodges, K. V. and Ghuma, M. A. (1980). Trace elements in continental margin magmatism. Part 2: Trace elements in Ben Ghnema batholith and nature of the Precambrian crust in central N. Africa. *Bull. geol. Soc. Amer.*, part 1, **41**, 445–447.

Rollinson, H. R. (1982a). Evidence from feldspar compositions of high temperatures in granite sheets in the Scourian complex, NW Scotland. *Mineral. Mag.*, **46**, 73–76.

Rollinson, H. R. (1982b). PT-conditions in coeval greenstone belts and granulites from the Archaean of Sierra Leone. *Earth Planet. Sci. Lett.*, **59**, 177–191.

Rollinson, H. R. and Windley, B. F. (1980a). An Archaean granulite-grade tonalite–trondhjemite–granite suite from Scourie, NW Scotland: geochemistry and origin. *Contrib. Mineral. Petrol.*, **75**, 265–281.

Rollinson, H. R. and Windley, B. F. (1980b). Selective elemental depletion during metamorphism of Archaean granulites, Scourie, NW Scotland. *Contrib. Mineral. Petrol.*, **72**, 257–263.

Rollinson, H. R., Windley, B. F. and Ramakrishnan, M. (1981). Contrasting high and intermediate pressures of metamorphism in the Archaean Sargur schists of S India. *Contrib. Mineral. Petrol.*, **76**, 420–429.

Romer, A. S. (1973). Permian reptiles. In A. Hallam (Ed.), *Atlas of Palaeobiogeography*, Elsevier, Amsterdam, pp. 159–167.

Romer, A. S. (1975). Intercontinental correlations of Triassic Gondwana vertebrate faunas. In K. S. W. Campbell (Ed.), *Gondwana Geology, 3rd Int. Gondwana Symp., 1973*, 469–473.

Rona, P. A. (1978). Criteria for recognition of hydrothermal mineral deposits in oceanic crust. *Econ. Geol.*, **73**, 135–160.

Rona, P. A. and Richardson, E. S. (1978). Early Cenozoic global plate reorganization. *Earth Planet. Sci. Lett.*, **40**, 1–11.

Rondot, J. (1978). Stratigraphie et metamorphismes de la region du Saint-Maurice. In *Metamorphism in the Canadian Shield, Geol. Surv. Can. Pap.*, **78-10**, 329–352.

Ronov, A. B. (1964). Common tendencies in the chemical evolution of the earth's crust, ocean and atmosphere. *Geochem. Int.*, **1**, 713–737.

Ronov, A. B. (1968). Probable changes in the composition of sea water during the course of geological time. *Sedimentology*, **10**, 25–43.

Ronov, A. B. (1972). Evolution of rock composition and geochemical processes in the sedimentary shell of the earth. *Sedimentology*, **19**, 157–172.

Ronov, A. B. and Migdisov, A. A. (1971). Geochemical history of the crystalline basement and the sedimentary cover of the Russian and North American platforms. *Sedimentology*, **16**, 137–185.

Ronov, A. B., Khain, V. E., Balukhovsky, A. N. and Seslavinsky, K. B. (1980). Quantitative analysis of Phanerozoic sedimentation. *Sed. Geol.*, **25**, 311–325.

Roobol, M. J., Ramsay, C. R., Jackson, N. J. and Darbyshire, D. P. F. (1983). Late Proterozoic lavas of the Central Arabian

Shield—evolution of an ancient volcanic arc system. *J. geol. Soc. Lond.*, **140**, 185–202.

Rose, E. R. (1973). Geology of vanadium and vanadiferous occurrences of Canada. *Rept. Geol. Surv. Canada*, **23**, 130 pp.

Rosencrantz, E. (1982). Formation of uppermost oceanic crust. *Tectonics*, **1**, 471–494.

Ross, C. P. (1970). The Precambrian of the United States of America: northwestern United States, the Belt Series. In K. Rankama (Ed.), *The Precambrian*, Wiley, New York, 145–252.

Roure, F. and Blanchet, R. (1983). A geological transect between the Klamath Mountains and the Pacific Ocean (southwestern Oregon): a model for palaeosubductions. *Tectonophysics*, **21**, 53–72.

Rowlands, N. J. *et al.* (1978). Gitological aspects of some Adelaidean stratiform copper deposits. *Mineral. Sci. Engng.*, **10**, 258–277.

Roy, J. (1980). Palaeomagnetism of the North American Precambrian. *Int. Geol. Cong. Paris, Abst.*, p. 613.

Roy, S. (1966). *Syngenetic Manganese Formations of India*, Jadavpur University, Calcutta, 219 pp.

Runcorn, S. K. (1961). Climatic change through geological time in the light of the palaeomagnetic evidence for polar wandering and continental drift. *Quart. J. roy. Meterol. Soc.*, **87**, 373, p. 282.

Rundle, C. C. and Snelling, N. J. (1977). The geochronology of uraniferous minerals in the Witwatersrand Triad: an interpretation of new and existing U–Pb age data on rocks and minerals from the Dominion Reef, Witwatersrand and Ventersdorp Supergroups. *Phil. Trans. R. Soc. Lond.*, **A286**, 567–583.

Russell, M. J. (1976). Incipient plate separation and possible related mineralization in lands bordering the North Atlantic. In D. F. Strong (Ed.), *Metallogeny and Plate Tectonics, Geol. Ass. Can. Sp. Pap.*, **14**, 339–349.

Russell, M. J. and Smythe, D. K. (1983). Origin of the Oslo graben in relation to the Hercynian-Allghenian orogeny and lithospheric rifting in the North Atlantic. *Tectonophysics*, **94**, 457–472.

Rutland, R. W. R. (1973). Tectonic evolution of the continental crust of Australia. In D. H. Tarling and S. K. Runcorn (Eds), *Implications of Continental Drift to the Earth Sciences*, vol. 2, Academic Press, London, pp. 1011–1033.

Rutland, R. W. R. (1981). Structural framework of the Australian Precambrian. In D. R. Hunter (Ed.), *Precambrian of the Southern Hemisphere*, Elsevier, Amsterdam, pp. 1–32.

Rutland, R. W. R. (1982). On the growth and evolution of continental crust: a comparative tectonic approach. *J. and Proc. R. Soc. New South Wales*, **115**, 33–60.

Ryan, W. B. F. and Cita, M. B. (1977). Ignorance concerning episodes of ocean-wide stagnation. *Marine Geol.*, **23**, 197–215.

Saager, R. and Muff, R. (1980). A new discovery of possible primitive life-forms in conglomerates of the Archaean Pietersburg greenstone belt, South Africa. *Geol. Rund.*, **69**, 179–185.

Saha, A. K. (1979). Geochemistry of Archaean granites of the Indian Shield: a review. *J. geol. Soc. India*, **20**, 375–392.

Saleeby, J. B. (1983). Accretionary tectonics of the North American Cordillera. *Ann. Rev. Earth Planet. Sci.*, **15**, 45–73.

Salop, L. J. (1972). A unified stratigraphic scale of the Precambrian. *24th Int. geol. Congr. Montreal. Sect. 1*, 253–259.

Salop, L. J. (1977). *Precambrian of the Northern Hemisphere*, Elsevier, Amsterdam, 378 pp.

Salop, L. J. (1983). *Geological Evolution of the Earth during the Precambrian*, Springer, Berlin, 459 pp.

Salop, L. J. and Scheinmann, Y. M. (1969). Tectonic history and structure of platforms and shields *Tectonophysics*, **7**, 565–597.

Sangster, A. L. and Bourne, J. (1982). Geology of the Grenville Province, and regional metallogenesis of the Grenville Supergroup. In R. W. Hutchinson, C. D. Spence and J. M. Franklin (Eds), *Precambrian Sulphide Deposits, Geol. Ass. Can. Sp. Pap.*, **25**, 91–125.

Sarkar, S. N. (1980). Precambrian stratigraphy and geochronology of Peninsula India: a review. *Indian J. Earth Sci.*, **7**, 12–26.

Sato, T. (1977). Kuroko deposits: their geology, geochemistry and origin. In *Volcanic Processes in Ore Genesis, Inst. Min. Metall./Geol. Soc. Lond. Sp. Publ.*, **7**, 153–161.

Savage, D. and Sills, J. D. (1980). High pressure metamorphism in the Scourian of NW Scotland: evidence from garnet granulites. *Contrib. Mineral. Petrol.*, **74**, 153–163.

Sawkins, F. J. (1976a). Massive sulphide deposits in relation to geotectonics. In D. F. Strong (Ed.), *Metallogeny and Plate Tectonics, Geol. Ass. Can. Sp. Pap.*, **14**, 221–242.

Sawkins, F. J. (1976b). Metal deposits related to intracontinental hotspot and rifting environments. *J. Geol.*, **84**, 653–671.

Sawkins, F. J. (1980). Single-stage versus two-stage re-deposition in subduction-related volcanoplutonic arcs. In J. D. Ridge (Ed.), *Proc. Fifth Quadrennial IAGOD Symp.*, vol. 1, Schweizerbart'sche Verlag, Stuttgart, pp. 143–154.

Sawkins, F. J. (1982). Metallogenesis in relation to rifting. In G. Palmason (Ed.), *Continental and Oceanic Rifts, Geodynamic Ser. 8, Amer. Geophys. Union*, 259–269.

Sawkins, F. J. and Burke, K. (1980). Extensional tectonics and mid-Palaeozoic massive sulfide occurrences in Europe. *Geol. Rund.*, **69**, 349–360.

Schenk, V. (1980). U–Pb and Rb–Sr radiometric dates and their correlation with metamorphic events in the granulite-facies basement of the Serre, S Calabria (Italy). *Contrib. Mineral. Petrol.*, **73**, 23–38.

Schermerhorn, L. J. G. (1975). Tectonic framework of late Precambrian supposed glacials. In A. E. Wright and F. Moseley (Eds), *Ice Ages: Ancient and Modern*, Seel House Press, Liverpool, pp. 241–274.

Schermerhorn, L. J. G. (1981). The West Congo orogen: a key to Pan-African thermotectonism. *Geol. Rund.*, **70**, 850–867.

Schermerhorn, L. J. G., Priem, H. N. A., Boelrijk, N. A. I. M., Hebeda, E. H., Verdvrmen, E. A. T. H. and Verschure, R. H. (1978). Age and origin of the Messejana dolorite fault–dyke system (Portugal and Spain) in the light of the opening of the North Atlantic ocean. *J. Geol.*, **86**, 299–309.

Schidlowski, M. (1980). The atmosphere. In O. Hutzinger (Ed.), *The Handbook of Environmental Chemistry*, vol. 1, Springer, Berlin, pp. 1–16.

Schidlowski, M. (in press). Early atmospheric oxygen levels. Constraints from Archaean photo-autotrophy. *J. geol. Soc. Lond.*

Schlanger, S. O. and Jenkyns, H. C. (1976). Cretaceous oceanic anoxic events: causes and consequences. *Geol. en Mijnb.*, **55**, 179–184.

Scholl, D. W., Marlow, M. S. and Cooper, A. L. (1977). Sediment subduction and offscraping at Pacific margins. In M. Talwani and W. C. Pitman (Eds), *Island Arcs, Deep Sea Trenches and Back Arc Basins, Amer. Geophys. Un.*, Maurice Ewing Ser. 1, pp. 199–210.

Scholl, D. W., von Huene, R., Vallier, T. L. and Howell, D. G. (1980). Sedimentary masses and concepts about tectonic processes at underthrust ocean margins. *Geology*, **8**, 564–568.

Schopf, J. W. (1972). Evolutionary significance of the Bitter Springs (late Precambrian) microflora. *24th Int. geol. Congr. Montreal., Sect. 1*, 68–77.

Schopf, J. M. (1973). Coal, climate and global tectonics. In D. H. Tarling and S. K. Runcorn (Eds), *Implications of Continental Drift to the Earth Sciences*, vol. 1, Academic Press, London, pp. 609–622.

Schopf, J. W. (1974). The development and diversification of Precambrian life. *Origins of Life*, **5**, 119–135.

Schopf, J. W. (1975). Precambrian palaeobiology: problems and perspectives. *Ann. Rev. Earth Planet. Sci.*, **3**, 212–249.

Schopf, J. W. (1977). Biostratigraphic usefulness of stromatolite Precambrian microbiotas: a preliminary analysis. *Precamb. Res.*, **5**, 143–173.

Schouten, H., Karson, J. and Dick, H. (1980). Geometry of transform zones. *Nature*, **288**, 470–473.

Schwab, F. L. (1978). Secular trends in the composition of sedimentary rock assemblages—Archaean through Phanerozoic time. *Geology*, **6**, 532–536.

Schwartz, D. J. and Fujiwara, Y. (1977). Komatiitic basalts from the Proterozoic Cape Smith range in northern Quebed, Canada. *Geol. Ass. Can. Sp. Pap.*, **16**, 193–201.

Sclater, J. G., Parsons, B. and Jaupart, C. (1981). Oceans and

384

continents: similarities and differences in the mechanisms of heat loss. *J. Geophys. Res.*, **86**, B12, 11535–11552.

Scotese, C. R., Bambach, R. K., Barton, C., van der Voo, R. and Ziegler, A. Y. (1979). Palaeozoic base maps. *J. Geol.*, **87**, 217–277.

Scott, B. (1976). Zinc and lead mineralization along the margins of the Caledonian orogen. *Trans. Inst. Min. Metall.*, **85B**, 200–213.

Scrutton, R. A. (1973). The age relationship of igneous activity and continental break-up. *Geol. Mag.*, **110**, 227–234.

Scrutton, R. A. (1976). Continental break-up and deep crustal structure at the margins of southern Africa. *An. Acad. bras. Ciênc.*, **48**, 275–286.

Scrutton, R. A. and Dingle, R. V. (1976). Observations on the processes of sedimentary basin formation at the margins of southern Africa. *Tectonophysics*, **36**, 143–156.

Sellwood, B. W. and Jenkyns, H. C. (1975). Basins and swells and the formation of an epeiric sea (Pliensbachian–Bajocian of Great Britain). *J. geol. Soc. Lond.*, **131**, 373–388.

Selverstone, J. (1982). Fluid inclusions as petrogenetic indicators in granulite xenoliths, Pali–Aike volcanic field, Chile. *Contrib. Mineral. Petrol.*, **79**, 28–36.

Selverstone, J. and Hollister, L. S. (1980). Cordierite-bearing granulites from the Coast Ranges, British Columbia; PT-conditions of metamorphism. *Can. Mineral.*, **18**, 119–129.

Semeneko, N. P., Rodionov, S. P., Usenko, I. S., Lichak, I. L. and Tsarovsky, I. D. (1960). Stratigraphy of the Precambrian of the Ukrainian Shield, *21st Int. geol. Congr. Norden*, part 9, 108–115.

Semikhatov, M. A. (1976). Experience in stromatolite studies in the U.S.S.R. In M. R. Walter (Ed.), *Stromatolites*, Elsevier, Amsterdam, pp. 337–357.

Semikhatov, M. A. (1981). New stratigraphic scale of the Precambrian in the U.S.S.R.: an analysis and its implications. *Int. geol. Rev.*, **23**, 193–154.

Sengör, A. M. C. (1979a). Mid-Mesozoic closure of Permo-Triassic Tethys and its implications. *Nature*, **279**, 590–593.

Sengör, A. M. C. (1979b). The North Anatolian transform fault: its age, offset and tectonic significance. *J. geol. Soc. Lond.*, **136**, 269–282.

Sengör, A. M. C. (1981). The geological exploration of Tibet. *Nature*, **294**, 403–404.

Sengör, A. M. C., Burke, K. and Dewey, J. F. (1978). Rifts at high angles to orogenic belts: tests for their origin and the upper Rhine graben as an example. *Amer. J. Sci.*, **278**, 24–40.

Sengör, A. M. C. and Kidd, W. S. F. (1979). Post-collisional tectonics of the Turkish Iranian plateau and a comparison with Tibet. *Tectonophysics*, **55**, 361–376.

Sengör, A. M. C., Yilmaz, Y. and Ketin, I. (1980). Remnants of a pre-late Jurassic ocean in northern Turkey: fragments of Permian-Triassic Palaeo-Tethys? *Bull. geol. Soc. Amer.*, **91**, part 1, 599–609.

Sepkoski, J. J., Bambach, R. K., Raup, D. M. and Valentine, J. W. (1981). Phanerozoic marine diversity and the fossil record. *Nature*, **293**, 435–437.

Sethuraman, K. and Moore, J. M. (1973). Petrology of metavolcanic rocks in the Bishop Corners—Donaldson area, Grenville Province, Ontario. *Can. J. Earth Sci.*, **10**, 589–614.

Seyfert, C. K. and Sirkin, L. A. (1979). *Earth History and Plate Tectonics*, 2nd edn. Harper and Row, New York, 600 pp.

Shackleton, R. M. (1976). Pan-African structures. *Phil. Trans. R. Soc. Lond.*, **A280**, 491–497.

Shackleton, R. M. (1981). Structure of southern Tibet: report on a traverse from Lhasa to Khatmandu organised by Academia Sinica. *J. Struct. Geol.*, **3**, 97–105.

Shackleton, R. M. (1984). Collision tectonics. In J. G. Ramsay, A. Ries and M. P. Coward (Eds), *Collision Tectonics*, Geol. Soc. London.

Shackleton, R. M., Ries, A. C., Coward, M. P. and Cobbold, P. R. (1979). Structure, metamorphism and geochronology of the Arequipa Massif of coastal Peru. *J. geol. Soc. Lond.*, **136**, 195–214.

Shackleton, R. M., Ries, A. C., Graham,. R. H. and Fitches, W. R. (1980). Late Precambrian ophiolitic mélanges in the eastern desert of Egypt. *Nature*, **285**, 472–474.

Shackleton, R. M., Ries, A. C. and Coward, M. P. (1982). An interpretation of the Variscan structures in SW England. *J. geol. Soc. Lond.*, **139**, 535–544.

Shams, F. A. (1980). Origin of the Shangla blueschists, Swat Himalaya, Pakistan. *Univ. Peshawar. Sp. Issue Geol. Bull.*, **13**, 67–70.

Shanks, W. C. (1977). Ore transport and deposition in the Red Sea geothermal system: a geochemical model. *Geochim. Cosmochim. Acta*, **41**, 1507–1519.

Shanti, M. and Roobol, M. J. (1979). A late Proterozoic ophiolite complex at Jabal Ess in N Saudi Arabia. *Nature*, **279**, 488–491.

Sharma, K. K., Bal, K. D., Parshad, R., Lal, N. and Nagpaal, K. K. (1980). Palaeo-uplift and cooling rates from various orogenic belts of India as revealed by radiometric ages. *Tectonophysics*, **70**, 135–158.

Sharpe, M. R. and Snyman, J. A. (1980). A model for the emplacement of the eastern compartment of the Bushveld Complex. *Tectonophysics*, **65**, 85–110.

Shegelski, R. J. (1980). Archaean cratonization, emergence and red bed development, Lake Shebandowan area, Canada. *Precamb. Res.*, **12**, 331–347.

Sheldon, R. P. (1981). Ancient marine phosphorites. *Ann. Rev. Earth Planet. Sci.*, **9**, 251–284.

Sheraton, J. W., Offe, L. A., Tingey, R. J. and Ellis, D. J. (1980). Enderby Land, Antarctica: an unusual Precambrian high-grade metamorphic terrain. *J. geol. Soc. Austral.*, **27**, 1–18.

Sheraton, J. W. and Black, L. P. (1981). Geochemistry and geochronology of Proterozoic tholeiite dykes of East Antarctica: evidence for mantle metasomatism. *Contrib. Mineral. Petrol.*, **78**, 305–317.

Sheridan, R. E. (1976). Sedimentary basins of the Atlantic margin of North America. *Tectonophysics*, **36**, 113–132.

Sheridan, R. E. and Gradstein, F. M. (1981). Early history of the Atlantic Ocean and gas hydrates in the Blake outer ridge. Results of the Deep Sea Drilling Project. *Episodes*, **2**, 16–22.

Shervais, J. W. (1979). Thermal emplacement model for the Alpine lherzolite massif at Balmuccia, Italy. *J. Petrol.*, **20**, 795–820.

Shimron, A. E. (1980). Proterozoic island arc volcanism and sedimentation in Sinai. *Precamb. Res.*, **12**, 437–458.

Shuldiner, V. I. (1982). The oldest high-grade terranes: possible relics of primeval earth crust. *Rev. Bras. Geoc.*, **12**, 45–52.

Sidorenko, A. V. and Borschchevskiy, Yu. A. (1979). General trends of evolution of isotopic composition of carbonates in the Precambrian and the Phanerozoic. *Dokl. Akad. Nauk. U.S.S.R.*, **234**, 96–98.

Sidorenko, Sv. A. and Sozinov (1982). Precambrian and carbonaceous formations and related areas. In *The Development Potential of Precambrian Mineral Deposits*. U.N. Nat. Res. Energy Div., Pergamon Press, New York, pp. 59–74.

Siedner, G. and Mitchell, J. G. (1976). Episodic Mesozoic volcanism in Namibia and Brazil: a K–Ar isochron study bearing on the opening of the South Atlantic. *Earth Planet. Sci. Lett.*, **30**, 292–302.

Sillitoe, R. H. (1976). Andean mineralization: a model for the metallogeny of convergent plate margins. In D. F. Strong (Ed.), *Metallogeny and Plate Tectonics*, Geol. Ass. Can. Sp. Pap., **14**, 58–100.

Sillitoe, R. H. (1979). Porphyry copper-type mineralisation in early Proterozoic greenstone belts, Upper Volta, West Africa. *Geol. Ass. Can. Prog. with Abst.*, **4**, 78.

Sillitoe, R. H. (1980). Are porphyry copper and Kuroko-type massive sulfide deposits incompatible? *Geology*, **8**, 1x–14.

Sillitoe, R. H. (1982). Extensional habitats of rhyolite-hosted massive sulfide deposits. *Geology*, **10**, 403–407.

Sills, J. D., Savage, D., Watson, J. V. and Windley, B. F. (1982). Layered ultramafic-gabbro bodies in the Lewisian of NW Scotland: geochemistry and petrogenesis. *Earth Planet. Sci. Lett.*, **58**, 345–360.

Simmons, E. C. and Hanson, G. N. (1978). Geochemistry and origin of massif-type anorthosites. *Contrib. Mineral. Petrol.*, **66**, 119–135.

Simmons, E. C., Hanson, G. N. and Lumbers, S. B. (1980). Geochemistry of the Shawmere anorthosite complex, Kapuskasing structural zone, Ontario. *Precamb. Res.*, **11**, 43–71.

385

Simonen, A. (1980). The Precambrian in Finland. *Geol. Surv. Finland Bull.*, **304**, 58 pp.

Simpson, P. R. and Bowles, J. F. W. (1977). Uranium mineralization of the Witwatersrand and Dominion Reef Systems. *Phil. Trans. R. Soc. Lond.*, **A286**, 527–547.

Simpson, P. R., Brown, G. C., Plant, J. and Ostle, D. (1979). Uranium mineralization and granite magmatism in the British Isles. *Phil. Trans. R. Soc. Lond.*, **291A**, 133–160.

Simpson, P. R. and Bowles, J. F. W. (1981). Detrital uraninite and pyrite: are they evidence for a reducing atmosphere? *U.S. Geol. Surv. Prof. Pap.*, **1161–S**, 1–12.

Sims, P. K. (1980). Subdivision of the Proterozoic and Archaean eons: recommendations and suggestions by the International Subcommission on Precambrian Stratigraphy. *Precamb. Res.*, **13**, 379–380.

Sinha-Roy, S. (1981). Metamorphic facies and inverted metamorphic sequences of the Eastern Himalayan crystalline rocks. In P. S. Saklani (Ed.), *Metamorphic Tectonites of the Himalaya*, Today and Tomorrow's Publ., New Delhi, pp. 279–302.

Sinha-Roy, S. (1982). Himalayan Main Central Thrust and its implications for Himalayan inverted metamorphism. *Tectonophysics*, **84**, 197–224.

Skinner, W. R. (1969). Geologic evolution of the Beartooth Mountains, Montana and Wyoming. Part 8. *Ultramafic Rocks in the Highline Trail Lakes area, Wyoming*. In L. H. Larsen (Ed.), *Geol. Soc. Amer. Mem.*, **115**, 19–52.

Sleep, N. H. (1979). Thermal history and degassing of the earth: some simple calculations. *J. Geol.*, **87**, 671–686.

Sleep, N. H. and Langan, R. T. (1981). Thermal history of the earth: some recent developments. *Adv. in Geophys.*, **23**, 1–24.

Sleep, N. H. and Windley, B. F. (1982). Archaean plate tectonics: constraints and inferences. *J. Geol.*, **90**, 363–379.

Smith, A. G. (1971). Alpine deformation and the oceanic areas of the Tethys, Mediterranean and Atlantic. *Bull. geol. Soc. Amer.*, **82**, 2039–2070.

Smith, A. G. and Hallam, A. (1970). The fit of the southern continents. *Nature*, **225**, 139–144.

Smith, A. G., Briden, J. C. and Drewry, G. E. (1973). Phanerozoic world maps. *Sp. Pap. Palaeont.*, **12**, 1–42.

Smith, A. G. and Woodcock, N. H. (1976). Emplacement model for some 'Tethyan' ophiolites. *Geology*, **4**, 653–656.

Smith, A. G., Hurley, A. M. and Bridan, J. C. (1981). *Phanerozoic Palaeocontinental World Maps*, Cambridge University Press, Cambridge, 102 pp.

Smith, A. G. and Woodcock, N. H. (1982). Tectonic syntheses of the Alpine-Mediterranean regions: a review. In *Alpine Mediterranean Geodynamics, Geodyn. Ser. 7, Am. Geophys. Un.*, 15–38.

Smith, D. B. (1979). Rapid marine transgression of the Upper Permian Zechstein Sea. *J. geol. Soc. Lond.*, **136**, 155–156.

Smith, J. V. (1981). The first 800 million years of Earth's history. *Phil. Trans. R. Soc. Lond.*, **A301**, 401–422.

Smith, R. L., Stearn, J. E. F. and Piper, J. D. A. (1983). Palaeomagnetic studies of the Torridonian sediments, NW Scotland. *Scott. J. Geol.*, **19**, 29–46.

Smith, T. E. and Noltimier, H. C. (1979). Palaeomagnetism of the Newark trend igneous rocks of the north central Appalachians and the opening of the central Atlantic Ocean. *Amer. J. Sci.*, **279**, 778–807.

Smythe, D. K., Dobinson, A., McQuillin, R., Brewer, J. A., Matthews, D. H., Blundell, D. J. and Kelk, B. (1982). Deep structure of the Scottish Caledonides revealed by the Moist reflection profile. *Nature*, **299**, 338–340.

Sonnenfeld, P. (1975). The significance of Upper Miocene (Messinian) evaporites in the Mediterranean Sea. *J. Geol.*, **83**, 287–311.

Soper, N. J. and Barber, A. J. (1979). Proterozoic folds on the NW Caledonian foreland. *Scott. J. Geol.*, **15**, 1–11.

Soper, N. J. and Barber, A. J. (1982). A model for the deep structure of the Moine thrust zone. *J. geol. Soc. Lond.*, **139**, 127–138.

Souther, J. G. (1972). Initial deposits in the Cordilleran geosyncline: evidence of a late Precambrian (>850 m.y.) continental separation. *Bull. geol. Soc. Amer.*, **83**, 1345–1360.

Speed, R. C. (1979). Collided Palaeozoic microplates in the western United States. *J. Geol.*, **87**, 279–292.

Spencer, A. M. (1971). Late Precambrian glaciation in Scotland. *Geol. Soc. Lond. Mem.*, **6**, 98 pp.

Spencer, A. M. (1975). Late Precambrian glaciations in the North Atlantic region. In A. E. Wright and F. Moseley (Eds), *Ice Ages: Ancient and Modern*, Seel House Press, Liverpool, pp. 217–240.

Spjeldnaes, N. (1978). Faunal provinces and the Proto-Atlantic. In D. R. Bowes and B. E. Leake (Eds), *Crustal Evolution in NW Britain and Adjacent Regions*, Seel House Press, Liverpool, pp. 139–150.

Spooner, E. T. C. (1977). Hydrothermal model for the origin of the ophiolitic cupriferous pyrite ore deposits of Cyprus. In *Volcanic Processes in Ore Genesis, Geol. Soc. Lond. Inst. Min. Metall.*, 58–71.

Spooner, E. T. C. and Bray, C. J. (1977). Hydrothermal fluids of sea-water salinity in ophiolitic ore deposits in Cyprus. *Nature*, **266**, 808–812.

Spray, J. G. and Williams, G. D. (1980). The sub-ophiolite metamorphic rocks of the Ballantrae igneous complex, SW Scotland. *J. geol. Soc. Lond.*, **127**, 359–368.

Srikantia, S. V. and Razdan, M. L. (1980). Shilakong ophiolite nappes of Zanskar mountains, Ladakh, Himalaya. *J. geol. Soc. India*, **22**, 277–234.

Stanley, D. J. (1974). Modern flysch sedimentation in a Mediterranean island arc setting. In R. H. Dott Jr and R. H. Shaver (Eds), *Modern and Ancient Geosynclinal Sedimentation, Soc. Econ. Pal. Mineral. Sp. Publ.*, **19**, 240–259.

Stanley, S. M. (1976). Fossil data and the Precambrian–Cambrian evolutionary transition. *Amer. J. Sci.*, **276**, 56–76.

Stanton, R. L. (1972). *Ore Petrology*. McGraw-Hill, New York, 713 pp.

Stanton, R. L. (1976). Petrochemical studies of the ore environment at Broken Hill, New South Wales. Part 4. Environmental synthesis. *Trans. Inst. Min. Metal. Sect. B*, **85**, 221–234.

Stauffer, M. R., Mukherjee, A. C. and Koo, J. (1975). The Amisk Group: an Aphebian? island arc deposit. *Can. J. Earth Sci.*, **12**, 2021–2035.

Steiger, R. H. and Jäger, E. (1977). Submission on geochronology: convention on the use of decay constants in geo and cosmochronology. *Earth Planet. Sci. Lett.*, **36**, 359–362.

Stein, S. (1978). A model for the relation between spreading rate and oblique spreading. *Earth Planet. Sci. Lett.*, **39**, 313–318.

Steiner, J. and Grillmair, E. (1973). Possible galactic causes for periodic and episodic glaciations. *Bull. geol. Soc. Amer.*, **84**, 1003–1018.

Stern, R. J. (1981). Petrogenesis and tectonic setting of late Precambrian ensimatic volcanic rocks, central eastern desert of Egypt. *Precamb. Res.*, **16**, 195–230.

Stewart, A. D. (1982). Late Proterozoic rifting in NW Scotland: the genesis of the Torridonian. *J. geol. Soc. Lond.*, **139**, 415–422.

Stewart, J. H. (1972). Initial deposits in the Cordilleran geosyncline: evidence of a late Precambrian (<850 m.y.) continental separation. *Bull. geol. Soc. Amer.*, **83**, 1345–1360.

Stewart, J. H. (1976). Late Precambrian evolution of North America: plate tectonic implications. *Geology*, **4**, 11–15.

Stöcklin, J. (1980). Geology of Nepal and its regional frame. *J. geol. Soc. Lond.*, **137**, 1–34.

St-Onge, M. R. (1981). 'Normal' and 'inverted' metamorphic isograds and their relation to syntectonic Proterozoic batholiths in the Wopmay Orogen, Northwest Territories, Canada. *Tectonophysics*, **76**, 295–316.

Stowe, C. W. (1974). Alpine-type structure in the Rhodesian basement complex at Selukwe. *J. geol. Soc. Lond.*, **130**, 411–426.

Strong, D. F. (1974). Plate tectonic setting of Appalachian–Caledonian mineral deposits as indicated by Newfoundland examples. *Trans. Ass. Inst. Min. Eng.*, **256**, 121–128.

Subrahmanyam, C. and Verma, R. K. (1982). Gravity interpretation of the Dharwar greenstone–gneiss–granite terrain in the southern Indian Shield and its geological implications. *Tectonophysics*, **84**, 225–245.

386

Suk, M. (1977). Le developpement métamorphique du Massif de Bohème. In *La Chaine Varisque d'Europe Moyenne et Occidentale*, Editions du CNRS No. 243, Paris, 341–348.

Sun Dazhong and Wu Changhua (1981). The principal geological and geochemical characteristics of the Archaean greenstone–gneiss sequences in N China. In J. E. Glover and D. I. Groves (Eds), *Archaean Geology, Geol. Soc. Australia, Sp. Publ.*, **7**, 121–132.

Sun, S. S. (1980). Lead isotope study of young volcanic rocks from mid-ocean ridges, ocean islands and island arcs. *Phil. Trans. R. Soc. Lond.*, **A297**, 409–445.

Sun, S. S. and Nesbitt, R. W. (1977). Chemical heterogeneity of the Archaean mantle, composition of the earth and mantle evolution. *Earth Planet. Sci. Lett.*, **35**, 429–448.

Sun, S. S. and Nesbitt, R. W. (1978). Petrogenesis of Archaean ultrabasic and basic volcanics: evidence from rare earth elements. *Contrib. Mineral. Petrol.*, **65**, 301–325.

Sun, S. S., Nesbitt, R. W. and Sharashin, A. Y. (1979). Geochemical characteristics of mid-ocean ridge basalts. *Earth Planet. Sci. Lett.*, **44**, 119–138.

Surlyk, F. (1980). The Cretaceous–Tertiary boundary event. *Nature*, **285**, 187–188.

Sutter, J. F. and Smith, T. E. (1979). $^{40}Ar/^{39}Ar$ ages of diabase intrusions from Newark trend basins in Connecticut and Maryland: initiation of central Atlantic rifting. *Amer. J. Sci.*, **279**, 808–831.

Sutton, J. (1976). Tectonic relationships in the Archaean. In B. F. Windley (Ed.), *The Early History of the Earth*, Wiley–Interscience, London, 99–104.

Swami Nath, J. and Ramakrishnan, M. (Eds) (1981). Early Precambrian supracrustals of southern Karnataka. *Mem. Geol. Surv. India*, **1112**, 350 pp.

Swett, K. and Smit, D. K. (1972). Cambro-Ordovician shelf sedimentation of western Newfoundland, northwest Scotland and central east Greenland, *24th Int. geol. Cong. Montreal, Sect. 6*, 33–41.

Swinden, H. S. and Strong, D. F. (1976). A comparison of plate tectonic models of metallogenesis in the Appalachians, the North American Cordillera and the East Australian Palaeozoic. In D. F. Strong (Ed.), *Metallogeny and Plate Tectonics, Geol. Ass. Can. Sp. Pap.*, **14**, 441–470.

Sykes, L. R., Oliver, J. and Isacks, B. (1970). Earthquakes and tectonics. In A. E. Maxwell (Ed.), *The Sea*, vol. 4, Wiley, New York, pp. 353–420.

Sylvester–Bradley, P. C. (1975). The search for protolife. *Proc. R. Soc. Lond.*, **B189**, 213–233.

Tahirkheli, R. A. K. (1974). Alluvial gold prospects in the north west, West Pakistan. *Nat. Inst. Geol. Mineral., Univ. Peshawar. Inf. Release*, **7**, 48 pp.

Tahirkheli, R. A. K. (1979). Geology of Kohistan and adjoining Eurasian and Into-Pakistan continent, N Pakistan. *Univ. Peshawar, Sp. Issue. Geol. Bull.*, **11**, 1–30.

Tahirkheli, R. A. K., Mattauer, M., Proust, F. and Tapponnier, P. (1979). The India–Eurasia suture zone in northern Pakistan: synthesis and interpretation of recent data at plate scale. In A. Farah and K. A. De Jong (Eds), *Geodynamics of Pakistan, Geol. Surv.*, Pakistan, Quetta, 125–130.

Talwani, M. and Pitman, W. (Eds), (1977). *Island Arcs, Deep Sea Trenches and Back Arc Basins*, Amer. Geophys. Un., Maurice Ewing Ser. 1, 470 pp.

Tankard, A. J., Jackson, M. P. A., Eriksson, K. A., Hobday, D. K., Hunter, D. R. and Minter, W. E. L. (1982). *Crustal Evolution of Southern Africa*, Springer, New York, 523 pp.

Tapponnier, P. and Molnar, P. (1976). Slip-line field theory and large scale continental tectonics. *Nature*, **264**, 319–324.

Tapponnier, P. and Molnar, P. (1977). Rigid plastic indentation: the origin of syntaxis in the Himalayan belt. In *Himalaya. Sciences de la Terre, Coll. Int.*, CNRS no. 268, Paris, pp. 431–432.

Tapponnier, P. *et al.* (29 authors) (1981a). The Tibetan side of the India–Eurasia collision. *Nature*, **294**, 405–410.

Tapponnier, P., Mercier, J. L., Armigo, R., Tonglin, H. and Zhou, J. I. (1981b). Field evidence for active normal faulting in Tibet. *Nature*, **294**, 410–414.

Tapponnier, P., Peltzer, G., Le Dain, A. Y., Armijo, R. and

Cobbold, P. (1982). Propagating extrusion tectonics in Asia: new insights from simple experiments with plasticine. *Geology*, **10**, 611–616.

Tarling, D. H. and Tarling, M. P. (1977). *Continental Drift*, Bell, London, 112 pp.

Tarney, J. (1976). Geochemistry of Archaean high-grade gneisses, with implications as to the origin and evolution of the Archaean crust. In B. F. Windley (Ed.), *The Early History of the Earth*, Wiley–Interscience, London, 405–417.

Tarney, J., Dalziel, I. W. D. and De Wit, M. J. (1976). Marginal basin 'Rocas Verdes' complex from S Chile: a model for Archaean greenstone belt formation. In B. F. Windley (Ed.), *The Early History of the Earth*, Wiley–Interscience, London, pp. 131–146.

Tarney, J. and Windley, B. F. (1977). Chemistry, thermal gradients and evolution of the lower continental crust. *J. geol. Soc. Lond.*, **134**, 153–172.

Tarney, J. and Saunders, A. P. (1979). Trace element constraints on the origin of Cordilleran batholiths. In M. P. Atherton and J. Tarney (Eds), *Origin of Granite Batholiths: Geochemical Evidence*, Shiva Publ., Kent, pp. 90–105.

Tarney, J., Weaver, B. and Drury, S. A. (1979). Geochemistry of Archaean trondhjemitic and tonalitic gneisses from Scotland and E Greenland. In F. Barker (Ed.), *Trondhjemites, Dacites and Related Rocks*, Elsevier, Amsterdam, 275–299.

Tarney, J. and Windley, B. F. (1979). Continental growth, island arc accretion and the nature of the lower crust—a reply to S. R. Taylor and S. M. McLennan. *J. geol. Soc. Lond.*, **136**, 501–504.

Tarney, J., Saunders, A. D., Mattey, D. P., Wood, D. A. and Marsh, N. G. (1981). Geochemical aspects of back-arc spreading in the Scotia Sea and Western Pacific. *Phil. Trans. R. Soc. Lond.*, **A300**, 263–285.

Tarney, J. and Windley, B. F. (1981). Marginal basins through geological time. *Phil. Trans. R. Soc. Lond.*, **A301**, 217–232.

Tarney, J., Weaver, B. L. and Windley, B. F. (1982). Geological and geochemical evolution of the Archaean continental crust. *Rev. Bras. Geoc.*, **12** (1–3), 53–59.

Tarling, D. H. A. (1974). A palaeomagnetic study of Eocambrian tillites in Scotland. *J. geol. Soc. Lond.*, **130**, 163–178.

Taylor, P. N., Moorbath, S., Goodwin, R. and Petrykowski, A. C. (1980). Crustal contamination as an indication of the extent of early Archaean continental crust: Pb isotope evidence from the late Archaean gneisses of W Greenland. *Geochim. Cosmochim. Acta*, **44**, 1437–1453.

Taylor, S. R. (1977). Island arc models and the composition of the continental crust. In M. Talwani and W. C. Pitman (Eds), *Island Arcs, Deep Sea Trenches and Back Arc Basins*, Amer. Geophys. Un., Maurice Ewing Ser. 1, pp. 325–335.

Taylor, S. R. (1979). Chemical composition and evolution of the continental crust: the rare earth element evidence. In M. W. McElhinny (Ed.), *The Earth: its Origin, Structure and Evolution*, Academic Press, London, 353–376.

Tenyakov, A. V., Koryakin, A. S., Kulish, E. E. and Predovsky, A. A. (1982). Sedimentary geology and metallogeny of strongly metamorphosed Precambrian complexes. In *The Development Potential of Precambrian Mineral Deposits*, U.N. Nat. Res. Energy Div., Pergamon Press, New York, pp. 41–58.

Thakur, V. C. (1977). Divergent isograds of metamorphism in some parts of Higher Himalaya zone. In *Himalaya. Science de la Terre, Coll. Int*, CNRS, no. 268, Paris, pp. 433–441.

Thakur, V. C. (1980). Tectonics of the Central Crystallines of western Himalaya. *Tectonophysics*, **62**, 141–154.

Thiede, J. and Suess, E. (1983). Sedimentary record of ancient coastal upwelling. *Episodes*, **v. 1983**, 15–18.

Thiessen, R., Burke, K. and Kidd, W. S. F. (1979). African hotspots and their relation to the underlying mantle. *Geology*, **7**, 263–266.

Thirlwall, M. F. (1981). Implications for Caledonian plate tectonic models of chemical data from volcanic rocks of the British Old Red Sandstone, *J. geol. Soc. Lond.*, **138**, 123–138.

Thomas, M. D. and Tanner, J. C. (1975). Cryptie suture in the eastern Grenville Province. *Nature*, **256**, 392–394.

Thomas, M. D. and Gibb, R. A. (197q). Gravity anomalies and deep structure of the Cape Smith foldbelt, northern Ungava, Quebec. *Geology*, **5**, 169–172.

Thomas, M. D. and Kearey, P. (1980). Gravity anomalies, block-faulting and Andean-type tectonism in the eastern Churchill Province. *Nature*, **283**, 61–63.

Thomas, P. R. (1979). New evidence for a central Highland root zone. In A. L. Harris, C. H. Holland and B. E. Leake (Eds), *The Caledonides of the British Isles—Reviewed*. Geol. Soc. Lond., Scottish Academic Press, Edinburgh, pp. 205–212.

Thorpe, R. S. (1978). Tectonic emplacement of ophiolitic rocks in the Precambrian Mona Complex of Anglesey. *Nature*, **276**, 57–58.

Thorpe, R. S. and Francis, P. W. (1979). Variations in Andean andesite compositions and their petrogenetic significance. *Tectonophysics*, **57**, 53–70.

Titley, S. R. (Ed.) (1982). *Advances in Geology of the Porphyry Copper Deposits*, University of Arizona Press, Tucson, Arizona, 560 pp.

Toksöz, M. N. and Bird, P. (1977). Formation and evolution of marginal basins and continental plateaus. In M. Talwani and W. C. Pitman (Eds), *Island Arcs, Deep Sea Trenches and Back Arc Basins*, Amer. Geophys. Un., Maurice Ewing Ser. 1, pp. 379–393.

Toksöz, M. N., Uyeda, S. and Francheteau, J. (Eds) (1980). *Oceanic Ridges and Arcs: Geodynamic Processes*, Elsevier, Amsterdam, 538 pp.

Torquato, J. R. and Cordani, U. G. (1981). Brazil–Africa geological links. *Earth Sci. Rev.*, **17**, 155–176.

Torske, T. (1977). The South Norway Precambrian region: a Proterozoic Cordilleran-type orogenic segment. *Norsk. geol. Tids.*, **57**, 97–120.

Touret, J. (1974). Facies granulite et fluides carboniques. *Bull. Ann. Soc. Geol. Belg.*, P. Michot vol., 267–287.

Toussaint, J. F. and Restrepo, J. J. (1982). Magmatic evolution of the northwestern Andes of Columbia. *Earth Sci. Rev.*, **18**, 205–213.

Towe, K. M. (1970). Oxygen-collagen priority and the early Metazoan fossil record. *Proc. N.Y. Acad. Sci.*, **65**, 781–788.

Tréhu, A. M. (1975). Depth versus (age)½: a perspective on mid-ocean rises. *Earth Planet. Sci. Lett.*, **27**, 287–304.

Trendall, A. F. (1968). Three distinct basins of Precambrian banded iron formations: a systematic comparison. *Bull. geol. Soc. Amer.*, **79**, 1527–1544.

Trendall, A. F. (1973). Iron formations of the Hamersley Group of western Australia: type examples of varved Precambrian evaporites. In *Genesis of Precambrian Iron and Manganese Deposits*, Proc. Kiev. Symp. 1970, UNESCO, Paris, 257–270.

Trommsdorff, V., Dietrich, V. and Honegger, K. (1982). The Indus suture zone: paleotectonic and igneous evolution in the Ladakh Himalayas. In K. J. Hsü (Ed.), *Mountain Building Processes*, Academic Press, London, pp. 213–220.

Trümpy, R. (1982). Alpine paleogeography: a reappraisal. In K. J. Hsü (Ed.), *Mountain Building Processes*, Academic Press, London, pp. 149–156.

Tuach, J. and Kennedy, M. J. (1978). The geologic setting of the Ming and other sulfide deposits, Consolidated Rambler Mines, NE Newfoundland. *Econ. Geol.*, **73**, 192–206.

Tugarinov, A. I. (1979). Evolution of the arc formation in the history of the earth. In L. H. Ahrens (Ed.), *Origin and Distribution of the Elements, Phys. Chem. of the Earth*, vol. II, Pergamon, Oxford, pp. 553–563.

Turcotte, D. L. (1980). On the thermal evolution of the Earth. *Earth Planet. Sci. Lett.*, **48**, 53–58.

Ukpong, E. E. and Olabe, M. A. (1979). Geochemical surveys for lead–zinc mineralization, southern Benue Trough, Nigeria. *Trans. Inst. Min. Metall.*, **88**, B81–92.

Umeji, A. C. (1983). Geochemistry and mineralogy of the Freetown layered basic igneous complex of Sierre Leone. *Chem. Geol.*, **39**, 17–38.

Underwood, M. B. and Bachman, S. B. (1982). Sedimentary facies associations within subduction complexes. In J. K. Leggett (Ed.), *Trench-Forearc Geology: Sedimentation and Tectonics in Modern and Ancient Active Plate Margins*, Geol. Soc. Lond., Blackwell, Oxford, 537–550.

Upadhyay, H. D. (1978). Phanerozoic peridotitic and pyroxenitic komatiites from Newfoundland. *Science*, **202**, 1192–1195.

Upton, B. G. J. (1974). The alkaline province of southwest Greenland. In H. Sørensen (Ed.), *The Alkaline Rocks*, Wiley, London, 221–237.

Upton, B. G. J., Aspen, P., Graham, A. and Chapman, N. A. (1976). Pre-Palaeozoic basement of the Scottish Midland Valley. *Nature*, **260**, 517–518.

Upton, B. G. J. and Blundell, D. J. (1978). The Gardar igneous province: evidence for Proterozoic continental rifting. In E. R. Neumann and I. B. Ramberg (Eds), *Petrology and Geochemistry of Continental Rifts*. Reidal, Holland, pp. 163–172.

Urabe, T. and Sato, T. (1978). Kuroko deposits of the Kosaka mine, NE Honshu, Japan—products of submarine hot springs on Miocene sea floor. *Econ. Geol.*, **73**, 161–179.

Urban, H. (1971). Zur Kenntnis der schichtgebundenen Wolfram–Molybdän-Vererzung im Örsdalen (Rogaland), Norwegen. *Mineral. Deposita*, **6**, 177–195.

Uyeda, S. (1981). Subduction zones and back-arc basins—a review. *Geol. Rund.*, **70**, 552–569.

Uyeda, S. (1982). Subduction zones: an introduction to comparative subductology. *Tectonophysics*, **81**, 133–159.

Uyeda, S. and Kanamori, H. (1979). Back arc opening and the mode of subduction. *J. Geophys. Res.*, **84**, B3, 1049–1061.

Vaasjoki, M., Äikäs, O. and Rehtijärvi, P. (1980). The age of mid-Proterozoic phosphatic sediments in Finland as indicated by radiometric U–Pb dates. *Lithos*, **13**, 257–262.

Vacquier, V. (1959). Measurement of horizontal displacements along faults in the ocean floor. *Nature*, **183**, 452–453.

Vail, J. R. (1970). Tectonic control of dykes and related irruptive rocks in eastern Africa. In T. N. Clifford and I. G. Gass (Eds), *African Magmatism and Tectonics*, Oliver and Boyd, Edinburgh, pp. 337–354.

Vail, J. R. and Hughes, D. J. (1977). Tholeiite derivative dyke swarms near Erkowit, Red Sea Hills, Sudan. *Geol. Rund.*, **66**, 228–237.

Vail, P. R., Mitchum, R. M. and Thompson, S. (1977). Seismic stratigraphy and global changes of sea level. Part IV. Global cycles of relative changes of sea-level. In Seismic stratigraphy—applications to hydrocarbon exploration. *Amer. Assoc. Petrol. Geol. Mem.*, **26**, 83.

Valencio, D. A. (1974). The South American palaeomagnetic data and the main episodes of the fragmentation of Gondwana, *Phys. Earth Planet. Inter.*, **9**, 221.

Valentine, J. W. and Moores, E. M. (1970). Plate tectonic regulation of faunal diversity and sea level: a model. *Nature*, **228**, 657–659.

Valentine, J. W. and Moores, E. M. (1972). Global tectonics and the fossil recvord. *J. Geol.*, **80**, 167–184.

Van Alstine, R. E. (1976). Continental rifts and lineaments associated with major fluorspar districts. *Econ. Geol.*, **71**, 977–987.

Van Andel, T. H., Thiede, J., Sclater, J. G. and Hay, W. W. (1977). Depositional history of the South Atlantic ocean during the last 125 million years. *J. Geol.*, **85**, 651–698.

Van Biljon, W. J. (1980). Plate tectonics and the origin of the Witwatersrand basin. In *Proc. 5th Quad. IAGOD Symposium*, Schweizerbart'sche Verlag, Stuttgart, pp. 217–226.

Van Breemen, O., Halliday, A. N., Johnson, M. R. W. and Bowes, D. R. (1978). Crustal additions in late Precambrian times. In D. R. Bowes and B. E. Leake (Eds), *Crustal Evolution in NW Britain and Adjacent Regions*, Seel House Press, Liverpool, pp. 81–106.

Van Breeman, O.,, Aftalion, M., Bowes, D. R., Dudek, A., Mísař, Z., Povondra, P. and Vrána, S. (1982). Geochronological studies of the Bohemian massif, Czechoslovakia and their significance in the evolution of Central Europe. *Trans. R. Soc. Edinb.: Earth Sci.*, **73**, 89–108.

Van Couvering, J. A., Berggren, W. A., Drake, R. E., Aguirre, E. and Curtis, G. H. (1976). The terminal Miocene event. *Mar. Micropal.*, **1**, 263–286.

Van der Voo, R. and Scotese, C. (1981). Paleomagnetic evidence for a large (2,000 km) sinistral offset along the Glen Glen fault during Carboniferous time. *Geology*, **9**, 583–589.

388

Van Eysinga, F. W. B. (compiler) (1975). *Geological Timetable*, 3rd edn, Elsevier, Amsterdam.

Van Houten, F. B. (1964). Origin of red beds—some unsolved problems. In A. E. M. Nairn (Ed.), *Problems in Palaeoclimatology*, Interscience, New York, pp. 647–659, 669–672.

Van Houten, F. B. (1974). Northern Alpine molasse and similar Cenozoic sequences of southern Europe. In R. H. Dott Jr and R. H. Shaver (Eds), *Modern and Ancient Geosynclinal Sedimentation*, Soc. Econ. Pal. Mineral. Sp. Publ., **19**, 260–273.

Van Houten, F. B. (1977). Trassic-Liassic deposits of Morocco and eastern North America: comparison. *Bull. Amer. Ass. Petrol. Geol.*, **61**, 79–99.

Vartiainen, H. and Wooley, A. R. (1974). The age of the Sokli carbonatite, Finland and some relationships of the North Atlantic alkaline igneous province. *Bull. geol. Soc. Finl.*, **46**, 81–91.

Veizer, J. (1973). Sedimentation in geologic history: recycling vs. evolution or recycling with evolution. *Contrib. Mineral. Petrol.*, **38**, 261–278.

Veizer, J. (1976a). $^{87}Sr/^{86}Sr$ evolution of seawater during geologic history and its significance as an index of crustal evolution. In B. F. Windley (Ed.), *The Early History of the Earth*, Wiley–Interscience, London, pp. 369–578.

Veizer, J. (1976b). Evolution of ores of sedimentary affiliation through geologic history: relations to the general tendencies in evolution of the crust, hydrosphere, atmosphere and biosphere. In K. H. Wolf (Ed.), *Handbook of Stratabound and Stratiform Ore Deposits*, vol. 3, Elsevier, Amsterdam, pp. 1–41.

Veizer, J. (1978). Secular variations in the composition of sedimentary carbonate rocks II Fe, Mn, Ca, Mg, Si and minor constituents. *Precamb. Res.*, **6**, 381–413.

Veizer, J. (1982). Geological evolution of the Archaean–early Proterozoic earth. In J. W. Schopf (Ed.), *Origin and Evolution of Earth's biosphere: an interdisciplinary study*, Princetown University Press.

Veizer, J. and Compston, W. (1976). $^{87}Sr/^{86}Sr$ in Precambrian carbonates as an index of crustal evolution. *Geochim. Cosmochim. Acta*, **40**, 905–914.

Veizer, J. and Hoefs, J. (1076). The nature of $^{18}O/^{16}O$ and $^{13}C/^{12}C$ secular trends in sedimentary carbonate rocks. *Geochim. Cosmochim. Acta*, **40**, 1387–1395.

Veizer, J. and Garrett, D. E. (1978). Secular variations in the composition of sedimentary rocks. I. Alkali metals. *Precamb. Res.*, **6**, 267–380.

Veizer, J. and Jansen, S. L. (1979). Basement and sedimentary recycling and continental evolution. *J. Geol.*, **87**, 341–370.

Veizer, J., Compston, W., Hoefs, J. and Nielsen, H. (1982). Mantle buffering of the early oceans. *Naturwissenschaften*, **69**, 173–180.

Venkatasubramaniam, V. S. (1974). Geochronology of the Dharwar craton—a review. *J. geol. Soc. India*, **15**, 463–468.

Vidal, G. (1972). Algal stromatolites from the late Precambrian of Sweden. *Lethaia*, **5**, 353–368.

Vidal, P., Cocherie, A. and Le Fort, P. (1982). Geochemical investigations of the origin of the Manaslu leucogranite (Himalaya, Nepal). *Geochim. Cosmochim. Acta*, **46**, 2279–2292.

Viljoen, M. J. and Viljoen, R. P. (1969). A collection of 9 papers on many aspects of the Barberton granite–greenstone belt, South Africa. *Geol. Soc. S. Afr. Sp. Publ.*, **2**, 295 pp.

Vine, F. J. and Matthews, D. H. (1963). Magnetic anomalies over ocean ridges. *Nature*, **199**, 947–949.

Vine, F. J., Poster, C. K. and Gass, I. G. (1973). Aeromagnetic survey of the Troodos igneous massif, Cyprus. *Nature Phys. Sci.*, **244**, 34–38.

Virdi, N. S. (1981a). Presence of parallel metamorphic belts in the northwest Himalaya: discussion. *Tectonophysics*, **72**, 141–154.

Virdi, N. S. (1981b). Chail metamorphics in the Himachal Lesser Himalaya. In P. S. Saklani (Ed.), *Metamorphic Tectonites of the Himalaya*, Today and Tomorrow's Publ., New Delhi, pp. 89–100.

Vitrac, A. M., Albarede, F. and Allègre, C. J. (1981). Lead isotopic composition of Hercynian granitic K-feldspars constrains continental genesis. *Nature*, **291**, 460–464.

Vlaar, N. J. and Wortel, M. J. R. (1976). Lithospheric aging, instability and subduction. *Tectonophysics*, **32**, 331.

Vogel, D. E. (1978). Polymetamorphism and structures in the Superior Province near the Grenville Front in southcentral Quebec. *Precamb. Res.*, **6**, 177–198.

Vokes, F. M. and Zachrisson, E. (1980). Review of Caledonian–Appalachian stratabound sulphides. *Sp. Pap. Geol. Surv. Ireland*, **5**, 81 pp.

Volbroth, A. (1962). Rapakivi-type granites in the Precambrian complex of Gold Butte, Clark Country, Nevada. *Bull. geol. Soc. Amer.*, **73**, 813–832.

Von Brunn, V. and Hobday, D. K. (1976). Early Precambrian sedimentation in the Pongola Supergroup of South Africa. *J. Sed. Pet.*, **46**, 670–679.

Vorma, A. (1976). On the petrochemistry of rapakivi granites, with special reference to the Laitila massif, SW Finland. *Geol. Surv. Finland, Bull.*, **285**, 98 pp.

Walker, J. C. G. (1976). Implications for atmospheric evolution of the inhomogeneous accretion model of the origin of the earth. In B. F. Windley (Ed.), *The Early History of the Earth*, Wiley–Interscience, London, pp. 543–546.

Walker, J. C. G. (1983). Possible limits on the composition of the Archaean ocean. *Nature*, **302**, 518–520.

Walker, R. N., Muir, M. D., Diver, W. L., Williams, N. and Wilkins, N. (1977). Evidence of major sulphate evaporite deposits in the Proterozoic McArthur Group, Northern Territory, Australia. *Nature*, **265**, 526–529.

Walter, M. R. (1970). Stromatolites used to determine the time of nearest approach of Earth and Moon. *Nature*, **170**, 1331–1332.

Walter, M. R. (1972). Stromatolites and the biostratigraphy of the Australia Precambrian and Cambrian. *Pal. Ass. Sp. Pap.*, **11**, 190 pp.

Walter, M. R., Buick, R. and Dunlop, J. S. R. (1980). Stromatolites 3400–3500 myr old from the North Pole area, W Australia, *Nature*, **284**, 443–445.

Wang Hong-Chen (1980). Megastages in the tectonic development of Asia. *Scientia Sinica*, **23**, 331–345.

Wang Liankui, Zhang Baogui and Cheng Jingping (1979). Geological and geochemical characteristic of Precambrian iron deposits at Anshan-Benxi area. NE China. *Ann. Rept. for 1978–70 Inst. Geochem. Acad. Sinica*, Guiyang, Guizhou, 22–24 (English abstract).

Wanless, R. K., Bridgwater, D. and Collerson, K. D. (1979). Zircon age measurements for Uivak II gneisses from the Saglek area, Labrador. *Can. J. Earth Sci.*, **16**, 962–965.

Wardle, R. J. (1981). Eastern margin of the Labrador Trough: an Aphebian proto-oceanic rift zone. *Geol./Min. Ass. Can. Abst.*, **6**, A59.

Watson, J. V. (1973). Influence of crustal evolution on ore deposition. *Trans. Inst. Min. Metall., Sect. B*, **82**, 107–114.

Watson, J. V. (1976). Mineralisation in Archaean provinces. In B. F. Windley (Ed.), *The Early History of the Earth*, Wiley–Interscience, London, pp. 443–453.

Watson, J. V. (1978). Precambrian thermal regimes. *Phil. Trans. R. Soc. Lond.*, **A288**, 431–440.

Watson, J. V. (1980). The origin and history of the Kapuskasing structural zone, Ontario, Canada. *Can. J. Earth Sci.*, **17**, 866–875.

Watson, J. V. and Dunning, P. W. (1979). Basement-cover relations in the British Caledonides. In A. L. Harris, C. H. Holland and B. E. Leake (Eds), *The Caledonides of the British Isles—Reviewed*, Scottish Academic Press, Edinburgh, pp. 67–92.

Watt, W. S. (1966). Chemical analyses from the Gardar igneous province, South Greenland, *Rap. Grønlands geol. Unders.*, **6**, 92 pp.

Watters, B. R. (1976). Possible late Precambrian subduction zone in south west Africa. *Nature*, **259**, 471–473.

Watterson, J. (1978). Proterozoic intraplate deformation in the light of SE Asian neotectonics. *Nature*, **273**, 636–640.

Watts, A. B., Weissel, J. K. and Larson, R. L. (1977). Sea floor spreading in marginal basins of the western Pacific. *Tectonophysics*, **37**, 167–181.

Weaver, B. L., Tarney, J., Windley, B. F., Sugavanam, E. B. and Venkata Rao, V. (1978). Madras granulites: geochemistry and PT conditions of crystallisation. In B. F. Windley and S. M. Naqvi (Eds), *Archaean Geochemistry*. Elsevier, Amsterdam, pp. 177–204.

Weaver, B. L. and Tarney, J. (1980). Rare earth geochemistry of Lewisian granulite–facies gneisses, NW Scotland: implications for the petrogenesis of the Archaean lower continental crust. *Earth Planet. Sci. Lett.*, **51**, 279–296.

Weaver, B. L. and Tarney, J. (1981a). The Scourie dyke suite: petrogenesis and geochemical nature of the Proterozoic subcontinental mantle. *Contrib. Mineral. Petrol.*, **78**, 175–188.

Weaver, B. L. and Tarney, J. (1981b). Lewisian gneiss geochemistry and Archaean crustal development models. *Earth Planet. Sci. Lett.*, **55**, 171–180.

Weaver, B. L., Tarney, J. and Windley, B. F. (1981). Geochemistry and petrogenesis of the Fiskenaesset anorthosite complex, SW Greenland: nature of the parent magma. *Geochim. Cosmochim. Acta*, **45**, 711–725.

Weaver, B. L. and Tarney, J. (1982). Andesitic magmatism and continental growth. In R. S. Thorpe (Ed.), *Andesites*, Wiley, London, pp. 639–661.

Weaver, B. L., Tarney, J., Windley, B. F. and Leake, B. E. (1982). Geochemistry and petrogenesis of Archaean metavolcanic amphibolites from Fiskenasset SW Greenland. *Geochim. Cosmochim. Acta*, **46**, 2203–2216.

Weiblen, P. W. and Morey, G. B. (1980). A summary of the stratigraphy, petrology and structure of the Duluth complex. *Amer. J. Sci.*, **280A**, 88–133.

Weissel, J. K. (1981). Magnetic lineations in marginal basins of the western Pacific. *Phil. Trans. R. Soc. Lond.*, **A300**, 223–247.

Wells, P. R. A. (1980). Thermal models for the magmatic accretion and subsequent metamorphism of continental crust. *Earth Planet. Sci. Lett.*, **46**, 253–265.

Wells, P. R. A. (1981). Accretion of continental crust: thermal and geochemical consequences. *Phil. Trans. R. Soc. Lond.*, **A301**, 347–357.

Wells, G., Bryan, W. D. and Pearce, T. H. (1979). Comparative morphology of ancient and modern pillow lavas. *J. Geol.*, **87**, 427–440.

Westbrook, G. K. (1982). The Barbados Ridge complex: tectonics of a mature forearc system. In J. K. Leggett (Ed.), *Trench-Forearc Geology: Sedimentation and Tectonics of Modern and Ancient Active Plate Margins*, Geol. Soc. Lond., Blackwell, Oxford, 275–290.

Wheeler, J. O. and Gabrielse, H. (1972). The Cordilleran structural province. In R. A. Price and R. J. W. Douglas (Eds), *Variations in Tectonic Styles in Canada*, Geol. Ass. Can. Sp. Pap., **11**, 1–82.

Wheeler, J. O., Charlesworth, H. A. K., Monger, J. W. H., Muller, J. E., Price, R. A., Ressor, J. E., Roddick, J. A. and Simony, P. S. (1974). Western Canada. In A. M. Spencer (Ed.), *Mesozoic–Cenozoic Orogenic Belts*, Geol. Soc. Lond. Sp. Publ., **4**, 591–624.

Whiteman, A. J., Ress, G., Naylor, D. and Pegrum, R. M. (1975). North Sea troughs and plate tectonics. *Norges geol. Unders.*, **316**, 137–161.

Whitney, J. A., Paris, T. A., Carpenter, R. H. and Hartley, M. E. (1978). Volcanic evolution of the southern slate belt of Georgia and S Carolina: a primitive oceanic island arc. *J. Geol.*, **86**, 173–192.

Wiebe, R. A. (1980). Anorthositic magmas and the origin of Proterozoic anorthosite massifs. *Nature*, **286**, 564–567.

Wiener, R. W. (1981). Tectonic setting, rock chemistry and metamorphism of an Archaean gabbro anorthosite complex, Tessinyakh Bay, Labrador. *Can. J. Earth Sci.*, **18**, 1409–1421.

Williams, D. (1969). Ore deposits of volcanic affiliation. In C. H. James (Ed.), *Sedimentary Ores, Ancient and Modern, Geology Dept. Univ. Leicester Sp. Publ.*, **1**, 197–206.

Williams, D. A. S. and Furnell, R. G. (1979). Reassessment of part of the Barberton type area, South Africa. *Precamb. Res.*, **9**, 325–347.

Williams, G. E. (1975). Late Precambrian glacial climate and the earth's obliquity. *Geol. Mag.*, **112**, 441–465.

Williams, G. E. (1981). Sunspot periods in the late Precambrian glacial climate and solar-planetary relations. *Nature*, **291**, 624–628.

Williams, H. (1977). Ophiolitic mélange and its significance in the Fleur de Lys Supergroup, N Appalachians, *Can. J. Earth Sci.*, **14**, 987–1003.

Williams, H. (1978a). Geological development of the northern Appalachians: its bearing on the evolution of the British Isles. In D. R. Bowes and B. E. Leake (Eds), *Crustal Evolution in NW Britain and Adjacent Regions*, Seel House Press, Liverpool, pp. 1–22.

Williams, H. (1978b). Tectonic lithofacies maps of the Appalachian orogen. Memorial University, Newfoundland. Scale 1:1.000,000.

Williams, H. (1979). Appalachian orogen in Canada. *Can. J. Earth Sci.*, **16**, 792–807.

Williams, H. (1980). Structural telescoping across the Appalachian orogen and the minimum width of the Iapetus ocean. In D. W. Strangway (Ed.), *The Continental Crust and its Mineral Deposits*, Geol. Ass. Can. Sp. Pap., **20**, 421–440.

Williams, H. and Smyth, W. R. (1973). Metamorphic aureoles beneath ophiolite suites and Alpine peridotites: tectonic implications with West Newfoundland examples. *Amer. J. Sci.*, **273**, 594–621.

Williams, H., Hibbard, J. P. and Bursnall, J. T. (1977). Geological setting of asbestos-bearing ultramafic rocks above the Baie Verte lineament, Newfoundland. *Geol. Surv. Can. Pap.*, **77-1A**, 351–360.

Williams, H. and Hatcher, R. D. (1982). Suspect terranes and accretionary history of the Appalachian orogen. *Geology*, **10**, 530–536.

Wilson, A. H. (1982). The geology of the Great Dyke, Zimbabwe: the ultramafic rocks. *J. Petrol.*, **23**, 240–292.

Wilson, H. D. B. (1982). Copper–nickel bearing rocks of the Precambrian and their current and potential development. In *The Development Potential of Precambrian Mineral Deposits, UN Nat. Res. Energy Div.*, Pergamon Press, New York, 95–114.

Wilson, J. T. (1963). Evidence from islands on the spreading of ocean floors. *Nature*, **197**, 536–538.

Wilson, J. T. (1965). A new class of faults and their bearing on continental drift. *Nature*, **207**, 343–538.

Wilson, J. T. (1966). Did the Atlantic close and then re-open? *Nature*, **211**, 676.

Wilson, J. T. (1973a). Continental drift, transcurrent and transform faulting. In A. E. Maxwell (Ed.), *The Sea*, vol. 4, Wiley, New York, pp. 623–644.

Wilson, J. T. (1973b). Mantle plumes and plate motions. *Tectonophysics*, **19**, 149–164.

Wilson, J. F., Bickle, M. J., Hawkesworth, C. J., Martin, A., Nisbet, E. G. and Orpen, J. L. (1978). Granite–greenstone terrains of the Rhodesian Archaean craton. *Nature*, **271**, 23–27.

Wilson, M. R. (1982). Magma types and the tectonic evolution of the Swedish Proterozoic. *Geol. Rund.*, **71**, 120–129.

Windley, B. F. (1973). Crustal development in the Precambrian. *Phil. Trans. R. Soc. Lond.*, **A273**, 321–341.

Windley, B. F. (1977). Timing of crustal growth and emergence. *Nature*, **270**, 426–428.

Windley, B. F. (1979). Tectonic evolution of the continents in the Precambrian. *Episodes*, **1979**, 12–16.

Windley, B. F. (1980). Evidence for land emergence in the early to middle Precambrian. *Proc. Geol. Ass.*, **91**, 13–23.

Windley, B. F. (1981). Phanerozoic granulites. *J. geol. Soc. Lond.*, **138**, 745–751.

Windley, B. F. (1982). Igneous rocks of the Lewisian complex. In D. S. Sutherland (Ed.), *Igneous Rocks of the British Isles*, Wiley, London, pi. 9–18.

Windley, B. F. (1983a). Banded iron formations in Proterozoic greenstone belts: call for further studies. *Precamb. Res.*, **20**, 585–588.

Windley, B. F. (1983b). Metamorphism and tectonics of the Himalayas. *J. geol. Soc. Lond.*, **140**, 849–866.

390

Windley, B. F. (1984). A tectonic review of the Proterozoic. In L. G. Medaris (Ed.), *Proterozoic Geology*, *Geol. Soc. Amer. Mem.*, **161**, 1–10.

Windley, B. F., Herd, R. K. and Bowden, A. A. (1973). The Fiskenaesset complex, West Greenland, Pt. 1: A preliminary study of stratigraphy, petrology and whole rock chemistry from Qeqertarssuatsiaq. *Grønlands geol. Unders. Bull.*, **106**, 1–80 (also *Meddr. Grønland*, **196**, 2).

Windley, B. F. and Smith, J. V. (1976). Archaean high-grade complexes and modern continental margins. *Nature*, **260**, 671–657.

Windley, B. F., Bishop, F. C. and Smith, J. V. (1981). Metamorphosed layered igneous complexes in Archean granulite–gneiss belts. *Ann. Rev. Earth Planet. Sci.*, **9**, 175–198.

Wise, D. U. (1974). Continental margins, freeboard and volumes of continents and oceans through time. In C. A. Burk and C. L. Drake (Eds), *The Geology of Continental Margins*, Springer. Berlin, pp. 45–58.

Wong, H. K. and Degens, E. T. (1983). Effects of CO_2–H_2O and oblique collision orogenesis—the European Hercynides as an example. *Tectonophysics*, **95**, 191–220.

Wood, D. S. (1973). Patterns and magnitudes of natural strain in rocks. *Phil. Trans. R. Soc. Lond.*, **A274**, 373–382.

Worst, B. G. (1960). The Great Dyke of southern Rhodesia. *Bull. geol. Surv. S. Rhod.*, 47.

Woussen, G., Dimroth, E., Corriveau, L. and Archer, P. (1981). Crystallisation and emplacement of the Lac St-Jean anorthosite massif (Quebec, Canada). *Contrib. Mineral. Petrol.*, **76**, 343–350.

Wright, J. B. (1976). Fracture systems in Nigeria and initiation of fracture zones in the South Atlantic. *Tectonophysics*, **34**, T43–T47.

Wright, J. B. and McCurry, P. (1973). Magmas, mineralizations and sea-floor spreading. *Geol. Rund.*, **62**, 116–125.

Wyman, W. F. (1980). Proterozoic komatiites from the Sangre de Cristo mountains, south-central New Mexico. *Geol. Soc. Amer. Abst. with Prog.*, **12**, 553.

Wynne-Edwards, H. R. (1976). Proterozoic ensialic orogenesis: the millipede model of ductile plate tectonics. *Amer. J. Sci.*, **276**, 927–953.

Wynne-Edwards, H. R. (1972). The Grenville Province. In R. A. Price and R. J. W. Douglas (Eds), *Variations in Tectonic Styles in Canada, Geol. Ass. Can. Sp. Pap.*, **11**, 263–334.

Young, G. M. (1970). An extensive early Proterozoic glaciation in North America? *Palaeogeog. Palaeoclimat. Palaeocol.*, **7**, 85–101.

Young, G. M. (1977). Stratigraphic correlation of upper Proterozoic rocks of NW Canada. *Can. J. Earth Sci.*, **14**, 1771–1787.

Young, G. M. (1978). Some aspects of the evolution of Archaean crust. *Geosci. Can.*, **5**, 140–149.

Young, G. M. (1980). The Grenville orogenic belt in the North Atlantic continents. *Earth Sci. Rev.*, **16**, 277–288.

Zeck, H. P. and Malling, S. (1976). A major global suture in the Precambrian basement of SW Sweden? *Tectonophysics*, **31**, T35–40.

Zeil, W. (1979). *The Andes: A Geological Review*, Gebrüder Borntraeger, Berlin, 260 pp.

Zeitler, P. K., Tahirkheli, R. A. K., Naeser, C. W. and Johnson, N. M. (1982a). Unroofing history of a suture zone in the Himalaya of Pakistan by means of fission-track annealing ages. *Earth Planet. Sci. Lett.*, **57**, 227–240.

Zeitler, P. K., Johnson, N. M., Naeser, C. W. and Tahirkheli, R. A. K. (1982b). Fission track evidence for Quaternary uplift of the Nanga Parbat region, Pakistan. *Nature*, **298**, 255–257.

Zen, E. A. (1981). An alternative model for the development of the allochthonous southern Appalachian Piedmont. *Amer. J. Sci.*, **281**, 1153–1163.

Zhang Ru-Yuan, Cong Bo-Lin, Ying Yu-Pu and Li Ji-Liang. (1981). Ferrifayalite-bearing eulysite from Archaean granulites. in Qianan county, Hebei, N China. *Tschermarks Min. Petr. Mitt.*, **28**, 167–187.

Ziegler, P. A. (1975). Geologic evolution of the North Sea and its tectonic framework. *Bull. Amer. Ass. Petrol. Geol.*, **59**, 1073–1097.

Ziegler, P. A. (1982a). Faulting and graben in western and central Europe. *Phil. Trans. R. Soc. Lond.*, **A305**, 113–142.

Ziegler, P. A. (1982b). Triassic rifts and facies patterns in western and central Europe. *Geol. Rund.*, **71**, 747–772.

Ziegler, A. M., Scotese, C. R., McKerrow, W. S., Johnson, M. E. and Bambach, R. K. (1979). Palaeozoic palaeogeography *Ann. Rev. Earth Planet. Sci.*, **7**, 473–502.

Zindler, A., Jagoutz, E. and Goldstein, S. (1982). Nd, Sr and Pb isotopic systematics in a three-component mantle: a new perspective. 298, 519–523.

Zonenshain, L. P. and Savostin, L. A. (1981). Movement of lithospheric plates relative to subduction zones: formation of marginal seas and active continental margins. *Tectonophysics*, **74**, 57–87.

Zwanzig, H. V., Syme, E. C. and Gilbert, H. P. (1979). Volcanic and sedimentary facies relationships at Lynn Lake, Manitoba: a reconstruction of parts of an Aphebian greenstone belt. *Geol. Ass. Can. Prog. with Abst.*, 134, 86.

Zwart, H. and Dornsiepen, U. F. (1978). The tectonic framework of central and western Europe. *Geol. en Mijnb.*, **57**, 627–654.

Zwart, H. J. and Dornsiepen, U. F. (1980). The Variscan and pre-Variscan tectonic evolution of central and western Europe: a tentative model. In J. Cogné and M. Slansky (Eds), *Geology of Europe*, B.R.G.M., Orléans, France, pp. 226–232.

Index

Abitibi greenstone belt, 35, 36, 44, 53, 54
Acadian orogeny, 181, 197, 201
Adelaide Basin, 117, 119
Adirondack massif, 99
Aldan Shield, 8, 12, 19, 27, 99
Alkaline igneous activity
  Alpine rifts, 304
  Anorogenic–late Proterozoic, 108, 113, 348
  Archaean, 37, 38
  aulacogens, 81, 88, 112
  continental margin orogenic belts, 283
  Early Proterozoic, 81, 88, 134
  East African rifts, 228–230
  Himalayan-type belts, 337
  in crustal evolution, 356
  island arcs, 256, 258, 259
  Late Proterozoic mobile belts, 140, 141
  mineralization, 114, 141, 231–232, 348
  oceanic islands, 246
  Pangaea breakup, 219, 228–230
  transform faults, 351
Alleghanian Orogeny, 181, 197, 201
Allochthonous terrains, 263, 267–272, 278
Almora thrust slab, 318
Alpine Fault, 239
Alpine flysch, 300
Alpine fold belt, 6, 292–308
  early rifting, 303–307
  evolution, 292
  magmatic activity, 304
  mineralization, 302–303
  oceanic crust, 304–306
  plate and microplate movements, 306–307
  rock units, 293–303
  synopsis, 293
  tectonic evolution, 303
  volcanics, 304
Altyn Tagh fault, 321
Amadeus Basin, 119, 131
Amapá Series, 95
Ameralik dykes, 12, 356
Amisk–Flin Flon belt, 147
Anabar Shield, 99
Andaman Sea, 321
Andean fold belt, 62, 274, 276–291
Andesites
  Archaean, 38, 52, 54, 55
  continental margin belts, 281 283, 284
  Early Proterozoic, 77
  island arcs, 256
Animikie Supergroup, 78
Animikie tillites, 89

Anorthosites
  Archaean, 9, 13–16, 26, 58
  in Late Proterozoic belts, 136
  in Pangaea, 214
  Mid-Proterozoic, 98–102, 112
Antarctica, 161, 170, 228
Aphebian Wollaston Group, 83
Appalachian fold belt
  ophiolites, 185–187
  mineralization, 197–199
  subdivisions, 193–194
  tectonic evolution, 193–197
Arabian–Nubian Shield, 139, 140, 348
Aravalli Group, 95
Archaean crustal evolution, 48–65
  Australia, 51
  Barberton, 50, 53
  Finland, 54
  India, 51
  Limpopo, 58, 59
  modern interpretation of, 61–65
  Ontario, 54
  Scotland, 60
  Zimbabwe, 50 51
Archaean granulite–gneiss belts, 8–27, 60–61, 358
  Africa, 24
  China, 10, 13
  comparative review of regions, 25–26
  Greenland, 8, 10, 12, 19–22, 25
  India, 10, 12, 13, 25
  Labrador, 8–10, 22, 25
  Limpopo, 25, 27
  Malagasy, 12, 27
  Scotland, 11, 12, 16, 23
  Selebi-Pikwe, 27
  Swaziland, 12
  USSR, 13
Archaean greenstone belts, 28–57, 60–61, 148, 344, 354
  Australia, 33
  Canada, 35
  classical models, 48–52
  general form and distribution, 28
  geochronology, 28
  India, 34–35
  Kambalda, 41
  metamorphism, 39
  mineralization, 39–44
  occurrences, 28
  Sierra Leone, 39
  stratigraphy, 30
  structure, 38–39
  Zimbabwe, 33, 42